Stomatal Function

Stomatal Function

Edited by
Eduardo Zeiger, G. D. Farquhar, and
I. R. Cowan

Stanford University Press, Stanford, California
1987

Stanford University Press, Stanford, California
© 1987 by the Board of Trustees of the Leland Stanford Junior University
Printed in the United States of America
CIP data appear at the end of the book

Preface

In April 1983, nearly forty scientists interested in stomatal function met at the East-West Center, Honolulu, Hawaii, under the auspices of the U.S.-Australia Cooperative Science Program, sponsored by the National Science Foundation of the United States and the Australian Department of Science and Technology. Although originally conceived as an Australian-American seminar, the meeting attracted a number of researchers from other countries and provided an international forum for an active exchange of data and ideas about the properties of stomata and their function in the plant.

The meeting also provided us with the opportunity to organize the assembly of a volume covering the main aspects of stomatal research. Despite the relevance of the subject for many disciplines of the plant sciences, there are few publications that provide a comprehensive view of stomatal function. The classic treatise *Physiology of Stomata*, by H. Meidner and T. A. Mansfield (New York: McGraw-Hill, 1968), is now nearly two decades old, and more recent publications like *Stomatal Physiology* (P. G. Jarvis and T. A. Mansfield; Cambridge: Cambridge University Press, 1981) or *Stomata* (C. M. Wilmer; White Plains, N.Y.: Longman, 1983) have either concentrated on selected themes or have been designed primarily as textbooks.

The present volume is intended as an up-to-date study of stomatal structure and function. It includes twenty chapters on different aspects of stomatal research. Some of these are original papers without precedent in the contemporary literature; others present reviews of their subjects, providing both background information and updates on recent conceptual and methodological developments. It is our hope that this volume will provide an introduction to stomata for the student and a valuable reference source for the researcher.

We thank the National Science Foundation and the Australian Department of Science and Technology for funding the Honolulu meeting on stomatal function. We are grateful as well to the contributing authors for their cooperative spirit—so often taxed throughout the process of putting this book together—and to Ms. Eleanor Crump for her superb editorial work.

Stanford and Canberra   E.Z.
March 1986   G.D.F.
                                                                                          I.R.C.

Contents

Contributors	ix
Introduction *Eduardo Zeiger*	1
1. Three Hundred Years of Research into Stomata *Hans Meidner*	7
2. The Evolution of Stomata *Hubert Ziegler*	29
3. The Development and Structure of Stomata *Fred D. Sack*	59
4. Stomatal Mechanics *Peter J. H. Sharpe, Hsin-i Wu, and Richard D. Spence*	91
5. An Introduction to Carbon Metabolism in Guard Cells *William H. Outlaw, Jr.*	115
6. Ionic Relations of Guard Cells *E. A. C. MacRobbie*	125
7. Guard Cell Bioenergetics *Sarah M. Assmann and Eduardo Zeiger*	163
8. Stomatal Responses to Light *Thomas D. Sharkey and Teruo Ogawa*	195
9. The Blue-Light Response of Stomata: Mechanism and Function *Eduardo Zeiger, Moritoshi Iino, Ken-Ichiro Shimazaki, and Teruo Ogawa*	209
10. Intercellular CO_2 Concentration and Stomatal Response to CO_2 *James I. L. Morison*	229
11. Action of Abscisic Acid on Guard Cells *Klaus Raschke*	253
12. Cytokinins and Stomata *L. D. Incoll and P. C. Jewer*	281
13. Auxins and Stomata *W. J. Davies and T. A. Mansfield*	293

Contents

14. Stomatal Responses to Air Humidity and to Soil Drought   311
    E.-D. Schulze, N. C. Turner, T. Gollan, and K. A. Shackel
15. Diurnal Variations in Leaf Conductance and Gas Exchange in Natural Environments   323
    J. D. Tenhunen, R. W. Pearcy, and O. L. Lange
16. Stomata in Plants with Crassulacean Acid Metabolism   353
    Irwin P. Ting
17. Leaf-Age Effects on Stomatal Conductance   367
    Christopher B. Field
18. Transfer Processes in Plant Canopies in Relation to Stomatal Characteristics   385
    J. J. Finnigan and M. R. Raupach
19. Breeding for Stomatal Characters   431
    Hamlyn G. Jones
20. Calculations Related to Gas Exchange   445
    J. Timothy Ball

Index of Subjects   477
Index of Authors   491

Contributors

Sarah M. Assmann received her Ph.D. at Stanford University in 1986. Her research covers studies of stomatal function at different levels of organization, including guard cell protoplasts, stomata in epidermal peels, and gas exchange in intact leaves. She is interested in the integration of basic biological processes with the structure and function of plants in their natural environments.

Timothy Ball is a doctoral candidate in biological sciences at Stanford University and is studying at the Department of Plant Biology, Carnegie Institution of Washington. His interests concern resource use by plants as determined by the interaction of plant form and function with the environment. Currently he is exploring the relationship of stomatal conductance to photosynthesis and the environment.

Ian Cowan received his Ph.D. from the University of London in 1966. His chief research interests include stomatal physiology and plant water use. He is Professor in the Department of Environmental Biology at the Research School of Biological Sciences of the Australian National University.

William Davies received his Ph.D. in botany and forestry from the University of Wisconsin, Madison, in 1974. He is now Senior Lecturer in Plant Physiology at the University of Lancaster. His research interests include plant water relationships and physiological plant ecology.

Graham Farquhar received his Ph.D. from the Australian National University in 1973. His research interests include stomatal physiology, isotopic composition of plants, and integration of photosynthesis with nitrogen and water use. He is a Senior Fellow in the Department of Environmental Biology at the Research School of Biological Sciences of the Australian National University.

Christopher B. Field received his Ph.D. from Stanford in 1981 and is currently a Staff Scientist at the Department of Plant Biology, Carnegie Institution of Washington. His primary research interests concern the relationship between the gas exchange of individual leaves and the growth and development of whole plants.

J. J. Finnigan received his Ph.D. from the Australian National University in 1978. His research interests include both theoretical and experimental aspects of turbulent transfer processes within plant canopies, interaction between internal gravity waves and turbulence, and wind flow over complex terrain. He is a Principal Research Scientist at the Division of Environmental Mechanics, CSIRO, Canberra, Australia.

Thomas Gollan studies biology at the University of Heidelberg and Bayreuth, West Germany. He received the degree of Diplom in biology in 1983, and has since then continued his studies toward a doctorate at CSIRO, Canberra, Australia, and at Bayreuth.

Moritoshi Iino is an Instructor in the Department of Biology, Tokyo Metropolitan University. He received his Ph.D. in 1982 from the Australian National University. From 1982 to 1984 he was a Postdoctoral Fellow in the Department of Plant Biology, Carnegie Institution of Washington. His major research interests are the hormonal and light regulations of plant growth and movement. His current research activity is centered on blue-light responses in plant movement, including phototropism of cereal coleoptiles and *Phycomyces* sporangiophores, and stomatal movement.

L. D. Incoll received his Ph.D. in botany from the University of Melbourne in 1968 for a dissertation on the regulation of photosynthesis in higher plants by the products of photosynthesis. After a postdoctoral fellowship with P. R. Day in Connecticut, he joined the Department of Plant Sciences at the University of Leeds, where he is presently Principal Research Officer. His current research interests include stomatal physiology and the role of temporary storage of assimilates in cereal stems during grain filling.

P. C. Jewer received his first degree in applied biology from the University of Bath. His Ph.D. was awarded by the University of London following work at East Malling Research Station, Maidstone, Kent, on the hormonal control of cambial activity in fruit trees. Since 1979 he has worked in the Department of Plant Sciences at the University of Leeds, first as a Postdoctoral Research Fellow, and currently as a Teaching Fellow. Regulation of stomatal movement, in particular by cytokinins, has been the main subject of his research at Leeds.

Hamlyn G. Jones is Principal Scientific Officer in the Physiology Department at East Malling Research Station, Maidstone, Kent. He holds an M.A. in natural sciences from Cambridge University and a Ph.D. from the Australian National University. His doctoral research was on the effects of water stress on photosynthesis. Since receiving his doctorate he has researched and taught plant water relations at the Plant Breeding Institute, Cambridge, and Glasgow University.

Otto L. Lange is Professor of Botany at the University of Würzburg, West Germany. His research has been devoted to assessing the gas exchange behavior of lower and higher plants in their natural habitats, and he has designed equipment for this purpose. Lange presently heads a research group investigating the behavior of plants under stress.

Enid MacRobbie is Reader in Plant Biophysics in the Botany School, University of Cambridge, and a Fellow of Girton College. She has worked on ion transport in giant algal cells and, more recently, on guard cells.

Terence A. Mansfield received his Ph.D. from the University of Reading, England, in 1961 and is now Professor of Plant Physiology at the University of Lancaster. In addition to stomatal physiology, he has worked on circadian rhythms, apical dominance, and the effects of SO_2 and NO_2 pollution on plants.

Hans Meidner earned his Ph.D. and D.I.C. at the University of London. For many years he worked on plant water relations and stomatal physiology at the University of Natal, South Africa. He then moved to the University of Reading, England, and later to the University of Stirling, Scotland, where he is Professor Emeritus and continues his experimental work with stomata.

James I. L. Morison received his Ph.D. from the University of Edinburgh in 1981, studying in the Forestry and Natural Resources Department. He is now a Lecturer in Agricultural Meteorology at the University of Reading, England, after spending four years in the Division of Plant Industry, CSIRO, Canberra, Australia, studying the effect of increasing atmospheric CO_2 on plant growth and water use. His interests include plant water relations and crop physiology.

Teruo Ogawa is Associate Director of the Solar Energy Research Group at the Institute of Physical and Chemical Research (RIKEN), Saitama, Japan. He received his Ph.D. from the Tokyo Institute of Technology under Kazuo Shibata. He studies photosynthesis in higher plants and algae, stomatal physiology, and CO_2-concentration mechanisms in cyanobacteria.

William H. Outlaw, Jr., is Professor of Biological Science at Florida State University, Senior U.S. Scientist at the University of Tübingen, and Associate Adjunct Professor in the Institute of Food and Agricultural Science at the University of Florida. He received his Ph.D. from the University of Georgia in 1974. Subsequently, he conducted postdoctoral research at Michigan State University and Washington University, St. Louis. The emphasis of his current work is guard cell biochemistry.

Robert W. Pearcy is Professor of Botany at the University of California at Davis. His research has concerned the physiological ecology of plants in diverse habitats. He has investigated mechanisms of photosynthetic temperature adaptation and the significance of C_4 photosynthesis. His current research is on the photosynthetic behavior of tropical-forest species in relation to their environment.

Contributors

Klaus Raschke received his Ph.D. in agricultural meteorology from the University of Poona, India, in 1955. After holding positions at two universities and in the agricultural chemicals industry, he became Professor in the MSU-DOE Plant Research Laboratory at Michigan State University, East Lansing, Michigan, in 1967. He is now at the Pflanzenphysiologisches Institut und Botanischer Garten of the University of Göttingen, West Germany, where he conducts research on the metabolism and function of abscisic acid in relation to ion transport and carbon metabolism in guard cells, photosynthetic tissue, and roots.

M. R. Raupach is Senior Research Scientist at the Division of Environmental Mechanics, CSIRO, Canberra, Australia. He has worked on the nature of air flow close to very rough surfaces, particularly vegetation canopies, and on dispersion in the atmospheric surface layer, with special emphasis on inhomogeneity and topographic effects. He received his Ph.D. from the Flinders University of South Australia in 1976.

Fred D. Sack is an Assistant Professor in the Department of Botany at Ohio State University in Columbus. He received his Ph.D. in plant biology from Cornell University in 1982.

Ernst-Detlef Schulze studied forestry at the University of Göttingen. He received the degree of Diplom in forestry in 1965, an M.A. in botany at the University of California, Los Angeles, in 1966, and the degree of Doctor Rerum Naturalium at Würzburg in 1969. He was formerly Associate Professor at Würzburg and Munich, and has been Professor and Director of the Lehrstuhl für Pflanzenökologie at the University of Bayreuth, West Germany, since 1975. He is currently Editor-in-Chief of *Oecologia* and a member of the review board of *Plant, Cell and Environment*. His areas of research include the ecophysiology of stomata, plant carbon, water and nutrient balance, and acid rain effects.

Kenneth Shackel is Assistant Professor in the Department of Viticulture and Enology at the University of California, Davis. He received his Ph.D. in plant physiology from the University of California, Riverside, in 1982.

Thomas D. Sharkey is Associate Director of the Biological Sciences Center, Desert Research Institute, Reno, Nevada. He received his Ph.D. from Michigan State University under Klaus Raschke. He studies photosynthesis in intact leaves, stomatal physiology, biochemistry of photosynthesis in leaves, and limitations to photosynthesis in CO_2-enriched air.

Peter J. H. Sharpe received his Ph.D. in biological sciences from the University of New South Wales in 1970. He is currently Professor in Bioengineering in the Biosystems Research Group of the Department of Industrial Engineering at Texas A&M University. He also serves as consultant to several research-oriented companies. His current areas of research interest include ecosystems modeling, physiological ecology, pest management, environmental impact, and space biotechnology.

Ken-Ichiro Shimazaki is a Research Worker in the Division of Environmental Biology, National Institute for Environmental Studies, Yatabi-machi, Japan. He received his Ph.D. in 1980 from Kyushu University at Fukuoka. He is interested in stomatal function, especially in the cooperation of guard cell organelles, and in the regulation of stomatal responses to environmental stimuli, including light and air pollutants.

Richard D. Spence is a Research Scientist in the Biosystems Research Group of the Department of Industrial Engineering at Texas A&M University. He received his Ph.D. in bioengineering from Texas A&M University in 1986. His research interests include cell-wall mechanics, stomatal physiology, and ecosystems water balance.

John D. Tenhunen is Adjunct Professor and Scientific Coordinator with the Systems Ecology Research Group at San Diego State University. He received his Ph.D. in physiological ecology from the University of Michigan in 1976. Tenhunen conducted research on the adaptation of plants to arid conditions at the University of Würzburg, West Germany, and in Portugal as a NATO Postdoctoral Fellow and a Research Fellow of the Deutsche Forschungsgemeinschaft.

Irwin P. Ting received his Ph.D. in 1964 from Iowa State University at Ames and is presently Professor of Biology at the University of California, Riverside, in the Department of Botany and Plant Sciences. His research interests have been largely in metabolism, especially crassulacean acid metabolism in succulent plants.

Neil C. Turner is Leader of the Biological Resources Group and Research Leader of the Dryland Crops and Soils Research Program in the CSIRO Laboratory for Rural Research, Perth, Western Australia. He has a Ph.D. from the University of Adelaide, South Australia, and a D.Sc. from the University of Reading, England. He was formerly a plant physiologist at the Connecticut Agricultural Experiment Station in New Haven, a research scientist in the CSIRO Division of Plant Industry, Canberra, Australia, an Alexander von Humboldt Research Fellow at the University of Bayreuth, West Germany, and a Fellow of the American Society of Agronomy. He has research interests in environmental effects on stomatal behavior and the water relations of plant communities.

Hsin-i Wu is Associate Professor of Bioengineering in the Biosystems Research Group of the Department of Industrial Engineering at Texas A&M University. He received his Ph.D. in mathematical physics from the University of Missouri, Columbia, in 1967. His current areas of research include cell-wall mechanics, pattern simulation using lattice-point processes, resource competition using classical field theory, and statistical ecophysics.

Eduardo Zeiger is Research Professor of Biology in the Department of Biological Sciences, Stanford University. He received his Ph.D. in 1970 from the University of California, Davis. He studies the function of stomata in the plant and the biological mechanisms regulating stomatal responses to light and other environmental stimuli.

Hubert Ziegler has been Professor and Chairman in the Department of Botany and Microbiology at the Technical University of Munich since 1970. He received his Ph.D. in botany from the University of Munich in 1950, and was Lecturer in the Department of Forest Botany at the University of Munich from 1951 until 1958, and Professor and Chairman in the Department of Botany and Director of the Botanical Garden at the Technical University of Darmstadt from 1959 until 1970. He has been a member of the Deutsche Akademie der Naturforscher Leopoldina since 1972 and of the Bayerische Akademie der Wissenschaften since 1973. He is currently Coeditor for the journals *Planta, Oecologia, Biochemie und Physiologie der Pflanzen, Trees, Naturwissenschaften*, and *Progress in Botany*. His research interests include transport in plants, tree physiology, biochemical ecology, and stable isotope discrimination in plant physiology and biochemistry.

Stomatal Function

Introduction

Eduardo Zeiger

Stomata, from the Greek for "mouth," are structural features of aerial organs in most plants. The term "stoma" denotes a microscopic pore or hole through the surface of the plant organ, which allows communication between the interior and the external environments, and a pair of specialized cells (the guard cells) that surrounds the pore. Guard cells respond to environmental signals by changing their dimensions and thereby regulating the size of the pore, or stomatal aperture. Stomatal movements, the continuously occurring changes in pore dimensions, are an intricate aspect of the physiology of the plant. The term "stomatal function" is meant to comprise the relationship between the structural and metabolic properties of stomata, the ensuing regulation of stomatal apertures and the gas exchange between the leaf and its environment, and the impact of these processes on the physiology, adaptation, and productivity of plants.

Inspection of the distribution of stomata in leaves growing in environments with different levels of available water gives us clues for the role of stomata in plant adaptation. Leaves from aquatic plants living underwater are devoid of stomata. Leaves that float in water, which are very common in ponds, have stomata in their upper surfaces but lack them in the surfaces in contact with water. Aerial leaves have stomata on either surface, although the frequency and distribution of the stomata vary substantially with phylogeny and environmental conditions. The requirement for stomata in all plant surfaces in contact with air is the consequence of an important adaptation to avoid desiccation. When aquatic plants invaded terrestrial habitats, they developed an impermeable cuticle, which prevents water loss. That adaptation, however, posed another problem for survival: any substance that effectively prevents the outward diffusion of water also acts as a barrier against the inward diffusion of carbon dioxide, an essential substrate for photosynthesis. In the absence of spatial selectivity for the diffusion of water and carbon dioxide, stomata provide a temporal adaptation. Stomata close when water is limiting and open under conditions favoring photosynthesis, thus preventing excessive, deleterious water loss. The fascinating subject of the evolution of stomata is covered in one of the papers in this volume (Ziegler); the role of stomata in an adaptation of plants to extreme environments, that of crassulacean acid metabolism, is the subject of a separate contribution (Ting).

Studies of stomata have been a classical undertaking of the botanical sciences; the history of such research is discussed in detail in one of the papers in this volume (Meidner). Guard cells are markedly specialized, and the process of their differentiation has often attracted the attention of developmental biologists. Of particular interest is the asymmetric division of epidermal cells, which leads to the differentiation of guard cells. That developmental sequence has been studied in detail in grasses because of its remarkable temporal and spatial regulation, but the mechanism controlling asymmetric division remains to be understood. Our current knowledge of guard cell structure and differentiation is presented in a paper in this volume (Sack); another paper (Field) deals with stomata in aging leaves, a different temporal dimension of cellular differentiation, and discusses how plant senescence, nutrient recycling, and stomatal function are interrelated.

Stomatal movements depend on turgor regulation in the epidermis. In the opening phase, increases in turgor generate mechanical forces on the thickened walls of the guard cells, resulting in larger pore apertures; closing ensues from the reverse process. The mechanical properties of the guard cells and epidermal cells are extensively discussed in a contribution in this volume (Sharpe, Wu, and Spence). Because water equilibrates rapidly between cells in the epidermis, steady-state turgor gradients in the epidermis cannot depend on water uptake alone; instead, osmotic gradients are generated by accumulation or loss of ions in the guard cells. The nature of the ion fluxes associated with stomatal movements (MacRobbie) and the concomitant metabolic processes (Outlaw) and their underlying energetics (Assmann and Zeiger) are the subjects of separate contributions in this volume.

The plant environment is continuously changing, and stomatal apertures adjust accordingly. Environmental cues are perceived by the guard cells and transduced into modulated ion fluxes and the concomitant changes in water content and turgor that result in aperture regulation. Under laboratory conditions, specific responses to environmental stimuli are studied in controlled experiments, in which the parameter under consideration is changed while all others are kept as nearly constant as possible. The remarkable sensitivity of guard cells to external and internal stimuli is evident in the extensive coverage given to stomatal responses in this volume, including responses to light (Sharkey and Ogawa), carbon dioxide concentrations (Morison), abscisic acid (Raschke), cytokinins (Incoll and Jewer), auxins (Davies and Mansfield), and air humidity and soil drought (Schulze et al.).

In nature, on the other hand, prevailing environmental cues are transduced into modulated stomatal responses, which represent an integration of all input signals. Most commonly, stomatal responses at the leaf level are measured with porometry or with gas exchange techniques and are expressed as stomatal conductances, the reciprocal of stomatal resistances to water diffusion from the leaf. Considerable advances in methodology have made gas exchange techniques readily available and have significantly expanded our understanding of stomatal function and its tight coupling with photosynthesis in the intact leaf. Details of gas exchange techniques and their computation are presented in this volume in a valuable paper (Ball) that provides the reader with the analytical tools required to run gas exchange experiments. Another contribution (Tenhunen, Pearcy, and Lange) describes the use of gas exchange to study stomatal behavior and its dependence on changes in natural environments. It is of interest that under unstressed conditions stomata retain their multisensory capacity but prevailing levels of stomatal conductances are primarily regulated by light and air humidity.

The valuable information on the ecophysiology of stomata supplied by single-leaf studies should not, however, overshadow the fact that in nature leaves occur in canopies, wherein the environment can be significantly different from that of the isolated leaf. The implications of the canopy environment for stomatal responses is the subject of a novel analysis included in this volume (Finnigan and Raupach).

Throughout the history of stomatal research, some of the basic ideas about stomatal function have been subjects of controversy. One long-controversial problem has been the role of light and carbon dioxide in stomatal responses. The question is complex: light stimulates both guard cell activity and photosynthesis in the underlying mesophyll; mesophyll photosynthesis leads to a reduction of the intercellular carbon dioxide concentrations and a concomitant stomatal response to carbon dioxide. Observations showing a specific effect of light on stomata date back to the early part of this century (see Meidner, this volume; Sharkey and Ogawa, this volume); but when a stomatal response to carbon dioxide was discovered, in the 1930's, the stomatal response to light was reinterpreted as an indirect response to carbon dioxide. Each of these points of view has been favored by a majority of researchers at one time or another, with the controversy extending well into the last decade. It is perhaps an indication of the considerable progress in stomatal research that debate on the question is virtually absent from this volume, superseded by an emerging consensus about the multisensory nature of guard cell responses and about the existence of a remarkable coupling between levels of stomatal conductances and prevailing photosynthetic rates in the mesophyll.

The attention of researchers is now shifting to other problems, and new controversies are emerging. Noteworthy are questions about the capacity of guard cell chloroplasts to photophosphorylate and to fix carbon dioxide photosynthetically, the relative role of ion influx and efflux in the control of steady-state apertures, and the mechanism facilitating coordinated responses of stomatal conductances and photosynthetic rates. Divergent, often opposite, points of view about these and other aspects of stomatal function can be found in different contributions to this volume: no uniformity has been imposed by the editors on currently controversial concepts; it is up to the reader to evaluate available evidence and, likewise, up to future research to establish more conclusive answers. Further, the reader will note that in different papers different units are employed in some of the typical measurements of stomatal responses. It is hoped that progress in stomatal research will result in uniform nomenclature and standard units.

The accelerated pace of stomatal research in the last decade has resulted in impressive methodological and conceptual progress, which will be apparent in the material covered in this volume. Microanalytic techniques to quantify metabolites in a few guard cells, development of methods using guard cell protoplasts for advanced physiological, biochemical, and electrophysiological studies, and availability of portable gas exchange equipment that incorporates state-of-the-art electronics and data acquisition have opened new research opportunities. Recent findings on the mechanism underlying the blue-light response of stomata (see Zeiger et al., this volume) illustrate the potential of this improved methodology, which has allowed the demonstration of a blue-light-activated proton pump at the guard cell plasmalemma.

It appears safe to predict that progress will continue at an accelerated pace in coming years. Emerging knowledge about sensory transduction during the light response could open the way for the isolation and characterization of the specific blue-light photoreceptor and for the elucidation of the molecular components of the proton pump and its dependence on the activity of the blue-light photoreceptor and the guard cell chloroplasts. A better understanding of the light response could also facilitate the characterization of sensory transduction during other stomatal responses and the way in which different environmental signals are integrated into modulated stomatal apertures. Areas in which progress is likely to be most rapid include the mode of action of phytohormones, such as abscisic acid and indoleacetic acid, and the role of calcium in regulating ion transport, particularly its effect on specific ion channels.

At the whole-leaf level, important open questions include an accurate partition between photosynthetic and stomatal responses in the leaf and the physiological basis of their coordination. Two existing messengers between stomata and mesophyll cells, abscisic acid and carbon dioxide, are already known to facilitate that coordination; it remains to be established whether their effects, added to the parallel responses of stomata and mesophyll to environmental stimuli, suffice to explain the observed phenomenology in the leaf, or whether other, hitherto undefined, coupling factors must be sought. An answer to that question is relevant for our understanding of the role of stomatal properties in leaf adaptation to different environments. Leaf adaptation could include specific changes in stomatal properties; or it could rely solely on the known adaptive changes of the photosynthetic apparatus, with stomata tracking those changes via their coupling with mesophyll activity.

These and other, unforeseeable, developments are likely to maintain and enhance the interest of biologists in studying stomata as a model system to explore basic cellular properties and that of agronomists and plant breeders in exploiting stomatal properties to improve carbon gain and plant adaptation to marginal lands.

1

Three Hundred Years of Research into Stomata

Hans Meidner

The gradual accumulation of information derived from experimental observations and testing of hypotheses constitutes the body of scientific knowledge. It will be instructive therefore to recall past investigations that form the background for today's research into stomatal physiology and function. This commentary focuses on those scientists whose work has stood the test of time as well as on more recent work that has given new direction to research on stomata. The account is chronological but refers in places to subsequent work in order to draw attention to the link between past and present and to the progress that has been made.

Discovery

The earliest known experimental measurements relating to stomata are those of Edmé Mariotte, who is reported by de Candolle (1852) to have used gravimetric methods to investigate the rate of evaporation from leaves in 1660. Mariotte was a mathematician and physicist but earned his living as a clergyman in Dijon. Despite his vocation he was a mechanist, as opposed to a vitalist, and explained all plant processes as resulting from the action of physical forces. He measured the pressure of sap in plant tissues and is of course best known for his work "De la nature de l'air." In his time plant physiology was not an established branch of science, and for about the next hundred years experiments on stomatal physiology and function were carried out by amateurs and scientists, other than botanists, who had an interest in plant life and who were looking for features common to animal and plant life and common to the physical and the living world. It is thus worthwhile to remember that from the earliest times right up to the present, contributions from chemists, physicists, and medical scientists frequently stimulated new botanical research into the role of stomata as gateways between the plant and its immediate environment, with which it continuously interacts.

Fifteen years after Mariotte's measurements, MARCELLO MALPIGHI (1628-94), professor of medicine at Bologna and Pisa and a specialist in skin tissue—best known for his discovery of the blood capillaries, extending and

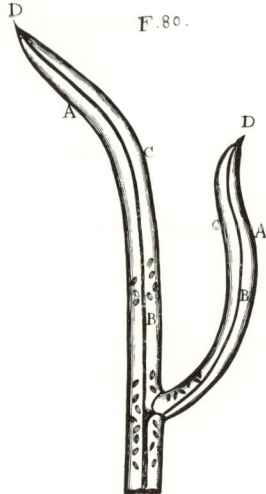
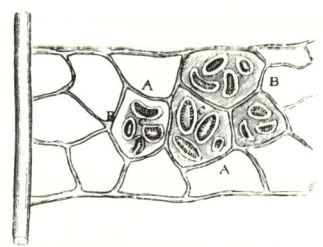

Fig. 1.1. *Left*, In plate 15, fig. 80, of Malpighi (1675), stomata are shown distinctly on the surface of the feathery leaf of *Foeniculum officinale*. *Right*, Plate 20, fig. 106, of Malpighi (1675): surface view of *Nerium oleander* leaf. "Intercostal areas" are shown at *A*, the "network of veins" at *B*, and "four or more apertures of cavities overarched with hairs in each area" at *C*. Malpighi did not in fact observe the stomata in the inner margins of these cavities, but he stated correctly that the cavities "form passages for breath or other humors." (I wish to thank Dr. Bert Drake of the Smithsonian for the reference to these figures.)

explaining Harvey's theory of the circulation of the blood—mentioned porelike structures he had observed in the surfaces of leaves with an early microscope (Malpighi, 1675). Within a few years the English physician and clergyman NEHEMIAH GREW (1641–1712) published drawings of the arrangements of such pores on different leaves, referring to them as "pass-ports" to the exterior in his *Anatomy of Plants* (Grew, 1682); on that account they were later referred to as "breathing pores." Grew laid the foundations for the development of scientific plant anatomy, convincingly postulated the sexuality of plants, and stimulated the study of plant physiology. It was Grew's detailed morphology and anatomy that caused Stephen Hales (1677–1761), the notable vicar of Teddington, to begin his classic experiments in 1700. Hales, however, did not study stomatal function specifically, but rather rates of water inflow and translocation, and of vapor loss from leaves.

Investigations into stomata did not begin until the middle of the eighteenth century, when their role in plant life began to be debated. The first signs of this were the variety of names proposed for these porelike structures. Though the terms "pass-ports" and "breathing pores" were already in use, JEAN ÉTIENNE GUETTARD (1715–86) referred to them as "miliary glands" (*glandes miliaires*, 1747), and HORACE-BENEDICT DE SAUSSURE (1740–99) called them "cortical glands" (*glandes corticales*, 1762; see Rehfous, 1917)—both names suggesting excretion as the function of the stomata. De Saussure is best known for constructing a magnetometer, an anemometer, and an instrument for measuring the intensity of blue in the sky. Professor of physics at Geneva, he introduced the term "geology" for the study of rocks and spent much of his time as an amateur botanist.

In the middle of the eighteenth century CHARLES BONNET (1720–93), also working at Geneva, examined leaves for the occurrence of stomata and, according to Rehfous (1917), arrived at the generalization that leaves of woody plants are hypostomatous and those of herbaceous plants are amphistomatous (Bonnet, 1754). However, only toward the end of the century was one of the functions of stomata truly appreciated when JOHANN HEDWIG (1730–99), expert microscopist and professor of botany at Leipzig, referred to them as "evaporatory pores" (*pores évaporatoires*, 1793) or "spiracula." Hedwig is most famous for his complete study of the life history of the mosses. (Did he examine stomata on the moss capsule, the least-evolved stage of plant life where they occur?) By giving stomata a functional designation Hedwig prompted the stomatal studies of water vapor loss from plants by de Candolle (1827), who by the beginning of the nineteenth century confirmed that the apertures of stomata were variable, as Hedwig had reported.

Structure and Function

AUGUSTIN PYRAMUS DE CANDOLLE (1778–1841) was professor of botany at Montpellier and worked chiefly on the classification of plants. A strict believer in the unity of structure and function, he contributed to shifting the interest in stomata toward their function in gas exchange. As reported by Rehfous (1917), it was HEINRICH FRIEDRICH LINK (1767–1851), professor of botany at Breslau, who introduced the term "stoma" in 1812, and by adopting this name, de Candolle (1827) ensured its continued use. At the end of the eighteenth century the fundamental discoveries of Jan Ingenhousz (1730–99) and Jean Senebier (1742–1809) had shown that CO_2 and oxygen entered into and escaped from plants via their leaves. De Candolle, aware of this new knowledge, had also benefited from the work of the ingenious GIOVANNI BATTISTA AMICI (1784–1860), mathematician, physicist, astronomer, rediscoverer of cyclosis, and inventor of the immersion lens. His detailed microscopic studies of the cuticles and epidermes of leaves (Amici, 1823) enabled de Candolle to consider the variable apertures of stomata as the path by which gases entered and escaped from leaves.

Williams (1841) pointed out that what was thought to be a membrane permanently sealing stomatal pores was an artifact resulting from "charring" leaves between two glass plates for examination. Curiously, a similar faulty method has led others, in our time, to suggest that leaves exude liquid droplets via their stomata (Kolpikov, 1964). Considerable progress was made when stomatal frequencies and rates of gas exchange were determined by M. GARREAU (1850). He found that quantities of water vapor lost from a leaf in daytime (*d'eau exhalée pendant le jour*) and the amount of CO_2 escaping from a leaf at night (*l'acide carbonique expirée pendant la nuit*) were proportional to stomatal numbers, confirming a shrewd notion that according to Rehfous (1917) had been advanced already by de Saussure (1762). By Garreau's time, hypotheses were formulated that connected the observed variability in the

BOTANISCHE ZEITUNG.

14. Jahrgang.   Den 3. October 1856.   **40. Stück.**

Inhalt. Orig.: H. v. Mohl, welche Ursachen bewirken d. Erweiterung u. Verengung der Spaltöffnungen. — C. Müller, Hal., üb. d. Pfl.zone zw. Cochinal u. Miguel-Diaz an d. peru.-chil. Küste. — Lit.: J. Giardini, Giornale d'Orticoltura Vol II, 2—7. — Koch, Synops. Fl. Germ. et Helv. I. 3te Ausg. — Bittcher, Pförtner-Album. — Jumpertz, de foecundatione plantarum. — Pers. Not.: Metsch. — Henschel.

— 697 —   — 698 —

Welche Ursachen bewirken die Erweiterung und Verengung der Spaltöffnungen?

Von

Hugo v. Mohl.

(Hierzu Taf. XIII.)

Dass sich die Spalte der Spaltöffnungen bald beim Mays und Zuckerrohr die Spaltöffnungen nur am frühen Morgen geöffnet zeigten, wenn die Sonne auf die noch thauigen Blätter schien, oder wenn der Verfasser das Blatt in ein Gefäss mit Wasser bog und es so dem Sonnenlichte aussetzte, also unter Umständen, unter welchen die Ausdünstung der Blätter zurückgehalten sei.

Fig. 1.2. The frontis page of the issue of *Botanische Zeitung* in which Hugo von Mohl's two papers appeared in 1856.

dimensions of stomatal pores with changes in the turgor of guard cells. The latter were thought to resemble turgor changes in cells of the pulvinus of leaflets, which John Ray (1627–1705) had considered over one hundred years previously as the mechanism of the sleep movements of leaflets.

By the middle of the nineteenth century the botanist Henri Joachim Dutrochet (1776–1847) showed that the substomatal cavity was continuous with the intercellular spaces of the leaf. More important, he had discovered and systematically investigated a process that he termed "osmosis" (Dutrochet, 1827): now progress could be made in stomatal physiology, and, indeed, professional plant physiologists began for the first time to advance the specific study of stomata.

The outstanding studies of the anatomy of guard cells, their turgor changes, and related deformations were those of Hugo von Mohl (1805–72), professor of botany at Tübingen and, incidentally, the inventor of the term "protoplasm." He published two papers (von Mohl, 1856) on stomatal structure and function, with drawings the quality of which served as model for N. J. C. Müller at Heidelberg and, a few years later, for Simon Schwendener at Basel. Their anatomical drawings could inspire us today, since a thorough knowledge of anatomy is a prerequisite and indispensable part of functional studies. Von Mohl documented practically everything that is known of guard cell anatomy, with the exception of fine-structural detail, which was to be revealed by electron microscopy, by dichroic staining, and by polarization microscopy. He described the outer ledges (*Vorsprünge*) forming the antechamber (*Vorhof*), and the inner ledges, where present, forming the rear chamber (*Hinterhof*) continuous with the substomatal cavity. He showed that guard cell lumina are frequently triangular in cross section when not under turgor and that the ventral and dorsal walls of guard cells are of unequal thickness. He drew attention to the thin portion of the dorsal wall of guard cells that connects them to neighboring cells, a thin-walled joint later termed *Hautgelenk* by Schwendener. Von Mohl discerned the thickening of the ledges and other cuticle features, and he discussed the importance of all these properties of guard cells for their particular deformations under turgor. He was also the first to puncture neighboring and guard cells and recognized thereby that the turgor of surrounding cells has a considerable influence on stomatal movements, as confirmed a hundred years later by Stålfelt (1966) and more quantitatively established in our time (Glinka, 1971). It was von Mohl, too, who first wrote of an "antagonism" between epidermal and guard cells (see Chap. 4). Heath (1938) used von Mohl's puncturing technique to show that guard cell deformations are not due to differential imbibition of layers of the cell wall as postulated in the "wall mechanism" hypothesis of Nadel (1935); and Mouravieff (1956) effectively applied this technique again to isolate the stomatal apparatus (*l'appareil stomatique isolé*).

During his studies with osmotic bathing solutions, von Mohl (1856) could control changes in the turgor of guard cells and could thereby confirm experimentally a hypothesis of his fellow botanist MATTHÄUS JACOB SCHLEIDEN (1804–81), of cell-theory fame, regarding the nature of guard cell deformation on a theoretical basis (Schleiden, 1842). The technique of altering cell turgor with osmotica was not advanced enough at the time to be used in determining the osmotic pressure of cell sap. This became possible only in our century, after HUGO DE VRIES (1848–1935) had discovered and named plasmolysis in 1884; the earliest published estimates of the osmotic pressures of guard cell sap are those of R. G. WIGGANS (1921). Finally, von Mohl introduced the use of epidermal peels for experiments on stomata and drew attention to the different properties of epidermal peels as compared with those of intact leaves.

Von Mohl's work provided the foundations for the vigorous studies about fifteen years later by N. J. C. MÜLLER, who discussed diverse mechanisms of guard cell deformations in different species (1870a, b). Müller detailed the best procedure for obtaining epidermal strips. In essence, it is the same as recommended by today's practitioners: score the opposite epidermis, kink or break the lamina through 120°, and pull the epidermis to be detached away from the mesophyll at right angles. In addition, Müller (1872) used a rocking movement with a curved blade to prepare serial sections (*Wiegenschnitte*) of single guard cells in order to ascertain the changes in their volume; he thus anticipated the use of microtome ribbons for that purpose (Raschke and Dickerson, 1973). Recalling Mariotte's work, Müller applied Graham's law of diffusion and Bunsen's gas volume calculations to his own studies of gas exchange through elegantly arranged epidermal tissue (see below).

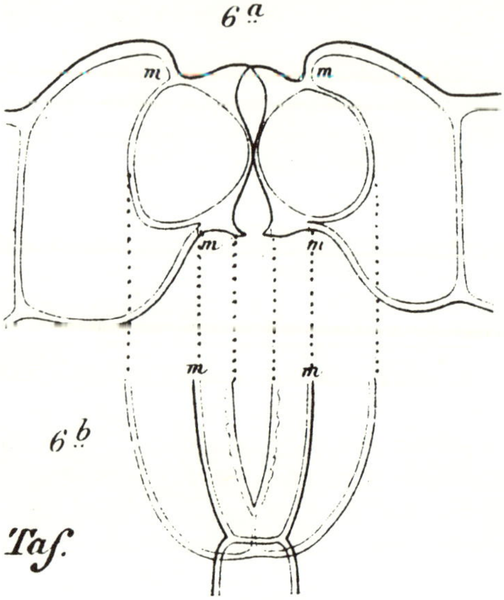

Fig. 1.3. N. J. C. Müller's diagram of a transverse section of a stoma of the *Amaryllis* type in *Clivia* spp. (Reproduced from Müller, 1872.)

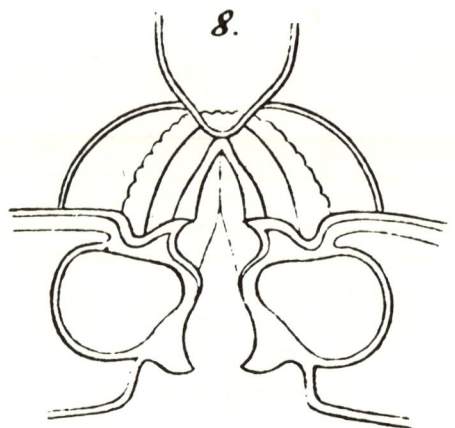

Fig. 1.4. Simon Schwendener's diagram of a transverse section of a stoma of *Amaryllis formosissima*. (Reproduced from Schwendener, 1881.)

The last and best-known anatomist of this era was SIMON SCHWENDENER (1829–1919), professor of botany at Basel. His work on the structure and mechanism of ellipse-shaped stomata and those found in the Gramineae and Cyperaceae has become the major reference work on this topic (Schwendener, 1881, 1889). Building on Schwendener's findings, later researchers observed that during the development of graminaceous stomata a stage occurs in which they are ellipse-shaped (Percival, 1921); the course of the changes in cell shape that this stage involves has lately been beautifully demonstrated by Palevitz (1981). Schwendener distinguished between stomatal types in *Helleborus*, *Amaryllis*, and *Mnium*, extending and refining von Mohl's observations. He emphasized the guard cell's *Hautgelenk* and the anchorage at the polar ends, writing about the elastic suspension of guard cells "as if hinged." This elastic suspension can be observed during manipulations with the pressure probe before the probe enters the lumen: both guard cells move elastically together as if loosely suspended between the surrounding cells. Changes in the orientation of the guard cells around Schwendener's *Hautgelenk* in the plane of the leaf affect both the dimensions of the outermost (eisodial) pore and the overall topography of the leaf. Thus, insofar as the boundary layer depends on surface features, it will change with stomatal movements.

Schwendener's anatomical studies were extended and complemented by Haberlandt (1887), who added several morphological features and introduced the term "stomatal complex"; HERMANN ZIEGENSPECK (1938, 1955) used dichroic stains to demonstrate the orientation of the micelles in guard cell walls, and FRIEDL WEBER (1937) made models of stomata from rubber tubes and distinguished various organelles by vital-staining techniques.

Salisbury (1928) drew attention to changes in stomatal frequencies due to the influence of water supply during leaf growth, height of insertion, and sun and shade. He also introduced the concept of *stomatal index*, $100[s/(s + e)]$, where s is the number of stomata and e is the number of epidermal cells per square millimeter. The stomatal index is fairly constant for leaves of a particular species.

Research Materials and Methods

With a sound knowledge of the morphology and anatomy of guard cells and an understanding of the elementary mechanisms of their movements, critical attention turned to methods and materials used in investigations into stomatal behavior. As already mentioned, von Mohl and Müller investigated aspects of stomatal and cuticular functioning in gas exchange. Von Mohl employed the most direct method of measuring changes in stomatal aperture, namely, microscopic observation on the intact leaf and on epidermal strips. Probably everyone who works with stomata has used microscopic measurements at some time. Stålfelt used this method almost exclusively, working always with intact leaves immersed in liquid paraffin. He painstakingly measured several fundamental stomatal responses, as will be detailed below. Microscopic measurements of pore widths on preserved material (Lloyd, 1908) and replica methods have been found to be beset with faults and are rarely used in stomatal work. The infiltration method was introduced by Molisch (1912) and much used in field work.

Mass flow porometry began with the instrument of Francis Darwin and D. F. M. Pertz (1911). Similar porometers were developed and refined during the first half of this century. Porometers overcome the problems posed by the variability of stomatal apertures in one sample and remain the most convenient method for estimating—albeit for mass flow—changes in stomatal-plus-leaf-airspace conductance. Measurements with mass flow porometers are reliable only for amphistomatous and heterobaric leaves—that is, for leaves with about equal numbers of stomata in either epidermis and with their stomata distributed in such a manner that the stomata are at both ends of airspaces at right angles to the plane of the leaf, an arrangement that allows for very little lateral air movement. Such leaves do not as a rule undergo marked changes in thickness, and hence their airspace conductance remains fairly constant. The complexity of measurements with mass flow porometers has been analysed by Penman (1942). With the reservations mentioned, mass flow measurements reflect most directly changes in pore dimensions and are thus suitable for the study of stomatal responses *per se*.

An ingenious type of water vapor diffusion porometer was used by N. J. C. Müller (1870a, b) when he fitted epidermal tissue over one end of a glass U-tube containing water and measured the volume of water that escaped as vapor with a capillary fitted to the other end of the U-tube. Such a system, using epidermal strips and with some necessary improvements, might well be tried again. Diffusion porometers using inert or radioactive gases are rather complex instruments developed since 1936. They estimate the diffusive conductance of the stomata plus that of the leaf airspace system, which is more relevant in studies of gas exchange capacity than estimates of mass flow conductance. Gases used were hydrogen (Gregory and Armstrong, 1936), nitrous oxide (Slatyer and Jarvis, 1966), argon (Moreshet, Stanhill, and Koller, 1968), and helium (Farquhar and Raschke, 1978).

After Müller, several forms of water vapor diffusion porometers were constructed. Stahl (1894) used cobalt chloride papers, and Darwin (1898) developed a horn hygroscope; in our time, electronic sensing devices were perfected (e.g. Wallihan, 1964; Beardsell, Jarvis, and Davidson, 1972). These instruments measure stomatal-plus-cuticular conductance and require accurate control of leaf temperature and atmospheric humidity; they assume vapor saturation of the leaf airspace. Thus, the shortcomings of mass flow porometers mentioned above and those of the water vapor diffusion porometers show that these instruments are appropriate to different kinds of investigations.

The Diffusion Path

Much of the early work on stomata had shown that there is a close relation between changes in rates of transpiration and CO_2 uptake and changes in stomatal aperture, so that it was a logical development for investigations into stomata to deal with that relationship in greater detail. The classic measurements were made by H. Brown and F. Escombe (1900) for both water vapor and CO_2; but because their measurements of water vapor diffusion through small pores in artificial membranes could not be reconciled with their measurements of transpiration from a sunflower leaf, much subsequent work was aimed at determining whether stomata really control the rate at which water vapor is lost from leaves.

F. E. LLOYD (1908), working at the Carnegie Institution of Washington, had reported that stomata did not regulate the rate of transpiration. In answer, Darwin (1916) wrote that Lloyd's conclusion was suspect because of his use of preserved epidermal peels, a material known to be unreliable. However, O. RENNER (1910), working in Graz, had already pointed out that Brown and Escombe failed to take into account the conductance of the leaf boundary layer; he thereby explained the discrepancy between theory and experimental results. A quantitative analysis of transpiration and the contribution of the leaf boundary layer was subsequently undertaken by Penman and Schofield (1951) and by Bange (1953).

FRANCIS DARWIN (1848–1925) concluded from the results of experiments he designed especially for this purpose that transpiration varies with and is regulated by stomatal aperture (Darwin, 1916). He was aware of the need for corrections to be made on account of cuticular vapor loss, which he attempted to estimate, and on account of changes in the intercellular volume of transpiring leaves, later experimentally quantified (Meidner, 1952, 1955a). Moreover, it was Francis Darwin who, on the advice of Sir Joseph Larmour, professor of mathematics at Cambridge, first suggested the square-root relation between diffusive flow and mass flow, which was further developed by Maskell (1928) and later tested experimentally (Meidner and Spanner, 1958).

Types, Phases, and Rhythms of Movements

J. V. G. LOFTFIELD (1921), also working at the Carnegie Institution of Washington, carried out his measurements of stomatal movements in the field. These are often referred to and are of historical interest because, among other things, Loftfield observed *midday closure* and the effects of temperature on rates of stomatal movements. Both of these came to be studied experimentally in the middle of the twentieth century.

Between 1927 and 1935, MARTIN GOTTFRIED STÅLFELT (1891–1968) published his papers on the relation between stomatal aperture and transpiration rate, which must be classed as of fundamental importance, together with those of Brown and Escombe, Renner, and Francis Darwin. Stålfelt (1927, 1928, 1929) identified stages and types of stomatal movement, assigning them terms that have endured and have lately received new justification: his *Spannungsphase* ("tension phase"), *Motorphase* ("motor phase"), and *hydroactive* and *hydropassive* (or "transient") *movements* have remained useful concepts in stomatal physiology. The reality of the *Spannungsphase* was later experimentally demonstrated with the pressure probe (Meidner and Edwards, 1975) by showing that pressures that were large enough to cause very slightly open stomata to open farther were insufficient to open originally closed stomata. It would seem therefore that a change, possibly connected with proton movements, has to occur in the properties of the guard cell wall before available turgor relations can initiate opening. Stålfelt's (1929) notion of an "optimum leaf water deficit" (*optimal Wasserdefizit*) has also remained a valuable concept, because it relates the degree of stomatal opening to the antagonism postulated by von Mohl and to a role of epidermal cells in stomatal mechanics. Before Stålfelt's studies, Darwin (1916) and R. C. KNIGHT (1916) had both noted transient stomatal movements and the influence of the water content of the leaf. Knight (1916) had recorded transient stomatal movements following leaf excision, which were later quantitatively analyzed (Meidner, 1965).

Stålfelt's data for transpiration in still air and moving air in relation to stomatal conductance provided the basis for detailed quantitative evidence showing that in leaves with an *adequate supply of water* transpiration rates in moving air are a function of stomatal aperture over the whole range of openings, whereas in still air the stomatal control of transpiration becomes less pronounced after stomata are about one-third open (Bange, 1953). In addition, to confirm his analysis, Bange examined the published results of Sierp and Seybold (1927) and of Huber (1930).

Diurnal rhythms in stomatal movement were discerned by Darwin and several other investigators who observed both changes in opening ability or rates of opening, normally found to be greater in the morning than in the afternoon, and the *night opening* occurring in many species before dawn. Rhythms in stomatal movements in continuous light and darkness were documented by Knight (1922), Sayre (1923), Maskell (1928), and Stålfelt (1929, 1963). That the true character of these rhythms is endogenous (circadian) and subject to phase shift was established with the systematic use of a recording porometer (Heath and Mansfield, 1962; Martin and Meidner, 1971; Heath and Meidner, 1983). Circadian stomatal rhythms are of ecological interest, and an awareness of their occurrence is a prerequisite for the controlled growth and preparation of experimental material.

Short-term oscillations in stomatal aperture were first observed by Stålfelt (1927, 1929). Gregory and Pearse (1937) found such oscillations superimposed on the diurnal rhythm they recorded. Stålfelt suggested that the gradual "pulsating" adjustment of the leaf water deficit under certain conditions was responsible for stomatal oscillations—via what today would be called a "feedback" system. Detailed investigations into short-term stomatal oscillations (Barrs and Klepper, 1968; Hopmans, 1971) were carried out mostly with leaves of rooted plants growing in hydroponic solutions and exposed to *sudden* changes in temperature, photon flux density, or atmospheric humidity. These experimental arrangements indicate that in a feedback system such oscillations originate in overshoot phenomena (Farquhar and Cowan, 1974).

Hypotheses of Mechanisms

Von Mohl (1856) hypothesized that the turgor of the guard cells provides the motive force for stomatal movement. In the first half of the twentieth century, other hypotheses were advanced. K. LINSBAUER (1917) postulated a *permeability theory*, according to which light-dependent changes of permeability in guard cells provide the motive force. G. W. SCARTH (1929) proposed in his *amphoteric colloid theory* that stomatal movement is mediated by pH-controlled swelling of colloids outside their isoelectric ranges. M. NADEL (1935) proposed a *wall mechanism*, whereby differential swelling of portions of the walls of the guard cells governed stomatal movement. However, the turgor hypothesis was generally favored; but the problem remained to identify the factors responsible for and the mechanism bringing about these turgor changes. Stomatal responses to illumination have been known since 1856 and were long thought to be related to the presence of chloroplasts in guard cells, supposedly involved in carbon assimilation. Lloyd (1908), advancing his *starch-sugar hypothesis*, postulated a light-stimulated conversion of starch to sugar; but when it was realized that illuminated stomata open also in CO_2-free air, it became clear that stomatal movement could not depend on any immediate product of carbon assimilation. It was also noted that if leaves were kept for prolonged periods in blue light and thus had destarched guard cells, their stomata nevertheless functioned normally. An extension of Lloyd's theory was therefore called for.

J. D. SAYRE (1923, 1926), working at Ohio State University, wrote of an indirect effect of light on stomata, operating via the intercellular concentration of CO_2, which would be reduced on account of photosynthesis in the mesophyll. Sayre referred to differences in pH that he had estimated with indicators in the sap of guard cells of open and closed stomata. He thought that such changes in pH might affect enzyme activity such as the postulated involvement of phosphorylase in the breakdown of starch. Although degradation of starch to glucose-1-phosphate would not bring about a change in osmotic pressure of guard cell sap, it had been reliably observed that the starch content of guard cells did decrease as stomata opened and increased when they closed. It was known, however, that starch-sugar conversion does not usually proceed so quickly as stomatal movement, although under certain circumstances, for instance, in wilting leaves, the formation of starch in guard cells can be astonishingly rapid (Iljin, 1922, 1930). But Heath (1949), working with *starch-free* guard cells of onion, showed that starch-sugar conversion was not essential to the stomatal mechanism; and I. MOURAVIEFF (1952), working in Lyon, showed that in any case light induced only very small changes in the quantity of starch in guard cells.

While the hypothesis of a pH-dependent conversion of starch to sugar was gradually being superseded, Scarth (1932) and Pekarek (1934) drew attention once more to the difference in the pH of guard cell sap of open and closed stomata. With today's knowledge of malate and proton metabolism in guard cells, these pH changes may be viewed in a different light.

Environmental Factors

Research had advanced rapidly with the development of refined techniques for measuring stomatal responses. After Darwin and Pertz's porometer of 1911, Laidlaw and Knight (1916) constructed a recording porometer, and Knight (1916) his double-cup porometer, at Imperial College of Science and Technology in London. There, too, the most prolific period of investigations into stomatal responses to environmental factors began about 1940: Gregory and Pearse (1934) designed the resistance porometer, and Heath and Russell (1951) the Wheatstone bridge porometer. These sensitive instruments, with detachable or ventilated leaf attachment cups, were suitable for statistically designed systematic investigations. A sustained research into stomata was thus started that by direct and indirect descent continues to this day.

The first series of investigations to be recalled includes the discovery of the influence of the intercellular concentration of CO_2 on stomata, a topic still at the forefront of research. As attention moved from starch-sugar conversion to light-dependent changes in the concentration of CO_2 in the leaf airspace, it was recalled that Linsbauer (1916) had noted an effect on stomata of changes in the intercellular concentration of CO_2 due to photosynthesis. A systematic investigation into the role of CO_2 in the stomatal mechanism had been made by H. FREUDENBERGER (1940) in Germany during the war, but because of the time and place of its publication this work was not widely known elsewhere.

Shortly after the war, O. V. S. HEATH (1948, 1950) observed that stomata in leaf areas permanently enclosed in unventilated attachment clamps had wider pores than those of the same leaf outside the clamps; this finding was the beginning of new experimentation to study the effects of intercellular CO_2 concentration on stomatal behavior. The decisive experiments were those of Heath and Russell (1954), who established that stomatal movement was affected by the interaction of light and the intercellular concentration of CO_2. Besides the vigorous introduction of experimental design to allow for the statistical analysis of results, these investigations included the control of the intercellular concentration of CO_2 (i.e. C_i), which is of such great interest today. This was achieved by passing air of known concentration of CO_2 through amphistomatous leaves of wheat, which are well suited for this technique both because the upper and lower stomatal densities are similar and because graminaceous leaves are not subject to substantial changes in thickness on account of their pattern of venation. At this time the heterobaric structure of leaves (see p. 13) like those of *Erithryna caffra* (Meidner, 1955b) and

Xanthium strumarium had not been discovered (Mansfield and Heath, 1963). Heath and Russell (1954) postulated the following direct and indirect effects of light on stomata:

First, an indirect light effect operating via the changes in the intercellular concentration of CO_2, beginning at 100 μmol mol^{-1} and extending beyond the normal concentration but becoming saturated at above 900 μmol mol^{-1} for C_3 leaves. When the zero CO_2-compensation point of C_4 leaves was discovered (Meidner, 1962), the range for these leaves was found to begin at zero CO_2 concentration.

Second, a direct effect thought to be due to photosynthetic events in guard cells that are now suspected of being exclusive of Calvin-cycle assimilation, but that are probably connected with ATP generation. This direct effect of light has now been confirmed (Sharkey and Raschke, 1981) and can be observed in epidermal strips and isolated protoplasts.

Two other effects, measurable in CO_2-free air, were also postulated: an indirect effect apparently transmitted from other cells by an agent not yet identified, and a direct effect perhaps related to the recently postulated blue-light receptor.

To these four must be added the effect of C_i itself. Most pronounced in the dark and with CO_2 concentrations between zero and 100 μmol mol^{-1} in the atmosphere surrounding the leaf, this effect has been attributed to respiratory CO_2 from epidermal cells and from guard cells.

Investigations into stomatal responses to colored light go back to F. G. KOHL (1895), but experiments attempting to keep the energy supply comparable in different wavebands began only with Sayre (1929), Paetz (1930), Sierp (1933), Harms (1936), Pyrkosch (1936), and Liebig (1942). Results obtained were contradictory, even as to whether blue or red light was the more effective for stomatal opening. However, Mouravieff (1958) concluded firmly that blue light caused greater stomatal opening than red and, incidentally, that it reduced starch in guard cells more than did red. This superiority of blue light over red was later confirmed and is no longer disputed (Kuiper, 1964; Meidner, 1968). In addition, a blue-light effect has been demonstrated in chlorophyll-free guard cells, so that the involvement of a blue-light receptor has been postulated (Zeiger and Hepler, 1977).

C. C. Wilson (1947), working at Duke University, documented the effects of light, temperature, and atmospheric vapor deficit on stomatal conductance, initially with measurements in the field and later in the laboratory. Wilson's work is noteworthy for three-dimensional graphs, so illuminatingly used by Heath and Russell (1954) for the analysis of results of laboratory experiments designed to elucidate the interaction of light and CO_2. Wilson used leaves with guard cells devoid of chloroplasts, investigated systematically later by Zeiger (1983) and by Jamieson and Willmer (1984).

Like Loftfield (1921), Wilson found that increases in temperature resulted in faster rates of stomatal opening to larger steady-state apertures. Subsequent work (Meidner and Heath, 1959; Stålfelt, 1962; Mansfield, 1965) confirmed Wilson's conclusion, with the added proviso that the intercellular concentration of CO_2 and the gradient in water vapor pressure between leaf and atmosphere must both remain constant to prevent midday closure. The temperature dependence of stomatal movements was later incorporated in a theoretical model of a feedback system of stomatal action (Raschke, 1975).

In respect of atmospheric vapor deficit, Wilson found that the largest stomatal openings occurred at low deficits and physiologically high temperatures (30° C) for the species used, provided that photon flux densities were moderate. At saturating photon-flux densities a plateau of widest stomatal opening was established, starting at medium atmospheric saturation deficits and temperatures; this plateau was maintained at high saturation deficits and temperatures. Systematic investigations of the humidity response of stomata have been undertaken since the late 1970's and are dealt with in this volume in Chapters 14 and 15.

Guard Cell Metabolism and Ion Transport

Although not specifically interested in stomata, MacCallum (1905) tested several plant tissues for the accumulation of potassium by staining with sodium cobaltinitrite and yellow ammonium sulphide. He discovered that guard cells were sites of considerable potassium concentrations, but he did not relate this finding to the state of stomatal opening. SHUN-ICHIRO IMAMURA (1943) published the results of a thorough investigation into the mechanism of changing turgor pressures in stomatal guard cells. He was primarily concerned with plasmolytic and deplasmolytic processes of guard cells bathed in salt solutions as compared with sucrose solutions and the effects of the presence of calcium ions in the incubation medium on these processes. His aim was the determination of limiting plasmolysis values for guard cells of stomata at different degrees of opening. Like Stålfelt before him, Imamura measured the influence of the state of turgor of the surrounding cells on the size of the stomatal aperture. During these investigations Imamura observed intracellular differences in guard cell structure between closed and open stomata and the changes that occurred during the *Spannungsphase*, which were later defined in more detail (Heller and Resch, 1967; Humbert and Guyot, 1972).

Imamura clearly discerned that the changes in the starch content of guard cells that occur during stomatal movements are not a primary occurrence that could explain guard cell deformations. But most important, it appeared to him that the primary event accompanying stomatal movement was a change in the concentration of potassium in the guard cells. However, his work did not immediately stimulate further research into ionic relations of guard cells, probably because he concentrated on attempting to demonstrate the validity of the amphoteric colloid theory (see p. 17), which was no longer in great favor. Ten years later, Yamashita (1952) showed that the potassium concentration in the sap of guard cells was highly correlated with stomatal aperture during diurnal cycles of opening and closing the stomatal pore.

It was not till 1967 that the ionic relations of guard cells became the focus of research, when MASAYOSHI FUJINO of the Faculty of Education, University of Nagasaki, published his account of the role of ATP and ATPase in stomatal movements. Although Fujino's emphasis was on a possible mechanism of energy availability for ion fluxes, he stated that "stomatal movement is caused by active transport of potassium between mesophyll and guard cells and the driving mechanism of potassium transport exists in the guard cells themselves" (Fujino, 1967). Subsequently, quantitative measurements of the potassium concentration in guard cells were related to degrees of stomatal opening (Fischer, 1968); and, at least for *Zea mays*, a "shuttle" of potassium between subsidiary and guard cells was found to occur during stomatal movements (Raschke and Fellows, 1971), while in *Commelina communis* a reversible gradient for potassium concentrations could be measured across the stomatal complex and the surrounding epidermal cells (Penny and Bowling, 1974).

Once the involvement of potassium ions in guard cell turgor relations had been recognized, proton movements with associated pH changes were investigated (Humble and Raschke, 1971). The study of the ionic movements (see Chap. 6), the biochemical changes in guard cells (see Chap. 5), as well as the behavior of isolated protoplasts (see Chap. 7), followed. Wright and Hiron (1969) and Mittelheuser and van Steveninck (1969) discovered that a hormone-like substance later identified as abscisic acid affects stomata, and many related research projects were soon under way. These are dealt with in this volume in Chapter 11, and work on other plant growth substances is discussed in this volume in Chapters 12 and 13.

Parallel with the research into ion transport and guard cell metabolism, another approach to the study of stomata has developed since the 1960's, namely, the application of control theory (Raschke, 1965). This served to clarify the interplay of factors affecting the turgor relations between epidermal and guard cells and the nature of the interactions between internal factors, such as epidermal and leaf water supply, transpirational vapor loss, intercellular concentration of CO_2, and photosynthesis, and, further, the dependence of these processes on external factors like photon flux density, atmospheric saturation deficit, and CO_2 concentration. New concepts, such as the optimization of stomatal conductance and water-use efficiency, were introduced.

The remarkable intensification of research into all aspects of stomatal physiology with an increased emphasis on their function in plant life has contributed to the emergence of a progressively more comprehensive scheme of stomatal physiology and function, which is further advanced by the contributions in this volume.

REFERENCES

Amici, G. B. 1823. Osservazione microscopiche sopra varie piante. *Societa italiana delle scienze*, Modena, 19: 20–26.
Bange, G. G. J. 1953. On the quantitative explanation of stomatal transpiration. *Acta botanica Neerlandica*, 2: 255–96.
Barrs, H. D., and E. Klepper. 1968. Cyclic variations in plant properties under constant environmental conditions. *Physiologia plantarum*, 21: 711–30.
Beardsell, M. F., P. G. Jarvis, and B. Davidson. 1972. A null balance diffusion porometer suitable for leaves of many shapes. *Journal of Applied Ecology*, 9: 677–90.
Bonnet, C. 1754. *Recherches sur l'usage des feuilles dans les plantes*. Göttingen: Luzac. [As quoted by Rehfous, 1917, q.v.]
Brown, H., and F. Escombe. 1900. Static diffusion of gases and liquids in relation to the assimilation of carbon and translocation in plants. *Philosophical Transactions of the Royal Society*, ser. B, 193: 223–91.
de Candolle, A. P. 1827. Chapter 6 in *Organographie végétale, ou description raisonée des organes des plantes*, 1: 78–88. Paris: Deterville.
_____. 1832. De l'émanation ou exhalation aqueuse des végétaux vasculaires. Chapter 4 in *Physiologie végétale*: 107–78. Paris: Béchet.
Darwin, F. 1898. Observations on stomata. *Philosophical Transactions of the Royal Society*, ser. B, 190: 531–621.
_____. 1916. On the relation between transpiration and stomatal aperture. *Philosophical Transactions of the Royal Society*, ser. B, 207: 413–37.
Darwin, F., and D. F. M. Pertz. 1911. On a new method of estimating the aperture of stomata. *Proceedings of the Royal Society*, ser. B, 84: 136–54.
Dutrochet, H. J. 1827. Nouvelles observations sur l'endosmose et sur la cause de ce double phénomène. *Annales de chimie et de physique*, 35: 393–400.
Farquhar, G. D., and I. R. Cowan. 1974. Oscillations in stomatal conductance. *Plant Physiology*, 54: 769–72.
Farquhar, G. D., and K. Raschke. 1978. On the resistance to transpiration of the sites of evaporation within the leaf. *Plant Physiology*, 61: 1000–1005.
Fischer, R. A. 1968. Stomatal opening: Role of potassium uptake by guard cells. *Science*, 160: 784–85.
Freudenberger, H. 1940. Die Reaktion der Schliesszellen auf Kohlensäure und Sauerstoffentzug. *Protoplasma*, 35: 15–54.
Fujino, M. 1967. Role of adenosine triphosphate and adenosine triphosphatase in stomatal movements. *Science Bulletin of the Faculty of Education, Nagasaki University*, 18: 1–47.
Garreau, M. 1850. Recherches sur l'absorption et l'exhalation des surfaces aériennes des plantes. *Annales des sciences naturelles*, 3d ser., 13: 321–40.
Glinka, Z. 1971. The effect of epidermal cell water potential on stomatal responses to illumination of leaf discs of *Vicia faba*. *Physiologia plantarum*, 24: 476–79.

Gregory, F. G., and J. I. Armstrong. 1936. The diffusion porometer. *Proceedings of the Royal Society*, ser. B, 121: 27–42.

Gregory, F. G., and H. L. Pearse. 1934. The resistance porometer and its application to the study of stomatal movement. *Proceedings of the Royal Society*, ser. B, 114: 477–93.

———. 1937. The effect on the behaviour of stomata of alternating periods of light and darkness of short duration. *Annals of Botany*, n.s., 1: 3–10.

Grew, N. 1682. *The anatomy of plants.* 2d ed. London: Royal Society.

Guettard, J. É. 1747. *Observations sur les plantes.* Paris: Durand.

Haberlandt, G. 1887. Zur Kenntnis des Spaltöffnungsapparates. *Flora*, 45: 97–109.

Harms, H. 1936. Beziehung zwischen Stomataweite und Lichtfarbe. *Planta*, 25: 155–93.

Heath, O. V. S. 1938. An experimental investigation of stomatal movement with some preliminary observations upon the response of the guard cells to "shock." *New Phytologist*, 37: 385–95.

———. 1948. Control of stomatal movement by a reduction in the normal [CO_2] of the air. *Nature*, 161: 179–81.

———. 1949. The role of starch in the light response of stomata. *New Phytologist*, 48: 186–211.

———. 1950. The role of carbon dioxide in the light response of stomata. *Journal of Experimental Botany*, 1: 29–62.

Heath, O. V. S., and T. A. Mansfield. 1962. A recording porometer with detachable cups operating on four separate leaves. *Proceedings of the Royal Society*, ser. B, 156: 1–13.

Heath, O. V. S., and H. Meidner. 1981. Feedback processes in the opening of leaf stomata in light. *Proceedings of the Royal Society*, ser. B, 213: 161–70.

Heath, O. V. S., and J. Russell. 1951. The Wheatstone bridge porometer. *Journal of Experimental Botany*, 2: 111–16.

———. 1954. An investigation of the light responses of wheat stomata with the attempted elimination of control by the mesophyll. *Journal of Experimental Botany*, 5: 1–15, 269–92.

Hedwig, J. 1793. *Sammlung seiner zerstreuten Abhandlungen* 1: 126. Leipzig: Crusius.

Heller, F. O., and A. Resch. 1967. Funktionell bedingter Strukturwechsel der Zellkerne in den Schliesszellen von *Vicia faba*. *Planta*, 75: 243–52.

Hopmans, P. A. M. 1971. Rhythms in stomatal opening of bean leaves. *Mededelingen van de Lanbouwhogeschool*, Wageningen, 71–73: 1–81.

Huber, B. 1930. Untersuchungen über die Gesetze der Porenverdunstung. *Zeitschrift für Botanik*, 23: 839.

Humbert, C., and M. Guyot. 1972. Modifications ultrastructurales des cellules stomatiques. *Comptes rendus de l'Académie des sciences*, Paris, 274: 380–82.

Humble, G. D., and K. Raschke. 1971. Stomatal opening quantitatively related to potassium transport. *Plant Physiology*, 48: 447–53.

Iljin, W. S. 1922. Über den Einfluss des Welkens der Pflanzen auf die Regulierung der Spaltöffnungen. *Jahrbücher für wissenschaftliche Botanik*, 61: 670–82.

———. 1930. Der Einfluss des Welkens auf den Ab- und Aufbau der Stärke. *Planta*, 10: 170–84.

Imamura, S.-I. 1943. Untersuchungen über den Mechanismus der Turgorschwankung der Spaltöffnungsschliesszellen. *Japanese Journal of Botany*, 12: 251–346.

Jamieson, A., and C. M. Willmer. 1984. Functional stomata in a variegated leaf chimera of *Pelargonium zonale* without guard cell chloroplasts. *Journal of Experimental Botany*, 35: 1053–59.

Knight, R. C. 1916. On the use of the porometer in stomatal investigations. *Annals of Botany*, 30: 57–76.
———. 1922. The interrelations of stomatal aperture, leaf water content and transpiration rate. *Annals of Botany*, 36: 1–83.
Kohl, F. G. 1895. Über Assimilationsenergie und Spaltöffnungsmechanik. *Botanisches Zentralblatt*, 64: 109–10.
Kolpikov, D. I. 1964. Observations on active non-stomatal gas and water exchange in higher plants. *Soviet Plant Physiology*, July/August 1964: 735–37. [Translated from *Fiziologia rastenii*, 11, no. 4, 1962.]
Kuiper, P. J. C. 1964. Dependence upon wavelength of stomatal movements in epidermal tissue of *Senecio ordoris*. *Plant Physiology*, 39: 952–55.
Laidlaw, C. G. P., and R. C. Knight. 1916. A description of a recording porometer and a note on stomatal behaviour during waiting. *Annals of Botany*, 30: 48–56.
Liebig, M. 1942. Untersuchungen über die Abhängigkeit der Spaltweite der Stomata von Intensität and Qualität der Strahlung. *Planta*, 33: 206–57.
Link, H. F. 1812. Recherches sur l'anatomie des plantes. *Annales du Muséum d'histoire naturelle*, 19: 307–44. [As quoted by Rehfous, 1917, q.v.]
Linsbauer, K. 1917. Beiträge zur Kenntnis der Spaltöffnungsbewegungen. *Flora*, 109: 100–143.
Lloyd, F. E. 1908. *The physiology of stomata*. Publications of the Carnegie Institution of Washington, no. 82. Washington, D.C.
Loftfield, J. V. G. 1921. *The behavior of stomata*. Publications of the Carnegie Institution of Washington, no. 314. Washington, D.C.
MacCallum, A.B. 1905. On the distribution of potassium in animal and vegetable cells. *Journal of Physiology*, 32: 95–128.
Malpighi, M. 1675. *Anatome plantarum*. London: Royal Society.
Mansfield, T. A. 1965. Stomatal opening in high temperature in darkness. *Journal of Experimental Botany*, 16: 721–31.
Mansfield, T. A., and O. V. S. Heath. 1963. Photoperiodic effects on rhythmic phenomena in stomata of *Xanthium strumarium*. *Journal of Experimental Botany*, 14: 334–52.
Martin, E. S., and H. Meidner. 1971. Endogenous stomatal movements in *Tradescantia virginiana*. *New Phytologist*, 70: 923–28.
Maskell, E. J. 1928. The relation between stomatal opening and assimilation—A critical study of assimilation rates and porometer rates in cherry laurel. *Proceedings of the Royal Society*, ser. B, 102: 488–533.
Meidner, H. 1952. An instrument for the continuous determination of leaf thickness changes in the field. *Journal of Experimental Botany*, 3: 319–25.
———. 1955a. Changes in the resistance of the mesophyll tissue with changes in leaf water content. *Journal of Experimental Botany*, 6: 94–99.
———. 1955b. The determination of paths of air movement in leaves. *Physiologia plantarum*, 8: 930–35.
———. 1962. The minimum intercellular space [CO_2], Γ, of maize leaves and its influence on stomatal movements. *Journal of Experimental Botany*, 13: 284–93.
———. 1965. Stomatal control of transpirational water loss. In G. E. Fogg, ed., *The state and movement of water in living organisms*: 185–203. Symposia of the Society for Experimental Biology, 19. Cambridge: Cambridge University Press.
———. 1968. The comparative effects of blue and red light on stomata of *Allium cepa* L. and *Xanthium strumarium*. *Journal of Experimental Botany*, 19: 146–51.
Meidner, H., and M. Edwards. 1975. Direct measurements of turgor pressure potentials of guard cells, I. *Journal of Experimental Botany*, 26: 319–30.

Meidner, H., and O. V. S. Heath. 1959. Stomatal responses to temperature and CO_2 in *Allium cepa* L. and their relevance to midday closure. *Journal of Experimental Botany*, 10: 206–19.

Meidner, H., and D. C. Spanner. 1959. The differential transpiration porometer. *Journal of Experimental Botany*, 10: 190–205.

Mittelheuser, C. J., and R. F. M. van Steveninck. 1969. Stomatal closure and inhibition of transpiration induced by RS-abscisic acid. *Nature*, 221: 281–82.

von Mohl, H. 1856. Welche Ursachen bewirken die Erweiterung und Verengung der Spaltöffnungen? *Botanische Zeitung*, 14: 697–704, 713–21.

Molisch, H. 1912. Das Offen- und Geschlossensein der Spaltöffnungen, veranschaulicht durch eine neue Methode (Infiltrationsmethode). *Zeitschrift für Botanik*, 4: 106–22.

Moreshet, S., C. S. Stanhill, and D. Koller. 1968. A radioactive tracer technique for the measurement of diffusion resistance. *Journal of Experimental Botany*, 19: 460–67.

Mouravieff, I. 1952. Variation de la quantité d'amidon dans les cellules stomatiques au cours de mouvements induits par la lumière. *Comptes rendus de l'Académie des sciences*, Paris, 234: 2637–39.

———. 1956. Action du gaz carbonique et de la lumière sur l'appareil stomatique isolé. *Comptes rendus de l'Académie des sciences*, Paris, 242: 926–27.

———. 1958. Action de la lumière sur la cellule végétale. *Bulletin de la Société botanique de France*, 105: 467–75.

Müller, N. J. C. 1870a. Über den Durchgang von Wasserdampf durch die geschlossene Epidermiszelle. *Jahrbücher für wissenschaftliche Botanik*, 7: 193–99.

———. 1870b. Untersuchungen über die Diffusion atmosphärischer Gase in der Pflanze und die Gasausscheidung unter verschiedenen Beleuchtungsbedingungen. *Jahrbücher für wissenschaftliche Botanik*, 7: 145–93.

———. 1872. Die Anatomie und die Mechanik der Spaltöffnungen. *Jahrbücher für wissenschaftliche Botanik*, 8: 75–116.

Nadel, M. 1935. On the influence of various fixatives on stomatal behaviour: A critical contribution to the theory of Lloyd's alcohol fixation method. *Palestine Journal of Botany*, 1: 22–42.

Paetz, K. W. 1930. Untersuchungen über die Zusammenhänge zwischen stomatärer Öffnungsweite and bekannter Intensitäten bestimmter Spektralbezirke. *Planta*, 10: 611–65.

Palevitz, B. A. 1981. The structure and development of stomatal cells. *Society of Experimental Biology, Seminar Series*, 8: 1–23.

Pekarek, J. 1934. Über die Aziditätsverhältnisse in den Epidermis- und Schliesszellen bei *Rumex acetosa* im Licht und im Dunkeln. *Planta*, 21: 419–46.

Penman, H. L. 1942. Theory of porometers used in the study of stomatal movements in leaves. *Proceedings of the Royal Society*, ser. B, 130: 416–33.

Penman, H. L., and R. K. Schofield. 1951. Some physical aspects of assimilation and transpiration. In J. F. Danielli and R. Brown, eds., *Fixation of carbon dioxide*: 115–29. Symposia of the Society for Experimental Biology, 5. Cambridge: Cambridge University Press.

Penny, M. G., and D. J. F. Bowling. 1974. A study of potassium gradients in the epidermis of intact leaves of *Commelina communis* L. in relation to stomatal opening. *Planta*, 119: 17–25.

Percival, J. 1921. *The wheat plant*. London: Duckworth.

Pyrkosch, G. 1936. Licht und Transpirationswiderstand, I. Die Transpirationswiderstände im monochromatischen Licht. *Protoplasma*, 26: 418–520.

Raschke, K. 1965. Die Stomata als Glieder eines schwingungsfähigen CO_2-Regelsystems. *Zeitschrift für Naturwissenschaften*, 20b: 1261–70.

———. 1975. Stomatal action. *Annual Review of Plant Physiology*, 26: 309–40.
Raschke, K., and M. Dickerson. 1973. Changes in shape and volume of guard cells during stomatal movement. *Plant Research*, 1972, 149–53.
Raschke, K., and M. P. Fellows. 1971. Stomatal movements in *Zea mays*. *Planta*, 101: 296–316.
Rehfous, L. 1917. Étude sur les stomates. *Bulletin de la Société botanique de Genève*, 2d ser., 9: 245–350.
Renner, O. 1910. Beiträge zur Physik der Transpiration. *Flora*, 100: 451–547.
Salisbury, E. J. 1928. On the causes and ecological significance of stomatal frequency with special reference to the woodland flora. *Philosophical Transactions of the Royal Society*, ser. B, 216: 1–65.
de Saussure, H. B. 1762. *Observations sur l'écorce des feuilles et des pétales*. Geneva: n.p. [As quoted by Rehfous, 1917, q.v.]
Sayre, J. D. 1923. Physiology of stomata of *Rumex patientia*. *Science*, 57: 205.
———. 1926. Physiology of stomata of *Rumex patientia*. *The Ohio Journal of Science*, 26: 233–67.
———. 1929. Opening of stomata in different ranges of wavelengths of light. *Plant Physiology*, 4: 323–28.
Scarth, G. W. 1929. The influence of H-ion concentration on the turgor and movement of plant cells with special reference to stomatal behaviour. In *Proceedings of the International Congress of Plant Sciences, Ithaca, N.Y.*, 2: 1151–62. Menasha, Wis.: Banta.
———. 1932. Mechanism of the action of light and other factors on stomatal movements. *Plant Physiology*, 7: 481–504.
Schleiden, M. J. 1842. *Grundzüge der wissenschaftlichen Botanik*, 3d ed., 1: 340. Leipzig: Engelmann.
Schwendener, S. 1881. Über Bau und Mechanik der Spaltöffnungen. *Monatsberichte der Königlichen preussischen Akademie der Wissenschaften zu Berlin, Physikalisch-mathematische Klasse*, July 1881: 833–67.
———. 1889. Die Spaltöffnungen der Gramineen und Cyperaceen. *Sitzungsberichte der Königlichen preussischen Akademie der Wissenschaften zu Berlin, Physikalisch-mathematische Klasse*, January 1889: 65–79.
Sharkey, T. D., and K. Raschke. 1981. Separation and measurement of direct and indirect effects of light on stomata. *Plant Physiology*, 68: 33–40.
Sierp, H. 1933. Untersuchungen über die Öffnungsbewegungen der Stomata in verschiedenen Spektralbereichen. *Flora*, 128: 269–85.
Sierp, H., and A. Seybold. 1927. Untersuchungen zur Physik der Transpiration. *Planta*, 3: 115–68.
Slatyer, R. O., and P. G. Jarvis. 1966. Gaseous diffusion porometer for continuous measurement of diffusive resistance of leaves. *Science*, 151: 574–76.
Stahl, F. 1894. Einige Versuche über Transpiration und Assimilation. *Botanische Zeitung*, 52: 117–23.
Stålfelt, M. G. 1927. Die photischen Reaktionen im Spaltöffnungsmechanismus. *Flora*, 121: 236–72.
———. 1928. Die Abhängigkeit der photischen Spaltöffnungsreaktionen von der Temperatur. *Planta*, 6: 183–96.
———. 1929. Die Abhängigkeit der Spaltöffnungsreaktionen von der Wasserbilanz. *Planta*, 8: 287–340.
———. 1962. The effect of temperature on opening of stomatal cells. *Physiologia plantarum*, 15: 772–79.
———. 1963. Diurnal dark reactions in the stomatal movements. *Physiologia plantarum*, 16: 756–66.

———. 1966. The role of the epidermal cells in the stomatal movements. *Physiologia plantarum*, 19: 241–56.
de Vries, H. 1884. Eine Methode zur Analyse der Turgorkraft. *Jahrbücher für wissenschaftliche Botanik*, 14: 427–601.
Wallihan, E. F. 1964. Modification and use of an electrical hygrometer for estimating relative stomatal aperture. *Plant Physiology*, 39: 86–90.
Weber, F. 1937. Ein Modell der Stomata-Bewegung. *Protoplasma*, 28: 119–23.
Wiggans, R. G. 1921. Variations in the osmotic concentrations of the guard cells during opening and closing of stomata. *American Journal of Botany*, 8: 30–41.
Williams, T. 1841. On the structure and uses of stomata. *Microscopic Journal and Structural Record*, 1: 118–21.
Wilson, C. C. 1947. The effects of some environmental factors on the movements of guard cells. *Plant Physiology*, 23: 5–37.
Wright, S. T. C., and R. W. P. Hiron. 1969. Abscisic acid, the growth inhibitor induced in wheat leaves by a period of wilting. *Nature*, 224: 719–20.
Yamashita, T. 1952. Influences of potassium supply upon various properties and movements of guard cells. *Sieboldia acta biologica*, 1: 51–70. [In Japanese, with English summary.]
Zeiger, E. 1983. The biology of stomatal guard cells. *Annual Review of Plant Physiology*, 34: 441–75.
Zeiger, E., and P. K. Hepler, 1977. Light and stomatal function: Blue light stimulates swelling of guard cell protoplasts. *Science*, 196: 887–89.
Ziegenspeck, H. 1938. Die Micellierung der Turgeszenzmechanismen, I. Die Spaltöffnungen. *Botanisches Archiv*, 39: 268–309.
———. 1955. Das Vorkommen von Fila in radialer Anordnung in den Schliesszellen. *Protoplasma*, 44: 385–97.

2

The Evolution of Stomata

Hubert Ziegler

When higher plants started to colonize land, only their subterranean organs (at first rhizomes, later on roots) found an environment with a relatively high water potential; the parts above ground were exposed to an atmosphere often at a very low water potential. To stabilize the high water potential necessary for active metabolism in the above-ground organs and to avoid rapid desiccation, terrestrial plants had to develop both an adequate system for absorbing and conducting water and a hydrophobic coating on their aerial parts. This coating is the cuticle in the primary, not secondarily thickened organs. Cutin and a structured cuticle evolved first among the Bryophyta, in the liverworts, the hornworts, and the mosses.

The main function of the cuticle is to decrease transpiration in aerial organs, and hence it must have a very low permeability to water vapor (Table 2.1). However, the cuticle also restricts the inward diffusion of CO_2 and, to a lesser extent, of oxygen. A cuticle without pores would allow at best for an adequate respiration rate, since the O_2 concentration gradient between the air and the interior of the respiring cell is steep, and since respiration needs less O_2 as fuel than photosynthesis does CO_2. Therefore, tissues that are respiring but not photosynthetically active, like those of fruits or whole parasitic higher plants (see pp. 39, 51), are able to grow with limited gas exchange and quite often either are devoid of stomata or have only nonfunctional stomata. A supply of CO_2 sufficient for intensively photosynthesizing tissues would, however, be impossible with an uninterrupted cuticular cover.

The need to balance CO_2 uptake and water vapor loss was achieved by the evolution of gas-permeable openings in the epidermis and its cuticle. In their simplest design these are small, permanently open pores; in more advanced designs they are hydraulically operated valves whose openings are adjustable depending on specific demands.

The balance of gas exchange is discussed in detail in Chapters 15 and 20. It involves mainly the regulation of stomatal action. Another serious problem

---

The author wishes to express his thanks to Professor Hans Meidner for critically reading the manuscript, improving the English, and making many valuable suggestions; to Professor Karl Mägdefrau for help in searching the literature; and to Professor Otto L. Lange for help in getting fresh *Anthoceros* material.

TABLE 2.1. Permeability coefficients of isolated cuticles of *Citrus aurantium* leaves and *Lycopersicon esculentum* fruits for different gases

Gas	Permeability coefficient (m s^{-1})	
	Citrus	*Lycopersicon*
H_2O	1.09×10^{-9}	2.47×10^{-8}
CO_2	5.40×10^{-8}	2.20×10^{-7}
O_2	3.05×10^{-7}	1.10×10^{-6}

SOURCE: Lendzian (1984).

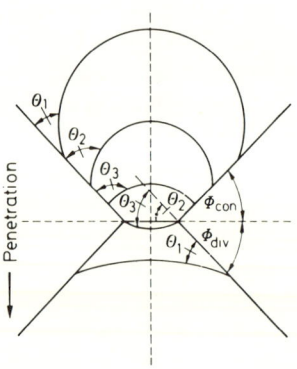

Fig. 2.1. Schematic drawing showing the ability of a liquid to rise in a conical capillary. The wall angles in the converging (Φ_{con}) and diverging (Φ_{div}) portions are both 45°. The menisci of three hypothetical liquids forming contact angles of 25° (Θ_1), 45° (Θ_2), and 90° (Θ_3) are indicated. Note that in the converging portion $\Phi_{con} + \Theta$ is always less than 180°; the penetrating pressure therefore is positive, as is indicated by the positive curvature of all three menisci, and all three liquids will rise in this portion of a capillary. At the constriction the capillary diverges, and only the liquid for which $\Theta = 25°$ can advance in this portion, because $\Phi_{div} \geqq \Theta_1$ and the pressure remains positive. The other two liquids cannot pass the constriction, since the penetrating pressure is either zero ($\Phi_{div} = \Theta$) or negative ($\Phi_{div} \leqq \Theta$). (Reproduced from Schönherr and Ziegler, 1975.)

for a plant with permanently or periodically open pores in the epidermis is the need to avoid the penetration of liquid water from outside into the subporous or substomatal cavity and the intercellular space, because this would have disastrous consequences for the gas exchange of the photosynthesizing cells. This requirement was met in the pores of the air chambers of some liverworts and in the stomata of the sporophytes of the bryophytes as well as of higher plants by specific characteristics of their architecture, as outlined below.

Principally, the penetration of a liquid through a capillary depends on the capillary pressure (Schönherr and Ziegler, 1975). In a capillary of circular cross section and radius r, the sign of this pressure, P, is determined by the contact angle, Θ, that the liquid of surface tension γ forms with the wall of the capillary:

$$P = \frac{2\gamma \cos \Theta}{r}$$

A liquid will rise in a capillary without applied pressure only if P is positive, that is, only if $\Theta < 90°$. Figure 2.1 shows the constricted portion of a conical capillary and the menisci of three hypothetical liquids. The penetrating pressure in the upper, converging portion is given by

$$P = \frac{2\gamma \sin(\Phi_{con} + \Theta)}{r}$$

and in the lower, diverging portion by

$$P = \frac{2\gamma \sin(\Phi_{div} - \Theta)}{r}$$

where Φ_{con} and Φ_{div} represent the wall angles of the converging and diverging portions, respectively.

The important difference between a cylindrical and a conical capillary is that the sign of P in the latter is determined by both the contact angle and the wall angle. In a converging capillary, P is positive when $(\Phi_{con} + \Theta) < 180°$. In a diverging capillary, P is positive when $0 < (\Phi_{div} - \Theta) < 180°$ or when $\Phi_{div} = 0$.

We conclude from this consideration that a proper air pore or stoma should have a diverging portion, thereby preventing the entrance of liquid water. The protection against infiltration is better, the smaller the wall angle Φ_{div}. A constriction, like an iris shutter protruding at right angles into the pore, has $\Phi_{div} = 0$ (see Fig. 2.2). This would optimize protection, since only liquids with zero contact angle could penetrate.

The contact angle itself is dependent on the nature of the surface of the pore wall and on the surface tension of the penetrating liquid. A criterion for the nature of a surface in this context is the critical surface tension; this term is defined as the surface tension of a liquid below which the contact angle with the surface becomes zero. Since water gives large contact angles on nonpolar surfaces, the pore and especially also the constriction should have a nonpolar coating.

All proper air pores and stomata of terrestrial plants have cutin as surface material; the critical surface tension of cutin is about $25-30 \times 10^{-5}$N cm^{-1}. Because of this low critical surface tension of the cutinous surface, and because of the geometry, and especially the cuticular ridges, of typical stomata, stomatal pores can be penetrated by liquids only if their surface tension is below about 30×10^{-5}N cm^{-1}. This is not the case for water, the surface tension of which in contact with air at 18° C is 73.05×10^{-5}N cm^{-1}, or for the aqueous solutions that may occasionally wet the aerial parts of terrestrial plants in nature.

We cannot question the view of Porsch (1905), who states, "Das älteste Stadium, gewissermassen der erste historische Schritt nicht nur zur Differenzierung eines Transpirationsregulators, sondern auch einer Einrichtung zur Aufnahme von Kohlensäure, wird wohl das Auftreten von Intercellularräumen

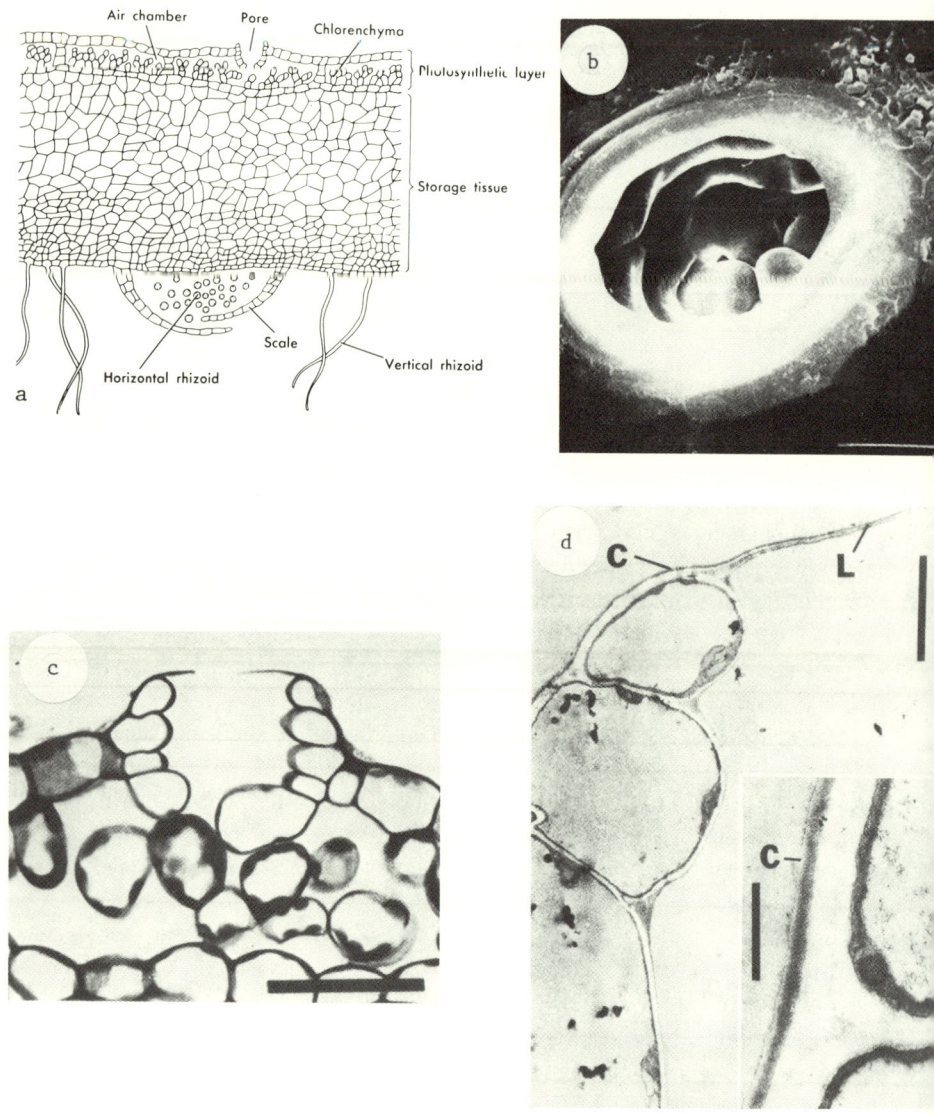

Fig. 2.2. Air chambers and air pores in *Marchantia* gametophytes. *a*, Cross section of *Marchantia polymorpha* thallus. (Reproduced with permission from Doyle, 1970.) *b*, Scanning electron micrograph of the air pore of *Marchantia paleacea*. Bar = 20 μm. *c*, Photomicrograph of cross section of an air pore in *Marchantia paleacea*. Bar = 10 μm. *d*, Transmission electron micrograph of an air pore of *Marchantia paleacea*. Note the cuticle (*C*) on the ledges (*L*). Bar = 20 μm, and 1μm in the inset, respectively. (*b*–*d* reproduced from Schönherr and Ziegler, 1975.)

zwischen beliebigen Epidermiszellen gewesen sein."* However, an opening only between the epidermal cells, as in the scent-producing spadix of some Araceae, is not a functioning air pore in the sense described above.

The first proper air pore, allowing gas exchange and preventing water entrance, had to have divergent portions (i.e. ridges) in the opening and a cuticle covering not only the surface but also the central pore. The first functioning stoma had to have in addition the ability to regulate the width of the central pore according to changing demand. Both these structures, proper air pores and functioning stomata, are found already in the Bryophyta.

The Air Pores of Liverwort Thalli

The fossil record of liverworts dates back at least to the Carboniferous, possibly to the Devonian (see Doyle, 1970). The gametophytes of these fossil liverworts were very similar to those of contemporary species; they developed a considerable complexity, the most advanced of all gametophytes in the plant kingdom. In *Marchantia* species, for example, a photosynthetic layer is situated in air chambers that are connected with the surrounding atmosphere by air pores (Fig. 2.2a–d). A closer look at the design of these pores shows that they are perfectly constructed according to the criteria described above for a proper air pore. They have protruding ridges, the surface of which has a critical surface tension of less than $30.4 \times 10^{-5} \mathrm{N} \, \mathrm{cm}^{-1}$, indicative of methyl or methylene groups. Evidence has been presented that the ledge is covered with a layer of cutin. No liverwort gametophyte has true stomata, not even in taxa in which the sporophyte has functioning stomata.

Pores and Stomata of Hornworts

There are no fossil records of the class Anthocerotae (hornworts), even though hornworts are found all over the world. The lower surface of the thallus in many species of hornworts has stomalike pores (Fig. 2.3), the development of which is similar to that of true stomata. However, the cavities below these pores are usually filled with slime, and the two cells surrounding each pore remain thin-walled and do not regulate the pore size; the pore remains completely open. According to Doyle (1970) this unique pore-chamber system may be interpreted as a vestige of an extensive aerating system that functioned either in the early but now-extinct hornworts or in the progenitors of hornworts.

Most remarkable is the structure of the hornwort sporophyte, which exhibits the greatest complexity of all bryophytes (Fig. 2.4a–c). The capsule contains a

---

*"The oldest stage, the first historical step, so to speak, not only toward the differentiation of a transpiration regulator but also of a setup for CO_2 uptake, was quite probably the formation of intercellular spaces between any epidermal cells."

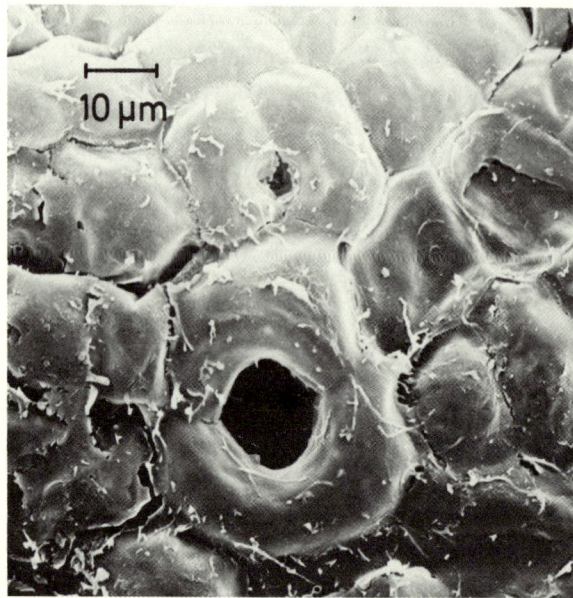

Fig. 2.3. *Anthoceros* sp. Pores on lower surface of gametophyte thallus. (Scanning electron micrograph courtesy of Jutta Künne.)

highly differentiated photosynthetic tissue and two out of five genera (*Anthoceros* and *Phaeoceros*) have true stomata and substomatal chambers (Fig. 2.4a, c). It is not surprising that the stomata of the hornwort sporophyte represent the most primary type, called "archetype" (Archetypus) by Kraus (1914), even in a specialized form (Fig. 2.5a). In this type the guard cells show thin walls on the ventral as well as on the dorsal side; an increase in turgor leads to a rounding of the lumina of the guard cells by vaulting the walls perpendicular to the plane of the organ surface without changing the positions of the dorsal wall.

The fact that even the most differentiated gametophytes did not develop stomata is interpreted by Porsch (1905), following ideas of von Wettstein (1901–3), as evidence that the gametophyte is phylogenetically the water generation, and the sporophyte the aerial generation. Stomata are assumed to be characteristic for the aerial generation. And in fact, the thalloid bryophytes grow mostly in wet biotopes.

Stomata of Mosses

In mosses, stomata occur only on the sporophyte; the occurrence of stomata on the calyptra (gametophytic tissue) of *Encalyptra ciliata* (Janzen, 1916) has not been confirmed. Stomata were first described in mosses by Treviranus (1811). Brown (1819) considered the very large stomata of *Lyellia crispa* as

Evolution of Stomata 35

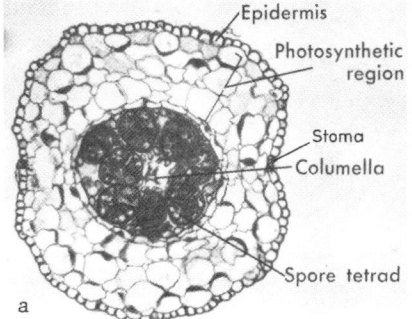

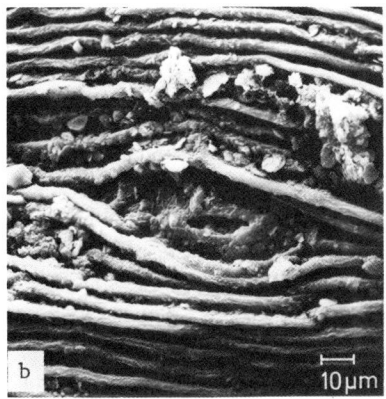
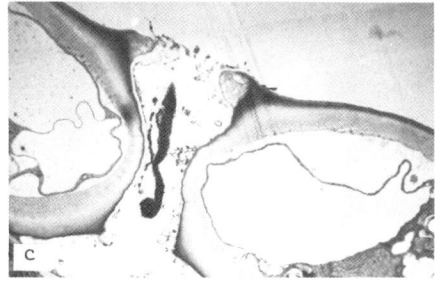

Fig. 2.4. Hornwort sporophyte. *a*, Cross section of the capsule with stomata. (Reproduced with permission from Doyle, 1970.) *b*, Scanning electron micrograph of a stoma from the capsule. *c*, Transmission electron micrograph of a cross section through a stoma from the capsule, × 2,000. (*b* and *c*, original pictures from *Anthoceros crispulus* [Montagne] Douin., collected in Ruppertshütte, Spessart, Germany, by Professor H. Volk, Würzburg.)

pores for spore distribution. Many famous botanists have since contributed to our knowledge of the anatomy, development, and phylogeny of moss stomata, including von Mohl (1838), Naegeli (1842), Schimper (1848), Strasburger (1867), Haberlandt (1886b), and Goebel (1906, 1908, 1915). Stomatal evolution in mosses was treated in detail by Porsch (1905) and Kuhlbrodt (1923). For additional historical details, see Bierschenk (1971).

Only in a few cases (e.g. all *Sphagnum* species) are stomata distributed around the surface of the whole capsule, and this pattern of distribution is regarded by Goebel (1906) as a primitive character. In most mosses, stomata are restricted to special regions of the capsule. In a few cases they are found near the peristome or in the middle of the capsule; normally, however, they are confined to the basal region, or apophysis. The apophysis is part of the theca, not of the seta, of the capsule (Haberlandt, 1886b), and it is in many cases the main photosynthetic tissue of the sporophyte. In connection with this function the apophysis became sterile and developed stomata with a different frequency and distribution. The distribution is strongly influenced by the special devel-

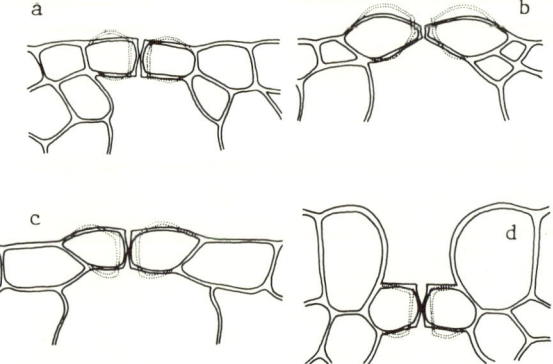

Fig. 2.5. Some types of stomatal movement in moss sporophytes. Dotted lines indicate the structure of the open stoma. *a*, Splachnaceae, etc.; *b*, *Polytrichum strictum*; *c*, Funariaceae and Bryaceae; *d*, Mniaceae. (Reproduced with permission from Bierschenk, 1971.)

opment and the final structure of the apophysis. Goebel (1930) postulated that initially the stomata became concentrated on an extending apophysis, and then the apophysis and its stomata were reduced again. The number of stomata per capsule shows some taxonomic regularities (Bierschenk, 1971), but it is also influenced by environmental factors (Schürmann, 1959).

Only a few moss capsules are devoid of stomata altogether (70 out of 900 European mosses, or 6.5%; see Bierschenk, 1971). Such species are found in all systematic groups (Acrocarpi and Pleurocarpi) and are especially frequent among the aquatic mosses. The absence of stomata is not paralleled by other reductions in capsule morphology, for example, of the peristome. Since there are frequently remnants of stomata on sporophytes that lack functioning stomata, it seems justified to assume that stoma-free moss capsules are evolutionarily advanced, not primitive (see Kuhlbrodt, 1923; Bierschenk, 1971). Stoma-free capsules have a thin-walled epidermis, which apparently allows adequate gas exchange, or else the epidermal cells themselves are photosynthetic.

Moss stomata resemble the archetype mentioned above or a modification of this type (Fig. 2.5). Porsch (1905) used the term "Muscineen-type" (Muscinean type) to characterize the form and function of the stomata of mosses and hornworts; at present, the term "*Mnium* type" (after Schwendener, 1881) is most often used in this sense. Within the moss stomata, Kuhlbrodt (1923) distinguished the archetype, the *Funaria* type, the *Polytrichum* type, and the *Mnium* type. Bierschenk (1971) emphasizes that large stomata with long pores have guard cells with a lumen that is flat and broad in cross section. When the stomata are smaller and the pores more rounded, the guard cell walls are thicker, and the nature of their movement is modified. Since such modifications can be present among the stomata of a single capsule, they cannot be considered a phylogenetic series.

Bierschenk (1971) distinguishes two types of stomatal movement in mosses (Fig. 2.5):

1. Unhindered vertical movement of outer and inner lateral walls that are long and have flat forms in cross section. Modifications are the *Polytrichum*, *Funaria*, and *Bryum* types (Fig. 2.5a–c), with some differences in the deformation of the outer and inner lateral walls during movement.

2. Restricted vertical deformation with the movement of only one (mostly the inner) wall (Fig. 2.5d), typical of guard cells that are oval or round in cross section. In this group we quite frequently find nonfunctional stomata with thick walls. These are usually combined with a reduced photosynthetic tissue and thus are considered a reduced, not a primitive, type.

Moss stomata are without morphologically or functionally specialized subsidiary cells ("Spaltöffnungen ohne Nebenzellen": Solereder, 1899). During stomatal development in mosses, the initial epidermal cell divides into one epidermal cell and the stoma mother cell. In contrast to those of most higher plants, the guard cells of the moss stomata are generally larger than or at least as large as the epidermal cells. This has consequences for the storage capacity of these cooperative cells during the shuttle of osmotic substances in the course of stomatal movement, if in fact such a shuttle occurs. We know very little about these physiological processes within the moss stomata (see Garner and Paolillo, 1973).

During the development of the moss stomata a repeated change of the micellar orientation in the guard cell walls can be observed with the polarizing microscope. Generally, in the moss stomata the micellar orientation does not play a role comparable to that in the stomata of higher plants (see Ziegenspeck, 1938–39), since the moss stomata do not bend their dorsal walls during movement and are embedded in a rigid epidermal frame.

Especially manifold are variations of the normal stoma type in the mosses. A special case of a still-functioning stoma is the one-celled stoma (Fig. 2.6), in which the pore is normally developed but the two guard cells are fused. It is typical for the Funariaceae but is present also in many Polytrichaceae and in *Buxbaumia aphylla*, for example. Haberlandt stated as early as 1886(b) that this fusion is achieved by the resorption of the dividing wall after division of the mother cell and after formation of the pore. It is interesting that the dividing wall is not complete before degradation starts but has openings at the polar ends, whence the resorption of the wall begins. This developmental process resembles that of the polar pores in the guard cells of the Gramineae.

Many reductions, tranformations, and malformations occur in moss stomata in connection with the reduction of the photosynthetic system of the

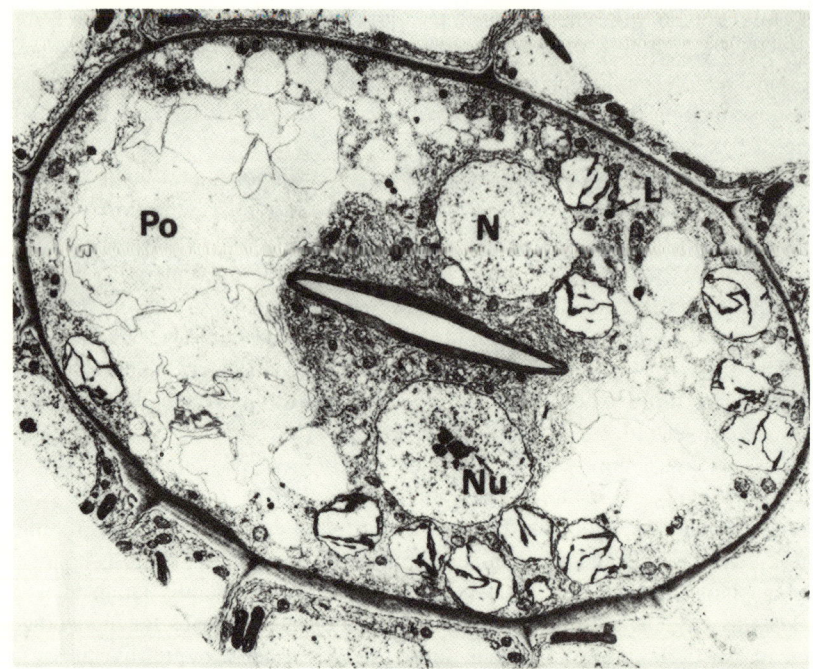

Fig. 2.6. *Funaria hygrometrica*. Paradermal section of mature one-celled stoma. *L*, lipid globule; *N*, nucleus; *Nu*, nucleolus; *Po*, polar region; × 3,000. (Reproduced with permission from Sack and Paolillo, 1983.)

sporophyte, within a single capsule as well as across species (Fig. 2.7). They are, therefore, clearly characterized as a declining series, starting with normal, functioning stomata, proceeding through many nonfunctioning malformations, and finally reaching sporophytes without any indication of stomata (see Kuhlbrodt, 1923). Very similar malformations can be found on special organs (e.g. cotyledons, fruits, petals; see Fig. 2.7) or in specialized higher plants (heterotrophic parasites and submerged aquatic plants) that are clearly derived from normal photoautotrophic plants.

All the observations mentioned so far indicate that the phylogenetically primitive mosses had a well-developed sporophyte with effective photosynthetic tissue and with well-developed, functioning stomata. This idea was emphasized by Haberlandt (1886a), who cultivated young sporophytes of *Funaria* and *Physcomitrium* in mineral solutions and achieved in these very first tissue-culture experiments the formation of capsules and spores. The strong dependence of the sporophyte in many mosses on the delivery of organic material from the gametophyte is, therefore, a secondary characteristic.

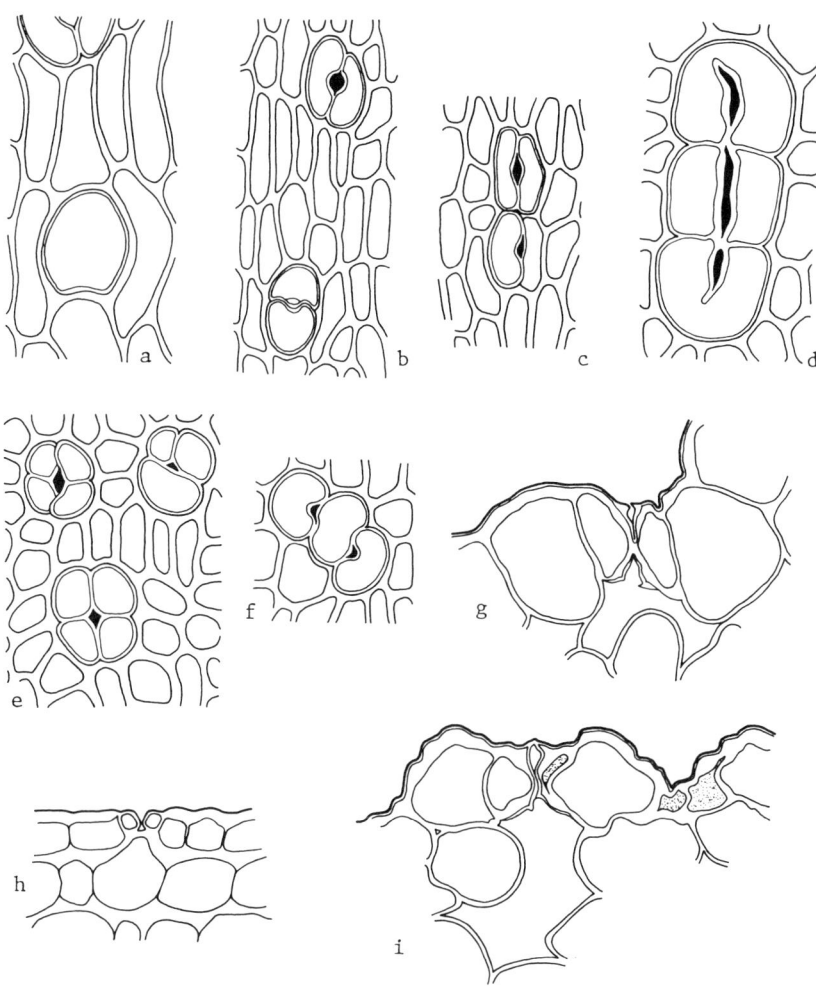

Fig. 2.7. Reductions and malformations of stomata. *a*, *Rhynchostegium murale* (moss): undivided stoma; *b*, *Platyhypnidium riparoides* (moss): normal and transversely divided stoma; *c*, *Anomodon attenuatus* (moss): serial twin stoma with one epidermal cell serving as guard cell; *d*, *Polytrichum alpinum* (moss): serial triplet stoma with degraded polar walls between the guard cells; *e*, *Dichodontium pellucidum* (moss): apophysis with three-celled and four-celled stomata; *f*, *Oxyrhynchium swartzii* (moss): triplet stoma. (*a–f* reproduced with permission from Bierschenk, 1971.) *g*, *Neottia nidus-avis*, Orchidaceae (parasite): stoma of the flower stalk, with ventral walls grown together. (Reproduced from Porsch, 1905.) *h*, *Atropa belladonna* (Solanaceae): epidermis of the fruit with closed substomatal cavity and fixed guard cells. (Reproduced from Fischer, 1929.) *i*, *Ruscus hypoglossum* (Liliaceae): reduced stomata on rudimentary leaves of the young plant. (Reproduced from Porsch, 1905.)

Stomata of the Pteridophyta

Members of the Pteridophyta were the first terrestrial plants in which the sporophyte (the aerial generation) dominated relative to the gametophyte. The first taxa colonizing the firm but still-swampy land were the Psilophytatae, which existed at the turn of the Silurian and Devonian (about 400 M.Y.B.P.); the Psilophytatae are considered the basic forms from which all other pteridophytes derived. The pteridophytes, as might be expected from their manifold morphological types of relative systematic independence, have developed a considerable diversity in stomatal form and ontogeny. We will consider the most important types, with special emphasis on their phylogenetic significance.

Psilophytatae. All the genera of this extinct taxon analyzed to date (*Rhynia*, *Hornea*, and *Asteroxylon*) exhibited stomata and cuticles. This is especially remarkable, since these plants did not have leaves or roots; stomata, then, are phylogenetically older than leaves! Whereas the *Hornea* stomata are not well preserved (they are reported to be similar to those of *Rhynia*), the stomata of *Rhynia* and *Asteroxylon* where intensively studied by Zimmermann (1927) and were found to be remarkably different.

The *Rhynia* stomata were situated at the leafless shoot. The cross section (Fig. 2.8a) shows uniformly thick outer and inner guard cell walls and, like most of the other pteridophyte or gymnosperm stomata, a well-developed ledge only on the outer wall. In general, this type of stoma is not far from the archetype mentioned above (e.g. the stoma of *Anthoceros*, Fig. 2.4). Since the cells neighboring the guard cells had no inside space to allow for deformation during stomatal opening (see p. 51), it can be assumed that only the guard cells themselves changed their shape during movements by bending their inner wall, as happens in some mosses (Fig. 2.5d).

The stoma of *Asteroxylon* (Fig. 2.8b) is characterized by a very thick ventral and a relatively thin dorsal wall. This type resembles the stomata of xerophytic angiosperms and may have functioned in a similar way, that is, by a bending of the dorsal walls against the neighboring cells during opening. This type is so specialized that it cannot be considered basic for other pteridophytes.

Psilotatae. The *Psilotum* species are still living and, in contrast to the Psilophytatae, develop true leaves. *Psilotum* has stomata on the surface of the stem (Fig. 2.8c), and the stem has a photosynthetic tissue and an extensive air-space system. The stomata are completely different from those of the Psilophytatae. The outer lateral and dorsal walls are quite thick, and the cross-sectional view of the stoma resembles strongly those of some gymnosperm stomata (e.g. *Bowenia*, Fig. 2.11c). This similarity is emphasized by the fact that the guard cell walls of *Psilotum* are partly lignified, as are those of

Evolution of Stomata   41

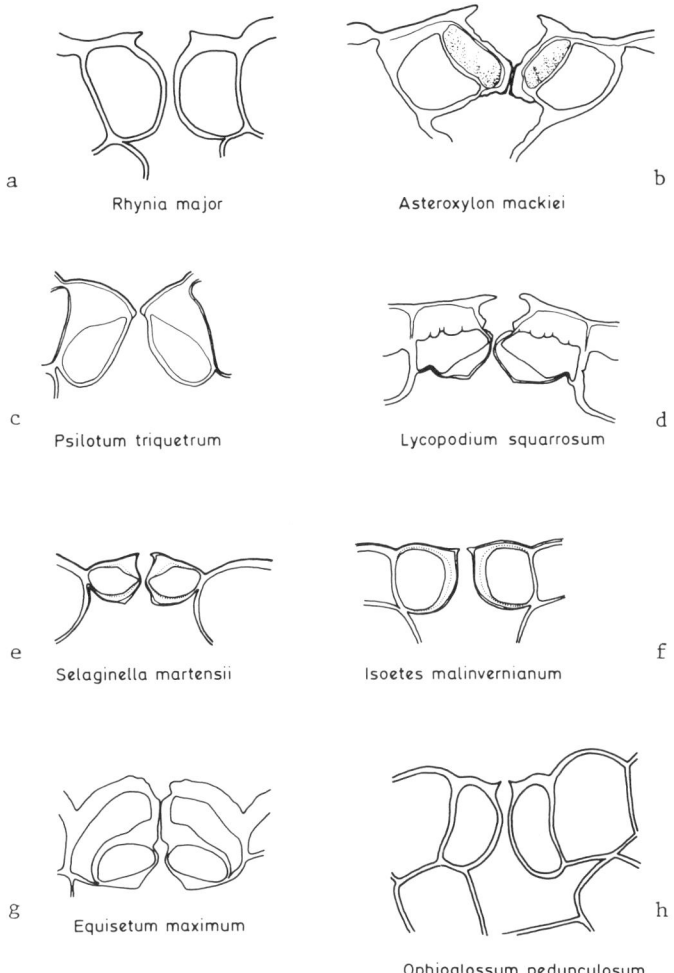

Fig. 2.8. Stomatal types of the Pteridophyta. *a–c*, Reproduced from Zimmermann (1927); stippled areas in *b* indicate lignified portions of cell wall. *d–g*, Reproduced from Riebner (1926). *h*, Adapted from Probst (1971).

Bowenia and other gymnosperms. Recall that lignin is not present in the Bryophyta, although there are a few cases of lignin-like compounds. After the appearance of lignin and lignification in the Pteridophyta, this substance was produced luxuriantly in the pteridophytes and gymnosperms (not so much in the angiosperms!) also for partial strengthening of guard cell walls. This lignification of the guard cell walls results in a distinct directional deformation of the guard cells during swelling and shrinking.

Lycopodiatae. Vascular plants of this class have been found in Lower Devonian strata contemporaneous with the oldest known Psilophytatae. Today, only five genera of lycopsids, all herbaceous, survive: *Lycopodium, Selaginella, Isoetes, Stylites,* and *Phylloglossum.* The stomata of the first three genera were studied in detail by Riebner (1926) and were found to be of quite different construction.

The stomata of the Lycopodiaceae (Fig. 2.8d) can be considered a special variation of the archetype. As in the *Mnium* type (*sensu stricto,* Fig. 2.5d), the outer lateral and dorsal walls of the guard cells are thick, and the ventral and inner lateral walls at least partially thin. In contrast, the well-developed outer ridge and the weakly expressed inner ridge are typical pteridophytic attributes. Conspicuous are wall ledges protruding from the outer and inner guard cell walls into the cell lumen. The guard cell walls are partly lignified. The opening of the *Lycopodium* stoma is achieved by a bending of the thin inner wall into the substomatal cavity, whereby the epidermal hinge ("Hautgelenk": Schwendener, 1881) at the dorsal wall of the guard cell acts to tilt the guard cell outward.

The stomata of the Selaginellaceae (Fig. 2.8e) represent the true pteridophytic type: the guard cells have thin outer and inner lateral walls and a thick dorsal wall. The outer ledge is again much better developed than the inner. The Selaginellaceae do not develop ledges on the inner side of the guard cell wall (which faces the substomatal cavity). The type of movement employed resembles that of Kraus's (1914) archetype: the outer and inner walls of the guard cell bend and widen the central pore.

In the Isoetaceae only the non-submerged species develop stomata. These are on the stems and resemble the floating-leaf type ("Schwimmblatt-Typus," Fig. 2.8f: Haberlandt, 1918).

Equisetatae. The Equisetaceae are characterized by a stomatal type unique in the plant kingdom (Fig. 2.8g). It consists of guard cells and subsidiary cells in an arrangement resembling that in the gymnosperms (see Fig. 2.11). The stomata of the Equisetaceae, however, have ledges on the wall of the subsidiary cell adjoining and external to the guard cell that radiate from the central pore. In addition, the guard cell walls are partly silicified. The opening of the stoma is achieved by a rounding-off of the guard cell lumen, followed by an inward bending of the thin ventral walls of the subsidiary cells, whereby the central pore (formed here by the ventral walls of the subsidiary cells) and the cuticular ridges of the guard cells below the pore are widened.

Filicatae. The first studies on fern stomata date back to Strasburger (1867) and Hildebrand (1866). The most detailed report was given by Probst (1971).

An equal distribution of stomata on all vegetative organs (except the root) is found in *Ophioglossum* species and can be considered phylogenetically primitive. A progressive attribute is the restriction of stomata to distinct regions

Evolution of Stomata   43

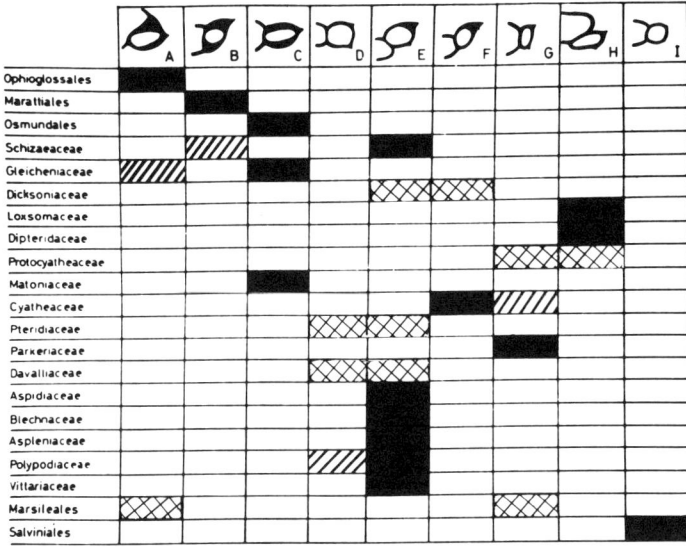

Fig. 2.9. Types of guard cells in the different families of ferns. *A, Ophioglossum* type (see Fig. 2.8h); *B, Anemia* type (see Fig. 2.10); *C, Osmunda* type (see Fig. 2.10); *D,* Archetype (see *Adiantum,* Fig. 2.13c); *E, Diplazium* type (Fig. 2.13d); *F,* Floating-leaf type (Fig. 2.15); *G, Ceratopteris* type; *H, Dipteris* type; *I, Azolla* type. *Solid blocks,* majority of fern species exhibit this type; *cross-hatched blocks,* about half of the ferns exhibit this type; *single-hatched blocks,* only a few species show this type. (Reproduced with permission from Probst, 1971.)

of the leaf (e.g. in *Schizaea, Lindsaya,* etc.) or even the complete absence of stomata (e.g. in the very thin-leafed Hymenophyllaceae and Hymenophyllopsidaceae). Especially remarkable is the occurrence of stomata on prothalloid outgrowths from isolated young leaves of the fern *Ceratopteris thalictroides,* since these regenerating tissues also develop antheridia and rhizoids, typical gametophytic structures (Goebel, 1908).

On the basis of the shape of the guard cells, Probst (1971) classified the fern stomata into nine groups (Fig. 2.9), which cannot be described here in detail. As Kraus (1914) showed, there are transitions from one of these types to another, even within one genus: for example, from the archetype to a floating-leaf type in the genera *Aspidium, Blechnum, Pteris,* and *Alsophila.*

The orders Ophioglossales, Marattiales, and Osmundales are very strictly characterized by special stoma types. The *Ophioglossum* type (Fig. 2.8h) has remarkable similarities with the *Psilotum* type and with the gymnosperm type. Ziegenspeck (1941) considers the *Marattia* type as primary and as the starting point for the evolution of the gymnosperm type and of the different types of the Filicales.

In many cases, the ontogeny of a structure is more helpful for phylogenetic evaluations than is the structure itself, since the ontogeny is more conservative.

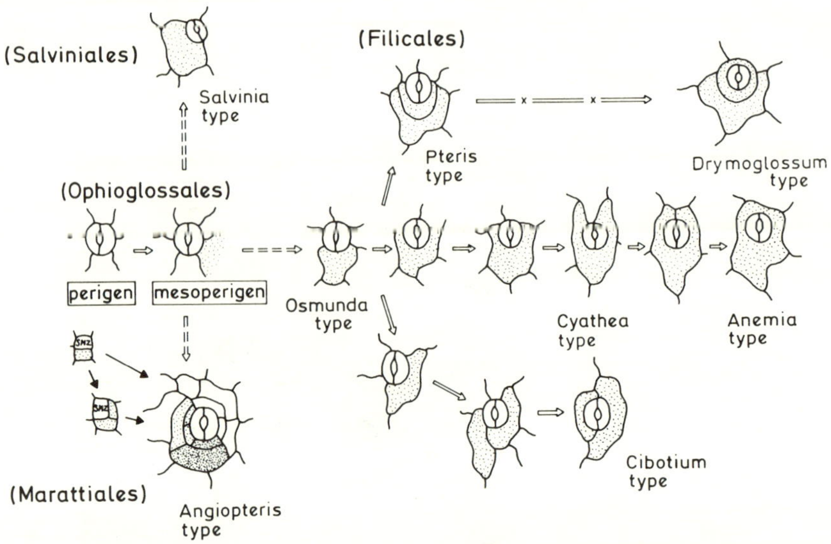

Fig. 2.10. Evolution of the types of stomatal development in ferns. *SMZ*, stoma mother cell. (Reproduced with permission from Probst, 1971.)

Stomatal ontogeny has proved valuable for phylogenetic characterization of the ferns and gymnosperms (for the latter, see p. 46).

The most primitive stomatal development is characterized by the appearance of a protodermal cell that becomes the guard-cell mother cell without an unequal cell division ("perigen" or "haplocheilic" development: Florin, 1931). Within the Filicatae this type was considered characteristic for the eusporangiate (primitive) Ophioglossales (Fig. 2.9; see also Maroti, 1961; Kondo, 1962; Pant and Khare, 1969). As Probst (1971) showed, the formation of the guard-cell mother cell in some *Ophioglossum* species is preceded by an unequal cell division ("meso-perigen"). To this meso-perigen type belongs the "vertical two-cutting-faces" type (*Angiopteris* type), characterized by two divisions of the primary mother cell ("Urmutterzelle": Probst, 1971) before formation of the guard-cell mother cell. Besides *Angiopteris*, other Marattiales belong to this type.

In most of the Filicales the marginal daughter cell of the primary mother cell develops into the stoma initial. Since the wall formed during the unequal cell division is bent, the sister cell encircles the guard cells like a horseshoe. This basipolar, horseshoelike sister cell often differs from other epidermal cells (smaller, nonundulated anticlinal walls: *Pteris* stomatal development, Fig. 2.10). It is characteristic of the Filicales, as Hildebrand (1866) emphasized, and it is not present in other Pteridophyta or in gymnosperms. We can assume that it arose from the more unspecific *Osmunda* type of stoma. From this, another kind of stoma with an annular sister cell can develop when the spindle apparatus during cell division forms an angle of more than 45° with the leaf plane (e.g. in species of *Aneimia*, *Cyclophorus*, *Drymoglossum*, etc.).

Another evolutionary series leading to an annular neighboring cell of the guard cells includes species still having one anticlinal wall in the annular neighboring cell. Yet another series includes several members with lateral neighboring cells only (e.g. *Cibotium*). A first step in this series could be an asymmetrical basipolar neighboring cell (e.g. *Cyathea capensis*). Within the Saviniales the guard-cell mother cell is mostly formed in one of the corners of the primary mother cell. As in the mosses, stomatal malformations occur frequently in the Filicatae. They are found in greater numbers on petioles and sporophylls. Most common are duplications or multiplications of stomata and transversely divided stomata (Probst, 1971).

Regulation of movement in pteridophyte stomata. Little work has been devoted to the study of the regulation of movement of pteridophyte stomata (Willmer and Pallas, 1973; Lösch and Bressel, 1979). There is some evidence that a K^+ shuttle may play a role similar to that in spermatophytes. A number of ferns show peculiar substomatal structures situated in an extracellular position between the pole of the guard cell complex and the polar subsidiary cell (Stevens and Martin, 1977; Lösch and Bressel, 1979). These discrete bodies are able to absorb ions (especially K^+) and may play a role in the ion exchange between the guard cells and the subsidiary cells. However, since similar structures have also been found in *Tradescantia*, they may not be specific for pteridophytic species and, therefore, not very useful as phylogenetic markers. Martin et al. (1984) showed that linear photosynthetic electron transport is carried out in the numerous and conspicuous guard cell chloroplasts of the ferns *Dryopteris ludoviciana* and *Platycerium bifurcatum*, just as it is in angiosperm guard cells, in which the number of chloroplasts is severely reduced.

Stomata of the Gymnosperms

There are numerous papers concerning gymnosperm stomata; see the extensive papers of Florin (1931, 1933) and of Napp-Zinn (1966) for detailed reviews. Only a few phylogenetically relevant data are mentioned here.

The stomata of gymnosperm leaves function as gas exchange pores, not as water pores (Lippmann, 1925). Sprecher (1907) considers the permanently open stomata with round pores on the cotyledons of *Ginkgo* as water pores; this remains to be proved.

In their guard cell structure, the main differences among the various gymnosperm taxa are development of antechambers and rear chambers formed by the ridges (if present), extension of the cuticle in the region of the central pore (throat), and pattern of lignification of the guard cell walls. It is striking that the gymnosperm stomata generally lack well-developed outer and inner

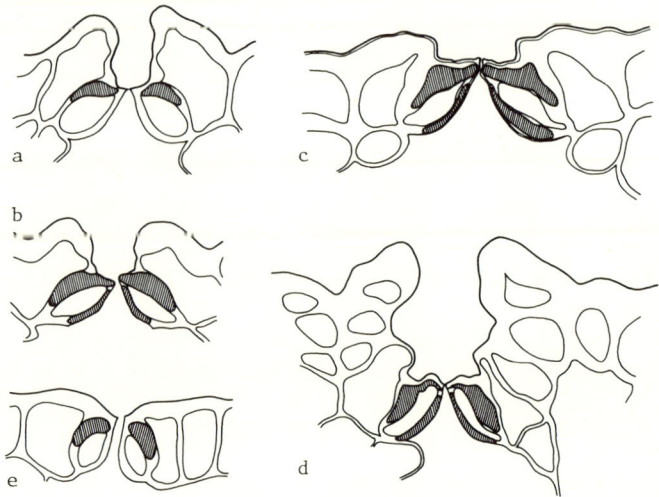

Fig. 2.11. Stomatal types in gymnosperms, with indication (shaded areas) of the lignification of the guard-cell walls. *a, Gingko biloba*; *b, Podocarpus neriifolia*; *c, Bowenia spectabilis*; *d, Dioon edule*; *e, Gnetum gnemon*. (Reproduced from Porsch, 1905.)

ridges, at least one of which, according to our introductory remarks, is essential for preventing the entrance of liquid water. In most cases the cuticle does not enter the substomatal cavity in the gymnosperms, although it is usually present up to the central pore (see Florin, 1931). It may be that the lack of developed structures for preventing entry of liquid water has some correlation with the well-known filling of the outer chamber of many gymnosperm stomata with the waxy, water-impermeable substance first reported by Link (1827). A detailed study is needed to test this interesting postulate.

As mentioned before (p. 41), like the guard cell walls of the pteridophytes, those of the gymnosperms are normally heavily lignified. The dorsal wall lignifies earlier than the ventral wall, and in a few cases the ventral wall remains unlignified: for example, in the stomata of the cotyledons of a few species and of the leaves of *Gingko* (Fig. 2.11a) and *Gnetum* (Fig. 2.11e; see Mahlert, 1885, and Florin, 1931).

The development of gymnosperm stomata has been intensively studied in many species (see Rehfous, 1923; Florin, 1933). The two main types of differentiation were briefly mentioned in connection with the stomata of the pteridophytes (p. 44). In one type the stoma mother cell produces not only the guard cells by a longitudinal division but also the lateral ("mesogenic") neighboring cells. This type was called "syndetocheilic" (compound labiate) differentiation by Florin (1931). The neighboring cells can develop directly into subsidiary cells, or they can by an additional division produce a subsidiary cell

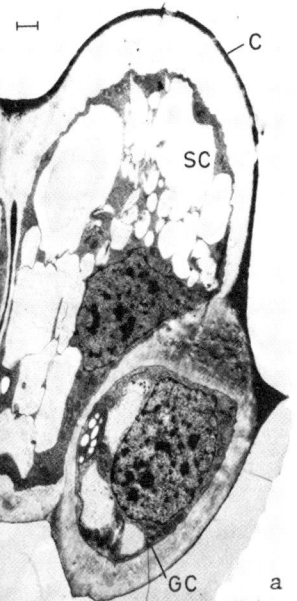 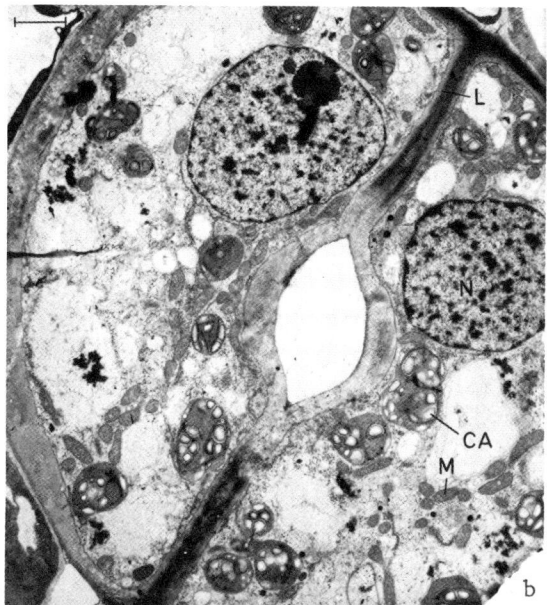

Fig. 2.12. Stomatal structure in *Welwitschia mirabilis*. *a*, Cross section showing half of the stoma: *GC*, guard cell; *SC*, subsidiary cell; *C*, cuticle. *b*, Paradermal section: *L*, lignified cell wall; *N*, nucleus; *CA*, chloroamyloplast; *M*, mitochondrion. Bar, 1 μm.

and a Kranz cell. Such syndetocheilic stomata are reported for the extinct Bennettitales and may be present also in the advanced gymnosperms *Welwitschia mirabilis* (Fig. 2.12a, b) and *Gnetum* (Florin, 1931). This would not allow great phylogenetic conclusions, especially since the presence of this stomatal type in the Gnetinae is doubted (Maheshwari and Vasil, 1961a, b).

All the other gymnosperm stomata develop by haplocheilic (simple labiate) differentiation, in which the lateral neighboring cells (perigenic cells) are not produced by the primary mother cell but by protodermis cells adjacent to the guard-cell mother cell. This developmental pattern is again divided into monocyclic and amphicyclic subtypes. The former is characterized by the development of the neighboring cells into subsidiary cells without further divisions and is represented by the genera *Torreya*, *Pseudotsuga*, *Pseudolarix*, *Larix*, *Pinus*, *Sciadopitys*, and *Thujopsis*. In the amphicyclic subtype the neighboring cells divide longitudinally, producing one layer of subsidiary cells and one to three layers of Kranz cells adjacent to the subsidiary cells. All other gymnosperm genera belong to this subtype.

There are some reports on malformations of stomata on gymnosperm leaves, but their phylogenetic significance is not clear.

Stomata of the Angiosperms

The greatest variety in the structure and function of stomata and their derivatives is found in the angiosperms. The first studies were carried out by Strasburger (1867) and by Vesque (1889), who recognized within the dicotyledons four categories based on the presence and arrangement of accessory cells and their mode of development. Only a few examples of phylogenetic trends can be touched on here.

There are three main stomatal types among the angiosperms shown with the other main stomatal types in the plant kingdom in Figure 2.13. These three are distinguished by the architecture of the guard-cell/subsidiary-cell complex, by the fine structure of the walls or part of the walls of the guard cells, by the manner in which the guard cells are connected to the subsidiary cells, and by the nature of the deformation resulting from these characteristics. In the *Amaryllis* type the guard cells move almost exclusively in a plane parallel to that of the organ. In the *Helleborus* type, which is widespread within the monocotyledons and dicotyledons, the increase in the volume of the guard cell extends the guard cell perpendicular as well as parallel to the surface of the leaf. Finally, in the stomata of the grasses the enlarged polar ends of the guard cells swell in such a way that the central parts move apart, forming the pore. This type is found in the Gramineae, Cyperaceae, Juncaceae, and Restionaceae.

Lignification of guard cell walls is rare in the angiosperms; it has been found in a few Gramineae and Pandanaceae and in *Laurus nobilis*, *Quercus ilex*, *Hedera helix*, and *Mahonia aquifolium* (Kaufmann, 1927).

All angiosperm stomatal types are characterized by a strong movement of the guard cells in the direction of the subsidiary cells during the opening process. This bending of the dorsal guard cell walls, or of parts of these in the stomata of grasses, is based on the special orientation of the micelles in these walls ("Radiomicellierung": Ziegenspeck, 1938–39). Another precondition for the opening of the angiosperm stomata is the substomatal cavity, which allows a deformation of the compressed subsidiary cell to the side, where there is no obstacle posed by a stiff cuticle.

Organization of the stomatal complex in monocotyledons. The different organizations of the stomatal complex in the leaves of monocotyledons and their bearing on phylogeny were studied by Stebbins and Khush (1961) and may serve here as an example for such considerations. The authors recognized

Fig. 2.13 (facing page). Stomatal types in angiosperms (*g–i*) in comparison with the most important types of stomata in lower taxa (*a–f*). Superimposed images and arrows indicate direction of and stoma shape after movement. *a*, Redrawn from Copeland, 1902. *b*, Redrawn from Haberlandt, 1886b. *c*, Redrawn from Kraus, 1914. *d, e, g, h*, Redrawn from Zimmermann, 1927. *f*, Redrawn from Klemm, 1886. *i*, Gramineaen stomatal apparatus: α, top view; β, bulbous ends of guard cells with open connection between the two cells: *left*, closed stoma; *right*, open stoma. (β redrawn from Schwendener, 1881.)

Mnium type (Archetype)

Anthoceros punctatus
a

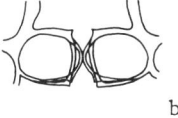
Mnium cuspidatum
b

Adiantum capillum veneris
c

Diplazium type

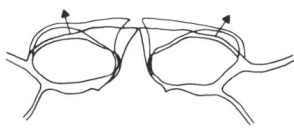
Diplazium celtidifolium
d

Psilotum type

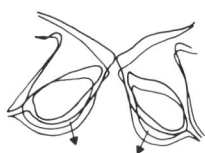
Psilotum sp.
e

Gymnosperm type

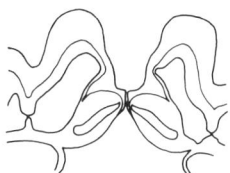
Juniperus macrocarpa   f

Angiosperm types

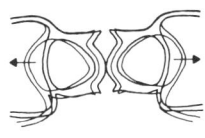
Amaryllis type   g

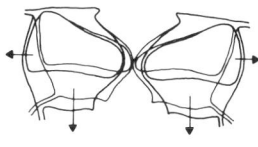
Helleborus type   h

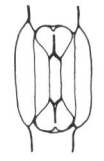
i

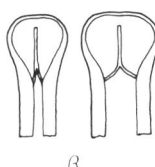

α   Graminean type   β

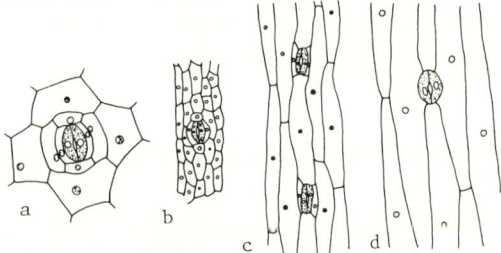

Fig 2.14. Types of stomatal complexes in monocotyledonous leaves. *a, Commelina communis* (Commelinaceae): 6 subsidiary cells; *b, Scheuchzeria palustris* (Scheuchzeriaceae): 5 subsidiary cells; *c, Juncus effusus* (Juncaceae): 2 subsidiary cells; *d, Nothoscordum inodorum* (Liliaceae): no subsidiary cells. (Reproduced with permission from Stebbins and Khush, 1961.)

four categories, two with four or more subsidiary cells surrounding the guard cells, one with two subsidiaries, and one with none. (Fig. 2.14). No correlation was found between these types of stomatal complex and xylem-vessel structure or leaf shape, but several correlations were found between stomatal type and type of seed germination, vascular anatomy of the seedling, growth habit of the mature plant, and geographic distribution. The authors assumed that the monocotyledons with more than two subsidiary cells are more primitive than those with none, for the following reasons: they occur in distantly related groups, some of which have relic characteristics (*Scheuchzeria*) and primitive flowers; they are associated with a low epidermal-cell/guard-cell ratio and hence with a smaller amount of differentiation in the epidermis; they are mostly evergreen phanerophytes, which are generally regarded as the most primitive life form; they are chiefly tropical in distribution, and a tropical origin is probable for the angiosperms; and their stomatal type with several subsidiaries is similar to the haplocheilic stomatal complex widespread in the gymnosperms (see p. 47).

Variations in structure and function of angiosperm stomata in connection with special adaptations. As is well known, all aquatic plant species within the angiosperms are derived from terrestrial relatives; they are secondary water plants, which still may have characteristics resembling those of their terrestrial ancestors. Schwendener (1881) has stated that several aquatic plants (*Alisma plantago-aquatica*, *Calla palustris*, and the fern *Salvinia natans*) never close their stomata. Haberlandt (1887) recognized that most phanerogamic floating plants have a uniform stomatal structure ("floating-leaf type"; see p. 42), in which the closing is achieved, if at all, not by contact of the protruded ventral walls but by contact of the outer cuticular ridges. The examples in Figure 2.15 demonstrate that the architecture of these floating-leaf stomata is suitable for preventing the entrance of water through the stomata, especially since the central pore is covered by a cuticle (Fig. 2.15).

Evolution of Stomata 51

The question arises whether floating leaves retain the ability to regulate stomatal width, since this ability could be considered a superfluous vestige of terrestrial ancestry. In this context, a loss in ability to regulate the size of permanently open stomata can be regarded as progressive. A study by Wagner (1973) revealed normal functioning stomata on the floating leaves of *Hydrocharis morsus-ranae*, *Polygonum amphibium*, *Potamogeton natans*, *Trapa natans* (Fig. 2.15d), *Callitriche palustris*, and *Spirodela polyrhiza* (Fig. 2.15b). Permanently open and nonfunctioning stomata were found in *Nymphaea alba*, *Nuphar lutea* (Fig. 2.15c), and *Lemna minor* (Fig. 2.15a); for *L. minor* Haberlandt (1887) already reported lack of functioning stomata.

It is not surprising that the stomata of *L. minor* cannot react, since the guard cells are dead and nearly empty (Fig. 2.16). It is remarkable, however, that we find such progressive (*Lemna*) and conservative (*Spirodela*) taxa in one family (Lemnaceae). Surprisingly, the guard cells of the nonfunctioning stomata of *Nymphaea* and *Nuphar* are cytologically intact, with chloroplasts, mitochondria, and so on. But because there is no substomatal cavity to allow the subsidiary cells to bend perpendicular to the plane of the leaf during stomatal opening, the guard cells are immobilized by the fixation of the subsidiaries.

As mentioned before (p. 29), there are many stomatal reductions and malformations in the angiosperms in connection with the loss or diminution of photosynthetic ability: for example, in parasitic higher plants (Fig. 2.7g), or in flower parts or fruits (Fig. 2.7h), or in rudimentary leaves (Fig. 2.7i). Special interest has been devoted to the differing stomatal structures on the different leaf-forms on a single plant (see Wassermann, 1924, and Fig. 2.17).

More fascinating are the cases in which stomalike pores, still partially able to regulate the pore width, serve functions other than gas exchange. This applies to some hydathodes in which the exuded water passes through the pores (Fig. 2.18a–c). Nectaries with stomalike pores (Fig. 2.18d) are considered

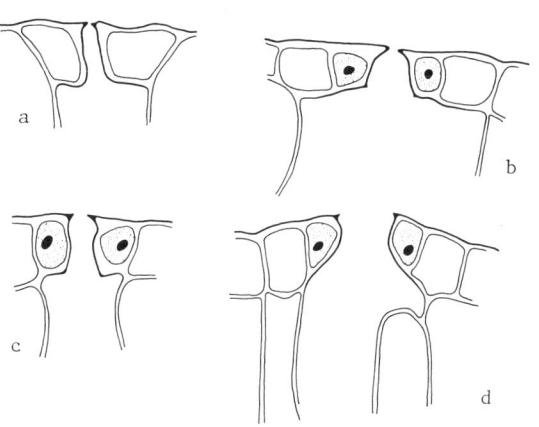

Fig. 2.15. Types of stomatal apparatus in floating leaves of angiosperms. *a*, *Lemna minor* (Lemnaceae); *b*, *Spirodela polyrhiza* (Lemnaceae); *c*, *Nuphar lutea* (Nymphaeaceae); *d*, *Trapa natans* (Trapaceae). (Reproduced with permission from Wagner, 1973.)

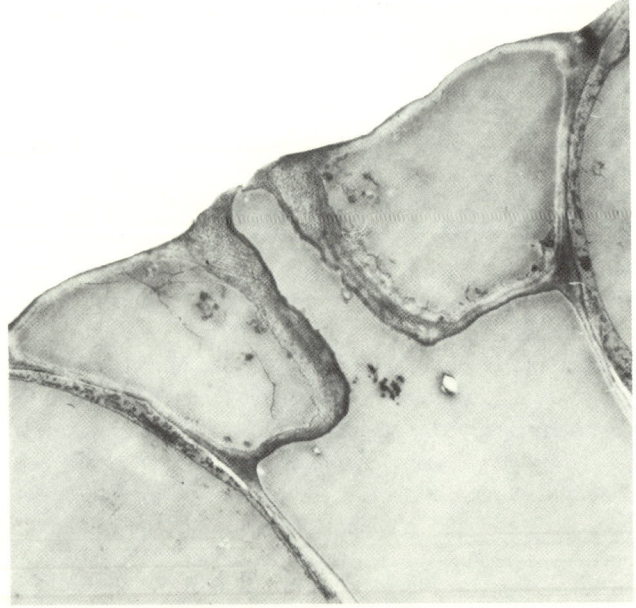

Fig. 2.16. Transmission electron micrograph of a cross section of an epistomatic stoma of *Lemna minor*. Note the empty guard cells. × 12,000. (Reproduced with permission from Wagner, 1973.)

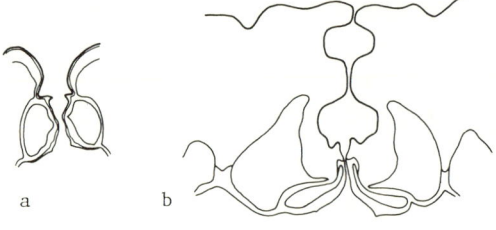

Fig. 2.17. Cross sections through the stomata of a primary leaf (*a*) and an adult pinnate leaf (*b*) of *Hakea suaveolens*. (Reproduced from Porsch, 1905.)

primitive and derived from hydathodes. Most spectacular is the transformation of a stomatal complex into a special structure assumed to serve the pitchers in carnivorous *Nepenthes* species to catch small animals (Fig. 2.18e). These were first identified as derived from stomata by Dickson (1883); together with waxy secretions, they constitute a slippery surface ("Gleitzone") inside the pitchers. For more details see Lloyd (1942).

Phylogenetic aspects of stomatal physiology in angiosperms. There are reports of only two important pecularities of stomatal physiology in angiosperms that can be evaluated for their phylogenetic meaning.

The first is the lack of chloroplasts in the functional guard cells of the genus

Evolution of Stomata 53

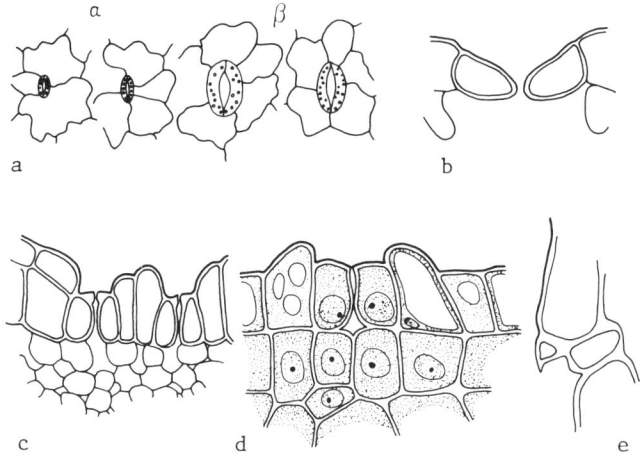

Fig. 2.18. Stomata with special functions. *a*, *Tropaeolum majus*: α, normal stomata; β, hydathodes of the same leaf. (Reproduced with permission from Burström and Odhnoff, 1963.) *b*, *T. majus*: cross section through the pore of the hydathode. (Reproduced from Haberlandt, 1918.) *c*, *Conocephalus ovatus*: cross section through the pore of the hydathode. (Reproduced from Haberlandt, 1918.) *d*, *Ribes rubrum*: cross section through the pore of the nectary in the flower cup. (Reproduced from Radtke, 1926.) *e*, Transformed stoma in the slippery surface of the pitcher of *Nepenthes* sp.

Paphiopedilum, Orchidaceae (Nelson and Mayo, 1975; Rutter and Willmer, 1979). We will not discuss here the question of how a stomatal apparatus without chloroplasts in its guard cells can function (see Chaps. 7–9), but we will briefly consider the phylogenetic significance of this condition.

It is assumed that a lack of guard cell chloroplasts is a progressive character, especially since the Orchidaceae are phylogenetically a rapidly evolving family. It is surprising, however, that this reduction seems to be restricted to a genus in the conservative subfamily Cypripedioideae, which has two inner stamens, and is not present in the most advanced subfamily, Orchidoideae, in which only the outer stamen is fertile. This is an indication that we cannot expect exclusively primitive characters in one taxon and exclusively progressive characters in another; rather, it is the sum of all characters that justifies the classification of a taxon as primitive or progressive.

The second case is the special mode of osmotic regulation in the guard cells of *Allium* species (Schnabl and Ziegler, 1977; Schnabl and Raschke, 1980). These plants lack starch in the guard cells and compensate the K^+ imported during the opening phase with coimported Cl^- rather than with starch-derived malate ions. This special feature is without doubt also a progressive character and may have been gained by halophilic species that had enough Cl^- for this regulation. In general, the Alliaceae are considered advanced within the Lilianae and especially within the order Asparagales (exhibiting, e.g., presence of bulbs, mustard oils, *Allium* type of gametophyte). The progressive physiology of the stomata, then, fits the general picture of this group.

Outlook

Our report makes it clear that the architecture of the stomatal apparatus, especially the form of the guard cells, the mode of the stomatal movement, and the development of the stomatal complexes, has unequivocal phylogenetic meaning, which in some cases is concealed by differences in various organs or by the influence of external factors. The phylogenetic significance of stomatal structure was used as a taxonomic tool for a long time, and stomata were among the most intensively studied plant parts during the pioneer times of plant anatomy.

There is a great gap in our knowledge about possible differences in the physiology and biochemistry of stomatal action in different plant groups. Our increased insight in this field has been mainly achieved with studies of a few angiosperm species and can now be used as a basis for looking intensively into the stomatal function of more primitive plants. It would not be too surprising if there should prove to be other mechanisms of osmoregulation, of ion transport between guard cells and neighboring cells, and of hormonal control than have yet been identified in higher plants.

REFERENCES

Bierschenk, K. 1971. Morphologie und Entwicklungsgeschichte der Laubmoos-Spaltöffnungen. Dissertation, University of Tübingen.
Brown, R. 1819. Character and description of *Lyellia*. *Transactions of the Linnean Society*, 12: 562–75.
Burström, H. G., and C. Odhnoff. 1963. *Vegetative anatomy of plants*. Stockholm: Svenska Bokförlaget.
Copeland, E. B. 1902. The mechanism of stomata. *Annals of Botany*, 16: 327–64.
Dickson, A. 1883. On the structure of the pitcher in the seedling of *Nepenthes* as compared with that of the adult plant. *Proceedings of the Royal Society of Edinburgh*, 1883–84: 381–85.
Doyle, W. T. 1970. *The biology of higher cryptogams*. London: Macmillan.
Fischer, M. 1929. Ergebnisse der Spaltöffnungsforschung an Früchten. *Botanisches Zentralblatt*, 49: 231–51.
Florin, R. 1931. *Untersuchungen zur Stammesgeschichte der Coniferales und Cordaitales*. Vol. 1, *Morphologie und Epidermisstruktur der Assimilationsorgane bei den rezenten Koniferen*. Kunglige svenska Vetenskapsakademiens Handlingar, ser. 3, 10, no. 1. Stockholm: Almquist & Wiksells.
———. 1933. *Studien über die Cycadales des Mesozoikums nebst Erörterungen über die*

Spaltöffnungsapparate der Bennettitales. Kunglige svenska Vetenskapsakademiens Handlingar, ser. 3, 12, no. 5. Stockholm: Almquist & Wiksells.

Garner, D., and D. J. Paolillo, Jr. 1973. On the functioning of stomates in *Funaria*. *Bryologist*, 76: 423–27.

Goebel, K. 1906. Archegoniatenstudien, X. *Flora*, 96: 1–200.

———. 1908. *Einleitung in die experimentelle Morphologie der Pflanzen.* Leipzig and Berlin: Engelmann.

———. 1915. *Organographie der Pflanzen.* 2d ed. Vol. 2, *Bryophyten.* Jena: Fischer.

———. 1930. *Organographie der Pflanzen.* 3d ed. Vol. 2, *Bryophyten, Pteridophyten.* Jena: Fischer.

Haberlandt, G. 1886a. Das Assimilationssystem der Laubmoos-Sporogonien. *Flora*, 44: 45–47.

———. 1886b. Beiträge zur Anatomie und Physiologie der Laubmoose. *Jahrbücher für wissenschaftliche Botanik*, 17: 457–77.

———. 1887. Zur Kenntnis des Spaltöffnungsapparates (v.a. bei Wasserpflanzen). *Flora*, 70: 97–110.

———. 1918. Physiologische Pflanzenanatomie. 5th ed. Leipzig: Engelmann.

Hildebrand, F. 1866. Entwicklung der Farnkrautspaltöffnungen. *Botanische Zeitung*, 24: 245–51.

Janzen, P. 1916. Eine Mooshaube mit Spaltöffnungen. *Hedwigia*, 57: 263–65.

Kaufmann, K. 1927. Anatomie und Physiologie der Spaltöffungsapparate mit verholzten Schliesszellmembranen. *Planta*, 3: 27–59.

Klemm, P. 1886. Über den Bau der beblätterten Zweige der Cupressineen. *Jahrbücher für wissenschaftliche Botanik*, 17: 489–54.

Kondo, T. 1962. A contribution to the study of fern stomata (II) with special reference to their structure and development. *Research Bulletin of the Faculty of Education of Shizuoka University*, 13: 239–61.

Kraus, F. 1914. Ein Beitrag zur Kenntnis der Anatomie und Physiologie der Pteridophytenspaltöffnungen. *Jahresberichte des fürstbischöflichen Gymnasiums Graz*, 1914: 7–50.

Kuhlbrodt, H. 1923. Über die phylogenetische Entwicklung des Spaltöffnungsapparates am Sporophyten der Moose. *Beiträge zur allgemeimen Botanik*, 2: 363–402.

Lendzian, K. J. 1984. Permeability of plant cuticles to gaseous air pollutants. In M. J. Koziol and F. R. Whatley, eds., *Gaseous air pollutants and plant metabolism: First international symposium on air pollution, Oxford, 1982*: 77–81. London: Butterworth.

Link, H. F. 1827. Über die Familie *Pinus* und die europäischen Arten derselben. *Abhandlungen der Königlichen preussischen Akademie der Wissenschaften zu Berlin, Physikalische Klasse*, 1827: 157–91.

Lippmann, G. 1925. Über das Vorkommen der verschiedenen Arten der Guttation und einige physiologische und ökologische Beziehungen. *Botanisches Archiv*, 11: 361–464.

Lloyd, F. E. 1942. *The carnivorous plants.* Waltham, Mass.: Chronica Botanica.

Lösch, R., and C. Bressel. 1979. Die Kaliumverteilung in den Schliesszellen leptosporangiater Farne unterschiedlicher Stomatypen. *Flora*, 168: 109–20.

Maheshwari, P., and V. Vasil. 1961a. The stomata of *Gnetum. Annals of Botany*, n.s., 25: 313–19.

———. 1961b. *Gnetum.* Botanical Monographs, 1. New Delhi: Council of Scientific and Industrial Research.

Mahlert, A. 1885. Beiträge zur Kenntnis der Anatomie der Laubblätter der Coniferen mit besonderer Berücksichtigung des Spaltöffnungs-Apparates. *Botanisches Centralblatt*, 24: 54–59, 85–88, 118–22, 149–53, 180–85, 214–18, 243–49, 278–82, 310–12.

Maroti, I. 1961. Untersuchungen der Entwicklung der Epidermis der *Psilotinae* und des *Filicinae*-Blattes und der Entwicklung des Stomas. *Acta biologica Szeged*, n.s., 7: 43-67.
Martin, G. E., W. H. Outlaw, Jr., L. C. Anderson, and S. G. Jackson. 1984. Photosynthetic electron transport in guard cells of diverse species. *Plant Physiology*, 75: 336-37.
von Mohl, H. 1838. Über die Entwicklung der Spaltöffnungen. *Linnaea*, 12: 544-48.
Naegeli, C. 1842. Botanische Beiträge, I. Entwicklung der Hautdrüsenzellen. *Linnaea*, 16: 237-40.
Napp-Zinn, K. 1966. *Anatomie des Blattes 1, Blattanatomie der Gymnospermen*. Handbuch der Pflanzenanatomie, vol. 8, part 1. Berlin: Bornträger.
Nelson, S. D., and J. M. Mayo. 1975. The occurrence of functional nonchlorophyllous guard cells in *Paphiopedilum* spp. *Canadian Journal of Botany*, 53: 1-7.
Pant, D. D., and P. K. Khare. 1969. Epidermal structure and stomatal ontogeny in some eusporangiate ferns. *Annals of Botany*, 33: 795-805.
Porsch, O. 1905. *Der Spaltöffnungsapparat im Lichte der Phylogenie*. Jena: Fischer.
Probst, W. 1971. Vergleichende Morphologie und Entwicklungsgeschichte der Spältöffnungen bei Farnen. Dissertation, University of Tübingen.
Radtke, F. 1926. Anatomisch-physiologische Untersuchungen an Blütennektarien. *Planta*, 1: 379-418.
Rehfous, L. 1923. Sur la phylogénie des stomates. *Comptes rendus de la Société de physique et d'histoire naturelle*, Geneva, 40: 68-78.
Riebner, F. 1926. Über Bau und Funktion der Spaltöffnungsapparate bei *Equisetinae* und *Lycopodinae*. *Planta*, 1: 260-300.
Rutter, J. C., and C. M. Willmer. 1979. A light and electron microscopy study of the epidermis of *Paphiopedilum* spp., with emphasis on stomatal ultrastructure. *Plant, Cell and Environment*, 2: 211-19.
Sack, F., and D. J. Paolillo, Jr. 1983. Structure and development of walls in *Funaria* stomata. *American Journal of Botany*, 70: 1019-30.
Schimper, W. Ph. 1848. Recherches anatomiques et morphologiques sur les mousses. Dissertation, University of Strasbourg.
Schnabl, H., and K. Raschke. 1980. Potassium chloride as stomatal osmoticum in *Allium cepa* L., a species devoid of starch in guard cells. *Plant Physiology*, 65: 88-93.
Schnabl, H., and H. Ziegler. 1977. The mechanism of stomatal movement in *Allium cepa* L. *Planta*, 136: 37-43.
Schönherr, J., and H. Ziegler. 1975. Hydrophobic cuticular ledges prevent water entering the air pores of liverwort thalli. *Planta*, 124: 51-60.
Schürmann, B. 1959. Über den Einfluss der Hydratur und des Lichts auf die Ausbildung der Stomainitialen. *Flora*, 147: 471-520.
Schwendener, S. 1881. Über Bau und Mechanik der Spaltöffnungen. *Monatsberichte der Königlichen preussischen Akademie der Wissenschaften zu Berlin, Physikalisch-mathematische Klasse*, July 1881: 833-67.
Solereder, H. 1899. *Systematische Anatomie der Dikotyledonen*. Stuttgart: Enke.
Sprecher, A. 1907. Le *Ginkgo biloba* L. Dissertation, University of Geneva.
Stebbins, G. L., and G. S. Khush. 1961. Variation in the organization of the stomatal complex in the leaf epidermis of monocotyledons and its bearing on their phylogeny. *American Journal of Botany*, 48: 51-59.
Stevens, R. A., and E. S. Martin. 1977. The morphogenesis of substomatal structures in *Polypodium vulgare*. *Canadian Journal of Botany*, 51: 37-42.

Strasburger, E. 1867. Ein Beitrag zur Entwicklungsgeschichte der Spaltöffnungen. *Jahrbücher für wissenschaftliche Botanik*, 5: 297–342.

Treviranus, L. C. 1811. *Beiträge zur Pflanzenphysiologie*. Göttingen: Dieterich.

Vesque, M. J. 1889. De l'emploi des charactères anatomiques dans la classification des végétaux. *Bulletin de la société botanique de France*, 36: 41–77.

Wagner, Th. 1973. Funktionieren die Stomata bei Schwimmblättern? Thesis (Zulassungsarbeit für das Lehramt an Gymnasien), Technical University of Munich.

Wassermann, J. 1924. Beiträge zur Kenntnis der Morphologie der Spaltöffnungen. *Botanisches Archiv*, 5: 26–69.

von Wettstein, R. 1901–3. *Handbuch der systematischen Botanik*. Leipzig and Vienna: Deuticke.

Willmer, C. M., and J. E. Pallas, Jr. 1973. A survey of stomatal movements and associated potassium fluxes in the plant kingdom. *Canadian Journal of Botany*, 51: 37–42.

Ziegenspeck, H. 1938–39. Die Micellierung der Turgeszenzmechanismen. *Botanisches Archiv*, 39: 268–309, 332–72.

———. 1941. Der Bau der Spaltöffnungen, III. *Repertorium specierum novarum regni vegetabilis*, Beiheft 123: 1–56.

Zimmermann, W. 1927. Die Spaltöffnungen der *Psilophyta* und *Psilotales*. *Zeitschrift für Botanik*, 19: 129–70.

3

The Development and Structure of Stomata

Fred D. Sack

Stomata have been described as "turgor-operated valves" for gas exchange (Raschke, 1979). Stomatal architecture and cytology are remarkably specialized for this role, but we are only beginning to understand the specific functional correlates of stomatal structure. This review will focus on the structural properties of the valve apparatus (i.e. of guard cell walls and shape), the morphological changes that accompany variations in turgor, the structure of the cuticle, the component organelles, and the role of the cytoskeleton in stomatal development. Ultrastructure and development will be emphasized.

Recent reviews of stomatal structure have been especially concerned with the relationship between the microtubules and the wall (Hepler, 1981; Palevitz, 1981a, 1982) or with the mechanics and changes in volume involved in stomatal movement (Raschke, 1979). General reviews include those of Allaway and Milthorpe (1976), who present an evolutionary and functional perspective, of Stevens and Martin (1979), and of Willmer (1983).

Terminology

Anticlinal walls and *anticlinal sections* are perpendicular to the plane of the epidermis; anticlinal sections include cross sections perpendicular to the long axis of the pore (see e.g. Figs. 3.1 and 3.9A, D) and longitudinal sections along the length of the stoma (see Fig. 3.9C, F). *Periclinal walls* and *paradermal sections* are parallel to the plane of the epidermis.

The term *stoma* (plural, *stomata*) generally designates not just the pore but also the surrounding *guard cells*. The *ventral wall* borders the pore (Fig. 3.1). The *dorsal wall* is the anticlinal wall adjacent to the surrounding epidermal cells. The *outer wall* and the *inner wall* are the periclinal walls bordering the atmosphere and the *substomatal cavity*, respectively. Generally, *ledges* overarch (*outer ledge*) and underarch (*inner ledge*) the pore (see Figs. 3.1, 3.7, 3.12). The pore is somewhat hourglass-shaped in cross section (see Fig. 3.7). The constricted mid-depth of the pore is termed the *central aperture*. The wider areas between this aperture and the ledges are the *front* (outer) *chamber* and the

---

The author wishes to thank Dr. David Priestley for Russian translation and A. Carl Leopold for support and encouragement during the writing of this review.

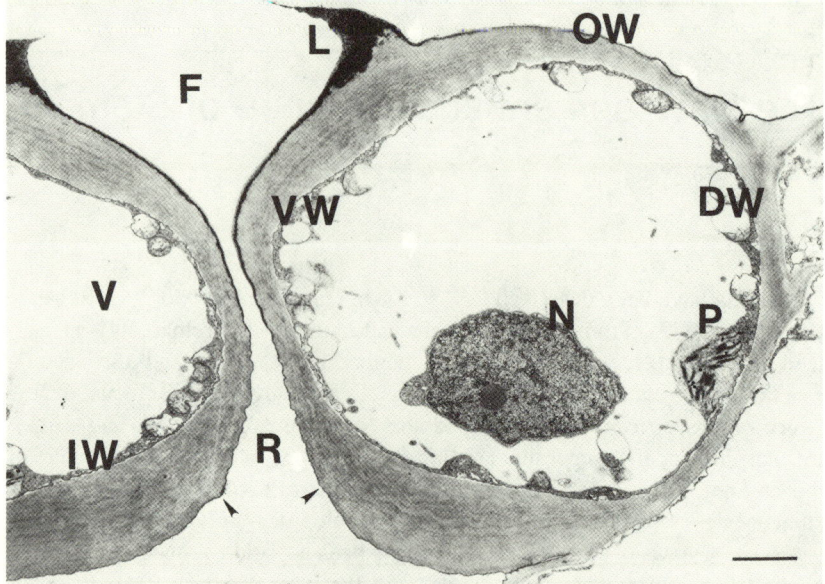

Fig. 3.1. Transmission electron micrograph of cross section through a pore of a mature *Nicotiana tabacum* stoma; atmosphere at top and substomatal cavity at bottom of micrograph. The well-developed outer ledges (*L*) are heavily cuticularized (electron-dense regions) and overarch the forechamber (*F*) of the pore. The inner ledges (arrowheads) are barely distinguishable (compare Fig. 3.7). Both guard cells are strongly vacuolate (*V*). *OW*, *IW*, *VW*, *DW*: outer, inner, ventral, and dorsal walls, respectively; *N*, nucleus; *P*, plastid; *R*, rear chamber of pore. Bar = 2 μm.

rear (inner) *chamber*. The areas between the ends of the pore and the ends of the stoma are the *polar regions* (see Fig. 3.4).

Subsidiary cells are usually defined as one or more cells surrounding a stoma that are distinct in size, shape, arrangement, and structure from other epidermal cells (Esau, 1977). Subsidiary cells are included in the assemblage designated by the terms *stomatal complex* or *stomatal apparatus*.

Stomatal Types and Distribution

Stomata are found on most aerial surfaces of plants, often including the adaxial epidermis (Fig. 3.2), stems, and flowers (Esau, 1977).

Guard cells typically are either dumbbell- or kidney-shaped when viewed from the surface. Dumbbell-shaped guard cells have bulbous ends containing most of the organelles, including the vacuoles (Fig. 3.3). The handle of the dumbbell usually encloses the elongate portion of the nucleus and surrounds the pore, which is typically slitlike. These guard cells adjoin much larger subsidiary cells. The structure of this stomatal type has been well described

Development and Structure of Stomata 61

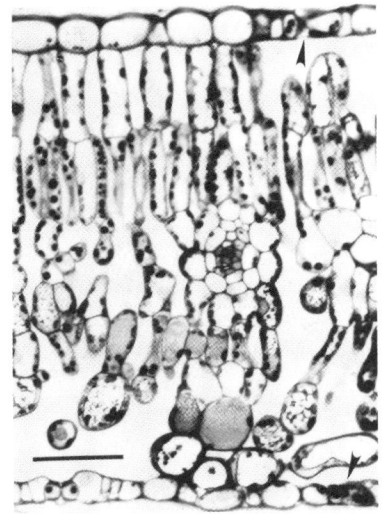

Fig. 3.2. Light micrograph of cross section of amphistomatous leaf of *Populus* hybrid (*P. maximowiczii* × *P. incrassata*). Abaxial stomata below scale bar and at lower right (arrowhead); adaxial stoma at upper right (arrowhead). Toluidine blue–stained thick section of fixed material. Bar = 20 μm.

(e.g. Srivastava and Singh, 1972; Galatis, 1980; Palevitz, 1981a, 1982). The dumbbell shape is characteristic of the guard cells in grasses and in a few sedges (Cyperaceae); other monocot families containing genera with similarly shaped guard cells include the Palmae, Flagellariaceae, Rapataceae, Marantaceae, Anarthriaceae, Restionaceae, Thurniaceae, Heliconiaceae, and Lowiaceae (Allaway and Milthorpe, 1976).

Kidney-shaped guard cells occur in most dicots, in many monocots, and in mosses, ferns, and gymnosperms. The overall outline of the two kidney-shaped guard cells is elliptical in paradermal section (Fig. 3.4). The ultrastructure of this type of guard cell has been studied also (e.g. Landré, 1972; Singh and Srivastava, 1973; Galatis and Mitrakos, 1980; Faraday, Thomson, and Platt-Aloia, 1982).

Although most guard cells can be classified as either kidney- or dumbbell-shaped, interesting variations occur. For example, stomata in *Azolla*, an aquatic fern, and in several mosses (including *Funaria*) have a roughly elliptical outline, as do stomata with kidney-shaped guard cells; but these stomata contain only one guard cell, which is continuous around the pore (Sack and Paolillo, 1983b; Busby and Gunning, 1984).

Stomatal shape may vary with development. As the pore forms, dumbbell-shaped stomata undergo a transient swelling, and the guard cells become kidney-shaped (Palevitz, 1981a). However, upon further wall deposition, the cells reconstrict to a dumbbell configuration.

Guard-cell and stomatal-complex structure vary greatly, presumably in concert with species habitat and leaf architecture. For example, the guard cells of xerophytes are typically sunken below epidermal cells (Fig. 3.5) or contain

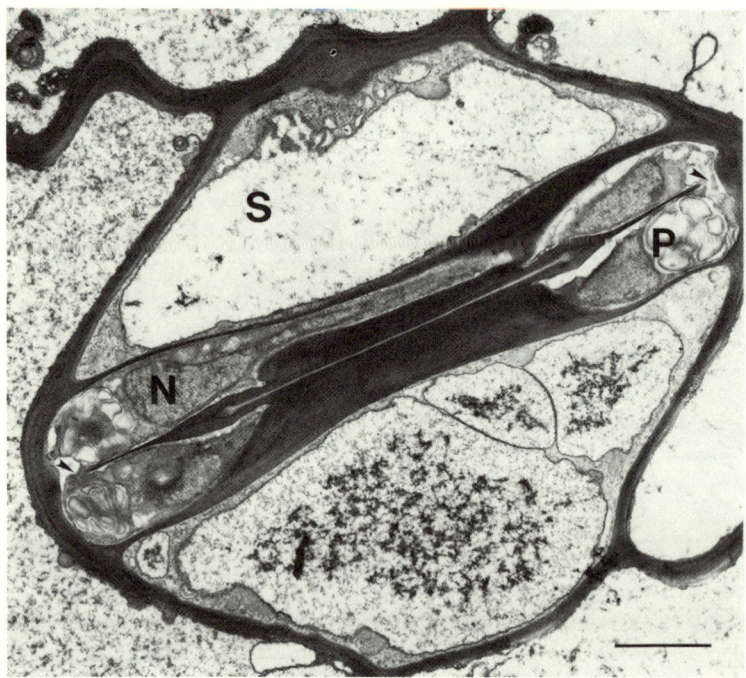

Fig. 3.3. Paradermal view of dumbbell-shaped stoma from *Oryza sativa*. Pore is narrow slit bisecting stoma. Ventral wall is perforated (arrowheads) at poles. Outer wall is sectioned in plane of wall to varying degrees in both guard cells near pore. Nucleus (*N*) follows shape of lumen and is also dumbbell-shaped. Large, starch-filled plastids (*P*) are seen at poles. Note substantial differences in volumes of guard cells and of two flanking subsidiary cells (*S*). Subsidiary cells are also much more vacuolate. Bar = 3 μm. (Electron micrograph by Dr. P. Dayanandan.)

prominent outer stomatal ledges (Carr and Carr, 1978). In the stomata of many genera, including gymnosperms and aquatic plants, the inner ledges are often small or absent (see Figs. 3.1, 3.9D, 3.10). Von Guttenberg (1971) and Napp-Zinn (1973) provide more extensive examples of structural variation. The number, origins, positions, and shapes of subsidiary cells also vary (Esau, 1977; Napp-Zinn, 1973).

Preprophase Band and Cytokinesis

The arrangement and sequence of cell divisions in the differentiation of the stomatal complex have been well described by light microscopists, who have produced an elaborate nomenclature for them (Esau, 1977). The stomatal complex has been a preferred object of ultrastructural studies aimed at determining how the positioning of the cell plate is related to the distribution of microtubules, particularly of the microtubules in the preprophase band (PPB).

Development and Structure of Stomata 63

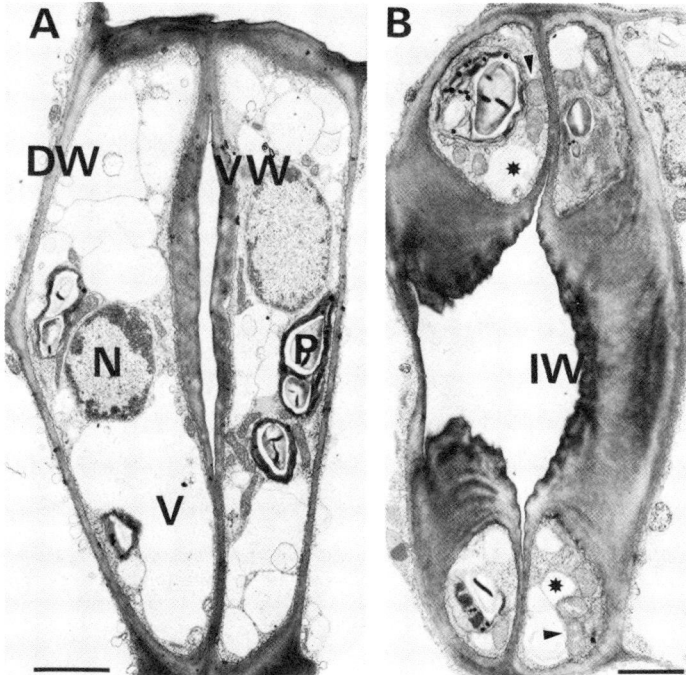

Fig. 3.4. Paradermal views of kidney-shaped stoma of *Lycoperisicon pennellii* (Plant Introduction No. 246502) at two different planes of sectioning. *A*, Sectioned at level of pore: section is oblique, so that length of stoma is not shown; most of lumen volume is occupied by vacuoles (*V*); nuclei (*N*) are positioned near pore; plastids (*P*) are filled with starch; *DW*, dorsal wall; *VW*, ventral wall; bar = 3 μm. *B*, Sectioned below pore in plane of inner wall (*IW*): note that guard cells are deeper at poles (*); arrowheads show mitochondria; bar = 2 μm.

Before prophase in most plant cells, a belt of microtubules appears around the circumference of the cell, usually perpendicular to the axis of the mitotic spindle. In most cases, this aggregation—the PPB—predicts the site where the cell plate will fuse with the parental walls (see Gunning, 1982). Since the PPB disappears before cytokinesis, however, positional correlations are difficult to study unless the new wall's site of insertion is predictable. Many stomatal complexes fulfill this criterion; in fact, one of the first accounts of PPBs was in guard-cell and subsidiary-cell parent cells (GPC, SPC; also known as "mother cells") of *Triticum* (Pickett-Heaps and Northcote, 1966).

In many genera, subsidiary cells are formed by asymmetrical division of the SPC. Before prophase, the nucleus migrates to a point near the adjacent GPC, positioning the mitotic apparatus, and thus the cell plate, near its final locus. During its last stage of growth, the cell plate either extends in its original plane or curves so that it joins the SPC wall at the former site of the PPB. This process has been well documented (e.g. Pickett-Heaps, 1969; Singh, 1977).

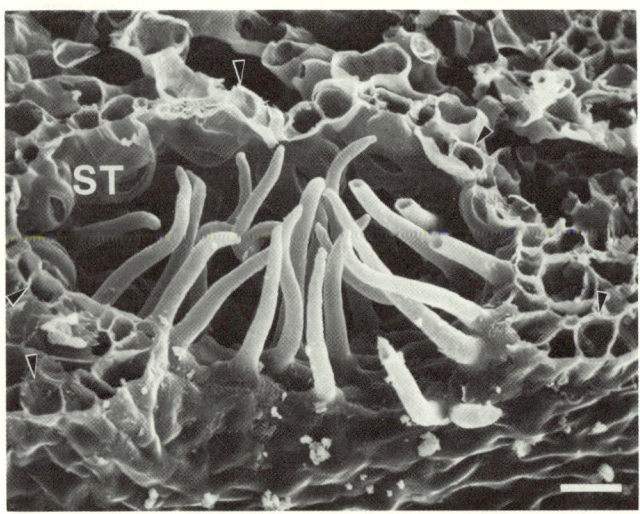

Fig. 3.5. Scanning electron micrograph of stomatal crypt of *Nerium oleander*. Leaf has been cut to reveal crypt, which is only exposed to atmosphere at opening (lower center). Arrowheads indicate epidermis, which is invaginated and includes stomata (*ST*). Epidermal hairs project into crypt. Bar = 20 μm. (Micrograph kindly furnished by M. K. Campenot.)

Preprophase bands have been reported in many GPCs (e.g. Palevitz and Hepler, 1974; Busby and Gunning, 1980, 1984). Some of these PPBs contain fewer microtubules and are less compact than the bands in SPCs (Galatis, 1982).

Because the PPB is not present during cell-plate growth, the PPB is probably only a marker of specialized cortical cytoplasm in this region, an area Gunning (1982) termed the "division site." The PPB or the division site may be of great importance in the final stages of cell-plate positioning (Galatis, Apostolakos, and Katsaros, 1984). This hypothesis is supported by the apparent absence of a PPB in *Funaria* GPCs, which undergo incomplete cytokinesis, and by the presence of a PPB in adjacent epidermal cells with normal cytokinesis (Sack and Paolillo, 1985). The one-celled stomata of *Azolla* do have a PPB in the GPC; in *Azolla*, however, the cell plate fuses with the parent-cell wall, and the ends of the plate are then resorbed to create the one-cell arrangement (Busby and Gunning, 1984). Cell plates in dumbbell-shaped stomata are also reported to show some resorption (Palevitz, 1981a), resulting in perforations in the mature septa; PPBs are present in these GPCs.

Evidence from GPCs in *Allium* suggests that the division site controls final cell-plate orientation. The metaphase and the cell plates are usually oriented obliquely, but the cell plate eventually rotates to a longitudinal position where a PPB was located (Palevitz and Hepler, 1974).

In many GPCs in the Leguminosae, the presence of a PPB is related to a precocious local thickening of the cell wall (Galatis et al., 1982). When the band appears, the wall thickens; it continues to do so even after the microtubules depolymerize during prophase. During guard cell growth, the walls all thicken, and differential thickening can no longer be detected at the former site of the PPB.

Differentiation of the stomatal complex often involves multiple asymmetric divisions at precise places and times. As indicated, GPCs and meristemoids may control differentiation in neighboring cells by inducing nuclear movement in SPCs, perhaps through some field effect. The stomata of *Vigna sinensis* can induce divisions in up to 25 surrounding protodermal cells (Galatis and Mitrakos, 1980). Because of such characteristics, stomatal complexes will undoubtedly remain valuable for studying the mechanisms of cytokinesis and differentiation in plants.

Microtubules and Wall Building

Because guard cell walls are highly ordered, stomata are also useful for studying the control of oriented wall synthesis (see Palevitz, 1981a, 1982; Hepler, 1981). Two relationships between microtubules and the wall have been consistently documented in developing guard cells: microtubules have been found in greatest numbers where the wall is thickest, and they are oriented parallel to newly synthesized wall microfibrils.

In all stomata studied, the ventral wall thickens shortly after the guard cell forms, so that the wall looks biconvex in median paradermal section (Fig. 3.6). Before this thickening occurs, one to three rows of anticlinally oriented microtubules appear along the midlength of the wall (e.g. Kaufman et al., 1970; Peterson and Hambleton, 1978). In kidney-shaped guard cells, the outer paradermal wall thickens at the same time as the ventral wall, and its fibrils are for the most part radially oriented about the pore at all stages of development (Ziegenspeck, 1955); the microtubules also show a radial orientation (Landré, 1969b; Palevitz and Hepler, 1976). In dumbbell-shaped guard cells, the fibrils in the outer wall and the underlying microtubules are radially oriented until the cells start to reconstruct after the transient swelling that accompanies pore formation. During reconstriction, newly synthesized wall fibrils take on a net axial orientation (Galatis, 1980; Mishkind, Palevitz, and Raikhel, 1981).

Most of the microtubules are found close to the plasmalemma, and many show bridges to other microtubules and to the cell membrane (Galatis, 1980; Palevitz, 1982). Smooth vesicles of apparent dictyosomal origin are frequently associated with these microtubules, and the contents of the vesicles may stain positively for polysaccharides with the periodic-acid–thiocarbohydrazide–silver-proteinate test (Landré, 1970; Galatis, Apostolakos, and Katsaros, 1983). Observations of coalignment of the wall microfibrils and microtubules that are associated with these vesicles are widespread in the literature and have

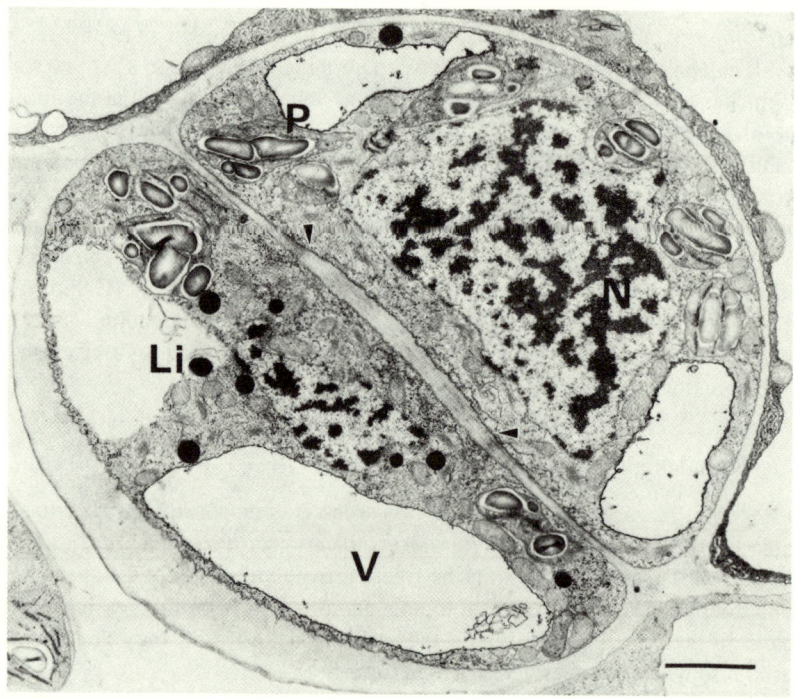

Fig. 3.6. Paradermal view of developing stoma of *Pisum sativum*. Ventral wall has thickened where pore will form (arrowheads). Vacuoles (*V*) are already differentiated. Section is oblique. *Lower left*, inner wall; *upper right*, dorsal wall; *Li*, lipid globules; *N*, nucleus; *P*, plastids. Bar = 2 μm.

led to several hypotheses regarding microtubule orientation of wall synthesis (Palevitz, 1982).

Careful analysis, especially of serial sections, has detected microtubular foci in developing guard cells (Galatis, 1980; Galatis and Mitrakos, 1980; Galatis, Apostolakos, and Katsaros, 1983; Busby and Gunning, 1984). During early stomatal growth, the microtubules converge at several foci along the cell edges at the junctions of the ventral wall with the two periclinal (outer and inner) walls. Microtubular foci appear concentrated along the midlength of the ventral wall, although they have also been detected close to the middle of the outer wall with high voltage electron microscopy (Palevitz, 1981b). The outer and inner edges of the ventral wall are preferentially thickened. Microtubular foci and many dictyosome vesicles are often found along the edges of the ventral wall. Galatis and co-workers have hypothesized that these microtubule-vesicle complexes play a role in localized wall deposition and are microtubular "nucleating sites" (Galatis, Apostolakos, and Katsaros, 1983, with references). Microtubule-vesicle complexes have also been described for developing PPBs in GPCs in *Zea* (Galatis, 1982).

Some GPCs in grasses contain a microtubular band located at right angles to the PPB. This "side band" was originally considered a PPB (Singh, 1977); however, Busby and Gunning (1980) and Galatis (1982) demonstrated that it forms before the PPB. The side band apparently contributes to an oriented thickening of the lateral wall of the elongating GPC. Wall sites adjacent to the side band and to the PPB seem predisposed to thicken and will do so even when the microtubules are depolymerized with colchicine (Galatis, 1982; Galatis, Apostolakos, and Katsaros, 1984).

Stomatal-complex microtubules are generally located close to the plasmalemma but can be found in other areas within the cytoplasm. For example, side-band microtubules may extend inward and be closely associated with proplastids (Galatis, 1982). Other microtubules project from the PPB, from the division site, or from microtubule foci to the nucleus (Galatis, 1982; Galatis, Apostolakos, and Katsaros, 1983; Busby and Gunning, 1984). It is not known whether organelles are being anchored or transported by such associations.

It is clear that the microtubular cytoskeleton of the stomatal apparatus is highly ordered in timing and distribution.

Formation of the Substomatal Cavity and of the Pore

Data indicate that the processes of wall separation that create the substomatal cavity and the pore differ in timing and probably in mechanism.

Formation of the substomatal cavity seems to be a special case of intercellular space formation. DeChalain and Berjak (1979) contend that the substomatal cavity of *Avicennia* develops lysigenously, that is, by lysis of the mesophyll cell or cells directly below the stomata. However, their study does not show intermediate stages or wall remnants lining the cavity; the cell debris in the cavity that has been interpreted as evidence of lysis could be an artifact of cutting the tissue during fixation. In *Equiseteum* (Laroche et al., 1976, fig. 24), *Polypodium* (Stevens and Martin, 1978, figs. 2, 3), *Zea* (Galatis, 1982, figs. 16, 24), and *Funaria* (Sack and Paolillo, 1983b) the cavity clearly originates through cell separation (schizogenously). In *Funaria* the first sign of cavity formation is a stretching of the middle lamella at the cell junctions. It is not known whether lytic enzymes dissolve the middle lamella. Mechanical forces are likely to affect the process, since the junctions are probably under tension because of differential expansion of the cells and tissues. The middle lamella may then be stretched into strands and sheets and, in *Funaria*, into a fibrillar meshwork as well (Sack and Paolillo, 1983b). A cuticle is not discernible on the inner wall (ie. the wall facing the cavity) in *Funaria* until the pore forms.

In most plants, the pore is the only place where the anticlinal walls detach from each other, a fact that suggests that the process is specific, localized, and

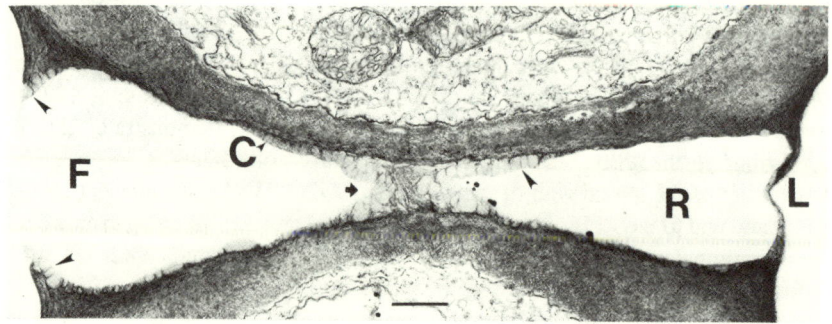

Fig. 3.7. Cross section of pore during formation in *Funaria*. At left, the outer ledges are detached (arrowheads), and the fore and rear chambers (*F*, *R*) have formed. The inner ledges are still attached (*L*). The ventral wall is almost separated at the central aperture, except for a bridge of material (arrow) that is being organized into the nascent cuticle (*C*). Plates and globules are involved in the development of the cuticle. The formed cuticle has lamellae and electron-dense fibrils that sometimes reach the cuticular surface (arrowheads). Bar = 0.5 μm.

tightly regulated. The pore typically develops after the substomatal cavity. Separation is rarely simultaneous throughout the depth of the ventral wall. In most cases, first the rear and fore chambers form, then the ledges, and finally the central aperture (Fig. 3.7; Carr and Carr, 1978; Galatis, 1980; Galatis and Mitrakos, 1980; Krarup-Hjort, 1983; Sack and Paolillo, 1983a). Along the length of the ventral wall, as seen in paradermal sections, separation begins in the middle of the thickened biconvex wall (e.g. Ziegler, Shmueli, and Lange, 1974; Peterson and Hambleton, 1978; Galatis, 1980). The separation sequence may be different in ferns. In *Adiantum* and *Polypodium*, the ventral wall first separates throughout its mid-depth and remains temporarily attached at the outer and inner walls (Stevens and Martin, 1978; Galatis, Apostolakos, and Katsaros, 1983).

Three elements may contribute to the separation of the ventral wall: mechanical stress-strain relationships, enzymatic hydrolysis of the middle lamella, and formation of the cuticle.

Mechanical relationships. No direct measurements of turgor have been reported for young stomata, but it is likely that guard cells increase in turgor before or during formation of the pore. Potassium and chloride levels increase substantially in *Phleum* and *Zea* guard cells when they swell before the pore forms (Palevitz, 1981a, 1982). These levels decrease as the cells reconstrict and elongate. In *Polypodium* and *Funaria*, the starch content of the plastids decreases upon pore formation (Stevens and Martin, 1978; Sack and Paolillo, 1983a).

Before the pore forms, the ventral and periclinal walls thicken, especially at their intersections, to form triangular corners (see illustrations in Stevens and

Martin, 1978; Galatis, 1980; Galatis and Mitrakos, 1980; Galatis, Apostolakos, and Katsaros, 1983). The ventral wall may appear biconcave in cross section (Sack and Paolillo, 1983b), or the thickenings may be greatest on the periclinal walls (Carr and Carr, 1978). Sometimes the periclinal walls are thinnest adjacent to the triangular thickenings (Galatis, Apostolakos, and Katsaros, 1983); according to Stevens and Martin (1978) this region may function as a hinge, about which the thickenings rotate to form the pore. In *Adiantum*, however, the mid-depth of the pore forms precociously, just after cytokinesis and considerably before the triangular thickenings and hinge develop, although the ledges remain attached until the thickenings appear (Galatis, Apostolakos, and Katsaros, 1983). In all these cases, it is plausible that an increase in turgor in the round guard-cell lumen (i.e. the intracellular space bounded by the cell wall) would set up forces that favor detachment at the corners of the ventral wall, which are square in outline.

The epidermis also may be under tension as a result of the expansion of the underlying tissues and of the organ. In addition, the radial orientation of the wall fibrils is attained in most guard cells prior to separation; it has been pointed out that this could contribute mechanically to pore formation, as it probably does to stomatal movement (Galatis, 1980; Galatis and Mitrakos, 1980; Busby and Gunning, 1984). In the atypical stomata of *Azolla*, this radial micellation may in turn favor pore elongation perpendicular to the stomatal axis but parallel to the axis of leaf expansion (Busby and Gunning, 1984). Similarly, when the guard cells reconstrict in grasses this radial pattern enables them to elongate parallel to the axis along which the stoma and the leaf will expand (Galatis, 1980; Palevitz, 1982).

Hydrolysis of the middle lamella. Tension on the ventral wall, even if greater than on other anticlinal walls, may be insufficient to cause separation. Enzymatic hydrolysis of the middle lamella may also occur, as has been widely assumed in the literature (Kaufman et al., 1970; Ziegler, Shmueli, and Lange, 1974; Stevens and Martin, 1978; Peterson and Hambleton, 1978), but little evidence for hydrolysis has been presented. Landré (1969a) claimed that the increased electron density of the middle lamella of the ventral wall originated from lysis, whereas Srivastava and Singh (1972) suggested that hydrolysis causes an absence of electron density. An electron-lucent middle lamella has been found prior to pore formation in several genera, and the middle lamella may not stain with the periodic-acid–thiocarbohydrazide–silver-proteinate test for polysaccharides (Singh and Srivastava, 1973; Galatis, 1980; Galatis and Mitrakos, 1980). But the lucency is not universal (Krarup-Hjort, 1983; Sack and Paolillo, 1983a), and, when it does occur, it can extend beyond the region where the ventral wall separates (Singh and Srivastava, 1973; Galatis and Mitrakos, 1980). Cytochemical tests appear necessary for definitive evidence on the nature of the electron lucency.

Formation of the cuticle. Several workers have noted that a cuticle lines the newly formed pore (Ziegler, Shmueli, and Lange, 1974; Galatis, 1980; Palevitz, 1981a), but the details of cuticle formation in the pore have been described only for *Funaria* and *Azolla* (Sack and Paolillo, 1983a; Busby and Gunning, 1984). In both these genera, structures involved in the organization of the cuticle are visible at the middle lamella before the ventral wall separates completely (Fig. 3.7). These structures include lamellae (and globules, in *Funaria*) whose contents have the same electron density as the matrix of the cuticle. They are present between the separating walls, and they coalesce to form the cuticle. Since these structures can be observed at the first detectable signs of separation, cuticular precursors must accumulate at the middle lamella beforehand. This accumulation could affect the binding properties of the middle lamella and thereby establish a weak point that facilitates wall separation.

Additional evidence that cuticular precursors may accumulate at the middle lamella prior to pore formation includes the infiltration of an electron-dense substance (Galatis and Mitrakos, 1980) and the appearance of membrane-like inclusions (Sack and Paolillo, 1983a). Stevens and Martin (1978) noted that "a cuticular layer is laid down adjacent to the middle lamella" before pore development.

It makes sense that formation of the pore and of the cuticle may be coupled to insure protection against water loss as soon as the surface is exposed to the atmosphere. Nevertheless, more genera need to be studied to determine whether cuticle formation is a general feature of stomatal pore development.

In *Funaria*, formation of the cuticle is simultaneous with that of the pore but not with that of the substomatal cavity. This difference underscores the specificity of the mechanisms involved in creating separation in two adjacent walls at different times.

Although the stoma is presumed to be functional after the pore forms, many developmental changes, such as thickening of the cuticle and of the walls (e.g. Galatis, 1980; Palevitz, 1981a; Sack and Paolillo, 1983b) and changes in organelle morphology, continue to occur after maturity.

Cell Shape and Thickening of the Wall in Guard Cells

Dumbbell-shaped guard cells. As indicated earlier, grass-type guard cells appear dumbbell-shaped in paradermal view (Fig. 3.3). Their shape is similar in longitudinal anticlinal sections because the poles are twice as deep as the canal (compare Fig. 3.8A and E). The pattern of wall thickening has been well depicted by Brown and Johnson (1962), Srivastava and Singh (1972), and Galatis (1980).

The narrower portion of the *Zea* guard cell, surrounding the pore, is elongate and has much thicker walls than do the bulbous ends. A cross section of the region surrounding the pore reveals that the inner and outer walls are

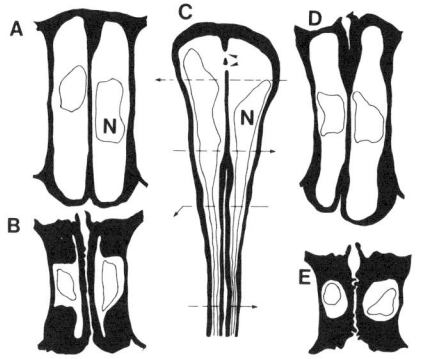

Fig. 3.8. Stomatal shape and the pattern of wall thickening in *Zea mays*. Only walls, lumen, and nuclei (*N*) are shown. Stoma and guard cells are dumbbell-shaped because the polar region is deeper and wider than the region around the pore. Arrowheads mark perforations in ventral wall at poles (*C*). Arrows across paradermal view (*C*) indicate level of corresponding cross section. See text for detailed description of pattern of wall thickening. Bar = 3 μm. (Drawn from electron micrographs of Srivastava and Singh, 1972.)

considerably thicker than the ventral and dorsal walls (Fig. 3.8E). Close to the end of the pore, the periclinal walls are thin where they join the ventral wall, so that the lumen appears flanged at this level (Fig. 3.8B). The periclinal walls are even thinner at the poles (Fig. 3.8A, D). The dorsal wall next to the subsidiary cell is thinner than the other walls and is especially thin at the level of the pore; this thinness could allow rapid passage of osmotica and flexible movement for equilibration. The dorsal wall thickens differentially near the end of the pore (Fig. 3.8D). The ventral wall is thinner at the poles than at the pore, and it thickens at its junctions with the periclinal walls.

Dumbbell-shaped stomata in grasses and sedges (e.g. Brown and Johnson, 1962; Galatis, 1980; Palevitz, 1981a) and stomata in *Equisetum* (Dayanandan and Kaufman, 1973; Laroche et al., 1976; Dayanandan, 1977) contain perforations in the polar parts of their ventral walls that can be large enough for plastids to pass through (Figs. 3.3, 3.8C). It is not known if members of other plant families with dumbbell-shaped stomata also have perforate septa. Raschke (1979) has shown that dumbbell-shaped guard cells are more efficient in opening than kidney-shaped ones; that is, they require a much smaller change in volume to produce a unit change in the width of the aperture.

Kidney-shaped guard cells. The three-dimensional pattern of wall thickenings and the details of cell shape have not been described so precisely for normal kidney-shaped guard cells as they have for the dumbbell-shaped cells of corn. Light-microscope descriptions (e.g. von Guttenberg, 1971; Singh and Srivastava, 1973; Raschke, 1979) indicate that kidney-shaped guard cells have several similarities to and differences from their dumbbell-shaped counterparts in corn. Noticeable similarities are a deeper lumen at the poles than at the middle of the pore (Figs. 3.4, 3.9C, E) and the fact that near the end of the pore the lumen is deepest close to the ventral wall (compare Figs. 3.8B and 3.9B). In longitudinal anticlinal sections through this region, the lumen of kidney-shaped guard cells may have an almost dumbbell-shaped profile (Fig. 3.8C).

72 Fred D. Sack

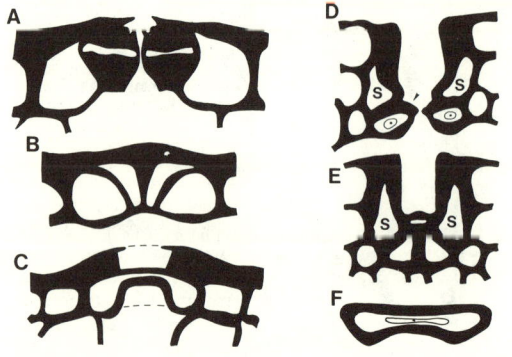

Fig. 3.9. Stomatal shape and the pattern of wall thickening in *Quercus ilex* (A–C) and *Pinus merkusii* (D–F). A, B, D, E, cross sections at the pore (A,D) and at the poles (B,E); C, F, longitudinal anticlinal sections close to the pore. Note that the poles are deeper than the areas along the pore. The *P. merkusii* stomatal apparatus is representative of the gymnosperm type, with the guard cells sunken below the epidermis and below subsidiary cells (S). Arrowhead in D indicates region shown in Fig. 3.11. Bar = 20 μm (*Pinus*). (Redrawn from von Guttenberg, 1971 [*Quercus*], and Esau, 1977 [*Pinus*], and based on light microscopic observations.)

The dumbbell-shaped guard cell, however, exhibits this shape in paradermal section as well. In both types of guard cells, the nucleus is centered in the shallower region, along the pore, whereas the vacuoles are found primarily at the deeper, polar regions (Figs. 3.3, 3.4).

A further similarity is found in the deformations accompanying movement. Raschke (1979) measured changes in volume during movement in living guard cells and concluded for both stomatal types that an increase in volume in the polar regions causes the midlength of the guard cell to bulge into the subsidiary cell, thus causing the pore to widen. Guard cells are much smaller than other epidermal cells or mesophyll cells, and small changes in volume may be effective in regulating the aperture. The changes in the volume of subsidiary cells (Couot-Gastelier, Laffray, and Louguet, 1984) and the structural properties of the entire stomatal apparatus also need to be included in discussions of the mechanics of movement (see Sharpe, Wu, and Spence, this volume).

Patterns of wall thickening are presumably related to changes in the shape and in the position of guard cells during movement (Allaway and Milthorpe, 1976), although differential thickenings may not be so important as some models of stomatal mechanics assume (Raschke, 1979).

Many variations in the patterns of wall thickening in kidney-shaped guard cells have been described with the light microscope (von Guttenberg, 1971). Generally, at least part of the dorsal wall is thinner than the other walls (Figs. 3.1, 3.9A). The ventral wall typically is thinnest where it forms the central aperture, and this wall thickens at its junctions with the periclinal walls (Figs. 3.1, 3.7, 3.9A).

In the elliptical stomata of *Funaria* the dorsal wall is extremely thin at its outer end, but the wall of the adjoining subsidiary cell is too thick to allow any movement of the dorsal wall (Sack and Paolillo, 1983b). The outer wall, however, is thin close to the subsidiary cell, and this area likely functions as a hinge during movement. Similar hinges, or "Hautgelenke," have been described in numerous other species, such as some gymnosperms (Chabot and Chabot, 1977).

The basic structure of gymnosperm stomata is consistent throughout the taxon. The guard cells are usually sunken below subsidiary cells, which are often arranged in a ring. The outer and dorsal walls are heavily thickened (see Figs. 3.9D, 3.11), except for the hinge at the base of the dorsal wall (Walles, Nyman, and Alden, 1973; Vassilyev and Vassilyeva, 1976; Chabot and Chabot, 1977). Another hinge is present below the outer ledge (Fig. 3.9D). The subsidiary-cell wall next to the guard cell is also heavily thickened. The guard cells are rotated inward with respect to the plane of the epidermis, so that the outer ledge forms the main aperture.

The stomata of *Equisetum* resemble those of gymnosperms; they have a similar position, except that protrusions on the wall of the subsidiary cell form the main aperture. The subsidiary-cell wall has prominent thickenings characteristic of the genus (Fig. 3.10; Dayanandan and Kaufman, 1973; Laroche et al., 1976; Dayanandan, 1977).

In *Polypodium*, Stevens and Martin (1977) have reported potassium-adsorbent sacs or bodies at the poles between the inner guard-cell wall and the cuticle lining the substomatal cavity; they claim that similar bodies occur in the intercellular space between the guard cells and subsidiary cells of *Tradescantia*. The identification of these structures by transmission electron microscopy may eliminate the possibility that they are artifacts of preparation for scanning electron microscopy. Carr, Carr, and Jahnke (1980) have described potassium-rich intercellular pectic strands that they believe are the "trabeculae" Stevens and Martin (1977) identified.

In summary, there are several major types of stomatal structure, with extensive variation in the patterns of wall thickening in the apparatus. Wall configurations allowing increases of volume in the polar regions may be critical for pore opening in all types.

Layering and Composition of the Wall

From polarized-light microscopy it has long been known that the fibrils in the periclinal walls of elliptical stomata show a radial orientation with respect to the pore (Ziegenspeck, 1955); this has been confirmed with electron microscopy (Singh and Srivastava, 1973; Palevitz and Hepler, 1976; Galatis and Mitrakos, 1980). However, the innermost layer of the periclinal wall in *Funaria* has sublayers with fibrils either radially or axially oriented (Sack and Paolillo,

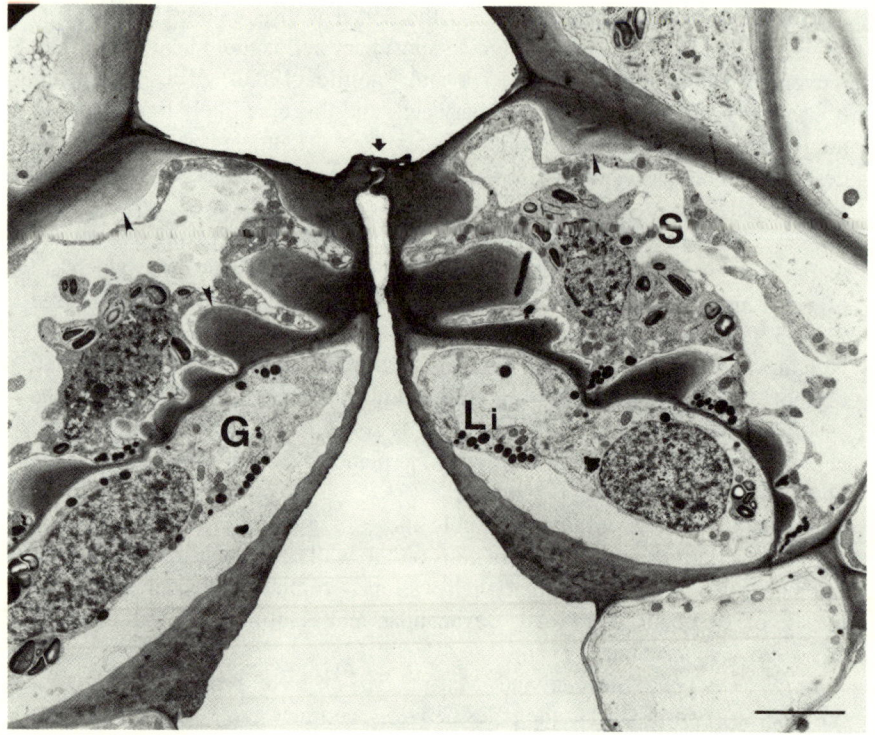

Fig. 3.10. Cross section of stomatal apparatus of *Equisetum hyemale*. As in *Pinus* (Fig. 3.9D), the guard cells (*G*) subtend subsidiary cells (*S*) that are in turn sunken below the epidermis. However, here ledges (arrow) are located on the subsidiary cells, not on guard cells, and the ledges have protuberances that interdigitate. Subsidiary-cell walls have massive thickenings (arrowheads). Stomatal walls are thin next to the subsidiary cells. Spherical, electron-dense lipid globules (*Li*) are present throughout. Bar = 5 μm. (Micrograph kindly provided by Drs. P. Dayanandan and P. B. Kaufman, from Dayanandan, 1977.)

1983b), and in *Allium*, fibrils that diverge slightly from a strict radial pattern can be seen in layers with an overall radial orientation (Palevitz, 1982). Since polarized-light microscopy reveals only net fibrillar orientation, there may be populations of microfibrils oriented otherwise within the dominant radial pattern; that is, the complexity of the walls may be greater than polarized-light studies can detect.

The net orientation of wall fibrils in mature dumbbell-shaped guard cells is axial, except when sensitive species are subjected to environmental manipulation or when guard cell shape is transient, during formation of the pore (Mishkind, Palevitz, and Raikhel, 1981; Palevitz, 1981a, 1982). As in the radial pattern in kidney-shaped guard cells, however, the wall fibrils in mature grass stomata may not be strictly axial but rather arranged in a steep crisscross pattern with a net axial orientation (Palevitz, 1982).

Wall layers identified by differing fibrillar orientations have been described in the guard cells of the gymnosperms *Araucaria* and *Agathis* (Vassilyev and Vassilyeva, 1976). The dorsal wall has an unusual secondary thickening composed of alternating narrow lamellae with axial fibrils and broad lamellae with transverse and oblique fibrils.

Layering in guard cell walls is suggested by ultrastructural features other than fibrillar orientation. In *Funaria* the wall layer deposited after the substomatal cavity forms has a mottled texture caused by electron-dense blotches connecting the wall fibrils (Sack and Paolillo, 1983b). By the time the pore has formed, a new, primarily fibrillar layer starts to be deposited. Cork and Nelmes (1979) reported electron-lucent blotches in the outer and ventral walls of guard cells from several genera.

Apart from these few studies, it is uncertain whether stomatal walls are generally unlayered; perhaps other studies simply have not examined the ultrastructure of the walls for this feature. Layering could be significant for stomatal mechanics if the different layers have different physical properties. As indicated, stomatal walls may thin during opening (Raschke, 1979), and the elasticity of the entire wall and of its layers could affect or effect the change in thickness. Layers could also vary in their ion-storage and ion-exchange capacities, which may be especially important, given the symplastic isolation of guard cells.

The chemical composition of the wall would also affect its mechanical and storage properties. Lignification would increase the wall's rigidity. Tests for lignin were positive in the guard cell walls of corn (Srivastava and Singh, 1972) and of several ferns (Waterkeyn and Bienfait, 1979) but negative in other elliptical stomata (Singh and Srivastava, 1973; Peterson, Firminger, and Dobrindt, 1975; Sack and Paolillo, 1983b; Gedalovich and Fahn, 1983). Guard cells in gymnosperms apparently have a regular pattern of lignification, with lignin conspicuously absent near the hinge regions (Fig. 3.11; Vassilyev and Vassilyeva, 1976; Chabot and Chabot, 1977; Appleby and Davies, 1983). Recently, Palevitz (1981a) found that in the stomatal walls of grasses and of some sedges only some of the phenolics constitute lignin; the remaining phenolics are probably esterified to carbohydrates.

Guard cell walls are thought to be rich in pectins (Singh and Srivastava, 1973; Raschke, 1979), although only the middle lamella stains positively with ruthenium red in the stomata of several genera (Srivastava and Singh, 1972; Sack and Paolillo, 1983b). Guard cell walls may react more intensely than adjacent cell walls with the carbohydrate stain silver methenamine; Wille and Lucas (1984) hypothesize that this indicates an abundance of pectins in the guard cell walls. The stomatal walls of ferns may contain callose, especially in immature cells (Peterson, Firminger, and Dobrindt, 1975; Waterkeyn and Bienfait, 1979).

Plasmodesmata

Mature stomata lack plasmodesmata. Early in development, plasmodesmata are present in all anticlinal stomatal walls. As the wall thickens, the plasmodesmata break and become embedded in the wall. Apart from one report of their presence (Pallas and Mollenhauer, 1972b), complete plasmodesmata have not been found in mature stomata (see literature cited in Willmer and Sexton, 1979; and in Wille and Lucas, 1984). Using serial thick sections viewed by electron microscopy, Wille and Lucas (1984) have convincingly demonstrated the general absence of plasmodesmata. This con-

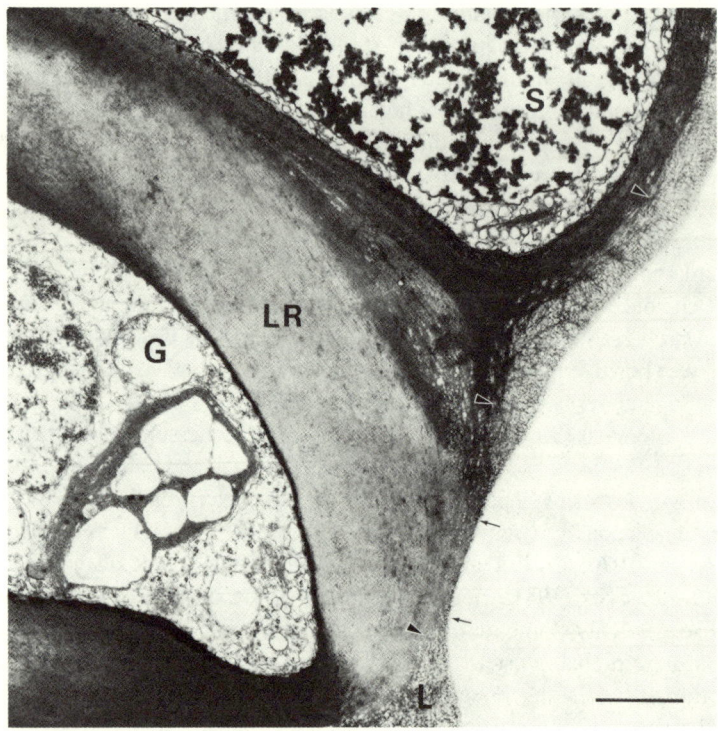

Fig. 3.11. Wall structure and cuticle thickness in stomatal apparatus of *Larix leptolepis*. Cross section showing part of subsidiary cell (*S*) and guard cell (*G*) in region comparable to that indicated by arrowhead in Fig. 3.9D. Gymnosperm stomatal walls are thought to be lignified in the region that is less electron-dense (*LR*) than other parts of the guard cell wall (e.g. bottom left; Chabot and Chabot, 1977). The cuticle is much thinner over the lignified region (between arrows) than over the subsidiary cell and the outer stomatal ledge (*L*). The approximate inner limits of the cuticle are denoted by arrowheads. Note that here, compared with *Equisetum* (Fig. 3.10), the guard cell wall is much thicker than the subsidiary wall where they adjoin. Bar = 1 μm.

Development and Structure of Stomata 77

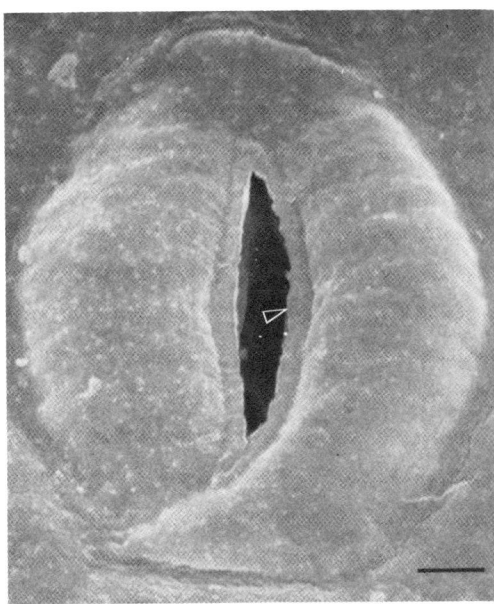

Fig. 3.12. Scanning electron micrograph of mature stoma of *Funaria hygrometrica* Hedw., showing outer ledge (arrowhead) overarching pore. Bar = 5 μm.

clusion has been reinforced by the finding that fluorescent symplasm markers microinjected into epidermal cells do not pass between the stomata and the surrounding cells (Erwee, Goodwin, and Van Bel, 1985; Palevitz and Hepler, 1985). The symplastic isolation of guard cells may make it easier to maintain differential pressures in turgor between guard cells and adjacent cells.

Interestingly, plasmodesmata are never found in the newly formed, incomplete ventral wall in *Funaria* (Sack and Paolillo, 1983c), but they are present in the young, perforated ventral walls of grasses (Kaufman et al., 1970; Srivastava and Singh, 1972). Extensive cytoplasmic continuity presumably makes plasmodesmata unnecessary in *Funaria*.

Membrane-like inclusions have been found in the guard cell walls of several species. In *Funaria*, they are concentrated where separation of the ventral wall begins, and they may play a role in pore formation (Sack and Paolillo, 1983a). Cork and Nelmes (1979) reported the existence of membrane-bound "vesicles" in stomatal walls; these vesicles may be cytoplasmic outgrowths continuous with the plasmalemma.

Structure of the Pore

Most stomatal pores have three apertures, two formed by the inner and outer ledges, and one, the central aperture, bounded by the bulging ventral wall at its mid-depth (Figs. 3.1, 3.7, 3.9A, 3.12). The inner aperture does not usually

78 Fred D. Sack

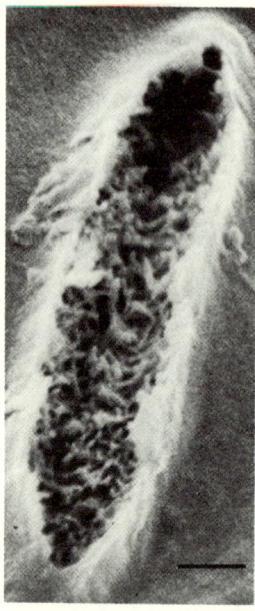

Fig. 3.13. Scanning electron micrograph of stoma of *Funaria hygrometrica* Hedw., showing plugged pore from older, presumably nonfunctional guard cell. Bar = 2 μm.

close completely, and either the outer or the central aperture may be narrowest at different stages of pore opening (Saxe, 1979). Thus, in any study that involves measuring stomatal aperture, it is important always to quantify the same aperture. Pore width may change upon fixation (Raschke, 1979).

As stomata age in *Funaria*, cuticular accumulations at the ends of the pore may shorten the aperture, and the pore itself may become filled with wax or cuticular tubules (Fig. 3.13). In many xerophytes and in gymnosperms, the pore may become permanently plugged with wax (Chabot and Chabot, 1977).

Cuticle

Several lines of evidence suggest that the cuticle of guard cells is more permeable to water than the cuticle of other epidermal cells. Stomata can open and close in response to changes in humidity that do not substantially affect the bulk water potential of the leaf (Schulze et al., this volume). Also, though most water exits the leaf through the pore (evapotranspiration), some may evaporate from exposed guard cell surfaces as well (peristomatal transpiration). Furthermore, stomatal walls are thought to be rich in ectodesmata, wall areas that may lie beneath sites where the cuticle is permeable to polar (charged) substances (see Sack and Paolillo, 1983a). These observations suggest that either the stomatal cuticle is more permeable to water, or else some areas of the guard cell wall are directly exposed to the atmosphere or are covered by only a thin cuticle.

Despite the importance of the stomatal cuticle as a barrier to water loss, there have been few studies of its fine structure and distribution. The cuticle is often thickest over the outer wall and the outer ledge (e.g. Carr and Carr, 1978; Galatis and Mitrakos, 1980; Appleby and Davies, 1983), and it is continuous through the pore and the inner wall (Saxe, 1979; Galatis, 1980; Palevitz, 1981a; Sack and Paolillo, 1983a).

Several workers have held that regions of the pore wall either are not covered by a cuticle or have only an extremely thin cuticle (Edwards and Meidner, 1978; Appleby and Davies, 1983), but definitive evidence, such as from high-magnification electron microscopy, is not yet available. There is evidence that the pore wall and outer ledge are the sites of greatest water loss (Maier-Maercker, 1983), but tracers of water movement are also found in the outer wall, and the degree of tracer accumulation is not necessarily proportional to the thickness of the cuticle (Appleby and Davies, 1983; Maier-Maercker, 1983). Furthermore, Mérida, Schönherr, and Schmidt (1981) found that cuticular permeability to water correlates more with chemical composition than with thickness.

Knowledge of the fine structure and chemical composition of the stomatal cuticle is important for understanding its permeability. In *Funaria*, the stomatal apparatus exhibits three different cuticular morphologies (Sack and Paolillo, 1983a). Electron-dense fibrils throughout the guard cell cuticle occasionally reach its surface (Fig. 3.7), but the subsidiary-cell cuticle lacks fibrils entirely. Similar fibrils in the cuticles of other genera and cell types have been shown to be hydrophilic, but these fibrils seldom reach the cuticle surface (Sack and Paolillo, 1983a). In the stomata of *Polypodium*, the fibrils come close to the surface, but the magnification of the published micrographs is insufficient to determine whether they reach it (Edwards and Meidner, 1978).

Few other details are available on guard cell cuticles. In the water fern *Azolla*, the cuticle is primarily lamellate (Busby and Gunning, 1984). In *Phleum*, the guard cell cuticle is more electron-dense than that in subsidiary cells (Palevitz, 1981a). In *Larix*, the cuticle is much thinner over the lignified (and thus less permeable) region of the guard cell wall than over adjacent nonlignified areas (Fig. 3.11). Clearly, more information is needed about the ultrastructure of the cuticle on stomata and on subsidiary cells.

Organelles in Mature and Developing Stomata

Because guard cells are symplastically isolated, and because stomatal movements involve the transport of large amounts of ions, investigators have looked for cell specializations that increase the surface area of the plasmalemma, like the wall ingrowths in transfer cells. Stomatal walls have no ingrowths, but plasmalemmasomes (infoldings of the plasmalemma) have been found (e.g. Thomson and DeJournett, 1970; Ziegler, Shmueli, and Lange,

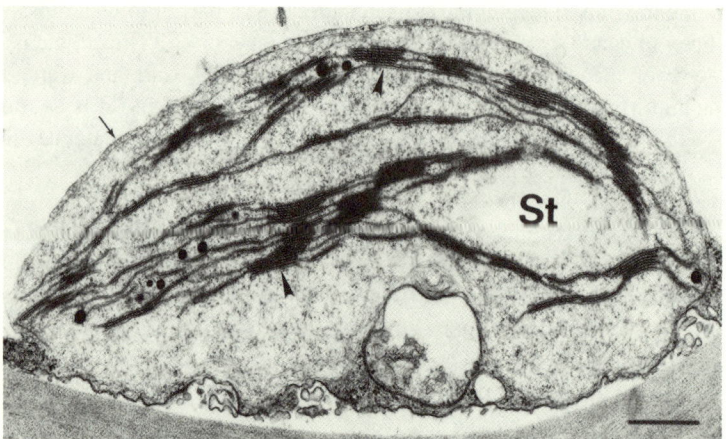

Fig. 3.14. Well-developed grana (arrowheads) in stomatal plastid of *Nicotiana tabacum*. Stomatal plastids in most genera are more filled with starch (*St*) than here. Spherical, electron-dense structures are plastoglobuli. Arrow marks tonoplast. Bar = 0.5 μm.

1974; Wille and Lucas, 1984). These may occur only rarely, however, and may be an artifact of fixation. Thus, it is likely that the guard cell membrane does not have an exceptional surface area.

The envelope of stomatal chloroplasts, though, is often extensively invaginated along its inner membrane. This peripheral reticulum could facilitate the import of starch precursors or the exchange of osmotica (Allaway and Setterfield, 1972; Pallas and Mollenhauer, 1972a; Srivastava and Singh, 1972; Singh and Srivastava, 1973; Allaway and Milthorpe, 1976; Stevens and Martin, 1979; Sack and Paolillo, 1983c). This reticulum can be more developed in the plastids of stomata than in those of mesophyll cells in the same species, and likewise in the plastids of open stomata in comparison with those of closed stomata (Allaway and Setterfield, 1972; Couot-Gastelier, Laffray, and Louguet, 1984).

Guard cell plastids generally contain chlorophyll, moderate-sized grana, and varying amounts of starch (Figs. 3.3, 3.4, 3.6, 3.11, 3.14; e.g. Whatley, 1972; Singh and Srivastava, 1973; Allaway and Milthorpe, 1976; Sanchez, 1977; Faraday, Thomson, and Platt-Aloia, 1982; Sack and Paolillo, 1983c). Plastids in the stomata of *Allium* lack starch and have moderately developed grana (Allaway and Setterfield, 1972; Palevitz and Hepler, 1976; Thorpe, 1980). Conversely, plastids from functional stomata of a variegated chimera of *Pelargonium zonale* and from guard cells of *Paphiopedilum* and *Haemeria discolor* (Orchidaceae) contain starch but lack grana and show either only weak chlorophyll fluorescence or none at all (Rutter and Willmer, 1979; Thorpe, 1980; Willmer, 1983; D'Amelio and Zeiger, unpublished). Stomata in other orchids mostly have plastids with chlorophyll and small grana; in some orchids and in *Nicotiana tabacum* and *Lycopersicon pennellii*, the grana are very well developed (Fig. 3.14; D'Amelio and Zeiger, unpublished). Osmiophilic, presumably lipid, globules can be found in most stomatal chloroplasts. Other, rarer inclusions in specific genera are fibrous (perhaps proteinaceous) bodies, paracrystalline structures (Fig. 3.15), and structures bound to or aggregated about the membrane (Rutter and Willmer, 1979; Thorpe, 1980; Faraday, Thomson, and Platt-Aloia, 1982; D'Amelio and Zeiger, unpublished).

Stomatal chloroplasts are typically smaller than plastids in mesophyll cells, and their grana are shallower (Allaway and Setterfield, 1972; Thorpe, 1980). Mesophyll cells of *Commelina* and *Vicia* have three times as many plastids per cell as guard cells do, but the plastids occupy a similar volume (4%) of the cytoplasm in both cell types (Allaway and Milthorpe, 1976). In some ferns, guard cells have 80 to 100 plastids, which occupy up to 40 percent of the cytoplasmic volume; nonstomatal epidermal cells also contain well-developed chloroplasts (Stevens and Martin, 1979). In other genera, plastids in nonstomatal epidermal cells are smaller, have shallower grana, and may not contain starch or chlorophyll (Allaway and Milthorpe, 1976).

Most stomatal chloroplasts are also amyloplasts (i.e. they are distended with starch), although some (e.g. those of *Allium*) lack starch entirely, and others may have only small amounts of it (Fig. 3.14). The cortical-cell plastids of *Funaria* initially contain much more starch than the guard cell chloroplasts do; this relationship is reversed by the time the pore forms (Sack and Paolillo, 1983c). The starch content decreases substantially when the pore is forming

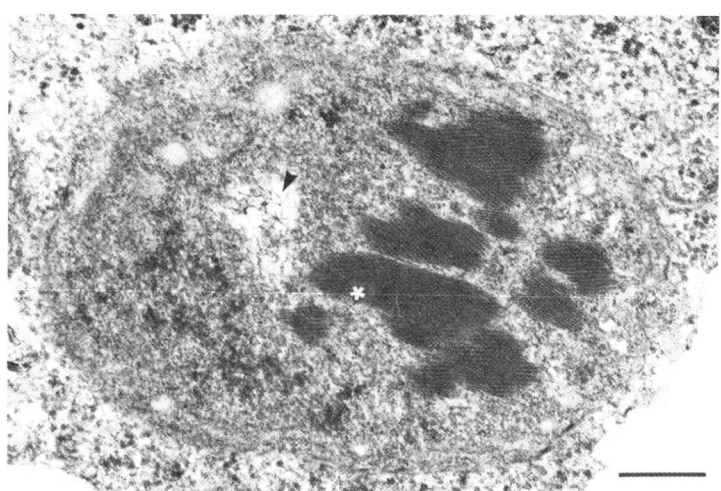

Fig. 3.15. Immature stomatal plastid from *Haemaria discolor* containing paracrystalline inclusions (*) and probable nucleoid region (arrowhead). Bar = 0.25 μm. (Courtesy of Drs. E. D'Amelio and E. Zeiger.)

and during stomatal opening (Outlaw, this volume; Stevens and Martin, 1978; Sack and Paolillo, 1983c).

Stomatal movement requires energy; thus, it is not surprising that guard cells contain many mitochondria. From counts of organelle profiles in micrographs, Allaway and Setterfield (1972) estimated that there are four to ten times as many mitochondria as chloroplasts in guard cells, but in mesophyll cells there are fewer mitochondria than plastids. Allaway and Setterfield (1972) also found mitochondria more abundant in the protoplast of guard cells than in that of mesophyll cells.

Lipid globules can frequently be observed in guard cells with light or electron microscopy (Figs. 3.6, 3.10; Sorokin and Sorokin, 1966; Pallas and Mollenhauer, 1972a; Dayanandan, 1977; Chabot and Chabot, 1977; Peterson and Hambleton, 1978). Even larger oil droplets may occur in stomata (Rutter and Willmer, 1979; Thorpe, 1980). Although the lipid globules are usually very electron-dense, they may become less so at developmental stages when lipid reserves might be depleted: for example, during rapid cuticle synthesis or as stomata age (Sack and Paolillo, 1983c). In *Adiantum*, lipid bodies (spherosomes) and microbodies preferentially accumulate near the regions of the stomatal wall that are thickening (Galatis, Apostolakos, and Katsaros, 1983).

Microbodies have been regularly found in guard cells, but usually they are sparse (Thomson and DeJournett, 1970; Allaway and Setterfield, 1972; Pallas and Mollenhauer, 1972a; Singh and Srivastava, 1973; Chabot and Chabot, 1977; Galatis and Mitrakos, 1980; Faraday, Thomson, and Platt-Aloia, 1982; Sack and Paolillo, 1983c). Histochemical tests suggest that their catalase levels are low (Thorpe, 1980); crystalline inclusions have not been observed in stomatal microbodies. The role of microbodies in stomatal metabolism is unknown.

Although the microtubular cytoskeleton in guard cells has been well described, either microfilaments are not abundant in stomata or else they are destroyed by fixation. Elements resembling microfilaments have been found near the plasmalemma of negatively stained guard cell protoplasts and in some thin sections (Doohan and Palevitz, 1980; Palevitz, 1982). Cytoplasmic streaming is common in stomata.

There are many different sorts of vesicles in the guard cell cytoplasm, including dictyosome vesicles containing carbohydrates that may be involved in wall biosynthesis (see "Microtubules and Wall Building"). Other vesicle types are distinguished by their size and by the electron density of their contents, and guard cells may have different sorts of vesicles than subsidiary cells do (Sack and Paolillo, 1983c). Coated vesicles and coated pits have often been found in developing and mature stomata (Singh and Srivastava, 1973; Galatis, 1980; Galatis and Mitrakos, 1980; Galatis, Apostolakos, and Katsaros, 1983; Sack and Paolillo, 1983c) and in guard cell protoplasts (Doohan and

Palevitz, 1980). Neither the function nor the direction of flow for any of these vesicles in stomata is known. Their numbers are much diminished in stomata that have been functional for some time; thus, they probably operate in cell development rather than directly in stomatal functioning.

One of the most dynamic components of the cytoplasm during stomatal development is the endoplasmic reticulum (ER). The ER may form a well-developed network of tubules, especially near the thickening ventral wall. In *Vicia faba*, this network appears to be the site of strong chlorotetracycline fluorescence, which indicates a high concentration of membrane-bound calcium (Palevitz and Hodge, 1984). The function of this network is uncertain. In *Funaria*, the ER is also particularly plentiful near the ventral wall until the pore forms, but in *Zea* and *Pisum* the ER is not preferentially localized (Singh and Srivastava, 1973; Galatis, 1980; Sack and Paolillo, 1983c). In all these studies, the ER was found closely associated with the plasmalemma.

The morphology of the ER may vary during development and between species. Before the ventral wall separates in *Funaria*, the ER may form arrays of parallel cisternae along this wall. About the time of pore formation, the ER becomes smooth, tubular, and strikingly abundant, practically filling the cytoplasm (Sack and Paolillo, 1983c). Apparently, a tubular ER also proliferates when the ventral wall finally separates in the stomata of *Adiantum* (Galatis, Apostolakos, and Katsaros, 1983). When the tubular ER appears in *Funaria*, the electron density of lipid globules decreases in the stomata but not in the subsidiary cells. Both organelles are possibly involved in the rapid cuticle synthesis accompanying pore formation (Sack and Paolillo, 1983c). There is less cisternal ER in mature stomata.

The ends of ER cisternae may be dilated; the swellings may contribute to the formation of vesicles and vacuoles (Sack and Paolillo, 1983c; Couot-Gastelier, Laffray, and Louguet, 1984; Palevitz and Hodge, 1984). The ER is a variable and complex component of the endomembrane system, and its structural complexity in guard cells probably reflects a multiplicity of functions.

The vacuolar system also undergoes complicated changes during stomatal development. Using videomicroscopy, Palevitz et al. (1981) took advantage of the blue-light fluorescence of guard cell vacuoles in *Allium* to follow vacuole differentiation in living cells. In the guard-cell parent cell, the vacuole is first large and globular but then subdivides into a reticulum that persists throughout cell division. The reticulum coalesces into larger, globular vacuoles as the stoma matures. The reticulum is extremely dynamic and changes position and interconnections over time.

Presumably the vacuolar volume increases in the mature guard cell during stomatal opening, but no such increase has been quantified. It is also uncertain whether the distribution and shape of the vacuole change during stomatal movement. Some observations indicate that open stomata have one or more large vacuoles in the poles and that closed stomata have many small vacuoles

(Miroslavov, 1972; Couot-Gastelier, Laffray, and Louguet, 1984, with references). Faraday, Thomson, and Platt-Aloia (1982), however, concluded that the shape and extent of vacuole dissection did not vary significantly with stomatal movement in *Vigna unguiculata*. Using semiserial sections, they demonstrated that what appeared to be separate vacuoles was actually confluent; this may account for some earlier reports of vacuole dissection. Observations of thick (10 μm) sections give the strong impression that closed stomata in *Funaria* do have many small vacuoles (Sack and Paolillo, 1983c), but it has not been determined whether these may result from the extended dark periods employed to induce closure rather than from the movement itself.

Guard cells are unlike most other mature plant cells in that their vacuole occupies a smaller portion of the protoplast volume (Faraday, Thomson, and Platt-Aloia, 1982; Couot-Gastelier, Laffray, and Louguet, 1984). The vacuole may exhibit convolution, which increases its surface-to-volume ratio and may facilitate the transport of osmotica and minimize the synthesis of the tonoplast during inflation (Singh and Srivastava, 1973; Faraday, Thomson, and Platt-Aloia, 1982).

Vacuolar inclusions exist in the stomata of several genera (Allaway and Setterfield, 1972; Peterson and Hambleton, 1978); in some cases, the inclusions change in electron density during stomatal movement (Faraday, Thomson, and Platt-Aloia, 1982). Small morphological variations in other organelles have been reported to occur during stomatal movement (Miroslavov, 1972; Humbert and Guyot, 1982).

Areas for Future Research

Studies of guard cell structure are highly relevant for an understanding of stomatal function as well as of basic mechanisms of cellular development. Many questions remain. Data on the location and role of microtubule foci during development should be expanded. It would be useful to determine whether a microfilamentous cytoskeleton exists in stomata. Information is also needed on the fine structure, distribution, and development of the cuticle of the stomatal apparatus, especially as it relates to pore formation and to cuticular permeability. Knowledge about wall substructure and histochemistry and about the changes in cell shape and in cell and cell-compartment volume that accompany movement is essential for any description of stomatal mechanics. Similarly, quantitative estimates of the changes in the volume of the cytoplasm and of the vacuoles during stomatal movement would be helpful for evaluating how osmotica are distributed.

To study changes in volume, data from several techniques should be compared in order to avoid, or at least in order to clarify, the artifacts introduced by conventional fixation for electron microscopy (e.g. shrinkage: Raschke, 1979). Freeze substitution and freeze fracture may be valuable in this regard. The only published freeze-fracture study of stomata used guard cell protoplasts (Schnabl, Vienken, and Zimmermann, 1980).

Light microscopists have been analyzing differences in the structure of the stomatal apparatus for over a century, yet we still know very little about ultrastructural variations among species. It will continue to be fascinating to uncover relationships between guard cell morphology and the physiology and ecology of the stoma and the plant.

REFERENCES

Allaway, W. G., and F. L. Milthorpe. 1976. Structure and functioning of stomata. In T. T. Kozlowski, ed., *Water deficits and plant growth* 4: 57–102. New York: Academic Press.
Allaway, W. G., and G. Setterfield. 1972. Ultrastructural observations on guard cells of *Vicia faba* and *Allium porrum*. *Canadian Journal of Botany*, 50: 1405–13.
Appleby, R. F., and W. J. Davies. 1983. The structure and orientation of guard cells in plants showing stomatal responses to changing vapour pressure difference. *Annals of Botany*, 52: 459–68.
Brown, W. V., and S. C. Johnson. 1962. The fine structure of the grass guard cell. *American Journal of Botany*, 49: 110–15.
Busby, C. H., and B. E. S. Gunning. 1980. Observations on preprophase bands of microtubules in uniseriate hairs, stomatal complexes, and *Cyperus* root meristems. *European Journal of Cell Biology*, 21: 214–23.
———. 1984. Microtubules and morphogenesis in stomata of the water fern *Azolla*: An unusual mode of guard cell and pore development. *Protoplasma*, 122: 108–19.
Carr, D. J., and S. G. Carr. 1978. Origin and development of stomatal microanatomy in two species of *Eucalyptus*. *Protoplasma*, 96: 127–48.
Carr, D. J., S. G. Carr, and R. Jahnke. 1980. Intercellular strands associated with stomata: Stomatal pectic strands. *Protoplasma*, 102: 177–82.
Chabot, J. F., and B. Chabot. 1977. Ultrastructure of the epidermis and stomatal complex of balsam fir (*Abies balsamea*). *Canadian Journal of Botany*, 55: 1064–75.
Cork, R. J., and B. J. Nelmes. 1979. Vesicles in guard-cell walls and their possible roles in the stomatal mechanism. *Journal of Cell Science*, 38: 83–95.
Couot-Gastelier, J., D. Laffray, and P. Louguet. 1984. Étude comparée de l'ultrastructure des stomates ouverts et fermés chez le *Tradescantia virginiana*. *Canadian Journal of Botany*, 62: 1505–12.
Dayanandan, P. 1977. Stomata in *Equisetum*: A structural and functional study. Ph. D. dissertation, University of Michigan, Ann Arbor.
Dayanandan, P., and P. B. Kaufman. 1973. Stomata in *Equisetum*. *Canadian Journal of Botany*, 51: 1555–64.
DeChalain, T., and P. Berjak. 1979. Cell death as a functional event in the development of the leaf intercellular spaces in *Avicennia marina* (Forsskål) Vierh. *New Phytologist*, 83: 147–55.
Doohan, M., and B. Palevitz. 1980. Microtubules and coated vesicles in guard cell protoplasts of *Allium cepa* L. *Planta*, 149: 389–401.
Edwards, M., and H. Meidner. 1978. Stomatal responses to humidity and the water

potentials of epidermal and mesophyll tissues. *Journal of Experimental Botany*, 29: 771–80.
Erwee, M. G., P. B. Goodwin, and A. J. E. Van Bel. 1985. Cell-cell communication in the leaves of *Commelina cyanea* and other plants. *Plant, Cell and Environment*, 8: 173–78.
Esau, K. 1977. *Anatomy of seed plants*. 2d ed. New York: Wiley.
Faraday, C. D., W. W. Thomson, and K. Platt-Aloia. 1982. Comparative ultrastructure of guard cells of C_3, C_4, and CAM plants. In I. P. Ting and M. Gibbs, eds., *Crassulacean acid metabolism*: 18–30. Rockville, Md.: American Society of Plant Physiologists.
Galatis, B. 1980. Microtubules and guard-cell morphogenesis in *Zea mays* L. *Journal of Cell Science*, 45: 211–44.
———. 1982. The organization of microtubules in guard mother cells of *Zea mays*. *Canadian Jornal of Botany*, 60: 1148–66.
Galatis, B., P. Apostolakos, and C. Katsaros. 1983. Microtubules and their organizing centres in differentiating guard cells of *Adiantum capillus veneris*. *Protoplasma*, 115: 176–92.
———. 1984. Experimental studies on the function of the cortical cytoplasmic zone of the preprophase microtubule band. *Protoplasma*, 122: 11–16.
Galatis, B., P. Apostolakos, C. Katsaros, and H. Loukari. 1982. Preprophase microtubule band and local wall thickening in guard cell mother cells of some Leguminosae. *Annals of Botany*, 50: 779–91.
Galatis, B., and K. Mitrakos. 1980. The ultrastructural cytology of the differentiating guard cells of *Vigna sinensis*. *American Journal of Botany*, 67: 1243–61.
Gedalovich, E., and A. Fahn. 1983. Ultrastructure and development of the inactive stomata of *Anabasis articulata* (Forsk.) Moq. *American Journal of Botany*, 70: 88–96.
Gunning, B. E. S. 1982. The cytokinetic apparatus: Its development and spatial regulation. In C. W. Lloyd, ed., *The cytoskeleton in plant growth and development*: 229–92. London: Academic Press.
von Guttenberg, H. 1971. Spaltöffnungen. In *Bewegungsgewebe und Perzeptionsorgane*: 203–19. Handbuch der Pflanzenanatomie, vol. 5, part 5. Berlin: Bornträger.
Hepler, P. K. 1981. Morphogenesis of tracheary elements and guard cells. In O. Kiermayer, ed., *Cytomorphogenesis in plants*: 327–47. Vienna: Springer-Verlag.
Humbert, C., and M. Guyot. 1982. Vacuome des cellules de garde et mouvements des stomates. *Physiologie végétale*, 20: 239–46.
Kaufman, P. B., L. B. Petering, C. S. Yocum, and D. Baic. 1970. Ultrastructural studies on stomatal development in internodes of *Avena sativa*. *American Journal of Botany*, 57: 33–49.
Krarup-Hjort, C. F. 1983. Ultrastructural development and lifetime functioning of stomata from squash (*Cucurbita maxima* Duch.) cotyledons. Ph.D. dissertation, University of California, Davis.
Landré, P. 1969a. Premières observations sur l'évolution infrastructurale des cellules stomatiques de la Moutarde (*Sinapis alba* L.) depuis leur mise en place jusqu'à l'ouverture de l'ostiole. *Comptes rendus des séances de l'Académie des sciences*, Paris, ser. D, 269: 943–46.
———. 1969b. Quelques aspects infrastructureaux des stomates des cotylédon de la Moutarde (*Sinapis alba* L.). *Comptes rendus des séances de l'Académie des sciences*, Paris, ser. D, 269: 990–92.
———. 1970. Activité golgienne en liaison avec celle du plasmalemme dans les cellules

stomatiques de la Moutarde (*Sinapis alba* L.) lors de la formation de l'ostiole. *Comptes rendus des séances de l'Académie des sciences*, Paris, ser. D, 271: 904–7.

———. 1972. Origine et développement des épidermes cotylédonaires et foliares de la Moutarde (*Sinapis alba* L.). Différenciation ultrastructurale des stomates. *Annales des sciences naturelles (botanique)*, 13: 247–322.

Laroche, J., D. Robert, C. Lecoq, and C. Guervin. 1976. Contribution à l'étude du transit du silicum chez l'*Equisetum arvense* L., I. Les stomates: étude ontogénétique et ultrastructurale. *Revue générale de botanique*, 83: 331–65.

Maier-Maercker, U. 1983. The role of peristomatal transpiration in the mechanism of stomatal movement. *Plant, Cell and Environment*, 6: 369–80.

Mérida, T., J. Schönherr, and H. Schmidt. 1981. Fine structure of plant cuticles in relation to water permeability: The fine structure of the cuticle of *Clivia miniata* Reg. leaves. *Planta*, 152: 259–67.

Miroslavov, E. G. 1972. Submicroscopic organization of guard cells in open and closed stomata. *Doklady Akademii nauk SSSR*, 203: 939–41.

Mishkind, M., B. A. Palevitz, and N. V. Raikhel. 1981. Cell wall architecture: Normal development and environmental modification of guard cells of the Cyperaceae and related species. *Plant, Cell and Environment*, 4: 319–28.

Napp-Zinn, K. 1973. *Anatomie des Blattes 2, Blattanatomie der Angiospermen*. Handbuch der Pflanzenanatomie, vol. 8, part 2A. Berlin: Bornträger.

Palevitz, B. A. 1981a. The structure and development of stomatal cells. In P. G. Jarvis and T. A. Mansfield, eds., *Stomatal physiology*: 1–23. Cambridge: Cambridge University Press.

———. 1981b. Microtubules and possible microtubule nucleation centers in the cortex of stomatal cells as visualized by high voltage electron microscopy. *Protoplasma*, 107: 115–25.

———. 1982. The stomatal complex as a model of cytoskeletal participation in cell differentiation. In C. W. Lloyd, ed., *The cytoskeleton in plant growth and development*: 346–76. London: Academic Press.

Palevitz, B. A., and P. K. Hepler. 1974. The control of the plane of division during stomatal differentiation in *Allium*, I. Spindle reorientation. *Chromosoma*, 46: 297–326.

———. 1976. Cellulose microfibril orientation and cell shaping in developing guard cells of *Allium*: The role of microtubules and ion accumulation. *Planta*, 132: 71–93.

———. 1985. Changes in dye coupling of stomatal cells of *Allium* and *Commelina* demonstrated by microinjection of Lucifer yellow. *Planta*, 164: 473–79.

Palevitz, B. A., and L. D. Hodge. 1984. The endoplasmic reticulum in the cortex of developing guard cells: Coordinate studies with chlorotetracycline and osmium ferricyanide. *Developmental Biology*, 101: 147–59.

Palevitz, B. A., D. J. O'Kane, R. E. Korbes, and N. V. Raikhel. 1981. The vacuole system in stomatal cells of *Allium*. Vacuole movements and changes in morphology in differentiating cells as revealed by epifluorescence, video and electron microscopy. *Protoplasma*, 109: 23–55.

Pallas, J. E., and H. H. Mollenhauer. 1972a. Physiological implications of *Vicia faba* and *Nicotiana tabacum* guard-cell ultrastructure. *American Journal of Botany*, 59: 504–14.

———. 1972b. Electron microscopic evidence for plasmodesmata in dicotyledonous guard cells. *Science*, 175: 1275–76.

Peterson, R. L., M. S. Firminger, and L. A. Dobrindt. 1975. Nature of the guard cell wall in leaf stomata of three *Ophioglossum* species. *Canadian Journal of Botany*, 53: 1698–1711.

Peterson, R. L., and S. Hambleton. 1978. Guard cell ontogeny in leaf stomata of the fern *Ophioglossum petiolatum*. *Canadian Journal of Botany*, 56: 2836–52.

Pickett-Heaps, J. D. 1969. Preprophase microtubules and stomatal differentiation in *Commelina cyanea*. *Australian Journal of Biological Sciences*, 22: 375–91.

Pickett-Heaps, J. D., and D. H. Northcote. 1966. Cell division in the formation of the stomatal complex of young leaves of wheat. *Journal of Cell Science*, 1: 121–28.

Raschke, K. 1979. Movements of stomata. In W. Haupt and M. E. Feinleib, eds., *Physiology of movements*, Encyclopedia of Plant Physiology, n.s., 7: 383–441. Berlin: Springer-Verlag.

Rutter, J. C., and C. M. Willmer. 1979. A light and electron microscopy study of the epidermis of *Paphiopedilum* spp. with emphasis on stomatal ultrastructure. *Plant, Cell and Environment*, 2: 211–19.

Sack, F. D., and D. J. Paolillo. 1983a. Stomatal pore and cuticle formation in *Funaria*. *Protoplasma*, 116: 1–13.

———. 1983b. Wall structure and development in *Funaria* stomata. *American Journal of Botany*, 70: 1019–30.

———. 1983c. Protoplasmic changes during stomatal development in *Funaria*. *Canadian Journal of Botany*, 61: 2515–26.

———. 1985. Incomplete cytokinesis in *Funaria* stomata. *American Journal of Botany*, 72: 1325–33.

Sanchez, S. 1977. The fine structure of the guard cells of *Helianthus annuus*. *American Journal of Botany*, 64: 814–24.

Saxe, H. 1979. A structural and functional study of the coordinated reactions of individual *Commelina communis* L. stomata (Commelinaceae). *American Journal of Botany*, 66: 1044–52.

Schnabl, H., J. Vienken, and U. Zimmermann. 1980. Regular arrays of intramembranous particles in the plasmalemma of guard cell and mesophyll cell protoplasts of *Vicia faba*. *Planta*, 148: 231–37.

Singh, A. P. 1977. The subcellular organization of stomatal initials in sugarcane leaves: The guard and subsidiary mother cells. *Canadian Journal of Botany*, 55: 2801–9.

Singh, A. P., and L. M. Srivastava. 1973. The fine structure of pea stomata. *Protoplasma*, 76: 61–82.

Sorokin, H., and S. Sorokin. 1966. The spherosomes of *Campanula persicifolia* L.: A light and electron microscope study. *Protoplasma*, 62: 216–36.

Srivastava, L. M., and A. P. Singh. 1972. Stomatal structure in corn leaves. *Journal of Ultrastructure Research*, 39: 345–63.

Stevens, R. A., and E. S. Martin. 1977. Ion-absorbent substomatal structures in *Tradescantia pallidus*. *Nature*, 268: 364–65.

———. 1978. Structural and functional aspects of stomata, I. Developmental studies in *Polypodium vulgare*. *Planta*, 142: 307–16.

———. 1979. The structure of guard cells and substomatal ion-adsorbent bodies. In D. N. Sen, ed., *Structure, function and ecology of stomata*: 7–21. Dehra Dun: Bishen Singh Mahendra Pal Singh.

Thomson, W., and R. DeJournett. 1970. Studies on the ultrastructure of guard cells of *Opuntia*. *American Journal of Botany*, 57: 309–16.

Thorpe, N. 1980. Accumulation of carbon compounds in the epidermis of five species with either different photosynthetic systems or stomatal structure. *Plant, Cell and Environment*, 3: 451–60.

Vassilyev, A., and G. Vassilyeva. 1976. (The ultrastructure of the stomatal apparatus in gymnosperms, with special reference to stomatal movements.) *Botanicheskij zhurnal*, Leningrad, 61: 449–65. [In Russian.]

Walles, B., B. Nyman, and T. Alden. 1973. On the ultrastructure of needles of *Pinus sylvestris*. *Studia forestalia suecica*, 106: 1–25.
Waterkeyn, L., and A. Bienfait. 1979. Production et dégradation de callose dans les stomates des Fougères. *La cellule*, 73: 81–97.
Whatley, J. M. 1972. The ultrastructure of guard cells of *Phaseolus vulgaris*. *New Phytologist*, 71: 175–79.
Wille, A. C., and W. J. Lucas. 1984. Ultrastructural and histochemical studies on guard cells. *Planta*, 160: 129–42.
Willmer, C. M. 1983. *Stomata*. London: Longman.
Willmer, C. M., and R. Sexton. 1979. Stomata and plasmodesmata. *Protoplasma*, 100: 113–24.
Ziegenspeck, H. 1955. Das Vorkommen von Fila in radialer Anordnung in den Schliesszellen. *Protoplasma*, 44: 385–88.
Ziegler, H., E. Shmueli, and G. Lange. 1974. Structure and function of the stomata of *Zea mays*, I. The development. *Cytobiologie*, 9: 162–68.

4

Stomatal Mechanics

Peter J. H. Sharpe, Hsin-i Wu, and Richard D. Spence

The role of guard cell mechanics in determining stomatal aperture has intrigued plant anatomists and physiologists since von Mohl (1856) proposed that pore opening and closing were mechanical processes. Since then, experimental observations and theoretical analyses have expanded our knowledge of the way stomata function. One of the major discoveries made during the ensuing years has been the extent to which the surrounding epidermal cells influence pore aperture. Despite more than a century of intense study, however, large gaps still exist in our knowledge of stomatal mechanics. Some aspects have solid mathematical foundations backed by firm empirical documentation; other mechanisms may only be theorized. This chapter presents a survey of what we have learned, hypothesized, and have yet to discover about stomatal mechanics.

Comprehension of the mechanical properties by which guard cells open and close the pore is of paramount importance in understanding the entire stomatal system. Good biomechanical analyses therefore provide frameworks into which theorists may fit their ideas and in which experimenters may test the validity of those ideas.

Conclusions from complex biomechanical models, unfortunately, often have not been interpreted in a form that has meaning for the stomatal physiologist. The material properties of the walls of guard cells are considerably more complex than those of most other plant cells. The level of mathematics required for a biophysical analysis is often unfamiliar to many biologists and inhibits the exchange of concepts and understanding concerning stomatal physiology. In this chapter, we attempt to provide a sufficient explanatory background to the biophysical principles involved.

The authors wish to thank the publishers of Springer-Verlag for permission to reproduce an illustration from Ziegenspeck (1955b); the editors of the Oxford University Press for permission to reproduce an illustration from Treloar (1949); the editors of *Plant, Cell and Environment* for permission to reproduce illustrations from Sharpe and Wu (1978), Wu and Sharpe (1979), and Wu et al. (1985); Dr. Ronald J. Newton, Department of Forest Science, Texas A&M University, and Profs. Jonathon D. B. Weyers and Neil W. Paterson, Department of Biological Sciences, University of Dundee, for their insight and suggestions regarding material in this paper; and Profs. Eduardo Zeiger, Ian Cowan, and Hans Meidner for their kind reviews of the manuscript.

Historical Review

The mechanical interpretation of stomatal function was begun by von Mohl (1856). He made six points, based on his observations, relating to stomatal function:

1. Guard cells swell by the inflow of water and contract by the loss of water. The pore opens when the guard cells become turgid and closes when they become flaccid.

2. Nonuniform thickening of the guard cell wall influences the changes in cell shape as the pore opens and closes.

3. The water-extracting power (differences in osmotic potential) of guard cells is greater than that of neighboring (epidermal) cells.

4. Guard cells may have a "favorable mechanical relationship" over their surrounding cells. (Our interpretation of this statement is that von Mohl believed the guard cell possessed a mechanical advantage over the epidermal cell, which he attributed to the central location of the guard cells, their thicker walls, and their higher turgor pressures. As was demonstrated later, both theoretically and experimentally [DeMichele and Sharpe, 1973; Edwards, Meidner, and Sheriff, 1976], von Mohl had this relationship exactly backwards.)

5. With the exception of the surface along which they are in contact with the epidermal cells, guard cells may expand without restriction. (Raschke [1975] interpreted von Mohl as saying that the effect of epidermal pressure on stomatal opening depends on the ratio of the surface areas on which guard cell and epidermal pressures act. We believe that von Mohl was less specific than Raschke indicated.)

6. Guard cells freed from the influence of surrounding epidermal cells, as in detached strips with ruptured epidermal cells, may not behave in the same way as on undamaged leaves. (See also Sack, this volume.)

These observations have been modified and restated over the years. Many studies in the literature during the late nineteenth and early twentieth centuries focused on von Mohl's second observation and attempted to relate changes in guard cell geometry during opening to the thickening pattern of cell walls. The pattern of differential wall thickening was postulated to be critical to stomatal function. Guard cells from different species have various patterns of wall thickening, but they typically have very thick inner and outer (lateral) walls, a thin dorsal wall, and a somewhat thickened ventral (pore) wall (Allaway and Milthorpe, 1976).

Geometric changes that occur during opening have been identified. Schwendener (1881) observed that the guard cell lumen has a triangular shape in cross section when the pore is closed. Ziegenspeck (1941) observed that guard cells bend into a half-moon shape during opening, with the thinner dorsal

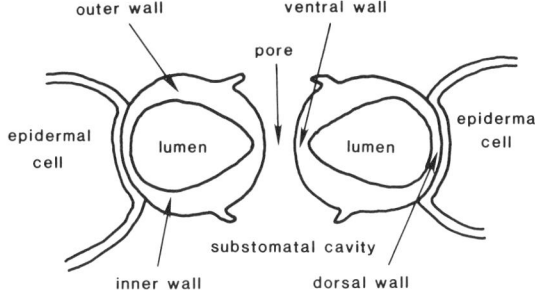

Fig. 4.1. Cross section through a typical non-grass stomatal system, showing the guard cell with respect to epidermal cells and the variable wall thickness on different surfaces of the guard cell.

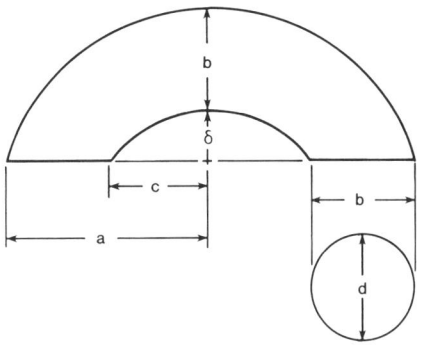

Fig. 4.2. Schematic sections of a typical, non-grass guard cell, illustrating the characteristic dimensions used in the text: a, semi-major axis of stomatal apparatus; b, width of guard cell; c, semi-major axis of pore; d, depth of guard cell; δ, semi-minor axis of pore. Aperture is 2δ. (Redrawn from Sharpe and Wu, 1978: 261.)

walls becoming extended and the thickened outer and inner walls (Fig. 4.1) bulging away from each other. Schwendener (1881) suggested that during opening the ventral wall bends less than the dorsal wall because the ventral wall is thicker. Haberlandt (1904) stated further that it is essential for the greater extension of the dorsal and ventral walls that they be thinner than the lateral walls. Otherwise, the counterpressure of the epidermal cells would cause the guard cells to extend vertically (bulge) from the plane of the epidermis.

Changes in stomatal dimensions during opening have received considerable attention. The characteristic dimensions of the stomatal apparatus are shown in Figure 4.2. The almost-universal conclusion of anatomical studies (see Haberlandt, 1904; Ziegenspeck, 1938) conducted during this century is that the length of the apparatus (Fig. 4.2, a) and of the pore (Fig. 4.2, c) remain essentially constant during opening. The guard cell (Fig. 4.2, b) first increases in width and then returns to approximately the original width during opening, reaching a maximum when the pore is about half open (Meidner and Mansfield, 1968). Raschke and Dickerson (1973) stated that the guard cell becomes circular in cross section as the pore opens, implying that b becomes equal to d, the depth of the guard cell. The models that follow explain how those dimensional changes occur.

Structure of the Guard Cell Wall

It is the unique structural arrangement of the guard cell walls that allows the pore to open. The shape deformation associated with guard cell expansion results from the rather complex interactions between the micellar bands and the nondirectional intermicellar (matrix) components (see also Sack, this volume). The micelles are cellulose-based microfibrils of indefinite length and low extensibility, arranged in parallel rows. Conceptually, the guard cell can be compared to a steel-belted radial tire. The micelles embedded in the wall matrix are oriented like belts wrapping radially around the cell, perpendicular to its longitudinal axis. The micellar bands restrict radial expansion so that the guard cell circumference remains approximately constant during opening and closing. Microfibrils arranged longitudinally within the thickened ventral wall cross-link the micellar bands, limiting longitudinal expansion along the ventral wall. Such unidirectional restrictive microfibrils are not observed elsewhere on the walls; as a result stretching primarily occurs laterally (i.e. the micellar bands move farther apart), so that the guard cell arches during opening. The stretching of the dorsal and ventral walls in conjunction with the micellar restrictions on the guard cell circumference forces the cell to bend outward into the epidermal cells as the stomate opens. The polar ends of the guard cells are anchored to the epidermal cells and to each other, so that the length of the apparatus remains essentially constant during opening.

Ziegenspeck (1938, 1941) conducted some of the most detailed studies of the micelles in guard cells. From his observations of micellar orientation and cell deformation, Ziegenspeck (1955a, b) postulated his "countermicellae elasticity law," which stated that the extensibility of the cell wall is greatest perpendicular to the plane of orientation of the micellar bands. This law forms the conceptual foundation of many recent models of stomatal mechanics.

Volume of the Guard Cell Lumen

Another unique feature of stomatal movement is the degree to which the volume of the guard cell lumen changes during opening: increases of up to 40 percent were measured on serial sections of *Vicia faba* by Humble and Raschke (1971). Such swelling, together with the restrictions that the micelles impose on the outside dimensions of the guard cell, must cause considerable thinning of the wall.

Volume changes during stomatal opening have attracted attention for more than half a century. In 1927, Stålfelt proposed that guard cells opened in two phases. These he described as *Spannungsphase* (tension phase) and *Motorphase* (strictly, *motorische Phase*, or pore opening phase). According to Stålfelt's observations, *Spannungsphase* is an inflation process preceding

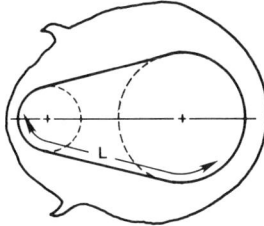

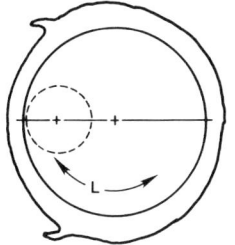

Fig. 4.3. Schematic cross section of a non-grass guard cell during the *Spannungsphase* movement. *top*, the cell in the relaxed state; *bottom*, the cell at the completion of *Spannungsphase*. As the cell expands, the flattened cross section, defined by the two centers of curvature (+ and dotted lines), assumes a circular shape. The cross-sectional lumen circumference, L, remains constant throughout *Spannungsphase*. (Redrawn from Sharpe and Wu, 1978: 262.)

Motorphase. In a common analogy, *Spannungsphase* may be likened to the inflation of a completely collapsed tire to a pressure at which it holds its shape but the walls are not stretched. In *Spannungsphase*, the guard cell lumen fills with fluid until its shape is circular in cross section.

Sharpe and Wu (1978) point out that at the end of *Spannungsphase* the turgor pressure may not be significantly greater than at the beginning. During *Spannungsphase*, guard cells change shape because of the application of force, either from inside or from outside the cell. Simultaneous changes in guard cell shape and osmotic potential can compensate for volume changes without increasing the lumen circumference (L), as shown in Figure 4.3. Increases in turgor pressure, therefore, occur only when compensatory changes in shape are no longer possible. This occurs at the transition between *Spannungsphase* and *Motorphase*. During *Motorphase*, an increase in turgor pressure results in wall extension and pore opening.

In the geometric model of Sharpe and Wu (1978), changes in volume during *Spannungsphase* were analyzed separately from *Motorphase*, since the two are mechanically distinct processes. *Spannungsphase* can be termed the swelling (inflating) stage (Fig. 4.3), while *Motorphase* can be described as the stretching (wall-thinning) stage. In their analysis, Sharpe and Wu assumed that volume increase during *Motorphase* occurred by two mechanisms: expansion by wall stretching, and polar expansion. Polar expansion is the considerable inflation of the rounded polar ends of the guard cells and occurs during the later stages of *Motorphase*. Polar expansion usually results in an extreme arching of the guard cells, so that the pore becomes rectangular as it opens to its widest aperture. The wall thinning caused by these two mechanisms results in an increased lumen cross section. As a result of their analysis, Sharpe and Wu (1978) concluded that volume increases of more than 100 percent are consistent with dimensions a, b, c, and d (Fig. 4.2) remaining approximately constant during opening.

Stomatal Models

The previously described anatomical observations formed the foundation from which a series of three different types of models of stomatal mechanics have been derived: bending-beam, finite-element, and volume-expansion models. Each successive type of model capitalized on new information and made improvements on preceding models.

Bending-beam models. The earliest mechanical models of stomatal motion treated the guard cell as a rigid beam subject to bending. The first such work was done by DeMichele and Sharpe (1971), and similar beam approaches were used by Aylor, Parlange, and Krikorian (1973), and by Shoemaker and Srivastava (1973). Aylor and his colleagues applied adhesive tape to a cylindrical balloon to simulate the micellar structure of the guard cell wall. The balloon was inflated to mimic opening. Beam-theory analysis was later extended by DeMichele and Sharpe (1973, 1974) to include the mechanical interactions with the adjacent epidermal cells. This particular topic will be treated later in this paper.

Each of these bending-beam models is open to criticism. They all assumed that the micellar structure of the thickened guard cell wall provided rigidity and constrained cellular expansion in certain directions. As pointed out by Cooke et al. (1976), these analyses treat the guard cell as a Bernoulli-Euler beam, although they use different assumptions with regard to both loading and boundary conditions. The DeMichele and Sharpe (1971) model does not allow for the stretching of the ventral wall, uses straight-beam theory for a curved member, and, like the model of Aylor, Parlange, and Krikorian (1973), allows rotation at the ends of the guard cell, resulting in overlaps and gaps. The Aylor, Parlange, and Krikorian (1973) balloon model ignores the essential geometry of the guard cell—a dorsal wall that is longer than the ventral wall. None of these models accounts for the observed large increase in lumen volume.

Finite-element models. Cooke et al. (1976) approached guard-cell and epidermal-cell wall mechanics by formulating a finite-element shell model. This model is computationally complex and embodies few biological or biophysical aspects of stomatal wall mechanics. Cooke et al. (1976) basically assumed linearity in their finite-element analysis, which led to the conclusion that stomatal aperture A was a linear combination of guard cell turgor p_g and epidermal cell turgor p_e. Thus

$$A = A_o + b_g p_g + b_e p_e \qquad A \geq A_o \qquad (1)$$

where the regression constant A_o is a nonzero intercept for $p_g = p_e = 0$, and b_g and b_e are termed "influence coefficients." A similar linear equation is implicit in the Cowan (1977) equation, linking changes in aperture to guard-cell and epidermal-cell pressures. Although this finite-element model follows stomatal displacements well, it is basically the result of a regression analysis and highlights little in the way of biological understanding.

Volume-expansion models. The relationship between the lumen volume, V_s (10^{-9} μl), of a guard cell pair and aperture, A (μm), in *Vicia faba* was found empirically by Raschke and Dickerson (1973) to be

$$V_s = V_{so} + 430A \qquad (2)$$

where V_{so} is the initial volume of the guard cell pair, and $5000 \leq V_{so} \leq 7300$. Equation 2 states that if the aperture increases to 12–15 μm, the volume of the cell lumen doubles. Such volume changes can be important because doubling the volume for a given solute in the cell causes a corresponding reduction in osmotic potential. Equation 2 does not explain how changes in volume occur. Geometric analyses were therefore undertaken by Sharpe and Wu (1978) to explain observed changes in guard cell volume during opening.

The pressure-volume relationship for guard cells measured experimentally by Raschke, Dickerson, and Pierce (1973) suggests that the material properties of guard cell walls are complex. These material properties do not appear to obey Hooke's law, a basic assumption of bending-beam (DeMichele and Sharpe, 1971, 1973, 1974; Aylor, Parlange, and Krikorian, 1973) and finite-element (Cooke et al., 1976) analyses. Wu and Sharpe (1979) used polymer elasticity theory to investigate the material properties of guard cell walls and to describe how they affect pressure-volume relationships in the guard cell.

Polymer Elasticity Theory

The structural organization and component material of cell walls, as with all bodies, to a large extent determine their mechanical properties. For small strains in one direction, materials such as steel and copper demonstrate the simple proportionality between stress and strain described by Hooke's law:

$$E = \text{stress/strain} = \frac{F/S}{\Delta \ell / \ell_o} \qquad (3)$$

where E is the constant of proportionality (Young's modulus), F is an applied force perpendicular to the cross-sectional area, S, and $\Delta \ell$ is the change in the original length, ℓ_o.

Cell walls differ mechanically from metallic materials in two important respects. First, during the growth phase, cell volume increases by plastic

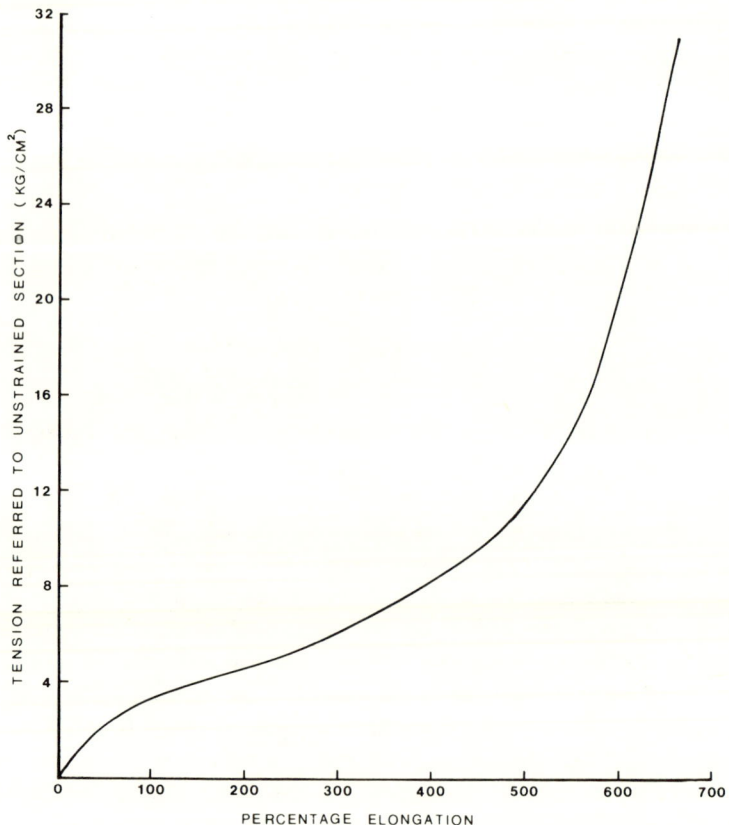

Fig. 4.4. Typical force-extension curve for a polymer-like substance. (Adapted from Treloar, 1949: 91.)

extension of the wall. Second, cell walls consist of a polymer structure made up of cellulose-based micellar microfibrils.

As shown by Treloar (1949), polymers do not generally obey Hooke's law. A typical stress-strain curve for a polymer substance is shown in Figure 4.4. Such a response is also observed in the guard cell wall (Raschke, Dickerson, and Pierce, 1973), which is basically a polymer structure. This is the essential drawback to the traditional approach of applying the constant bulk modulus to plant cell elasticity implied by Hooke's law. In the traditional approach, a small change in pressure ΔP brings about a corresponding change ΔV in volume V:

$$\Delta P = \epsilon \frac{\Delta V}{V} \qquad (4)$$

where ϵ is the bulk elastic modulus, a proportionality parameter between observed change ΔP and change in relative cell volume $\Delta V/V$.

This simple relationship is useful only if ϵ is a constant. It has been known for some years, however, that ϵ is not a constant but changes with volume and turgor (Hellkvist, Richards, and Jarvis, 1974; Cheung, Tyree, and Dainty, 1976; Steudle, Zimmermann, and Lüttge, 1977). As a result, Equation 4 is wholly inadequate to describe cell pressure-volume relations. A thorough analysis of the weaknesses of the use of ϵ may be found in Wu et al. (1985). The direct relationship between P and V is sufficient to describe cell mechanics and is intuitively more obvious. The current focus in the literature, however, is a restatement of complex pressure-volume relationships based on $\epsilon = VdP/dV$. Such a focus deflects attention from the fundamental relationship between pressure and volume in plant cells.

Cell wall elasticity has recently been approached using the shear-modulus property of the cell wall (Wu and Sharpe, 1979). The shear modulus is the ratio of the tangential force on a surface per unit area to its angular deformation and is a measure of the ability of a material to withstand deformation during stress. When force is applied to stretch the cell wall, the layers of polymer molecules making up the wall tend to slide over and past each other. The shear modulus measures the resistance of the cell wall material to this sliding.

Most of the theory of polymer elasticity has been derived for isotropic polymer materials (Treloar, 1949). Isotropic materials, when stressed by directional forces, show equal directional strain per unit force. Isotropic elasticity implies that the elastic stretching tendencies are nondirectional. The most common cause of isotropic strain is a completely random orientation of the polymer molecules. Many synthetic polymers are isotropic. Plant cell walls, muscle tissue, and some artificial polymers, however, show directional strain because of the nonrandom orientation of their polymer strands. When there is a tendency to stretch more in one direction than in another in response to the same force applied in each direction, the material is said to be anisotropic. As such materials stretch, the polymer strands change their orientation. In a plant cell, continued volume expansion eventually causes the microfibril network to become taut and distended (Fig. 4.5), causing directional changes in cell stretching and wall thinning.

Whenever a stretching force is generated within the cell wall, its resulting size and shape deformations are determined by the restrictions of the micellar network. The changes in shape can be elastic, plastic, or both (Tyree and Jarvis, 1982). In elastic deformation, the cell returns to its original shape after the stress is released. In plastic deformation, the cell retains the deformed shape. Plant growth requires plastic deformation, whereas guard cell function requires elastic deformation. The present study, therefore, is limited to elastic, nonpermanent changes in cell shape induced by changes in turgor pressure under standard conditions.

Fig. 4.5. The change in orientation of micellar strands during stretching of the guard cell wall. Stretched (A), partially stretched (B), and relaxed (C) micelles are shown. (From Ziegenspeck, 1955b: 387; reproduced by kind permission of the publishers.)

A   B   C

Consider a typical cell, with nonlignified walls containing micelles of the multinet arrangement (Taiz, 1984). If we assume it to be a cylinder of axial length ℓ_o, radius r_o, and cell wall thickness t_o, the extension ratio λ of its axial length when stretched is

$$\lambda = \frac{\ell}{\ell_o} \qquad (5)$$

where ℓ and ℓ_o are its stretched and relaxed lengths, respectively. The extension ratio of the cross-sectional dimension r/r_o is then some related function λ^q, where q is an anisotropic factor. If ℓ and r increase with cell expansion and the cell wall volume remains constant, then the relative change in cell wall thickness t must be a contraction described by $\lambda^{-(1+q)}$ (Wu et al., 1985). In terms of cell wall volume V_w, the resisting tensile force F_τ of such a cell under stress may then be described by

$$F_\tau = \frac{V_w}{\ell} G \left[\lambda^2 - \lambda^{-2(1+q)} \right] \qquad (6)$$

where G is the shear modulus of the cell wall. The expression $G[\lambda^2 - \lambda^{-2(1+q)}]$ is the tensile force in the wall per unit area.

Some materials change their volume during elastic deformation. The Poisson

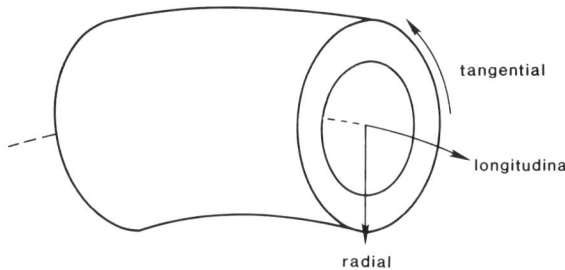

Fig. 4.6. A schematic representation of the guard cell coordinates: along the guard cell center line (longitudinal); along the guard cell circumference (tangential); along the guard cell radius (radial). (Adapted from Wu, Sharpe, and Spence, 1985: 270.)

ratio is an estimate of changes in volume, with a ratio of 0.5 indicating that volume does not change during stretching. Tobolsky (1960) lists Poisson ratio values for various materials. For most crystalline materials and glasses, the Poisson ratio varies from 0.2 to 0.33. The Poisson ratio for polymer or rubbery materials varies from 0.4996 to 0.5. It is therefore appropriate to assume that cell wall volume is conserved during guard cell expansion.

Guard Cell Wall Mechanics

As the guard cell stretches lengthwise along the center axis, the wall stretches in the circumferential direction by a factor λ^q and thins in the radial direction by $\lambda^{-(1+q)}$ (Fig. 4.6). Guard cell walls are designed to withstand the high internal pressures required to open the pore, which must overcome the resisting tensile force of the guard cell plus the opposing epidermal turgor-generated force.

For an isolated guard cell, where there is no epidermal counterforce, the turgor-generated opening force F is counterbalanced only by the resisting tensile force F_T. The guard cell turgor pressure p as a function of elongation λ may then be expressed (Wu and Sharpe, 1979)

$$p = \kappa(\lambda^{1-q} - \lambda^{-3(1+q)}) \tag{7}$$

where

$$\kappa = \frac{GV_w}{2\rho_o \ell_o (a_o - c_o)} \tag{8}$$

with the subscript o denoting values at transition between *Spannungsphase* and *Motorphase* (see Sharpe and Wu, 1978), and with ρ being the radius, ℓ the length of the inner lumen cavity, a the half-length of the entire stomatal apparatus, and c the half-length of the pore (see Fig. 4.2). Equation 7 relates internal pressure changes to guard cell elongation. Similarly, the volume of the cell lumen during elongation can be expressed

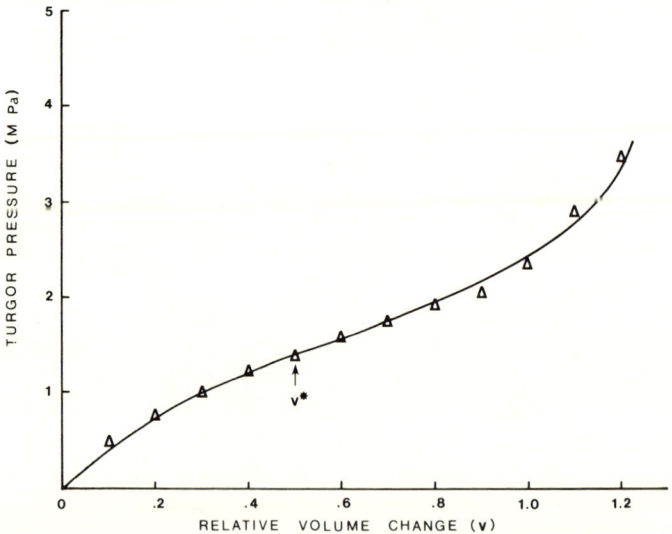

Fig. 4.7. Guard cell turgor pressure in relation to its relative volume increment. The values denoted by △ are from Raschke, Dickerson, and Pierce (1973). The solid line is from Wu and Sharpe (1979). The value v^* represents the relative volume increment that marks the transition between isotropic and anisotropic expansion.

$$V = \pi\rho^2\ell = \pi(\rho_o\lambda^q)^2(\lambda\ell_o)$$
$$= V_o\lambda^{1+2q} \qquad (9)$$

where V_o is the volume of the cell lumen at transition. By defining the relative volume of change $v = (V/V_o) - 1$, it is found that

$$\lambda = (1 + v)^{\frac{1}{1+2q}} \qquad (10)$$

and therefore guard cell turgor can be expressed in terms of the relative volume change v (Wu, Sharpe, and Spence, 1985):

$$p = \kappa\left[(1 + v)^{\frac{1+2q}{1-q}} - (1 + v)^{-\frac{3(1+q)}{1+2q}}\right]. \qquad (11)$$

The relationship between the pressure p and the relative volume change v given by Equation 11 is shown in Figure 4.7. The observed values of Raschke, Dickerson, and Pierce (1973) for *Vicia faba* are also shown. The guard cell turgor pressure can be expressed as

$$p = \psi_w - \frac{\psi_w^o}{1 + v} \qquad (12)$$

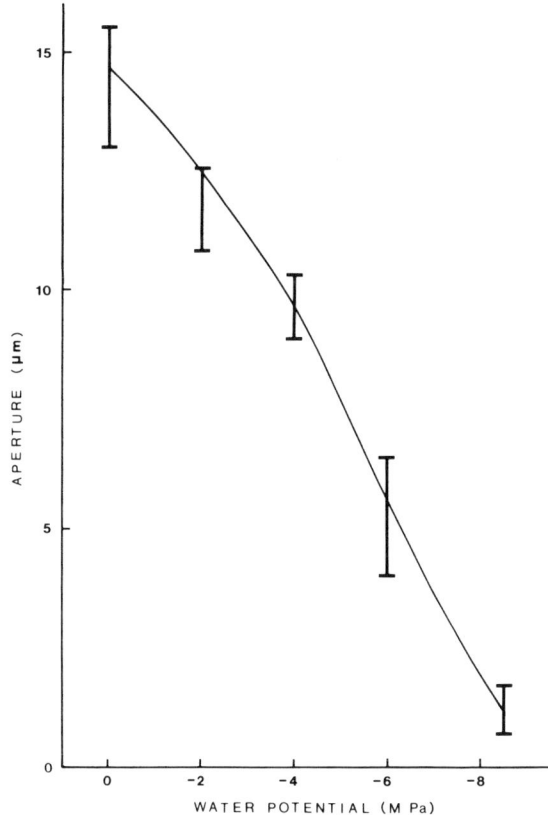

Fig. 4.8. Stomatal aperture in relation to water potential for *Vicia faba*. Experimental values are adapted from Raschke, Dickerson, and Pierce (1973). The solid line is from Wu and Sharpe (1979).

where ψ_w and ψ_w^o are the water potentials corresponding to *Motorphase* and transition, respectively. By equating Equations 11 and 12, a relation for the relative volume change and the water potential is obtained:

$$\psi_w = \frac{\psi_w^o}{1 + v} + \kappa \left[(1 + v)^{\frac{1-q}{1+2q}} - (1 + v)^{-\frac{3(1+q)}{1+2q}} \right] \quad (13)$$

where $q = q(v)$ (see Wu and Sharpe, 1979). Equation 13 can be used to describe *Motorphase* motion. Solving for v for a given water potential ψ_w is a rather complicated procedure. After v is obtained, however, the aperture-volume relationships can be used to compute the aperture for given values of ψ_w (Sharpe and Wu, 1978; Wu and Sharpe, 1979), as shown in Figure 4.8. Calculated values are compared with observed values for *Vicia faba* given by Raschke, Dickerson, and Pierce (1973).

Mechanical Interactions between Guard Cells and Epidermal Cells

Schäfer (1888) postulated that "the width of the stomatal pore at any given time is the result of two forces opposing each other, but working along the same straight line." He identified these forces as the guard cell turgor pressure and the opposing force of the neighboring epidermal cell turgor pressure. In his hypothesis, Schäfer confused forces and pressures. (Pressure is defined as force per unit area.)

DeMichele and Sharpe (1973, 1974) were the first to analyze mechanically the opposing forces operating on stomata. They found that the epidermal cells had what they termed a *mechanical advantage* over the guard cells. The mechanical advantage of the epidermal cell over the guard cell has been experimentally verified by Edwards, Meidner, and Sheriff (1976), who monitored simultaneously the pore aperture and the guard-cell and adjacent epidermal turgor pressures. A mechanical advantage defines a relationship between forces, in particular, the ratio of load to effort (Burns and MacDonald, 1970). The formulation as used by DeMichele and Sharpe (1973) reduces to a ratio between pressures. The advantage lies with the epidermal cells in that they counter the opening force of the guard cell with a lower turgor pressure.

The mechanical analysis of DeMichele and Sharpe (1973) was seriously limited because it ignored the tensile force in the stretched guard cell wall. They considered only the balance of forces generated by the epidermal and guard cell turgor pressures. The model is inconsistent physically for isolated guard cells. When no counterforce from the epidermal cells is considered, the isolated guard cell in the model, with no restrictive internal tensile force, would expand until it burst. This manifestly does not happen. Subsequent analyses of Cooke et al. (1976) and Cowan (1977) also do not explicitly include the tensile forces within the guard cell walls.

In addition, Cowan (1977) proposed a relationship between stomatal aperture, guard cell volume, guard cell turgor pressure, and epidermal cell turgor pressure. Although a number of relationships are presented, they are difficult to reconstruct, because the initial assumptions are not clearly stated.

Cooke et al. (1976) redefined the mechanical advantage in a partial derivative form:

$$\alpha = - \frac{\left(\frac{\partial A}{\partial p_e}\right)_{p_g}}{\left(\frac{\partial A}{\partial p_g}\right)_{p_e}}. \tag{14}$$

Stomatal Mechanics

The negative sign arises from the numerator, where an increase in epidermal pressure p_e causes a reduction in aperture A. The subscripts p_g and p_e in the expression of partial derivatives indicate that these quantities are held constant while the respective derivatives $\partial A/\partial p_e$ and $\partial A/\partial p_g$ are taken. Cooke et al. (1976) termed this formulation the *antagonism ratio*. While being a correct description of the mechanical interactions between guard cells and epidermal cells, the Cooke et al. (1976) model has some practical limitations. In comparing the DeMichele and Sharpe (1973) and Cooke et al. (1976) models, Meidner and Bannister (1979) found that the antagonism-ratio model gave a better fit to the data from which it was calculated, but failed entirely near zero aperture. Furthermore, to test the Cooke et al. (1976) model required "technically difficult measurements" (Meidner and Bannister, 1979).

Wu, Sharpe, and Spence (1985) presented a biophysical analysis demonstrating that the effective force in the stomatal system may be expressed in terms of simple stomatal geometry. They redefined and reinterpreted the mechanical advantage of the epidermal cell based upon simple geometric relationships calculated from measurable anatomical dimensions.

The force F_g generated by the guard cell turgor p_g is given by the equation (Wu, Sharpe, and Spence, 1985)

$$F_g = 4\rho(a_o - c_o)p_g \qquad (15)$$

where $\rho = \rho_o \lambda^q$ and guard cell dimensions a_o and c_o as defined in Figure 4.2 are assumed to be constant during pore opening. A similar relationship can be formulated to describe the force F_e generated by the epidermal turgor pressure p_e:

$$F_e = 4\chi a_o \rho p_e \qquad (16)$$

where $4\chi a_o \rho$ is the effective contact area of the epidermal cell, and the parameter χ is the effective epidermal cell depth, d_e, in relation to the depth of the lumen cavity, 2ρ, whose cross section is assumed to be circular:

$$\chi = d_e/2\rho. \qquad (17)$$

Anatomically, χ varies from about 2 to nearly zero. Figure 4.9 shows a range of geometric relationships between guard cells and epidermal cells. For an anatomical interpretation of the relation of guard-cell and epidermal forces, compare Equations 15 and 16, and see Figure 4.10.

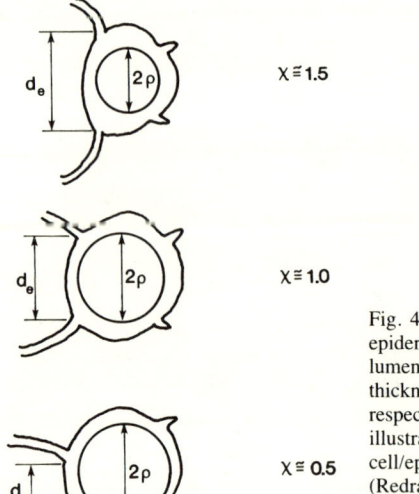

Fig. 4.9. Different ratios (χ) of effective epidermal lumen depth (d_e) to guard cell lumen depth depending on guard cell wall thickness and guard cell orientation with respect to the epidermal cell. These figures illustrate a range of hypothetical guard cell/epidermal cell configurations. (Redrawn from Wu, Sharpe, and Spence, 1985: 272.)

The resisting tensile force generated within the guard cell walls F_τ is given by

$$F_\tau = \frac{2V_w}{\ell_o} G(\lambda - \lambda^{-3-2q}). \tag{18}$$

The opening force F_g is balanced by the epidermal counterforce F_e and the resisting guard cell wall tensile force F_τ:

$$F_g - F_e = F_\tau. \tag{19}$$

This may be written

$$p_g - \sigma p_e = \kappa \left[(1 + v)^{\frac{1-q}{1+2q}} - (1 + v)^{-\frac{3(1+q)}{1+2q}} \right] \tag{20}$$

where $\sigma = a_o \chi/(a_o - c_o)$ and κ is as defined in Equation 8.

The right-hand side of Equation 20 is extremely cumbersome, and q is a complex function of volume v. Fortunately, Raschke, Dickerson, and Pierce (1973) proposed an empirical regression function that Wu and Sharpe (1979) have shown approximates the theoretical result. The Raschke, Dickerson, and Pierce (1973) equation is expressed

Stomatal Mechanics   107

$$\hat{p}_g = p_o v^n \tag{21}$$

where $\hat{p}_g$ is the isolated guard cell turgor pressure, and p_o and n are regression parameters. This form of the empirical relationship is extremely useful in that the necessary mathematical manipulations are quite tractable, while the regression parameters p_o and n drop out of the final result. Thus Equation 21 is used in place of the right-hand side of Equation 20:

$$p_g - \sigma p_e = \hat{p}_g = p_o v^n \tag{22}$$

or

$$v = \left(\frac{p_g - \sigma p_e}{p_o}\right)^{\frac{1}{n}}. \tag{23}$$

Sharpe and Wu (1978) have shown that the linear relationship between aperture A and guard cell volume v found by Raschke and Dickerson (1973) closely approximates the behavior of their geometric model. Written for an individual guard cell, it is

$$V = V_o + \eta^\dagger A \tag{24}$$

where $\eta^\dagger$ is a regression coefficient.

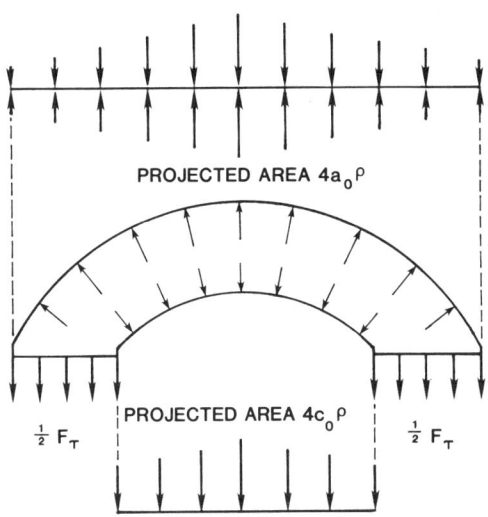

Fig. 4.10. Force balance between the guard cells and epidermal cells, showing how the force of the guard cell directed toward the pore partially nullifies that directed toward the epidermal cells. F_T, tensile force generated within the guard cell walls; a_0, half-length of the stomatal apparatus; c_0, half-length of the stomatal pore; ρ, cross-sectional radius of the guard cell lumen. (Redrawn from Wu, Sharpe, and Spence, 1985: 272.)

Substituting Equation 24 into the definition for relative volume change $v = (V/V_o) - 1$,

$$v = \frac{\eta^\dagger A}{V_o} \tag{25}$$

or

$$A = \frac{v}{\eta} \tag{26}$$

where $\eta = \eta^\dagger V_o$.

Substituting Equation 23 into Equation 26 yields an expression for aperture A in terms of guard-cell and epidermal-cell turgor, p_g and p_e, respectively. This relationship has the form

$$A = \frac{1}{\eta}\left(\frac{p_g - \sigma p_e}{p_o}\right)^{\frac{1}{n}}. \tag{27}$$

Equation 27 provides the biophysical basis for calculating a revised definition of the mechanical advantage.

The partial derivative expression of the mechanical advantage proposed by Cooke et al. (1976) in Equation 14 can be given a biophysical foundation by using Equation 27. Substituting Equation 27 into the numerator and denominator of Equation 14 gives

$$\left(\frac{\partial A}{\partial p_e}\right)_{p_g} = -\frac{\sigma}{n\eta p_o}\left(\frac{p_g - \sigma p_e}{p_o}\right)^{\frac{1}{n} - 1} \tag{28}$$

and

$$\left(\frac{\partial A}{\partial p_g}\right)_{p_e} = \frac{1}{n\eta p_o}\left(\frac{p_g - \sigma p_e}{p_o}\right)^{\frac{1}{n} - 1} \tag{29}$$

Therefore, we obtain the simple result (Wu, Sharpe, and Spence, 1985)

$$\alpha = \sigma = \frac{a_o \chi}{a_o - c_o}. \tag{30}$$

It can therefore be concluded that the mechanical advantage has a simple geometric interpretation. It indicates the ratio of guard cell turgor to epidermal turgor necessary to overcome the closing force and to induce pore opening. Thus, the magnitude of the pressure advantage of the epidermal cell is influenced largely by the shape of the guard cell.

The other influential anatomical feature is the relative epidermal cell contact depth χ. For example, when $\chi > 1$, the effective depth of the epidermal lumen through which the epidermal turgor can be applied is greater than the depth of the guard cell lumen (Fig. 4.9). When $\chi = 1$, a unit reduction in p has the same effect on aperture as increasing p_g by $a_o/(a_o - c_o)$.

The effective epidermal contact depth χ depends upon the geometric arrangement between the guard cells and epidermal cells, which can be very complicated. In defining the effective forces in the stomatal system, therefore, we have made two simplifying assumptions. First, the counterforce of the atmosphere that presses against the external walls of the guard cells and epidermal cells is small compared to the forces generated by their turgor pressures. Second, the part of the epidermal cell wall not in contact with the guard cell is not so rigid as to cause a significant transfer of force along the wall. Breaking the epidermal cell force into its component directional forces shows that only those acting directly against the guard cell influence aperture when the above assumptions are considered. Assuming that the epidermal cell wall is very stretchable, then the turgor-generated forces within the epidermal cell that are not exerted directly against the guard cell are absorbed by the bulging of the epidermal cell wall. These "ineffective" forces are counteracted by the elastic force of the epidermal cell walls.

These assumptions are rather ideal when the full range of stomatal anatomies is considered. If the epidermal cell wall facing the guard cell is ideally rigid, as in a steel plate, then all the forces directed toward the guard cell and pore are transferred along the wall. In such a case, however, if the inner and outer lateral walls were as rigid as that facing the guard cell, then there would be no cell expansion, because the walls would be unable to stretch; and therefore the epidermal cell would be ineffective in closing the pore. In many species the outer wall tends to be somewhat thicker and more rigid than the inner wall, which may cause the guard cell to rotate outward during pore closing. The effective lumen depth therefore depends on a number of factors, such as the rigidity of the walls involved and the spatial orientation of the guard cells and epidermal cells. These factors are species-specific and may be difficult to measure in routine anatomical studies. The influence of different anatomical structures on the mechanics of stomatal opening represents an intriguing area for future research.

Water-Stress, Hormonal, and Mineral Effects on Wall Material and Mechanical Properties

It has been observed that stomata on plants grown under water stress are smaller than those of well-watered plants (Cutler, Rains, and Loomis, 1977) and are better able to maintain open pores at low plant water potentials (Brown, Jordan, and Thomas, 1976). Plant responses to water stress, however, may not be entirely physiological. It has been speculated that the dimensions of the stomata and the mechanical properties of the guard cell walls contribute to a plant's response to water stress (McCree, 1974). By employing the geometric definition of the mechanical advantage (Wu, Sharpe, and Spence, 1985), Spence et al. (1986) determined that the smaller stomata observed on plants grown under water stress have the mechanical capability of opening and maintaining open pores at lower guard cell turgor pressures, relative to the turgor of the surrounding epidermal cells, than do larger guard cells. Their study provided a basis for explaining that at least a part of stomatal adaptation allowing pore opening during conditions of water stress results from anatomical adaptation during growth, the result of which is to affect the stomatal dimensions and thus the mechanical properties of the stomata.

There are numerous reports in the literature suggesting that the material properties of cell walls are influenced by hormones and minerals. Meidner and Edwards (1975) report that the potassium ions or protons that traverse the guard cell wall may change its elastic properties. Aluminum has been shown to increase wall rigidity by cross-linking pectins (Foy, Chaney, and White, 1978). Other metals believed to change the elastic properties of cell walls are silicon and zinc (R. J. Newton, personal communication). It is apparent that immature, expanding cells have walls that are stimulated or inhibited by hormones and minerals during plastic elongation. The sensitivity, however, of mature cell walls that no longer experience plastic elongation is not known. Virtually all hormonal and mineral-nutrition studies have been with immature or partially immature cell walls that exhibit both plastic and elastic deformation. No studies have been undertaken of cell walls that exhibit elastic deformation only.

Meidner and Edwards (1975) have speculated that changes in concentrations of cations or of hormones, or of both, may alter the elastic properties of guard cell walls during *Spannungsphase*. Their observations on cation concentration reveal a disparity between turgor pressure measurements made directly and those made from estimates of the internal osmotic pressure using the plasmolytic method. In particular, they found that the pressures in guard cells measured by microcapillaries inserted into the cells are somewhat less than expected from estimates of internal osmotic pressure using the plasmolytic method (Cowan, 1977). This disparity can be explained from the analyses of

Wu and Sharpe (1979). For example, data of Raschke, Dickerson, and Pierce (1973) indicate that a water potential of -9 MPa is required to plasmolyze guard cells of *Vicia faba*. The analysis of Sharpe and Wu (1978) indicates that during opening, from the beginning of *Spannungsphase* (which approximates plasmolysis) to fully open at the end of *Motorphase*, the lumen volume increases by 250 percent. Thus, the internal osmotic potential of -9 MPa measured at plasmolysis would be equal to -3.6 MPa when changes in volume are taken into account. It is therefore to be expected that direct and indirect measurements of guard cell pressure will not agree without accounting for adjustments in cell volume.

Problems for Further Consideration

In their chapter on stomatal structure and function, Allaway and Milthorpe (1976) regret the lack of information about essential stomatal dimensions and properties, including the areas of the various walls, osmotic potentials, and the modulus of elasticity over a range of apertures. Nearly a decade later, this lack of information still hinders efforts to test models of stomatal function. In light of new insights into stomatal mechanics and function, we can identify five areas that need attention:

1. The geometric expression of the mechanical advantage (Eq. 30) needs to be tested.

2. Because dimensional constancy is a major part of most models of stomatal mechanics, further investigation should be conducted to explore this phenomenon fully.

3. The degree of mechanical interaction between guard cells and epidermal cells must be determined. Edwards, Meidner, and Sheriff (1976) and Meidner and Bannister (1979) measured turgor pressures of guard cells and epidermal cells directly, and Spence et al. (1983) correlated aperture with degree of contact with live epidermal cells. However, more detailed studies of the turgor pressure–force interactions between guard and epidermal cells are needed.

4. The degree and magnitude of polar expansion in guard cells should be investigated. Rupturing the epidermal cells may tend to exaggerate polar expansion and change the aperture at which the transition from *Spannungsphase* to *Motorphase* occurs (Sharpe and Wu, 1978).

5. Hormonal and ionic effects on the elastic properties of mature walls should be investigated.

The influence of environmental growth conditions on the material properties of guard cell walls is an area for which experimental evidence either is lacking or has not been collected in a manner that can be theoretically analyzed. Such experiments may have consequences in ecology, plant physiology, and plant growth analysis.

REFERENCES

Allaway, W. G., and F. L. Milthorpe. 1976. Structure and functioning of stomata. In T. T. Kozlowski, ed., *Water deficits and plant growth* 4: 57–102. New York: Academic Press.

Aylor, D. E., J.-Y. Parlange, and A. D. Krikorian. 1973. Stomatal mechanics. *American Journal of Botany*, 60: 163–71.

Brown, K. W., W. R. Jordan, and J. C. Thomas. 1976. Water stress induced alterations of the stomatal response to decreases in leaf water potential. *Physiologia plantarum*, 37: 1–5.

Burns, D. M., and S. G. G. MacDonald. 1970. Chapter 9, section 9, in *Physics for biology and pre-medical students*: 153–55. London: Addison-Wesley.

Cheung, Y. N. S., M. T. Tyree, and J. Dainty. 1976. Some possible sources of error in determining bulk elastic moduli and other parameters from pressure-volume curves of shoots and leaves. *Canadian Journal of Botany*, 54: 758–65.

Cooke, J. R., J. G. DeBaerdemaeker, R. H. Rand, and H. A. Mang. 1976. A finite element shell analysis of guard cell deformation. *Transactions of the American Society of Agricultural Engineers*, 19: 1107–21.

Cowan, I. R. 1977. Stomatal behavior and environment. *Advances in Botanical Research*, 4: 117–228.

Cutler, J. M., D. W. Rains, and R. S. Loomis. 1977. The importance of cell size in the water relations of plants. *Physiologia plantarum*, 40: 255–60.

DeMichele, D. W., and P. J. H. Sharpe. 1971. A model of stomatal action: Its agronomic and ecological significance. Paper no. 71-589, winter meeting, American Society of Agricultural Engineers, Chicago, December 1971.

———. 1973. An analysis of the mechanics of guard cell motion. *Journal of Theoretical Biology*, 41: 77–96.

———. 1974. A parametric analysis of the anatomy and physiology of the stomata. *Agricultural Meteorology*, 14: 229–41.

Edwards, M., H. Meidner, and D. W. Sheriff. 1976. Direct measurements of turgor pressure potentials of guard cells, II. The mechanical advantage of the subsidiary cells, the *Spannungsphase*, and the optimum leaf water deficit. *Journal of Experimental Botany*, 27: 163–71.

Foy, C. D., R. L. Chaney, and M. C. White. 1978. The physiology of metal toxicity in plants. *Annual Review of Plant Physiology*, 29: 511–66.

Haberlandt, G. 1904. *Physiologische Pflanzenanatomie*. Leipzig: Engelmann.

Hellkvist, J., G. P. Richards, and P. G. Jarvis. 1974. Vertical gradients of water potential and tissue water relations in Sitka spruce trees measured with the pressure chamber. *Journal of Applied Ecology*, 11: 637–67.

Humble, G. D., and K. Raschke. 1971. Stomatal opening quantitatively related to potassium transport. *Plant Physiology*, 48: 447–53.

McCree, K. J. 1974. Changes in the stomatal response characteristics of grain sorghum produced by water stress during growth. *Crop Science*, 14: 273–78.

Meidner, H., and P. Bannister. 1979. Pressure and solute potentials in stomatal cells of *Tradescantia virginiana*. *Journal of Experimental Botany*, 30: 255–65.

Meidner, H., and M. Edwards. 1975. Direct measurements of turgor pressure potentials of guard cells, I. *Journal of Experimental Botany*, 26: 319–30.

Meidner, H., and T. A. Mansfield. 1968. *Physiology of stomata.* London; McGraw-Hill.
von Mohl, H. 1856. Welche Ursachen bewirken die Erweiterung und Verengung der Spaltöffnungen? *Botanische Zeitung,* 14: 697–704.
Raschke, K. 1975. Stomatal action. *Annual Review of Plant Physiology,* 26: 309–40.
Raschke, K., and M. Dickerson. 1973. Changes in shape and volume of guard cells during stomatal movement. *Plant Research,* 1972: 149–53.
Raschke, K., M. Dickerson, and M. Pierce. 1973. Mechanics of stomatal response to changes in water potential. *Plant Research,* 1972: 155–57.
Schäfer, R. F. C. 1888. Über den Einfluss des Turgors der Epidermiszellen auf die Funktion des Spaltöffnungsapparates. *Jahrbucher für wissenschaftliche Botanik,* 19: 188–205.
Schwendener, S. 1881. Über Bau und Mechanik der Spaltöffnungen. *Monatsberichte der Königlichen preussichen Akademie der Wissenschaften zu Berlin, Physikalisch-mathematische Klasse,* July 1881: 833–67.
Sharpe, P. J. H., and H. Wu. 1978. Stomatal mechanics, I. Volume changes during opening. *Plant, Cell and Environment,* 1: 259–68.
Shoemaker, E. M., and L. M. Srivastava. 1973. The mechanics of stomatal opening in corn (*Zea mays* L.) leaves. *Journal of Theoretical Biology,* 42: 219–25.
Spence, R. D., P. J. H. Sharpe, R. D. Powell, and C. A. Rogers. 1983. Epidermal and guard cell interactions on stomatal aperture in epidermal strips and intact leaves. *Annals of Botany,* 52: 1–12.
Spence, R. D., H. Wu, P. J. H. Sharpe, and K. G. Clark. 1986. Water stress effects on guard cell anatomy and the mechanical advantage of the epidermal cells. *Plant, Cell and Environment,* 9: 197–202.
Stålfelt, M. G. 1927. Die photischen Reaktionen im Spaltöffnungsmechanismus. *Flora,* n.s., 21: 236–72.
Steudle, E., U. Zimmermann, and U. Lüttge. 1977. Effect of turgor pressure and cell size on the wall elasticity of plant cells. *Plant Physiology,* 59: 285–89.
Taiz, L. 1984. Plant cell expansion: Regulation of cell wall mechanical properties. *Annual Review of Plant Physiology,* 35: 585–657.
Tobolsky, A. V. 1960. *Properties and structure of polymers.* New York: Wiley.
Treloar, L. R. G. 1949. *The physics of rubber elasticity.* Oxford: Clarendon Press.
Tyree, M. T., and P. G. Jarvis. 1982. Water in tissues and cells. In O. L. Lange, P. S. Nobel, C. B. Osmond, and H. Ziegler, eds., *Physiological plant ecology* 2, *Water relations and carbon assimilation,* Encyclopedia of Plant Physiology, n.s., 12B: 35–77. Berlin: Springer-Verlag.
Wu, H., and P. J. H. Sharpe. 1979. Stomatal mechanics, II. Material properties of guard cell walls. *Plant, Cell and Environment,* 2: 235–44.
Wu, H., P. J. H. Sharpe, and R. D. Spence. 1985. Stomatal mechanics, III. Geometric interpretation of the mechanical advantage. *Plant, Cell and Environment,* 8: 269–74.
Wu, H., R. D. Spence, P. J. H. Sharpe, and J. D. Goeschl. 1985. Cell wall elasticity, I. A critique of the bulk elastic modulus approach and an analysis using polymer elastic principles. *Plant, Cell and Environment,* 8: 563–70.
Ziegenspeck, H. 1938. Die Micellierung der Turgeszenzmechanismen, I. Die Spaltöffnungen (mit phylogenetischen Ausblicken). *Botanisches Archiv,* 39: 268–309, 332–72.
———. 1941. Der Bau der Spaltöffnungen, III. Eine phyletisch-physiologische Studie. *Repertorium specierum novarum regni vegetabilis,* Beiheft 123: 1–56.
———. 1955a. Die Farbenmikrophotographie, ein Hilfsmittel zum objektiven Nachweis

submikroskopischer Strukturelemente. Die Radiomicellierung und Filierung der Schliesszellen von *Ophioderma pendulum*. *Photographie und Wissenschaft*, 4: 19–22.

———. 1955b. Das Vorkommen von Fila in radialer Anordnung in den Schliesszellen. *Protoplasma*, 44: 385–88.

5

An Introduction to Carbon Metabolism in Guard Cells

William H. Outlaw, Jr.

Stomatal guard cells make up a disperse tissue system. They occur as somewhat autonomous pairs of cells lying parallel to each other, with an opening, or pore, between them. These pairs are sprinkled about in the otherwise relatively impervious leaf surface and represent less than 1 percent of the total leaf volume. Yet, by regulating the size of the aperture between them, guard cells control the entrance into the leaf of CO_2, which subsequently is reduced to organic compounds. Thus, the functioning of these cells sets the magnitude of the first in a series of resistances to the photosynthetic rate of the leaf. Since water vapor is lost from the leaf through these same pores, CO_2 entry into the leaf and water loss from the leaf are obligatorily coupled. To prevent desiccation, the aperture size is varied from moment to moment to effect a compromise between water loss and CO_2 entry. Guard cells have evolved a specialized metabolism in order to fulfill their crucial role in the plant's physiology. The purpose of this chapter is to present an overview of the unique biochemical adaptations that distinguish guard cells from other leaf cells. Although I have attempted to incorporate the most recent confirmed research results, the treatment of controversial areas will reflect my personal bias, which comes from my experience with *Vicia faba*. For more comprehensive reviews with citations of primary literature, the reader is referred to Hsiao (1976), Raschke (1979), Outlaw (1982, 1983), and Zeiger (1983).

Aperture Size and Guard Cell Turgor

The size of the stomatal aperture depends on the integration of several physiological, environmental, temporal, and historical parameters. Despite these complexities, the biochemical and biophysical bases for changes in aperture size are similar among diverse species. Guard cell walls distend asymmetrically as their internal volume and pressure change. The pore enlarges if water moves into the cells. Water moves into guard cells if their water

---

The author wishes to acknowledge research support from the United States Department of Energy and the National Science Foundation. Travel funds were furnished by the United States–Australia Science and Technology Exchange Program. This paper was submitted in November 1983.

potential is lower than that of the surrounding solution. In other words, water moves into guard cells, increasing their volume and turgor and the size of the aperture, when they accumulate solutes. Thus, the unique aspects of carbon metabolism in guard cells are related to their enhanced ability to osmoregulate. As will be seen later, this simple statement belies the extent to which guard cell metabolism differs from that of other cells.

Interest in the identification of the solutes that fluctuate in guard cells has surged and ebbed. The observation that starch content is often inversely related to the size of the stomatal aperture was the basis for an earlier theory that starch-to-sugar interconversions were responsible for the solute fluctuations that cause stomatal movements (Scarth, 1927). Although doubts occasionally surfaced when only weak correlations between guard cell starch content and stomatal aperture size were reported, this hypothesis was accepted by most physiologists until the mid-1960's. In the following years, several investigators (e.g. Fischer and Hsiao, 1968) reported that changes in the concentration of potassium salts alone could account for the changes in osmolarity during stomatal movements. (Actually, potassium involvement in stomatal movements had been implicated much earlier by Japanese scientists [e.g. Imamura, 1943; Fujino, 1967], but their reports went unnoticed by Western contemporaries.) Although the importance of potassium salts as osmoregulatory agents is still not in question, doubt has been raised that fluctuations in the levels of these salts account entirely for fluctuations in osmolarity, especially at some ranges of stomatal aperture size (e.g. MacRobbie, 1981). Nevertheless, at present, potassium salts are the only significant osmoregulatory solutes to have been identified unequivocally as important in stomatal movements. Therefore, the thrust of the following discussion will be restricted to carbon metabolism in the accumulation and dissipation of potassium salts.

Metabolic Implications of Potassium Influx into and Efflux from Guard Cells

Most likely, potassium moves between guard cells and adjacent cells via liquid in the apoplast. Stomatal opening, and probably closing also, is an active process and therefore requires energy. Because stomatal movements may occur in darkness, the simplest interpretation is that guard cells must be able to derive the required energy from carbon oxidation. Consistent with this interpretation are: the results of inhibition studies (see review by Outlaw, 1983); the presence of high activity levels of marker enzymes of glycolysis, of the tricarboxylic acid cycle, and of the oxidative pentose phosphate pathway (see Hampp, Outlaw, and Tarczynski, 1982); and the presence of many mitochondria in guard cells (e.g. Allaway and Setterfield, 1972).

Net cation uptake into guard cells is electrically balanced by net proton extrusion. Regardless of a co-transported species, stoichiometry, electrogenicity, energy coupling, or spatial arrangements, the overall increase in potassium-ion concentration must be electrically balanced. In a typical case, guard cell potassium contents fluctuate by about 300 mM. Of this fluctuation, about 100 mM is "offset," in effect, by chloride-ion movements. In contrast, the concentration of protons in guard cells is always at least three orders of magnitude lower than that of these other ions. Furthermore, cells lack sufficient buffering capacity (i.e. concentrations of ionizable groups near pH 7) to stabilize pH against massive (about 200 mM) proton effluxes. Thus, the first consideration is for a mechanism to stabilize cellular pH.

Control of Cellular pH During Stomatal Movements

Most plant cells regulate pH coarsely by proton exchange with the surroundings. This mechanism, of course, cannot operate if the cell must primarily regulate large potassium-ion fluxes. Thus, guard cells regulate cellular pH by the synthesis and degradation of organic anions, which are considered mechanisms for fine control of pH in most other plant cells (see Smith and Raven, 1979). The specific mechanism, as it applies to guard cells, is shown in the left panel of Figure 5.1. It is useful to interpret this outline from different perspectives. First, in the variation shown, a divalent anion (malate) is synthesized from a neutral precursor (starch). Thus, the key steps are the oxidation of the aldehyde to the carboxylate (phosphoglyceraldehyde → phosphoglycerate), which releases a proton, and the hydration of CO_2, which releases another proton; the resulting HCO_3^- is used in the carboxylation of phosphoenolpyruvate (PEP). There are other metabolite interconversions in this scheme that involve H^+ also (e.g. the reduction of oxaloacetic acid requires two electrons and two protons, but the redox couple [NADH → NAD^+] supplies two electrons but only one proton). These other reactions, however, are balanced elsewhere in the cell (e.g. by the reduction of NAD^+). Second, I feel certain that the concepts presented in this outline (Fig. 5.1) will prove essentially correct. The evidence is very strong (Outlaw, 1982): malate concentration in the guard cells of open stomata is about six times higher than in those of closed stomata; at least initially, the starch content of guard cells declines when stomata open; PEP carboxylase (PEPCase), which catalyzes the reactions at the metabolic branch point, is present at much higher specific activity in guard cells than in other leaf cells. Third, most of the steps shown in this outline (Fig. 5.1) have not been verified experimentally. For example, we do not know whether starch is degraded by an amylase or by a phosphorylase, or which metabolite is transported from the plastid, or anything of the nature of the transport processes at the tonoplast.

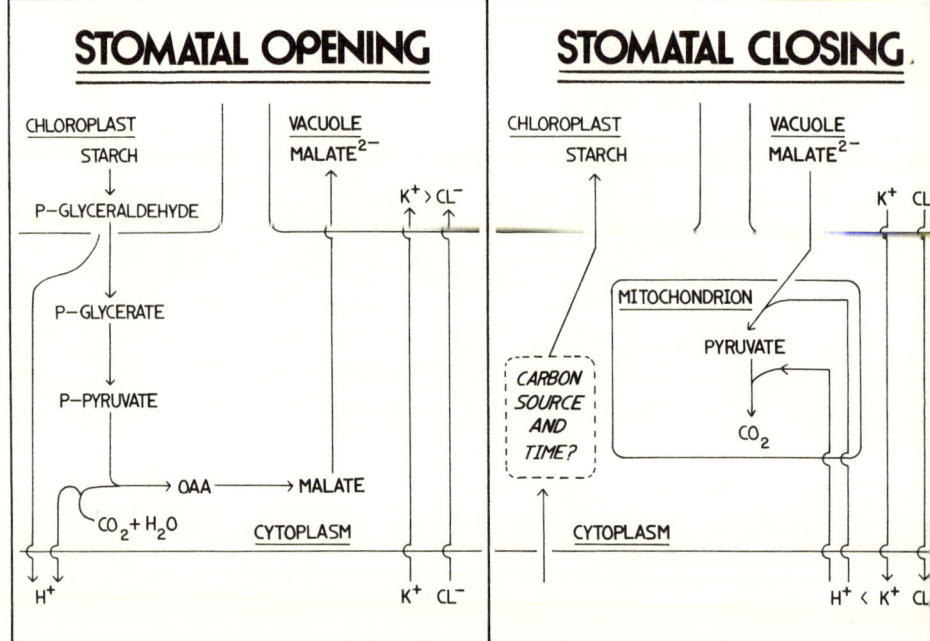

Fig. 5.1. Outline of metabolic events postulated to occur in guard cells during stomatal movements.

A description of the metabolic events that occur during stomatal closing must be equivocal. There are three possibilities. First, malate may leave the guard cells "with" potassium. Indeed, there is one report (Van Kirk and Raschke, 1978) that malate efflux occurs during rapid stomatal closure. Second, carbons 1, 2, and 3 of malate could be converted back to starch by PEP carboxykinase or pyruvate orthophosphate dikinase. Reports to date have indicated the lack of significant levels of PEP carboxykinase. Despite earlier controversy, current reports indicate the absence of metabolically significant levels of pyruvate orthophosphate dikinase (see discussion in Outlaw, Springer, and Tarczynski, 1985). Third, malate may be depleted by complete oxidation in mitochondria, either in guard cells or in adjacent epidermal cells. This third possibility, which I favor, is supported by high levels of malic enzyme in both cell types (see Outlaw, 1982). Although the loss of total leaf carbon would be small, complete malate oxidation would place an additional demand on carbon import into guard cells, since, as discussed later, these cells lack the ability to reduce CO_2 photosynthetically. However, this additional demand would not seem to place a constraint on stomatal functioning, because sucrose in the mesophyll exchanges fairly rapidly with that in the epidermal layer (e.g. Outlaw and Fisher, 1975; Outlaw, Fisher, and Christy, 1975).

Regulation of Carbon Metabolism in Guard Cells

The regulatory aspects of carbon metabolism in guard cells have been studied in independent laboratories at only two metabolic positions, PEPCase and ADP-glucose pyrophosphorylase.

Phosphoenolpyruvate is at a metabolic branch point. It may be oxidized via pyruvate or consumed during gluconeogenesis. Alternatively, it may be carboxylated to form oxaloacetate (Fig. 5.1). Thus, the metabolic fate of PEP obviously must be regulated. Even considered alone, the carboxylation of PEP may have diverse functions. One function is to "trap" CO_2 as in the mesophyll cells of C_4 plants. A compound resulting from this carboxylation is transported to other cells, where the CO_2 is released and subsequently reduced. A similar function is to "trap" CO_2 as in CAM plants. A resulting acid is stored intracellularly until conditions are favorable for CO_2 reduction. As shown in Figure 5.1, another function is the release of protons for pH regulation; in this case, a resulting anion is stored temporarily in the cells. To fulfill the diverse metabolic functions of PEP carboxylation, several unique forms of PEPCase have evolved, with different properties. Even in single organs, distinct forms of this enzyme have been reported (e.g. Ting and Osmond, 1973; Amagasa, 1984; Goatly, Coombs, and Smith, 1975). Although the overall distribution of PEPCase activity among different cell types is known, little is known about the properties of its activity in guard cells. Furthermore, the foregoing overview indicates that it is not prudent to extrapolate to the guard cell PEPCases from properties of other PEPCases. In brief, this enzyme is potentially regulated by two non–mutually exclusive mechanisms. The first is alteration of the chemical environment—for example, an influence on K_m (Mg·PEP) by pH (see Smith and Raven, 1979); despite several reports, including one from my laboratory, little of relevance is known about this mechanism. A second mechanism of regulating this enzyme is through reversible modification of the enzyme itself, similar to the mechanisms operating in CAM plants (e.g. Kluge, Brulfert, and Queiroz, 1981; Winter, 1982). Two reports (Schnabl, Elbert, and Krämer, 1982; Rao and Anderson, 1983) indicate that the level of activity of PEPCases that can be extracted from stomatal systems depends on the physiological state of the tissue. These reports are particularly exciting, but more detailed work is needed.

Starch degradation is the first step in the mobilization of carbon skeletons that can be used as a source for PEP. Stomatal opening, and therefore guard cell starch degradation, normally occurs when a leaf is illuminated. By contrast, starch accumulates then in photosynthetic parenchyma. Thus, one challenge has been to understand how starch biosynthesis can be regulated so differently in adjacent cells. This specific aspect of starch metabolism in guard cells is now understood. The regulated step in starch biosynthesis is ADP-glucose pyrophosphorylase. In leaf cells, including guard cells, this enzyme is regulated by the ratio of the concentration of 3-phosphoglycerate to that of orthophosphate (Robinson, Zeiger, and Preiss, 1983; Outlaw and Tarczynski, 1984). In photosynthetic parenchyma, 3-phosphoglycerate concentration is elevated in light. The dominant and perhaps sole cause of elevated 3-phosphoglycerate concentration is photosynthetic carbon metabolism, the carboxylation of ribulose-1,5-bisphosphate (RuBP) to two molecules of 3-phosphoglycerate being activated (albeit indirectly) by light. In contrast to photosynthetic parenchyma, guard cells do not carry out photosynthetic carbon metabolism,* and the concentration of 3-phosphoglycerate in these cells is low and unaffected by illumination. Thus, it appears that the absence of the Calvin-Benson cycle from guard cells effectively uncouples starch biosynthesis from the effects of illumination in these cells. Of course, many questions about starch metabolism in guard cells remain; we do not know, for example, when guard cells synthesize starch or how starch biosynthesis in these cells is regulated. In my opinion, answering such questions is critical to our understanding of guard cell function.

I hope a summary of the regulation of carbon metabolism in guard cells soon may be more comprehensive. Aside from the two metabolic steps I have outlined above, several others are potentially regulated: for example, malate catabolism, cytoplasmic osmoregulation, carbon import, and tonoplast transport. From one perspective, there are reasons to expect progress to be rapid; it has been only a decade since the publication of the first quantitative study of guard cell carbon metabolism. During this time, considerable effort has been invested in the development of appropriate methodology. Currently, many methods are in hand, and the field has attracted many investigators with diverse talents.

Photosynthetic Electron Transport in Guard Cells

In photosynthetic parenchyma, about 80 percent of the products of electron transport (reducant and ATP) are metabolized in the reduction of CO_2, nitrogen reduction being the other major sink. However, although guard cells lack the photosynthetic carbon-reduction pathway, they conduct linear electron trans-

*There are unpublished indications from immunochemistry that the protein RuBP carboxylase (RuBPCase) is present in guard cells. However, the bulk of the evidence indicates that the Calvin-Benson cycle does not operate in these cells (see Outlaw, 1982). That evidence includes: the virtual absence of RuBPCase activity in extracts of guard cell protoplasts from different species and assayed in different laboratories, the virtual absence of 3-phosphoglycerate as a product of CO_2 incorporation by guard cell protoplasts, the insignificant levels of RuBPCase in guard cells of more than 20 species as assayed by a single-cell approach, the virtual absence of phosphoribulokinase and β-hydroxypyruvate reductase activities in guard cells, the lack of detection of RuBPCase in extracts of guard cell protoplasts by rocket electrophoresis, and the absence of an effect of light on the size of the 3-phosphoglycerate pool in guard cells (Outlaw and Tarczynski, 1984).

port (e.g. Outlaw et al., 1981; Zeiger, Armond, and Melis, 1981). Stated more conservatively, the standard biophysical correlates of photophosphorylation and photosynthetic $NADP^+$ reduction have been documented for guard cells, although neither ATP synthesis nor $NADP^+$ reduction has been directly documented. To put the situation into proper perspective, additional facts must be considered somewhat independently. First, guard cell turgor is increased by photosynthetically active radiation (PAR), even when indirect influences (e.g. CO_2 concentration and blue-light effects) are constant. Alone, this fact is not surprising; photophosphorylation may "fuel" an ion-transporting ATPase. It is more challenging to explain simply why PSII inhibitors, which would abolish the generation of reductant but not photophosphorylation, moderate the PAR effect. It is tempting to speculate that some reductant (e.g. ferredoxin) may be involved in the activation of some metabolic processes. The second consideration is quantitative. Because guard cells lack the Calvin-Benson cycle but seem to reduce $NADP^+$ photosynthetically, a central problem is to identify the Hill oxidant.* Whether this problem is important depends upon one's viewpoint. For *Vicia faba*, guard cell chlorophyll content (on a per-cell basis) is less than 5 percent of that of a palisade parenchyma cell. On the other hand, the chlorophyll/protein ratio (on a mass basis) in guard cells (1:35) is similar to that for palisade parenchyma cells (1:20).

From the foregoing discussion, it is obvious that our understanding of how the light reactions of photosynthesis influence carbon metabolism in guard cells is rudimentary. This state of affairs is understandable; reliable reports of PSII in guard cells were published only in 1981. At this point, I wish to warn against complacency in drawing strict analogies between the function of chloroplasts in guard cells and that in other cells. In general, guard cell chloroplasts and their mesophyll counterparts differ in morphology. For example, guard cell chloroplasts are smaller, less abundant, and paler, have fewer thylakoids per granum, have areas of peripheral reticulum, and contain relatively more starch. One biochemical difference (fluorescence kinetics) underlying these morphological differences has already been identified. In my opinion, the relationship between the size of the stomatal aperture and the functioning of photosynthetic pigments in guard cells deserves intensive investigation by stomatal physiologists.

*My earlier report that guard cells lack $NADP^+$-triose-P dehydrogenase was in error (see Outlaw, 1982). The specific activity (on a mass basis) of this enzyme in guard cells of *Vicia faba* is about ten percent of that in the palisade cells (Outlaw, Springer, and Tarczynski, 1985).

REFERENCES

Allaway, W. G., and G. Setterfield. 1972. Ultrastructural observations in guard cells of *Vicia faba* and *Allium porrum*. *Canadian Journal of Botany*, 50: 1405–13.

Amagasa, T. 1984. Enzymatic properties of phosphoenolpyruvate carboxylase isoforms in the CAM plant *Kalanchoe daigremontiana*. *Plant and Cell Physiology*, 25: 625–33.

Fischer, R. A., and T. C. Hsiao. 1968. Stomatal opening in isolated epidermal strips of *Vicia faba*, II. Response to KCl concentration and the role of potassium absorption. *Plant Physiology*, 43: 1953–58.

Fujino, M. 1967. Role of adenosine triphosphate and adenosine triphosphatase in stomatal movement. *Science Bulletin of the Faculty of Education, Nagasaki University*, 18: 1–47.

Goatly, M. B., J. Coombs, and H. Smith. 1975. Development of C_4 photosynthesis in sugar cane: Change in properties of phosphoenolpyruvate carboxylase during greening. *Planta*, 125: 15–24.

Hampp, R., W. H. Outlaw, Jr., and M. C. Tarczynski. 1982. Profile of basic carbon pathways in guard cells and other leaf cells of *Vicia faba* L. *Plant Physiology*, 70: 1582–85.

Hsiao, T. C. 1976. Stomatal ion transport. In U. Lüttge and M. G. Pitman, eds., *Transport in plants* 2, Encyclopedia of Plant Physiology, n.s., 2B: 195–221. Berlin: Springer-Verlag.

Imamura, S.-I. 1943. Untersuchungen über den Mechanismus der Turgorschwankung der Spaltöffnungsschliesszellen. *Japanese Journal of Botany*, 12: 251–346.

Kluge, M., J. Brulfert, and O. Queiroz. 1981. Diurnal changes in the regulatory properties of PEP-carboxylase in crassulacean acid metabolism (CAM). *Plant, Cell and Environment*, 4: 251–56.

MacRobbie, E. A. C. 1981. Ionic relations of stomatal guard cells. In P. G. Jarvis and T. A. Mansfield, eds., *Stomatal Physiology*: 51–70. Cambridge: Cambridge University Press.

Outlaw, W. H., Jr. 1982. Carbon metabolism in guard cells. *Recent Advances in Phytochemistry*, 16: 185–222.

———. 1983. Current concepts of the role of potassium in stomatal movements. *Physiologia plantarum*, 59: 302–11.

Outlaw, W. H., Jr., and D. B. Fisher. 1975. Compartmentation in *Vicia faba* leaves, I. Kinetics of ^{14}C in the tissues following pulse labeling. *Plant Physiology*, 55: 699–703.

Outlaw, W. H., Jr., D. B. Fisher, and A. L. Christy. 1975. Compartmentation in *Vicia faba* leaves, II. Kinetics of ^{14}C-sucrose in the tissues following pulse labeling. *Plant Physiology*, 55: 704–11.

Outlaw, W. H., Jr., B. C. Mayne, V. E. Zenger, and J. Manchester. 1981. Presence of both photosystems in guard cells of *Vicia faba* L. Implications for environmental signal processing. *Plant Physiology*, 67: 12–16.

Outlaw, W. H., Jr., S. A. Springer, and M. C. Tarczynski. 1985. Histochemical technique. A general method for quantitative enzyme assays of single cell "extracts" with a time resolution of seconds and a reading precision of femtomoles. *Plant Physiology*, 77: 659–66.

Outlaw, W. H., Jr., and M. C. Tarczynski. 1984. Guard cell starch biosynthesis regulated by effectors of ADP-glycose pyrophosphorylase. *Plant Physiology*, 74: 424–29.

Introduction to Carbon Metabolism in Guard Cells

Rao, I. M., and L. E. Anderson. 1983. Light and stomatal metabolism, I. Possible involvement of light modulation of enzymes in stomatal movements. *Plant Physiology*, 71: 451–55.

Raschke, K. 1979. Movements of stomata. In W. Haupt and M. E. Feinleib, eds., *Physiology of movements*, Encyclopedia of Plant Physiology, n.s., 7: 383–441. Berlin: Springer-Verlag.

Robinson, N. L., E. Zeiger, and J. Preiss. 1983. Regulation of ADP-glucose synthesis in guard cells of *Commelina communis*. *Plant Physiology*, 73: 862–64.

Scarth, G. W. 1927. Stomatal movement: Its regulation and regulatory role. A review. *Protoplasma*, 2: 498–511.

Schnabl, H., C. Elbert, and G. Krämer. 1982. The regulation of the starch-malate balances during volume changes of guard cell protoplasts. *Journal of Experimental Botany*, 33: 996–1003.

Smith, F. A., and J. A. Raven. 1979. Intracellular pH and its regulation. *Annual Review of Plant Physiology*, 30: 289–311.

Ting, I. P., and C. B. Osmond. 1973. Photosynthetic phospho*enol*pyruvate carboxylases. Characteristics of alloenzymes from leaves of C_3 and C_4 plants. *Plant Physiology*, 51: 439–47.

Van Kirk, C. A., and K. Raschke. 1978. Release of malate from epidermal strips during stomatal closure. *Plant Physiology*, 61: 474–75.

Winter, K. 1982. Properties of phosphoenolpyruvate carboxylase in rapidly prepared, desalted leaf extracts of the crassulacean acid metabolism plant *Mesembryanthemum crystallinum* L. *Planta*, 154: 298–308.

Zeiger, E. 1983. The biology of stomatal guard cells. *Annual Review of Plant Physiology*, 34: 441–75.

Zeiger, E., P. Armond, and A. Melis. 1981. Fluorescence properties of guard cell chloroplasts. Evidence for linear electron transport and light-harvesting pigments of photosystems I and II. *Plant Physiology*, 67: 17–20.

6

Ionic Relations of Guard Cells

E. A. C. MacRobbie

With hindsight it is clear that the importance of potassium salts (K-salts) in the function of guard cells should have been recognized long before it was. The first demonstration of very high concentrations of K-salts in open, but not in closed, guard cells was given by the histochemical studies of Macallum (1905), and it is now forty years since evidence for the special role of K-salts in the opening and closing of stomata was provided by the extensive studies of Imamura (1943). Although Yamashita (1952) showed a correlation between the K^+ content of guard cells and stomatal aperture, the hypothesis that the extent of K-salt accumulation in guard cells determines their turgor, and hence stomatal aperture, was not developed until the late 1960's. This came with the first publication in English of Fujino's work (1967), and from independent work by Fischer (1968a; Fischer and Hsiao, 1968). It has now been well established, by several methods and in different plants, that K^+ levels are very high indeed in open, but much lower in closed, guard cells; there is, however, very little understanding of how this is achieved.

In the forty years since Imamura established the importance of K^+ in guard cell function, other researchers have provided a quantitative description of the changes in net salt accumulation associated with changes in stomatal aperture in a few selected species. However, we have a very limited understanding of the control of salt accumulation in guard cells, particularly in the intact leaf. In addition, and perhaps more important, we need to understand the means by which guard cells, as distinct from nearly all other plant cells, sustain very large fluctuations in their salt levels as part of their normal function.

Attention has concentrated on the mechanisms of ion uptake in guard cells, yet a more interesting question may be the reason for the failure of such accumulation in response to quite ordinary signals, such as darkening, or increased CO_2 levels. It is this failure that distinguishes these cells, and I believe that more attention should be focused on the mechanisms involved in the loss of the ability to maintain the very high level of vacuolar salt accumulation in closing conditions.

There are two distinct problems. The first is to establish the differences in ionic state beween open and closed guard cells and to describe the steady-state conditions at a range of apertures. It is important that such a description be dynamic and include values not only for the steady-state ion contents of the guard cells at different apertures but also for the steady-state ion fluxes at the plasmalemma and tonoplast. The aim must then be to identify the nature of the fluxes, particularly the active fluxes, by which steady non-equilibrium states are maintained.

The second problem is to identify the transients involved in stomatal movements, the changes in ionic state in response to environmental signals for opening and closing that result in flux changes and transitions to new steady states with different apertures and contents. The critical issues are the identification of the ion fluxes that are sensitive to such signals and therefore control stomatal aperture, their nature (K^+, H^+, Cl^-, malate), their location (plasmalemma, tonoplast, or both), and their direction (influx, efflux, or both). We need also to distinguish between primary fluxes responding to the opening and closing signals, and the secondary fluxes that follow. Most important, we need to describe two transients, one involved in the initiation of opening and the other in the initiation of closing; it seems likely that the nature of the flux changes will be different for the two transitions.

One difficulty is immediately obvious: we have methods for the determination of ion contents, but not fluxes, in guard cells in their normal state in the intact leaf. For the measurement of ion fluxes we need an experimental system in which the extracellular medium is defined and in which the ion fluxes into and out of guard cells are not masked by the contributions from the subsidiary and epidermal cells. These requirements demand the use of epidermal strips and the direct exposure of the guard cells to the bathing medium. For meaningful flux measurements it is also essential that the epidermal strips contain "isolated" guard cells, obtained by a pretreatment that kills all epidermal and subsidiary cells but leaves guard cells alive and functional. It is clear that the information required for an understanding of the control of ion transport in guard cells cannot be acquired under normal physiological conditions. However, we may hope that an understanding of the ionic responses in experimentally accessible guard cells may throw light on events occurring in the normal state in the intact leaf.

The evidence for the role of K-salts in guard cells has been extensively considered in earlier reviews and need not be repeated. Outlaw (1983) provides the most recent review of K^+ in guard cells. Zeiger (1983) assesses the evidence that in guard cells, as in other plant cells, salt accumulation is achieved by primary pumping of protons out of the cell, setting up proton gradients that drive the secondary transport of other ions. In this review I shall concentrate on the two species in which a reasonable amount of quantitative information is available, *Vicia faba* and *Commelina communis*. My first aim is to present an integrated picture of the ionic relations of guard cells of these two species, and my second aim is to review progress toward an understanding of the processes involved and to assess directions for future research.

Ion Contents of Open and Closed Guard Cells

Although the accumulation of K^+ in guard cells seems to be a universal process in stomatal opening, quantitative data are available for very few species, and only in *Vicia faba* and *Commelina communis* have fluxes been measured. Table 6.1 presents values for the K^+ contents in open and closed guard cells in these species in "intact" and "isolated" conditions. The term "intact" refers to guard cells of stomata opened or closed on the intact leaf in which content or concentration measurements were made before significant changes in ion levels took place. Intact guard cells swell against the back pressure from subsidiary or epidermal cells and require enough turgor to overcome the resistance to expansion provided both by the guard cell wall and by the turgor of surrounding cells. The term "isolated" refers to guard cells of stomata opened in epidermal strips in which only guard cells are alive. These stomata tend not to close completely in the absence of back pressure, and openings of 4 to 6 μm, or more, are commonly found in what would be closing conditions in the intact leaf; this need not imply an intrinsic anomaly in the closing response. The difference in the guard cell turgor necessary to open the pore in intact and isolated conditions is clearly reflected in the difference in K^+

TABLE 6.1. Potassium contents of open and closed guard cells

Tissue and reference	K content (pmol per guard cell)		K concentration (mmol l^{-1})	
	Open	Closed	Open	Closed
Vicia faba				
Intact				
a	2.12	0.1	880	77
b	2.72	0.55	552	112
c	2.25	0.41	460–760	80
Isolated				
d	1.38	0.26	276	52 (7 μm)
e	1.78	0	357	0 (4 μm)
f	1.61	0.40	322	80 (6 μm)
Commelina communis				
Intact				
g*	3.1	0.4	448	95
h*	2.7–4.2	0.3–0.36	385–600	75–90
Isolated				
i*	1.1	0.2	157 (15 μm)	51 (5 μm)
i	1.2	0.1–0.2	167–174 (15 μm)	20–56 (5 μm)

SOURCES: *a*, Humble and Raschke (1971): electron microprobe analysis. *b*, Allaway and Hsiao (1973): aperture set on intact leaf, then epidermal peel rolled to kill epidermal cells; flame photometry. *c*, Outlaw and Lowry (1977): dissection of guard cell pairs from frozen-dried tissue; enzymic assay for K. *d*, Fischer (1972): $^{42}K^+$ uptake. *e*, Pallaghy and Fischer (1974): $^{42}K^+$ uptake. *f*, Fischer (1968a), Fischer and Hsiao (1968): $^{86}Rb^+$ uptake. *g*, Penny and Bowling (1974): K-sensitive electrode. *h*, MacRobbie and Lettau (1980b): K-sensitive electrode. *i*, MacRobbie and Lettau (1980a): K-sensitive electrode (*i**) or $^{86}Rb^+$ uptake from RbCl.

NOTE: The calculated K^+ concentrations (*a–f, i*) use the guard cell volumes measured by the original authors; allowance is made for the volume change during opening.

*Measurement of potassium activity rather than total content.

contents. Thus, in *Vicia*, guard cell K$^+$ concentrations of 460 to 880 mmol l^{-1} are associated with opening against turgid epidermal cells, whereas opening is achieved in isolated guard cells with only 276 to 357 mmol l^{-1} K$^+$. In *Commelina*, open intact guard cells contain 385 to 600 mmol l^{-1} K$^+$, compared with only 157 to 174 mmol l^{-1} in open isolated guard cells. For *Commelina*, values are available that allow the comparison of this difference with the difference that would be expected on osmotic grounds. The contribution of the subsidiary-cell turgor to the water relations of the guard cell (P_{gs}) may be estimated by determining the osmotic pressure of sucrose that has to be added to the external solution to bring the guard cells back to the initial aperture after the removal of subsidiary-cell turgor (MacRobbie, 1980). The osmotic difference associated with the difference in K-salt accumulation can be estimated (approximately) as the difference in the osmotic pressures of corresponding solutions of KCl (Π_{KCl}). The results of this comparison showed that over the aperture range of 6 to 14 μm, $\Delta\Pi_{KCl}$ rose from 110 to 420 mOsmol kg^{-1}, and P_{gs} rose from 89 to 355 mOsmol kg^{-1}. Thus the extra K-salt observed in intact open guard cells is close to the calculated amount necessary to overcome P_{gs}.

Although flux measurements are only possible in isolated guard cells, we may consider the net ion differences between open and closed intact guard cells in relation to the time required for opening and closing. Over the range of stomatal movement the intact guard cells of *Vicia* gain or lose an estimated 1.8 to 2.2 pmol K$^+$ per cell, and those of *Commelina* an estimated 2.4 to 3.8 pmol K$^+$ per cell. Guard cell volumes and areas are difficult to estimate accurately, but available measurements suggest areas of 17 × 10^{-6} cm^2 per cell in *Vicia* (Fischer and Hsaio, 1968), and 25 × 10^{-6} cm^2 per cell for *Commelina* (MacRobbie, 1981a). Thus the net K$^+$ differences expressed on the basis of guard cell area are about 120 nmol cm^{-2} in both species. If stomatal movements involved only the fluxes of K-salts, then opening over 1 to 2 hours would require a flux of about 30 to 15 pmol cm^{-2} s^{-1}, but closing over 20 minutes would require a flux of about 100 pmol cm^{-2} s^{-1}. These are very large fluxes, particularly that for closing, and are higher than the measured fluxes discussed in a later section. But the fact that stomata can close in 20 minutes suggests that closure does not result simply from switching off the process of accumulation, but involves a stimulation of efflux in response to the closing signal. There is also evidence that during opening the maximum aperture is reached before the buildup of K$^+$ is complete (Laffray, Louguet, and Garrec, 1982). This suggests that other osmotica may also be involved, and in this case the true fluxes during opening may be somewhat lower than those calculated above.

The very high levels of K$^+$ accumulated in open intact guard cells, much higher than the maximum value reached in other plant cells, are apparent in Table 6.1. In many systems salt accumulation is limited by feedback signals that regulate either turgor or internal ionic concentrations (see reviews by Cram, 1976; Zimmermann, 1977; Gutknecht, Hastings, and Bisson, 1978). In many marine algae the active ion influx primarily responsible for the generation of turgor seems to be inhibited by increasing turgor; in some instances this is active K^+ influx, and in others, active Cl^- influx. Cram (1980) has compared the regulation of salt uptake in carrot and in beet. In carrot (perhaps representative of glycophytes) influx of KCl is regulated by internal Cl^- or NO_3^- concentrations, but not by turgor; whereas the accumulation of K-carboxylate may be turgor-regulated. By contrast, in beet (perhaps representative of halophytes), Cl^- influx is inhibited both by increasing turgor and by high internal levels of Cl^- or NO_3^-. Maximum levels of KCl accumulated in carrot are about 160 μmol g^{-1}, which may be compared with 80 μmol g^{-1} in barley roots (Pitman, 1969). In beet, turgor seems to be regulated to about 0.64 MPa. If guard cells have such controls, their set points must be at much higher levels. Osmotic pressures measured in open guard cells are very high indeed, provided that plasmolytic measurements are taken either before appreciable solute leakage has vitiated the results or else in the presence of high K^+ in the plasmolysing solution. At incipient plasmolysis, values of 4.7 MPa in *Vicia*, 6 MPa in *Commelina*, and 7 to 9 MPa in *Zebrina* were found by Imamura (1943); Raschke (1979) quotes values of 7 to 9 MPa in *Vicia*. Thus, even allowing for a one-and-one-half- to twofold increase in volume during opening, the osmotic pressure in guard cells must be of the order of 4 to 6 MPa. The nature of the controls on salt accumulation in guard cells remains one of the interesting and unstudied problems of their physiology.

Behavior of Guard Cells in Epidermal Strips

The ionic conditions of bathing solution required for opening of isolated guard cells of *Vicia* and of *Commelina* are qualitatively similar but quantitatively different.

Vicia faba. In light, stomatal opening in *Vicia* is achieved with an external concentration of 1 to 10 mmol l^{-1} KCl or RbCl, but concentrations of 100 mmol l^{-1} or more are required with NaCl, LiCl, or CsCl (Humble and Hsiao, 1969); this specificity is lost in Ca^{2+}-free solutions. Stomata open less effectively in the dark, but will open if given 100 mmol l^{-1} of any of these cations. Increasing Ca^{2+} concentrations above 0.1 mmol l^{-1} reduces aperture, but opening can be achieved in 10 mmol l^{-1} KCl in the presence of 1 mmol l^{-1} Ca^{2+} (Willmer and Mansfield, 1969; Fischer, 1972). The opening in 10 mmol l^{-1} KCl plus 0.1 mmol l^{-1} $CaCl_2$ is not affected by the pH of the bathing solution in the range 4 to 7 (Fischer, 1972).

Commelina communis. *Commelina* differs from *Vicia* in requiring higher concentrations of KCl for opening and in being much more sensitive to Ca^{2+}.

TABLE 6.2. Steady-state apertures in light or darkness in isolated guard cells of *Commelina communis*

Solution	Aperture (μm)	
	Light	Dark
KCl	18.4 ± 0.4 (13, 187)	11.3 ± 0.3 (16, 186)
KBr	16.1 ± 0.5 (9, 101)	6.1 ± 0.8 (4, 49)
RbCl	13.7 ± 0.2 (14, 223)	7.4 ± 0.6 (11, 176)

SOURCE: MacRobbie (1984).
NOTE: All solutions contained 30 mmol l^{-1} of the salt indicated and 10 mmol l^{-1} MES, pH 3.9. Data presented as mean ± S.E.M. (number of strips, number of stomata).

Isolated guard cells open with about 20 to 30 mmol l^{-1} KCl in the medium, although with intact subsidiary and epidermal cells an external concentration of 60 to 100 mmol l^{-1} KCl could be required for the same degree of opening (Willmer and Mansfield, 1969; MacRobbie and Lettau, 1980a). Opening is abolished by the addition of 1 mmol l^{-1} Ca^{2+} and severely inhibited even at 0.1 mmol l^{-1} Ca^{2+} (Fujino, 1967; Willmer and Mansfield, 1969). As in *Vicia*, there appears to be little effect of pH; pH 3.9 is used to isolate guard cells, and at this pH they appear able to survive for long periods and to open as effectively as at pH 6.7 (MacRobbie and Lettau, 1980a; Clint and MacRobbie, 1984). The osmotic dependence of the aperture was not significantly different in intact and in isolated guard cells, and it therefore appears that the properties of the guard cell wall are not altered by the treatment at low pH (MacRobbie, 1980).

Isolated guard cells are still sensitive to light and dark; the steady-state apertures reached after incubation in 30 mmol l^{-1} in light or dark, in KCl, KBr, or RbCl, and the apparent order of effectiveness are shown in Table 6.2. There is a clear effect of light, although, as would be expected in the absence of back pressure, the dark opening is appreciable, as Fischer (1972) also found in *Vicia faba*.

In the absence of added Ca^{2+}, opening in *Commelina* also occurs in the presence of NaCl (Willmer and Mansfield, 1969), although Jarvis and Mansfield (1980) showed that Na-opened guard cells were much less sensitive to closing stimuli, such as dark, high CO_2, and abscisic acid (ABA), than were those opened in KCl.

In isolated guard cells there is little difference between the degree of opening at pH 3.9 and at pH 6.7 (MacRobbie and Lettau, 1980a; Clint and MacRobbie, 1984). However, in epidermal strips in which there are turgid subsidiary and epidermal cells, the opening in any given condition is significantly less than that in isolated guard cells. If it is assumed that ion contents are equal in intact and in isolated guard cells after incubation in a given external concentration, then the aperture of the intact stomata fits well with what would be predicted for that ion content (see also "Aperture Dependence of Ion Contents").

One interesting question concerning these aperture values is what limits opening in isolated guard cells. In *Commelina* bathed with 30 mmol l^{-1} K(Rb)Br(Cl) we observe apertures of 14 to 18 μm, depending on the salt, with intracellular ion concentrations of about 160 to 200 mmol l^{-1}. Wider openings can sometimes be observed and are therefore mechanically possible, and intact guard cells can accumulate much higher levels of K$^+$, 400 mmol l^{-1} or more, with a resulting higher turgor. The same is true in *Vicia*: isolated guard cells achieve only about 300 mmol l^{-1} K$^+$, compared with 460 to 880 mmol l^{-1} in intact cells. We can therefore ask why the isolated guard cells do not continue to accumulate salt and open further.

Isolated guard cells may fail to open further because their transport processes are impaired by the isolation procedure, a possibility that could be tested with isolated guard cells in solutions of high osmotic pressure. Alternatively, isolated guard cells could leak excessively to the now-infinite bathing medium, an explanation implying that in the leaf the K$^+$ concentration outside the guard cell reaches very high levels. This is argued by Maier-Maercker (1979, 1983) but is not generally accepted, and we have no experimental evidence bearing on the question. However, flux measurements by Fischer (1972), to be discussed later ("Flux Measurements in Guard Cells of *Vicia*"), do not show very high effluxes and therefore suggest that this is not the reason. A third possibility is that ion accumulation in guard cells is regulated neither by turgor nor by internal concentration, but by volume. Such control is found in wall-less algae, such as *Poterio-ochromonas* (Kauss, 1979), and volume changes in guard cells are large enough for this to be a feasible hypothesis.

Mechanism of Ion Uptake in Guard Cells

The evidence available suggests that the mechanism of salt accumulation in guard cells is, in fact, very similar to that in plant cells in general, differing in quantity but not in kind.

It is now generally accepted that salt accumulation in plant cells is driven by primary proton extrusion at the plasmalemma, powered by ATP. Proton extrusion generates an electrical driving force (inside negative) for cation entry, and a pH gradient across the plasmalemma (cytoplasm more alkaline); the pH gradient may then be dissipated by synthesis of malic acid in the cytoplasm or by chloride uptake by co-transport with one or more protons. Part of the evidence for a primary role of a proton pump lies in the sensitivity of K$^+$ and Cl$^-$ uptake to the fungal toxin fusicoccin. Marrè's group has shown that fusicoccin stimulates active proton extrusion by many plant cells or releases an inhibition by an endogenous controlling factor, and as a result, stimulation of a process by fusicoccin has become a diagnostic for a link with the proton pump (see review by Marrè, 1979). The fusicoccin-stimulated opening of stomata and associated changes in ion contents (Turner and Graniti, 1969; Turner, 1973; Squire and Mansfield, 1972, 1974) therefore constitute evidence for the role of a proton pump in initiating the events leading to salt accumulation in guard cells.

There is also some direct evidence for proton extrusion from guard cells. Raschke and Humble (1973) measured proton extrusion associated with guard cells of *Vicia faba* as the excess protons excreted into solutions of K- as compared with Ca-iminodiacetate, an impermeant anion. Stomata opened in K-iminodiacetate but not in Ca-iminodiacetate. The amount of proton extrusion was consistent with a one-to-one exchange of K^+ and H^+. The initial release of H^+ into Ca^{2+} solutions or from frozen thawed tissue was very high, about 2.7 pmol per guard cell. This implies a large exchange capacity and an initially low pH in the guard cell wall. A one-to-one exchange of K^+ for H^+, if continued throughout the course of stomatal opening, would involve a very large release of H^+ (about 2 pmol per cell) into a very limited cell-wall space, with restricted diffusion away from that site. The result would be large changes in pH in the cell wall and would require the guard cell membrane to be very resistant to low pH. We can speculate that such falling pH in the wall might then lead to solute leakage from the neighboring subsidiary or epidermal cells, which are known to be less tolerant of low pH than are guard cells. Thus proton extrusion from guard cells might tilt the competitive balance, leading to uptake by the guard cells of ions lost from subsidiary cells rendered leakier by the falling pH.

Gepstein, Jacobs, and Taiz (1982) measured rates of H^+ release from epidermal strips of *Vicia faba*, in which more than 90 percent of epidermal cells were dead, on 10 mmol l^{-1} KCl plus 0.1 mmol l^{-1} $CaCl_2$. The rates were 5 to 18 nmol h^{-1} for 10 to 12 strips of area 20 to 25 mm^2 each. If we use a median figure of 250 mm^2 for the total area, with 60 stomatal complexes per square millimeter and single-guard-cell areas of 17×10^{-6} cm^2 per cell, then these rates convert to 3 to 10 pmol cm^{-2} s^{-1} and are comparable with the K^+ fluxes in *Vicia* to be discussed later. The proton efflux was inhibited in the dark, and by 1 mmol l^{-1} vanadate in the light; it was stimulated by fusicoccin. Abscisic acid prevented K^+ accumulation but only partially inhibited H^+ release. Reduced proton efflux was associated with reduced opening, both in the dark and in the vanadate treatment. Vanadate is a well-recognized inhibitor of active proton efflux in other plant cells and of the plasmalemma ATPase, which is held to be responsible for such transport (Bowman et al., 1978; Bowman and Slayman, 1979; Cocucci, Ballarin-Denti, and Marrè, 1980; Jacobs and Taiz, 1980; DuPont, Burke, and Spanswick, 1981). The results therefore give experimental support to the hypothesis that proton pumping is the primary process in the ion accumulation associated with stomatal opening.

There is early evidence from work with dyes that the vacuolar pH is higher in open than in closed guard cells (Scarth, 1932; Pekarek, 1934; Small and Maxwell, 1939). Direct intracellular measurements with a pH-sensitive microelectrode gave values of 5.6 in open guard cells and 5.19 in closed guard cells (Penny and Bowling, 1975). These changes are similar to those observed during KCl accumulation in low-salt barley roots (Hiatt, 1967; Hiatt and Hendricks, 1967). Changes in guard cell fluorescence have also been taken as indicators of vacuolar pH (Zeiger, 1981, 1983), since increased fluorescence can be induced by conditions likely to produce intracellular alkalinity, such as treatment with ammonium chloride at pH 8; increased fluorescence after the addition of fusicoccin suggests alkalinization attributable to the proton pump.

In many plant cells the membrane potential is a sensitive indicator of the activity of the electrogenic proton pump, and membrane potentials much more negative than any possible diffusion potential have been observed. When the activity of the pump is inhibited, the marked hyperpolarizations disappear, and potentials are close to the diffusion potentials. Since membranes are more permeable to K^+ than to any other ion, this value is generally close to E_K, the K^+ equilibrium potential. In fact, guard cells seem to have rather low membrane potentials, although it is difficult to define the extracellular K^+ concentration representative of the conditions in intact leaves. A number of measurements in defined conditions on epidermal strips may be noted. In onion guard cells, with 1 mmol l^{-1} KCl outside, the potential in dark-adapted cells was -72 mV, and the potential in light was more negative, on average by 10 to 23 mV, though sometimes up to 50 mV (Zeiger et al., 1977; Moody and Zeiger, 1978). Saftner and Raschke (1981), working with guard cells of several species bathed in 30 mmol l^{-1} KCl, measured membrane potentials of about -40 mV in open guard cells and -30 mV in closed. The potential changed by about 50 mV per tenfold change in external concentration, with little specificity for the particular cation. Thus guard cell potentials seem to have only a small contribution from the electrogenic pump and to behave more like diffusion potentials. At first sight, this may seem surprising for a cell with such a massive capacity for salt accumulation, but a case can be made that it is a consequence of this very property.

Th evidence for gradient-coupled Cl^- influx in a range of plant cells is reviewed by Sanders (1984). In *Chara*, for which the stoichiometry has been established by measuring the electrical effects of initiating high Cl^- influx in chloride-starved cells, it appears to be $2H^+:1Cl^-$; hence the process leads to depolarization of the membrane potential (Sanders, 1980; Beilby and Walker, 1981). The observed kinetics of chloride influx in *Chara* and the detailed dependence on internal and external pH and chloride concentration can be fitted closely by a theoretical model in which two H^+ are transported with one Cl^-, with Cl^- first on and first off the carrier and with charge carried on the loaded carrier on the inward translocation (Sanders and Hansen, 1981). Although we lack detailed information on Cl^- fluxes in other plant cells, including guard cells, we can speculate about the existence of co-transport mechanisms of similar stoichiometry. It follows that if guard cells had a very high activity of a $2H^+:1Cl^-$ co-transport system, then we might predict the electrical behavior seen. The combined entry of K^+ and ($2H^+ + 1Cl^-$) would short out the electrogenic component, and although the proton pump would contribute greatly to the proton flux and current, its effect on the membrane potential would be relatively small. Clearly there is a need for a coordinated study of electrical effects and fluxes in experimental conditions that result in different rates of transport.

Zeiger (1983) suggests that a direct stimulation of the proton motive force is the most plausible mode of action of light in stomatal responses. This hypothesis must be tested experimentally, although there are indications that the control may be much more complex. The flux measurements in *Commelina*, to be discussed in a later section, show that dark-induced closing of stomata is achieved not by inhibition of ion influxes but by stimulations in ion effluxes in a complex transient. The effects of light on two-way ion fluxes in guard cells need to be clearly established by experimental measurements before we can assess possible mechanisms of photocontrol of aperture.

In fact, the mechanism of the light effect on the proton pump and on other transport activities remains obscure, even in the better-understood cell systems. It is thought unlikely that it results from a large reduction in the available supply of ATP. Recent measurements of ATP levels by NMR spectroscopy (Roberts, 1984) suggest that some earlier determinations should be reexamined. Nevertheless, the measurements of adenine nucleotides in subcellular fractions of wheat protoplasts (prepared within 0.1 sec after disruption of the protoplasts) meet such criticism; this work shows that illumination is not accompanied by a general increase in ATP or in the ratio of ATP to ADP outside the chloroplast, in the cytosol (Stitt, Lilley, and Heldt, 1982). In *Chara*, which has a very marked light activation of its proton pump, Penth and Weigl (1971) found no effect of light on the ATP level in the cell; and later work, in which transport or ATP supply was manipulated in other ways, shows a similar lack of correlation between transport activity and ATP levels. Two specific examples may be cited. A comparison of the detailed time courses of ATP changes and of changes in electrical properties after adding carbonylcyanide *m*-chlorophenylhydrozone (CCCP) to *Chara* cells showed a sudden switch-off of the pump at a threshold ATP concentration of about 0.5 to 1.5 mmol l^{-1} in the cytoplasm (Spanswick, 1980). The results suggested activation of the proton pump by ATP, rather than a substrate effect of ATP, as the energy source. Similarly the inhibition of proton pumping in *Chara* by CO_2 was not accompanied by a change in the ATP levels measured in the cell (Spanswick and Miller, 1977). Work with perfused, tonoplastless cells of *Chara*, where the internal concentration is under experimental control, emphasizes the lack of any explanation for the light effects on transport. Such work suggests that the effect is not mediated through changes in either ATP or internal pH (Kikuyama et al., 1979; Fujii, Shimmen, and Tazawa, 1979). We should therefore be cautious in interpreting light effects on proton transport in guard cells in terms of changes in available ATP.

Zeiger (1983) has considered the possible sources of energy for proton pumping in guard cells and the evidence that oxidative phosphorylation in the dark or photophosphorylation in guard cell chloroplasts in light can supply the ATP required. He argued that the wide opening induced by fusicoccin in the dark provides evidence that oxidative phosphorylation is capable of producing adequate ATP to support opening, though the transport is inhibited in the dark in normal conditions. Schwartz and Zeiger (1984) suggested that opening in light depends on ATP from photophosphorylation and that in the absence of fusicoccin oxidative phosphorylation does not substitute for photophosphorylation as the energy source. However, in my view, it is more likely that light activation of the transport mechanism is involved, rather than gross differences in cytosolic ATP levels in light and dark conditions.

Recent observations of fluorescence transients in guard cell chloroplasts have provided an elegant method of following changes in guard cell metabolism in the intact leaf (Melis and Zeiger, 1982; Ogawa et al., 1982). Of particular interest is the slower fluorescence transient thought to be associated with photophosphorylation; its existence in guard cells is taken as evidence for their capacity for photophosphorylation. Ogawa et al. (1982) showed that in epidermal peels the magnitude of this transient was increased by K^+ and reduced by ABA in the medium. Melis and Zeiger (1982) found that in guard cell chloroplasts, as in mesophyll chloroplasts, this transient was reduced by CO_2. The explanation given for the effect of CO_2 in mesophyll chloroplasts is a dissipation of the high energy state by CO_2 fixation (Krause, 1973), but in the absence of carbon fixation in guard cells (Outlaw et al., 1979; Vaughn and Outlaw, 1983), this explanation is not available. However, if, as in *Nitella*, the guard cell proton pump is inhibited by CO_2, and if ion transport is a major dissipator of energy in the guard cell, then an effect of CO_2 on the high energy state of the chloroplast might be explained. The general picture for energy relations in plant cells points to phosphorylation geared to demand, with the processes that consume ATP setting the rates of production. It is further recognized that, in general, changes in the rate of the ATP-consuming processes, including transport, are more likely to be achieved by specific regulation of these processes, rather than by gross changes in the pool of available energy. Further studies should reveal if this is also true of guard cells.

Aperture Dependence on Ion Contents

It is important to look at the dependence of aperture on ion content over the full range of apertures, and not only in open and closed stomata. The critical question is the extent to which changes in K-salt content can account for

TABLE 6.3. Aperture dependence of steady-state ion content in isolated guard cells

		Change of content with aperture (dQ/dA) (pmol mm^{-2} μm^{-1})	
Variable	Reference	^{86}Rb$^+$	^{82}Br$^-$
External concentration	a	6.1, 6.6, 9.7, 10.6, 13.4*, 14.8	13.6, 16.2*
Light/dark	b	5.4, 6.5	9.3, 13.9
ABA	c	12.5, 18.3	
FC 0–20 min	d	19.9, 45.2	
FC 0–135 min	d	12.5, 14.0	

SOURCES: a, MacRobbie and Lettau (1980a), MacRobbie (1981a): external concentration 10 to 90 mmol l^{-1} at pH 3.9; values quoted are slopes of regressions of tracer content on aperture. b, MacRobbie (1983, 1984): incubation in 30 mmol l^{-1} ^{86}RbCl or K^{82}Br, pH 3.9, in light or dark; values quoted are $\Delta Q^*/\Delta A$, from the difference in means of aperture and content in light (L) and dark (D). c, MacRobbie (1981b): change in aperture and tracer content after 3 to 4 h in 20 μmol l^{-1} ABA, added to steady-state, light-loaded tissue; 20 or 60 mmol l^{-1} RbCl, pH 3.9. d, Clint and MacRobbie (1984): change in aperture and content measured over 20 or 135 min after adding 30 μmol l^{-1} fusicoccin (FC) to steady-state, light-loaded tissue; 10 mmol l^{-1} RbCl pH 3.9.

NOTE: 120 guard cells per mm^2 of epidermis; single guard cell area about 25 × 10^{-6} cm^2. Values marked with an asterisk (*) will be used for calculation of $d\Pi_{KCl}/dA$ in Table 6.5.

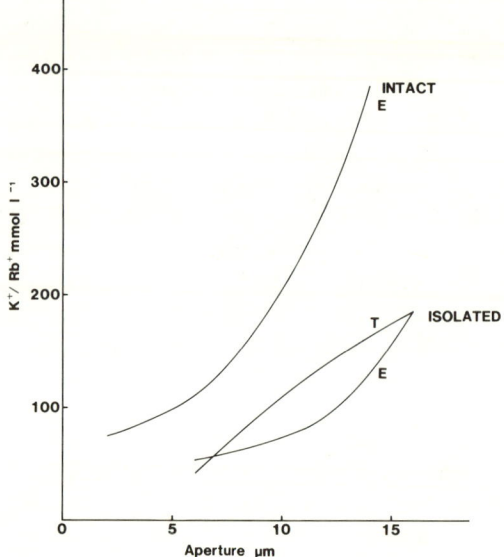

Fig. 6.1. Ion concentrations in guard cells of *Commelina communis*. *Intact*, guard cells in freshly removed epidermal strips with live subsidiary cells. *Isolated*, guard cells in epidermal strips in which only the guard cells are alive. E, from K$^+$ activity, measured with K-sensitive electrode. T, tracer measurements, from steady-state levels of ^{86}Rb$^+$ (from RbCl), and estimates of guard cell volumes. Smooth curves were drawn through the experimental points. (Data from MacRobbie and Lettau, 1980a, b.)

the osmotic changes required to change the aperture over every stage of the stomatal movement. Values allowing this comparison over a range of apertures are available for *Commelina*. In this species K$^+$ changes have been estimated either from measurements of K$^+$ activity with an intracellular K-sensitive microelectrode or from measurements of the steady-state tracer levels in isolated epidermal strips.

The tracer measurements have been made using $^{86}RbCl$ and $K^{82}Br$: $^{86}Rb^+$ was used as a tracer for Rb^+, as an analogue for K^+; and $^{82}Br^-$ acted as tracer for Br^-, as an analogue for Cl^-. Because of the slow cation exchange in the free space at pH 6.7, the cation experiments were done at pH 3.9, and to allow direct comparison this pH was also used for the anion experiments. Opening appears normal at this pH. Different apertures were produced by varying the external concentration (C_o) of K(Rb)Br(Cl), by incubation in light or dark, or by adding ABA or fusicoccin (FC) to guard cells already opened by incubation in light. (MacRobbie and Lettau, 1980a; MacRobbie, 1981a, b, 1983, 1984; Clint and MacRobbie, 1984). The values from tracer measurements for the increase in ion content as a function of aperture are shown in Table 6.3. Estimates of K^+ concentrations in guard cells using a K-sensitive microelectrode for both isolated and intact guard cells, together with one set of $^{86}Rb^+$ measurements in isolated guard cells, are shown in Figure 6.1 (MacRobbie and Lettau, 1980a, b).

Contribution of Potassium Salts to the Osmotic Change Required for Stomatal Opening

The osmotic effects of changes in ion content can be estimated by using values of Π_{KCl} for equivalent concentrations. Since Rb^+ and Br^- show the same relation to aperture, it appears that, in peels floating on K(Rb)Br(Cl), the uptake of $K^+(Rb^+)$ is balanced by $Br^-(Cl^-)$, and Π_{KCl} should therefore provide a good estimate. The measured $d\Pi_{KCl}/dA$, the increase with aperture of the osmotic contribution of K-salts, can then be compared with the osmotic requirement for opening, $d\Pi/dA$, estimated by aperture measurements as sucrose is added to the bathing solution.

From the results with *Commelina* shown in Figure 6.1 and Table 6.3, values of Π_{KCl} were estimated at different apertures; the results of the comparison of $d\Pi_{KCl}/dA$ with the required $d\Pi/dA$ are shown in Table 6.4 for intact guard cells and in Table 6.5 for isolated guard cells.

In intact guard cells it is clear that the osmotic contribution of the increase in K-salt is comparable with the required osmotic change only in the stage of wide opening. For openings below 10 μm, the discrepancy between $d\Pi_{KCl}/dA$ and $d\Pi/dA$ is large. Thus in the early stages of opening of intact guard cells, up to about 10 μm, the K-salt changes are much too small to account for the osmotic changes required to open the pore, and other changes in the guard cell must also contribute significantly to the opening. Above 10 μm the extent of K-salt accumulation does appear to determine the degree of opening achieved.

In isolated guard cells, the salt changes are much too small to account for the osmotic changes, over the whole range of aperture observed, irrespective of the

TABLE 6.4. Comparison of $d\Pi_{KCl}/dA$ with $d\Pi/dA$ in intact guard cells

| | | Change with aperture (mOsmol kg^{-1} μm^{-1}) | | |
| | | Aperture range | | |
Parameter	Reference	7–9 μm	11–13 μm	Above 15 μm
$d\Pi_m/dA$, intact	a	91 ± 6 (33)	137 ± 7 (20)	162 ± 17 (4)
$d\Pi/dA$	a	116	162	225
$d\Pi_{KCl}/dA$	b	34–58	110–225	
$d\Pi_{KCl}/dA$	b	38–50	73–85	

SOURCES: a, MacRobbie (1980): $d\Pi_m/dA$ is the measured osmotic slope; $d\Pi/dA$ allows also for the increase in P_{gs} with aperture; $d\Pi/dA$ is therefore the osmotic change required to open against the resistance to expansion from both guard cell and subsidiary cell. $d\Pi_m/dA$ given as mean ± S.E.M. (n); $d\Pi/dA$ given as value calculated from mean $d\Pi_m/dA$. b, MacRobbie and Lettau (1980b): data given as calculated values over the given aperture range.

TABLE 6.5. Comparison of $d\Pi_{KCl}/dA$ with $d\Pi/dA$ in isolated guard cells

Variable	Reference	Aperture range (μm)	$d\Pi_{KCl}/dA$ (mOsmol kg^{-1} μm^{-1})	$d\Pi/dA$
^{86}Rb$^+$				
External concentration	a	6–14	27.5	
Light/dark	b	6–14	11	
Light/dark	b	10–14	31	
ABA	c	7–9	28	
ABA	c	8.5–10.5	42.5	
FC 0–20 min	d	6–7	62, 124	
FC 0–135 min	d	6–12	32, 35	
^{82}Br$^-$				
External concentration	a	6–15	34	
Light/dark	b	6–17	16.5	
Light/dark	b	11–14	27	
K$^+$ electrode				
	e	7–9	10	
	e	11–13	20–40	
Osmotic requirement				
	f	7–9		74 ± 6
	f	11–13		121 ± 8
	f	above 15		188 ± 21

SOURCES: a–d, References and details as in Table 6.3. e, MacRobbie and Lettau (1981a). f, MacRobbie (1980).

opening stimulus. However, since isolated guard cells will, under the same conditions, open much wider than intact guard cells, opening also against the back pressure of turgid subsidiary cells, their status over the whole range of aperture observed is likely to correspond to that of intact guard cells below 10 μm. It is therefore likely that in isolated guard cells we are looking at the processes by which intact guard cells would achieve the earlier stages of their opening, up to about 10 μm. For this reason it is not surprising that the discrepancy between $d\Pi_{KCl}/dA$ and $d\Pi/dA$ should appear over the whole range of aperture in isolated guard cells.

However, in one case, that is, in the first 20 minutes after adding fusicoccin, the K-salt changes are adequate to explain the osmotic changes. Thus the initial effects of fusicoccin are consistent with stimulation of KCl accumulation (Marrè, 1979), but this is followed by further opening, in which the continuing KCl accumulation is not adequate to account for the osmotic changes. The effects of ABA are less well defined, as the time course is not well established; the decrease in ^{86}Rb$^+$ content after adding ABA gives values of the change of content with aperture (dQ/dA) of 12.5 and 18.3 pmol mm^{-2} μm^{-1}, over 3 to 4 hours, compared with a value of only 6.1 pmol mm^{-2} μm^{-1} in the same tissue when external concentration was the variable setting the aperture. Thus the primary effect of ABA also may be on the component of the osmotic change that can be attributed to salt.

Thus in *Commelina* the accumulation of K-salts makes a major contribution to the generation of turgor in the guard cells, and aperture is therefore sensitive to factors affecting the level of such accumulation. On the other hand, in opening up to 10 μm in intact guard cells, osmotic changes other than K-salt accumulation play a major role.

Although it has been widely accepted that the accumulation of K-salts can account, quantitatively, for the osmotic changes in guard cells, there are considerable uncertainties in the values on which this conclusion is based. This view is based on the estimation of the osmotic requirements in *Vicia* by plasmolytic techniques. Humble and Raschke (1971) measured the K$^+$ change as 0.2 pmol cell^{-1} μm^{-1}, with a guard cell volume of only 2.4 pl open and 1.3 pl closed; the osmotic requirement was estimated as 1.6 atm μm^{-1}, based on 30 minutes' exposure to the plasmolyzing solution. The K-salt changes then appear adequate, but as Raschke (1979) points out, the osmotic requirement may be underestimated. He showed that considerable solute leakage occurs under these conditions (as did Imamura, 1943) and quotes a value of 4.8 atm μm^{-1} as a revised osmotic requirement. Potassium changes have also been estimated by other authors, using a variety of methods; values have been collected by Outlaw (1983) and are generally in the range 0.1 to 0.25 pmol cell^{-1} μm^{-1}. However, the quoted guard cell volumes are larger, at about 5 pl in open stomata, reducing the estimates of $d\Pi_{KCl}/dA$. Thus in *Vicia*, although it is established that K-salt accumulation makes a major contribution to the increased osmotic value of guard cells as they open, the data do not establish that amounts of K-salt are adequate.

Additional evidence for the involvement of other solutes comes from the balance sheet for solute changes in *Vicia* guard cells as they open, obtained in the microanalytical studies of Outlaw and co-workers. Their measurements suggest that opening to 10 μm involves increases of 1.6 pmol K$^+$, 0.42 to 0.72 pmol hexose, and 0.14 pmol sucrose (Outlaw and Lowry, 1977; Outlaw and Manchester, 1979). Furthermore, Outlaw and Kennedy (1978) suggested that little malate was synthesized in the early stages of opening in *Vicia*. Since in the intact leaf of *Vicia* K$^+$ is almost exclusively balanced by malate rather than by chloride, this suggests that the early stages of opening are achieved by processes other than salt accumulation.

Thus it would appear that in *Vicia*, as in *Commelina*, K-salt accumulation plays a major role in turgor generation in guard cells, but some other process is also involved. There remain unsolved problems, therefore, even in identifying the end results of the processes responsible for stomatal movements.

Flux Measurements in Guard Cells of *Vicia*

Fischer made the first measurements of ion fluxes in "isolated" guard cells, using epidermal strips of *Vicia faba* in which other cells had been disrupted (Fischer, 1972; Pallaghy and Fischer, 1974). The 1972 paper has important implications. The isolated guard cells had initial apertures, in the absence of epidermal cell turgor, of about 7 μm. After incubation for 4 to 23 hours on 10 mmol l^{-1} KCl plus 0.1 mmol l^{-1} $CaCl_2$, the steady-state conditions of K^+ content and aperture were 7 nmol cm^{-2} and 7 to 9 μm in normal air in the dark, and 18 nmol cm^{-2} and 12 to 13 μm in CO_2-free air in light. The ^{42}K influx at the start of the opening in CO_2-free air in light was much higher than the fluxes in the near-steady state achieved after 3 to 5 hours. Thus the initial influx was 11.0 and 16.5 pmol cm^{-2} s^{-1} in two separate experiments, whereas after 200 to 300 minutes on the solution in light, when the aperture would have increased to about 13 μm, influx and efflux had both fallen to about 2 to 4 pmol cm^{-2} s^{-1}. (Fluxes are expressed on the basis of guard cell area.) In contrast, after incubation for 200 to 300 minutes on non-radioactive solutions in the non-opening conditions in the dark, the aperture had changed little; and on transfer of this tissue to CO_2-free air in light the influx was also high, at 8.5 and 11.0 pmol cm^{-2} s^{-1} in the two experiments. After opening, the internal K^+ would have reached about 280 mmol l^{-1}, well below the level for the same aperture in the intact leaf. It is clear from these experiments that the submaximal steady-state content in isolated guard cells is not the result of excessive efflux and that the fluxes must therefore be regulated in some way. Fischer speculated on the possiblity of inhibition of influx by increasing turgor but showed that the steady-state influx was not significantly increased by the addition of 0.2 mol l^{-1} sucrose to the incubation medium, arguing against this possibility. As neither turgor nor internal concentration have reached their potential maxima, the results may suggest, instead, a regulation based on cell volume.

Flux Measurements in Guard Cells of *Commelina*

Fluxes have now been measured in the isolated guard cells of *Commelina* in a range of conditions, both in the steady state and in the transitions between different states. In most experiments $^{86}RbCl$ and $K^{82}Br$ were used, but a few measurements using K^{36}Cl have also been taken (MacRobbie, 1981a, b, 1983, 1984; Clint and MacRobbie, 1984).

Provided that the tissue is in a steady state, analysis of the kinetics of efflux may be used to measure cytoplasmic and vacuolar contents, and fluxes at plasmalemma and tonoplast separately. For the kinetics of the two phases to be distinguishable, the tonoplast fluxes should not be too high, relative to the plasmalemma fluxes. This condition is not met for ^{86}Rb$^+$ fluxes, but in most experiments with ^{82}Br$^-$ the plasmalemma fluxes were high enough relative to those at the tonoplast for the two phases to be distinguished in the efflux, allowing cytoplasmic and vacuolar contributions to be separately assessed. Isolated epidermal strips were equilibrated overnight in light or dark and at various external concentrations. Measurements of influx and efflux were made over the subsequent 4 to 8 hours, either in constant conditions for the analysis of efflux kinetics, or before and after a change of conditions, to examine the flux transients. The results are therefore relevant to the questions of what determines the ability of guard cells to stay open, or to open wider, and of what flux changes are associated with a closing transient. The quite different question of the control of ion fluxes in the initial stages of opening remains for future work, but the closing process is more likely to hold the key to the specific function of guard cells.

Steady-state fluxes. At pH 3.9 the ^{86}Rb$^+$ influx rose with external concentration and could be fitted by a Michaelis-Menten relation with a maximum velocity (V_{max}) of about 34 pmol mm^{-2} h^{-1}, on the basis of area of epidermal strip, and a Michaelis constant (K_m) of about 32 mmol l^{-1}. The values are approximate, since they pool the results of different batches of tissue. Expressed on a basis of guard cell area, the value of V_{max} converts to about 3.5 pmol cm^{-2} s^{-1}. In the efflux experiments the ^{86}Rb$^+$ exchanged as a single compartment after the free-space exchange, with an apparent rate constant, $\bar{k}$ (= rate of loss/tissue content), of 0.25 h^{-1}. This corresponds to a half-time of 2.8 hours and gave efflux values that were comparable to the measured influxes. The rate constant $\bar{k}$ appeared to be independent of external concentration and was equal in steady-state tissue in light and dark; it is therefore independent of aperture (MacRobbie, 1981a, 1983). These results therefore suggest that the steady-state apertures are determined by the balance of the influx and an efflux that is proportional to content, but we have no measurements allowing comparison of these fluxes with those existing at the start of the opening process.

Plasmalemma fluxes of ^{82}Br$^-$ were similar to those measured with ^{86}Rb$^+$, though somewhat higher, and increased steeply with external concentration. Values in the range 7 to 112 pmol mm^{-2} h^{-1} were measured in different conditions, converting to fluxes of about 1 to 12 pmol cm^{-2} s^{-1} on the basis of guard cell area.

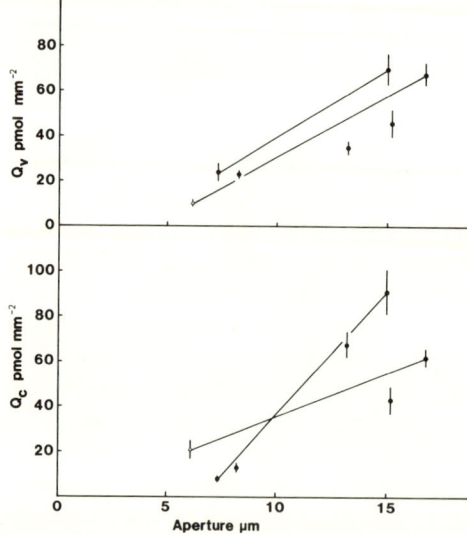

Fig. 6.2. Br^- contents of cytoplasm and vacuole of guard cells of *Commelina* estimated from steady-state efflux analysis, based on the area of epidermal strips. Aperture varied by external concentration or light (●) or dark (△). Lines join values for replicate tissues at two different apertures, at 20 and 60 mmol l^{-1} KBr in light, or at 30 mmol l^{-1} KBr in light or dark. All solutions also contained 10 mmol l^{-1} MES, pH 3.9. Q_c, cytoplasmic content; Q_v, vacuolar content. Light intensity, 135–145 μmol m^{-2} s^{-1}.

Efflux analysis: intracellular distribution of $^{82}Br^-$. Because the plasmalemma fluxes of $^{82}Br^-$ were in general higher than the tonoplast fluxes, efflux analysis was possible, yielding more information on the intracellular distribution of ions in the cell. The efflux curve could be split into two exponential terms: the faster-exchanging component had a half-time of about 38 minutes, independent of aperture; whereas the half-time of the slower component depended on aperture, with values of 2.6 to 5.3 hours at 15 to 17 μm, but only 16 to 35 hours at 6 to 8 μm aperture. Thus the exchange of vacuole and cytoplasm were kinetically distinguishable, and values for the cytoplasmic and vacuolar contents and plasmalemma and tonoplast fluxes could be calculated from the efflux curves. It is important to note the assumptions implicit in this analysis. The calculation assumes that the cytoplasm and vacuole are in series and that during the efflux (but not during all the loading period) the contents of cytoplasm and vacuole, and the fluxes, are constant. It also assumes that at the start of the efflux the internal specific activity is equal to that outside, an assumption that is justified because the loading time has been long enough to reach this state and because there was little or no Br^- in the tissue at the start of the loading.

The most important consequence of such calculations is that it is possible to assess separately the changes with aperture of the concentrations in the cytoplasm and in the vacuole. Figure 6.2 shows the calculated Br^- contents of cytoplasm and vacuole (expressed per unit area of epidermis), from different experiments, including two experiments in which the same tissue was examined at two different apertures, at two different external concentrations, or in light and dark. Clearly both cytoplasmic content (Q_c) and vacuolar content (Q_v) increase with aperture, but, surprisingly, the slope of the cytoplasmic relation is comparable with or greater than that of the vacuolar component, in spite of the much larger vacuolar volume in open guard cells. There is uncertainty in trying to convert these figures to concentration changes, since the intracellular distribution of volume is even more uncertain than the cell volume itself. Two sets of calculations were done, assuming that the cytoplasm occupied either 50 or 70 percent of the volume of closed guard cells, and that its volume remained constant during opening, the volume increase going to the vacuole. (Such constancy of cytoplasmic volume during swelling of *Albizzia* motor cells was shown by Campbell and Garber [1980], and it is certain that the cytoplasmic volume fraction is much higher in closed than in open guard cells.) The value of 70 percent in closed guard cells is likely to be an overestimate and therefore gives an upper limit for the changes in vacuolar concentrations. Figure 6.3 shows estimates of cytoplasmic and vacuolar concentrations (C_c and C_v) as a function of aperture in the combined experiments, calculated on the basis of 70 percent cytoplasmic volume in closed guard cells. Table 6.6 shows values for $\Delta Q_c/\Delta A$ and $\Delta Q_v/\Delta A$, estimates of $\Delta C_c/\Delta A$ and $\Delta C_v/\Delta A$ in the two sets of replicate tissue, and the slopes of the linear regressions of C_c and C_v against aperture for the combined results from Figure 6.3; values are given for both volume fractions. Depending on the conditions and on the assumptions made, estimates of $\Delta C_c/\Delta A$ are in the range 11 to 45 mmol l^{-1} μm^{-1}, whereas those of $\Delta C_v/\Delta A$ are only in the range 6.3 to 8.6 mmol l^{-1} μm^{-1}. The osmotic requirement for the change in aperture is about 112 mOsmol kg^{-1} μm^{-1}, averaged over 7 to 15 μm. Thus, allowing also for the accompanying K^+, the higher estimates of Br^- increase in the cytoplasm can account for a

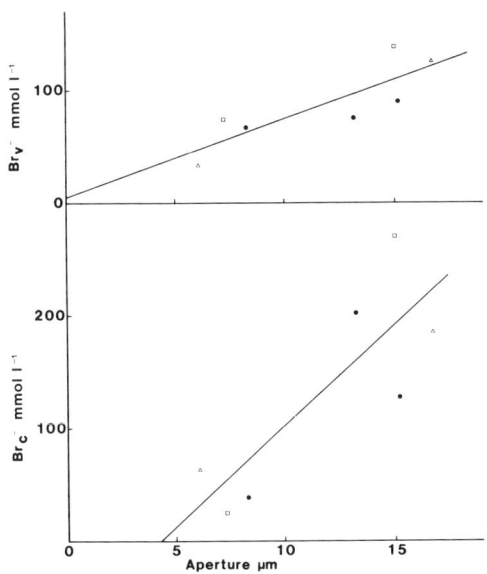

Fig. 6.3. Cytoplasmic and vacuolar concentrations in guard cells of *Commelina* calculated from amounts Q_c and Q_v from efflux analysis, assuming constant cytoplasmic volume, equal to 70 percent of the volume of closed guard cells, with the increase in volume with opening attributed to vacuolar volume. Values collected from different experiments. △ Replicate tissue in light or dark, in 30 mmol l^{-1} KBr. □ Replicate tissue in light, in 20 or 60 mmol l^{-1} KBr.

TABLE 6.6. Aperture dependence of cytoplasmic and vacuolar Br⁻ in isolated guard cells of *Commelina*

Variable	$\frac{\Delta Q_c}{\Delta A}$	$\frac{\Delta Q_v}{\Delta A}$	$\frac{\Delta C_c}{\Delta A}$ $V = 0.5$	$\frac{\Delta C_v}{\Delta A}$	$\frac{\Delta C_c}{\Delta A}$ $V = 0.7$	$\frac{\Delta C_v}{\Delta A}$
Light/dark	3.9	5.5	16	7.5	11	8.6
External concentration	10.3	5.9	45	8	32	8.4
Overall			25	6.3	18	6.9

NOTE: A, aperture, μm; C_c, C_v, estimated cytoplasmic (c) and vacuolar (v) concentrations (C), mmol l⁻¹; Q_c, Q_v, cytoplasmic (c) and vacuolar (v) contents (Q), pmol mm⁻²; v, value assumed for fraction of cytoplasmic volume in closed guard cells, cytoplasmic volume assumed constant with aperture (with increase in cell volume attributed to the vacuole); *Overall*, slopes of regressions of content or concentration on aperture for combined experiments. Standard errors of quoted values: 12–13% in light/dark and for the cytoplasmic values in external concentration; 25–29% for the vacuolar values in external concentration; 22–25% in overall regressions.

large part of the increased osmotic pressure. With slightly different estimates of the cytoplasmic volume, for example, allowing for a smaller cytoplasm or a fraction of non-osmotic volume (Weyers and Fitzsimons, 1982), the higher estimates for Br_c^- increase could satisfy the osmotic requirement. However, in the light/dark experiment the cytoplasmic salt increase seems to be too small to match the required osmotic change. In contrast, when we consider the vacuolar changes, it is clear that none of the estimates of increasing salt with aperture can be stretched to meet the osmotic requirement.

There are uncertainties in these figures, but the essential conclusions are nevertheless clear and will not be affected by any reasonable manipulation of the volume estimates or by the uncertainties in fitting of exponentials in the efflux analysis. The most important conclusion is that the increase in vacuolar salt concentration is relatively small, less than that in the cytoplasm and much less than that required osmotically. The discrepancy between measured salt accumulation and required solute accumulation is primarily a vacuolar shortfall. As the aperture increases at higher external concentrations, the cytoplasmic salt increase can account for a major fraction, and perhaps all, of the osmotic increase. Thus it should be emphasized that this is not a situation in which salt accounts for the vacuolar increase and some other, compatible solute accounts for that in the cytoplasm, as has been suggested by Outlaw (1983); the determination of the intracellular distribution by the efflux analysis shows that the reverse is true. Further work is necessary to establish the nature of the other vacuolar solute or solutes involved.

Plasmalemma and tonoplast fluxes: $^{82}Br^-$. The influx at the plasmalemma can be measured by short-term loading, and the efflux analysis allows calculation of steady-state fluxes at the plasmalemma and tonoplast separately; in steady-state tissue influx and efflux are equal. Figure 6.4 shows the fluxes determined in a number of experiments plotted as a function of aperture. The plot is drawn in this form (implying aperture is the independent variable) for comparison with the previous graphs, but this may be misleading. It is more likely that aperture is set by the level of plasmalemma influx (ϕ_p), which determines the level of cytoplasmic Br^-, which in turn determines the tonoplast flux (ϕ_t) and the level of vacuolar Br^-; Br_c^- and Br_v^- will then in part determine the aperture.

It is worth considering the ratios of flux to content or concentration in the same tissue in two different conditions, as shown in Table 6.7. Aperture was varied by changing the external concentration of KBr (20 or 60 mmol l^{-1}) in one experiment and by incubation in light or dark in the other. With concentration as the variable, both ϕ_p/Q_c and ϕ_t/Q_c are similar in the two states, and the aperture appears to be set by the very considerable increase in plasmalemma flux at high C_o. It should, however, be noticed that ϕ_p/C_o has increased considerably, suggesting that the internal state of the guard cell and its capacity for ion uptake are affected by external concentration. Similarly in the comparison of steady-state tissue in light and dark, the ratio ϕ_p/C_o is much higher in the light than in the dark. The values also show that the tonoplast fluxes are much lower in the dark, implying that transfer from cytoplasm to vacuole is significantly inhibited in the dark.

The important point to be made is that factors other than internal and external concentrations may control the fluxes at both plasmalemma and tonoplast. These results suggest that the transport capacities, and therefore the aperture, are set not only by the steady-state ion levels but also by the internal state of the tissue, in ways that remain to be identified. In this connection it is surprising

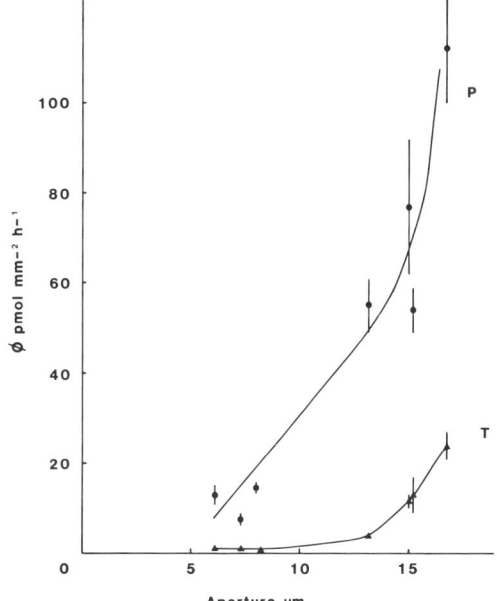

Fig. 6.4. Steady-state fluxes of Br^- (ϕ) at the plasmalemma (P) and tonoplast (T) of guard cells of *Commelina*, estimated from efflux analysis. Aperture varied by changing external concentration (20 to 60 mmol l^{-1} KBr), or by incubation in light or dark. All solutions also contained 10 mmol l^{-1} MES, pH 3.9.

TABLE 6.7. Ratios of $^{82}Br^-$ fluxes to contents at different apertures

Parameter	Variable external concentration		Variable light/dark, 30 mmol l^{-1}	
	20 mmol l^{-1}	60 mmol l^{-1}	Dark	Light
A (μm)	7.3 ± 0.3 (73)	15.0 ± 0.4 (58)	6.1 ± 0.4 (49)	16.7 ± 0.4 (46)
Q_c	7.9 ± 0.9	91 ± 10	21 ± 4	62 ± 4
Q_v	24 ± 4	70 ± 7	10.2 ± 0.7	68 ± 5
C_c	24	271	63	185
C_v	75	139	34	125
ϕ_p	7.4 ± 1.0	77 ± 15	13 ± 2	112 ± 12
ϕ_t	1.2 ± 0.3	11.5 ± 1.4	1 ± 0.2	24 ± 3
ϕ_p/Q_c	0.95 ± 0.11	0.82 ± 0.08	0.62 ± 0.05	1.8 ± 0.2
ϕ_t/Q_c	0.15 ± 0.03	0.13 ± 0.01	0.05 ± 0.01	0.39 ± 0.05
ϕ_t/Q_v	0.05 ± 0.01	0.165 ± 0.006	0.10 ± 0.02	0.35 ± 0.05

SOURCE: MacRobbie (1981a, 1984).
NOTE: A, aperture, mean ± S.E.M. (number of stomata); C_c, C_v, cytoplasmic (c) and vacuolar (v) concentrations (C), mmol l^{-1}; Q_c, Q_v, cytoplasmic (c) and vacuolar (v) contents (Q), pmol mm^{-2}; ϕ_p, ϕ_t, plasmalemma (p) and tonoplast (t) fluxes (ϕ), pmol mm^{-2} h^{-1}; ϕ/Q in units of h^{-1}. Values calculated from efflux analysis of steady-state tissue. Each figure is the mean ± S.E.M. of four strips.

that the effect of external concentration should be so dramatic; it suggests that the cytoplasmic ion concentration plays a very critical role in determining the state of the transport systems. This is reminiscent of Imamura's (1943) demonstration of threshold concentrations of external K^+ that determined the capacity of guard cells to retain high levels of solute, presumably by setting cytoplasmic K^+ levels. Above about 80 mmol l^{-1} external KCl, open guard cells of *Zebrina* retained solutes and maintained osmotic pressures of 7 to 9 MPa; whereas if external K^+ was reduced, there was extensive solute leakage, and the osmotic pressure fell to about 0.8 to 1 MPa within 15 to 30 minutes. Similar effects were also seen with *Vicia* and *Commelina*. The flux experiments suggest that control of the plasmalemma influx may be a key factor in these changes, indicating the need for detailed studies of the effect of internal salt concentration on the transport capacity at the plasmalemma.

Flux Transients

Perhaps the most important thing we need to understand about guard cells is the nature of the flux transients in response to opening or closing signals. There are two quite distinct questions involved: first, the mechanism by which ion accumulation is switched on in response to an opening signal, such as light or reduced CO_2; and second, the mechanism by which the ability to maintain high solute concentrations is lost in response to a closing signal, such as light-off, high CO_2, or exposure to ABA. It is therefore necessary to record flux transients in such conditions by measuring both steady-state fluxes before the signal and the immediate response of both influx and efflux to the change.

While it is likely that opening is initiated by switching on the proton pump, with consequent stimulation of secondary transports of K^+ and Cl^-, it is unclear how closing is initiated. Inhibition of ion uptake has been assumed as the basic mechanism (Raschke, 1977, 1979; Zeiger, 1983), but consideration of the early flux measurements in *Vicia* (Fischer, 1972) might suggest that this would be inadequate, and Hsiao (1976) speculated on the possible involvement of an efflux pump in stomatal closure. As was discussed in an earlier section ("Flux Measurements in Guard Cells of *Vicia*"), Fischer found that in steady-state open guard cells influx and efflux of K^+ at the plasmalemma were both very low, only 2 to 4 pmol cm^{-2} s^{-1}, compared with 17 pmol cm^{-2} s^{-1} in the early stages of opening. Thus, a simple inhibition of influx would produce only very slow closing, and a marked stimulation of efflux would seem to be necessary for a reasonably fast response. These fluxes should be compared with the net efflux calculated from the change in content in the intact leaf; the net efflux required for closing over 20 minutes is about 100 pmol cm^{-2} s^{-1}, based on guard cell area. Even allowing for possible differences in fluxes between guard cells in the intact leaf and in epidermal strips, it is hard to avoid the conclusion that more than switching off the uptake mechanism will be required for a reasonably fast closing response.

Results on flux transients in *Commelina* in response to turning the light off or adding ABA show that the closing response is indeed achieved by stimulation of efflux, not by inhibition of influx.

Effects of light/dark on fluxes in Commelina. The effect of transfer from light to dark on the two-way fluxes of both anions and cations have been studied in open isolated guard cells (MacRobbie, 1983, 1984). Such guard cells respond to transfer to the dark by reductions in content and aperture, though, as might be expected in the absence of subsidiary-cell back pressure, more sluggishly and to a lesser extent than do guard cells in the intact leaf.

The values in Table 6.8 show no significant effect of transfer of open guard

TABLE 6.8. Effect of transfer to dark on ion influx in *Commelina communis*

Initial aperture (μm)	Influx (pmol mm^{-2} h^{-1})		
	Light L*	Dark D*	Dark DD*
$^{86}Rb^+$			
15.1 ± 0.3 (14, 149)	52.4 ± 3.2 (7)	58.3 ± 2.6 (7)	
15.3 ± 0.2 (12, 122)	44.6 ± 1.1 (6)	45.5 ± 2.3 (6)	
11.7 ± 0.3 (8, 128)	14.7 ± 1.3 (4)	12.3 ± 1.8 (4)	
9.3 ± 0.1 (34, 545)	23.6 ± 1.5 (12)	21.1 ± 2.3 (13)	26.1 ± 2.4 (9)
$^{82}Br^-$			
14.8 ± 0.4 (3, 29)	48.1 ± 2.0 (10)	52.6 ± 1.8 (11)	
14.3 ± 1.0 (1, 12)	43.4 ± 2.8 (8)	41.5 ± 2.7 (8)	

SOURCE: MacRobbie (1983, 1984).

NOTE: Influxes were measured over 27 to 33 min in light (L*) or in dark after incubation overnight in non-radioactive solution in light (D*), or over 30 min in dark after 1-h pretreatment in non-radioactive solution in the dark (DD*). Apertures were measured in light before the influx period. Light intensity was 135–145 μmol m^{-2} s^{-1} (400–700 nm). The solutions were 30 mmol l^{-1} RbCl or 25 mmol l^{-1} KBr; both contained 10 mmol l^{-1} MES, pH 3.9. Apertures presented as mean ± S.E.M. (number of strips, number of pores). Influx data presented as mean ± S.E.M. (number of strips).

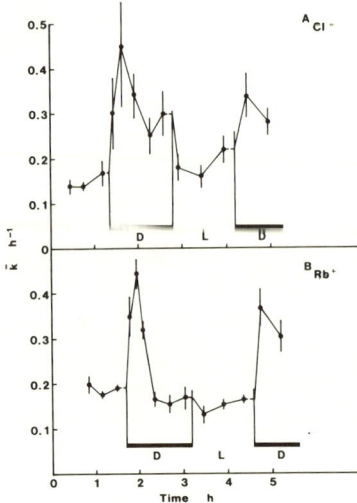

Fig. 6.5. Effect of transfer to the dark and return to light on the efflux of $^{36}Cl^-$ and $^{86}Rb^+$ from guard cells of *Commelina* after labeling to steady state in light. A, $^{36}Cl^-$: 20 mmol l^{-1} KCl, 4 mmol l^{-1} Na$_2$SO$_4$, 10 mmol l^{-1} MES, ph 3.9; B, $^{86}Rb^+$: 30 mmol l^{-1} RbCl, 10 mmol l^{-1} MES, pH 3.9; $\bar{k}$, the apparent rate constant for exchange, = (rate of tracer loss)/(tissue content).

cells to the dark on the influx of either $^{86}Rb^+$ or $^{82}Br^-$. Thus reduction in influx is not responsible for the initiation of the closing response after a light-off signal.

In contrast, there is a very marked effect on efflux of $^{86}Rb^+$, $^{82}Br^-$, and $^{36}Cl^-$, with a very sharp transient stimulation of efflux on transfer to the dark. This is best seen in the plot of the apparent rate constant for exchange, $\bar{k}$ (= rate/content), against time. Typical results are shown in Figure 6.5A and B for $^{36}Cl^-$ and $^{86}Rb^+$, respectively; results with $^{82}Br^-$ were very similar.

The interpretation is simpler for $^{86}Rb^+$, which exchanges as a single compartment. Efflux of $^{86}Rb^+$ was stimulated by a light-off signal by a factor of 2.27 ± 0.09 (n = 30), but the stimulation was transient and $\bar{k}$ returned to close to the previous level in light within about an hour. There was no significant difference between $\bar{k}$ in steady-state light and steady-state dark. The effect of reillumination after darkness depended on the length of the dark period. As shown in Figure 6.5B, after 89 minutes in the dark, there was little change in $\bar{k}$ on return to light; but in other experiments (not shown), after shorter times in the dark (28 or 51 minutes, for example) $\bar{k}$ did fall again when the light was turned on: it looked as if $\bar{k}$ decreased in cells returned to light only during the period of stimulated efflux and not after about 1 to 1.5 hours in darkness. The ability to generate a stimulated efflux in response to a light-off signal could be restored by a second period (about 30 min) in light; this regeneration took place during efflux and therefore involved internal changes in the tissue, and did not require the further uptake of labeled ions from the solution. The amount of extra ion loss during the whole transient is difficult to estimate, since the specific activity becomes uncertain after the transfer, but it was about 11 ± 1 ($n = 12$) pmol mm^{-2}, with a half-time of about 12 to 20 minutes. This is a relatively small fraction of the total ion content under these conditions.

The effects clearly represent a specific transient state of high efflux from the open guard cells, induced by transfer to the dark, rather than a simple difference between efflux in light and dark. The response to reillumination depends on the previous history of the tissue and on whether the dark transient is complete.

The effects of light-off on the efflux of ^{82}Br$^-$ and ^{36}Cl$^-$ were similar to those seen with ^{86}Rb$^+$. Figure 6.5A shows an experiment in which steady-state tissue was loaded with ^{36}Cl$^-$ in light and was transferred to dark during the efflux; the effects with ^{82}Br$^-$-loaded tissue were similar. With both anions $\bar{k}$ after the peak remained higher during the period in the dark and decreased on return to the light, even after long periods in the dark. This finding differs from experiments with Rb$^+$, where $\bar{k}$ fell to the previous light level after about an hour in the dark and did not then change on return to light. In the anion experiments $\bar{k}$ increased by a factor of 2.1 ± 0.2 ($n = 11$) in response to the first light-off; on return to light after the transient, $\bar{k}$ fell by 1.58 ± 0.07 ($n = 8$). A second transfer to the dark gave a second transient, smaller than the first; however, because of the changing chemical content in the cells direct comparison of the ratios of $\bar{k}$ before and after the transfers may be misleading.

The interpretation of the ^{82}Br$^-$ transients is more complex than those of ^{86}Rb$^+$, since two intracellular phases are kinetically distinguishable and $\bar{k}$ is therefore falling with time in constant conditions. The factor of increase in $\bar{k}$ after the first transfer to the dark was independent of the stage of efflux at which the transfer was made (from zero to 291 minutes); this implies that the effect reflects a stimulation of the efflux at the plasmalemma. The subsequent time course of $\bar{k}$ will, however, be affected by changes in specific activity in the cytoplasm and may not reflect accurately the time course of the changes in total ion efflux. As with Rb$^+$, the total amount of extra Br$^-$ loss during the transient is difficult to estimate accurately, but it was about 10 to 25 pmol mm^{-2}, similar to the amount of excess Rb$^+$ loss, and also a small fraction of the total content.

Borle, Uchikawa, and Anderson (1982) pointed out that a peak in tracer efflux in such circumstances need not reflect a similar peak in total ion efflux but may reflect a transient in the specific activity in the phase delivering tracer to the medium, with a constant stimulated efflux. It was therefore necessary to examine the detailed time course of $\bar{k}$ before concluding that the transient represented a true transient in efflux. Simulation of the time course showed that it could not be produced by a new constant efflux, but that there must be a genuine transient in the stimulation of plasmalemma efflux after the transfer to the dark.

Ion contents and fluxes were no longer constant after the transfer of cells from light to dark, invalidating the use of the fit to two exponentials in the efflux analysis to calculate fluxes and contents, and allowing only qualitative

TABLE 6.9. Efflux from light-loaded isolated guard cells of *Commelina* in light and dark

Parameter	Light	Dark
k_1	2.2 ± 0.1	1.73 ± 0.07
k_2	0.27 ± 0.01	0.096 ± 0.019
A_1	39 ± 5	67 ± 13
A_2	91 ± 7	42 ± 4
Q_T	130 ± 7	109 ± 16
Q_c	62 ± 4	
Q_v	68 ± 5	

SOURCE: MacRobbie (1984).

NOTE: Contents A_1, A_2, Q_T, Q_c, and Q_v expressed in pmol mm^{-2}; rate constants k_1 and k_2 expressed in h^{-1}; Q_T, the total tracer in the tissue, is expressed in pmol mm^{-2} and fitted to the sum of two exponentials ($A_1 e^{-k_1 t} + A_2 e^{-k_2 t}$). Tissue was loaded in 30 mmol l^{-1} K^{82}Br for 15 to 16 h in light, then washed out in light (4 strips) or dark (3 strips). The light washout is steady-state and can be used to estimate cytoplasmic and vacuolar contents, Q_c and Q_v. Light intensity: 135–145 μmol m^{-2} s^{-1}. All data presented as mean ± S.E.M.

indications of the nature of the flux changes. Table 6.9 compares the efflux kinetics of tissue loaded in the light but washed in light or dark. It is clear that in the dark wash, much more of the total Br$^-$ content appears in the faster-exchanging component and that the rate constants, particularly the slow-rate constant, k_2, are both reduced. Thus the flux from cytoplasm to vacuole is much reduced in the dark, and indeed it appears that in the early stages of the washout in the dark all of the cytoplasmic tracer seems to be lost to the solution, with little or no movement to the vacuole.

Hence if we transfer steady-state open guard cells from light to dark, we see no effect on the influx, but a marked transient stimulation of the plasmalemma efflux and an inhibition of the flux from cytoplasm to vacuole (MacRobbie, 1984).

If, however, we compare these effects with the comparison of steady-state Br$^-$ fluxes in light and dark tissue, a very striking distinction emerges. The steady-state fluxes at both plasmalemma and tonoplast in cells kept in darkness are much lower than those in cells kept in light (Fig. 6.4). Indeed the data in Table 6.7 show that not only the fluxes ϕ_p and ϕ_t, but also all the apparent rate constants, ϕ_p/C_o, ϕ_p/Q_c, ϕ_t/Q_c, and ϕ_t/Q_v, were reduced in the dark-treated tissue. It would appear therefore that the low fluxes in the dark steady state are not a direct effect of darkness, but rather a consequence of changes in the internal state of the tissue and its capacity for transport. The influx in the dark is high in open guard cells but low in cells with low apertures that have been kept in the dark; this suggests an effect not of light but of internal ion concentration on the transport ability, as was also argued (see "*Plasmalemma and tonoplast fluxes*," above) for the reduction in ϕ_p/C_o at low external KBr concentration. These observations are consistent with the hypothesis that the state of guard cell transport processes is very sensitive to the level of cytoplasmic salt.

In relation to light/dark sensitivity of guard cells it is interesting to consider events in Na^+-loaded cells. Jarvis and Mansfield (1980) found that although NaCl and KCl were equally effective in promoting the opening of *Commelina* guard cells in light and CO_2-free air, the NaCl-loaded guard cells were insensitive to the usual closing signals, dark, high CO_2, or ABA. The efflux transients in Na^+-loaded tissue on transfer to the dark were therefore measured and compared with those in replicate Rb^+-loaded tissue (MacRobbie, 1983). The very marked transient stimulation of $^{86}Rb^+$ efflux was not paralleled in the $^{22}Na^+$ efflux. In two experiments the ratios of $\bar{k}_{max}$ after the transfer to dark to $\bar{k}_L$ in light before the change were 2.55 ± 0.14 and 3.0 ± 0.4 for Rb^+, compared with 1.37 ± 0.08 and 1.12 ± 0.08, respectively, for Na^+ in the same tissue (each figure is the mean of four strips). Thus Na^+-loaded tissue does not show the characteristic flux changes on darkening and would not be expected to show an aperture response. The response was determined by the presence of Na^+ or Rb^+ within the cells, not by the external ion; thus Rb^+-loaded cells showed the transient high efflux on darkening whether washing into RbCl or into NaCl. However, it does not necessarily follow that the extrusion process is specific for Rb^+ (or K^+) and cannot deal with Na^+, since guard cells opened in NaCl look much less healthy than those opened in KCl, KBr, or RbCl. Thus aperture is not the only criterion on which the state of guard cells should be judged. It is possible that the metabolic competence of Na^+-loaded tissue is quite different from that of Rb^+(K^+)-loaded cells. The ability to transfer salt to the vacuole is unimpaired, but this may not be true of all metabolic activities. The results do not settle the question of whether the cation efflux or the anion efflux is primary in the stimulated salt extrusion.

Effects of ABA on ion fluxes in Commelina. The effect of adding ABA to open guard cells (MacRobbie, 1981b) is similar to that of light-to-dark transitions and suggests that the two closing responses share a common mechanism.

When ABA was added during the efflux to steady-state cells in light, the time courses for both $^{86}Rb^+$ and $^{82}Br^-$ efflux were similar to those seen for transfer to the dark. The data in Figures 6.6A and B show the effect of adding 20 µmol l^{-1} ABA on $\bar{k}$ for $^{86}Rb^+$ and $^{82}Br^-$, respectively. In the Rb^+ experiment, where only one phase is apparent in the efflux kinetics, $\bar{k}$ is constant before the addition; whereas with Br^- the value of $\bar{k}$ falls steadily before the addition of ABA, reflecting the biphasic time course of efflux. In both cases addition of ABA results in a marked transient stimulation of efflux, lasting about 30 to 60 minutes. The data summarized in Table 6.10 show the ratio of the maximum $\bar{k}$ after the ABA addition to that before ABA addition. The stimulation is somewhat higher for Br^- than for Rb^+ and is also higher at lower external salt concentrations. Again, simulation of the Br^- transient showed that it could not be produced by a constant stimulated efflux and

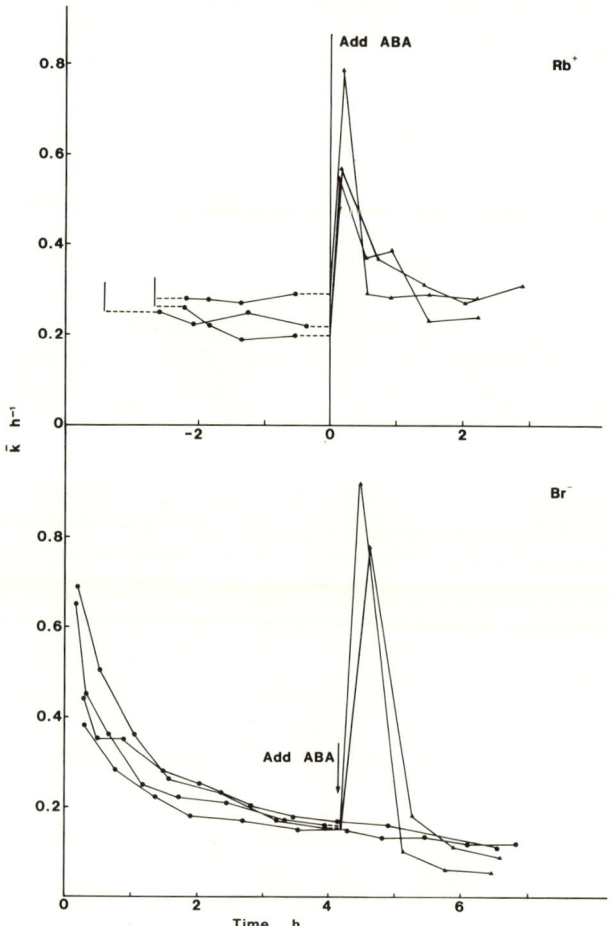

Fig. 6.6. Effect on rate constant ($\bar{k}$) of adding 20 μmol l^{-1} ABA during the efflux of Rb$^+$ and Br$^-$ from steady-state guard cells of *Commelina* in light. ABA added at time 0 in the Rb$^+$ plot. Rb$^+$: 25 mmol l^{-1} RbCl, 10 mmol l^{-1} MES, pH 3.9; Br$^-$: 60 mmol l^{-1} KBr, 10 mmol l^{-1} MES, pH 3.9; ●, in absence of ABA; ▲, after addition of ABA.

changing specific activity but rather must represent a genuine transient in total ion efflux (MacRobbie, unpublished).

As with darkening, the addition of ABA did not inhibit the influx of either ^{86}Rb$^+$ or ^{82}Br$^-$. The apparent influx in the presence of ABA, measured over 0.5 to 1 hours, was 75 to 89 percent of the control influx, but this reduction can be accounted for by the observed increase in efflux after the addition of ABA. An enhanced leakage of ^{86}Rb$^+$ from intact epidermal strips of *Commelina* was also reported by Weyers and Hillman (1980). Their use of autoradiography did

TABLE 6.10. Effects of ABA on efflux of ^{86}Rb$^+$ and ^{82}Br$^-$ in isolated guard cells of *Commelina*

External concentration (mmol l^{-1})	$\bar{k}_{max}/\bar{k}_C$	
	^{86}RbCl	K^{82}Br
10	1.6, 7.4, 6.1	
25	2.8, 2.8, 2.4, 2.6	
20		6.0, 16.3
60		4.1, 4.8

SOURCE: MacRobbie (1981b).
NOTE: All solutions contained 10 mmol l^{-1} MES, pH 3.9. Tissue was loaded overnight in light, in the absence of ABA, then washed out in light, with ABA (20 μmol l^{-1}) added during the course of the efflux. $\bar{k}_C$ is the value of $\bar{k}$ (rate/content) before the addition of ABA, and $\bar{k}_{max}$ is the maximum value during the transient.

not allow either calculation of fluxes or the separation of the guard cell contribution to the efflux from that of the subsidiary cells, but their results suggest that the effects seen in isolated strips do in fact reflect processes responsible for the ABA-induced closure in the intact leaf.

Nature of the closing transient. These results show that the control of plasmalemma efflux is complex, and rule out a number of simple mechanisms. The response cannot be explained as an effect of changes in membrane potential, because the efflux of both anions and cations is stimulated and because there is no effect on influx. In addition the measured depolarizations in the dark are small, about 10 to 20 mV at lower K$^+$ concentrations than used in the flux work (Zeiger et al., 1977; Moody and Zeiger, 1978; Zlotnikova, Gunar, and Panichkin, 1977); at higher C_o the potential changes are likely to be even smaller.

The important observations are that on darkening the stimulation is strictly transient and that the response to a transfer from dark to light depends on the previous history of the cells. Thus we are not dealing with a change in membrane properties, but rather with the control of a specific process of salt excretion. This is consistent with the observations of Weyers et al. (1982) that the ABA-induced closure of guard cells was inhibited by azide, cyanide, or hypoxia and that metabolism was required to achieve a net reduction in ion content. A number of authors have found that dark-induced closure is prevented by a variety of metabolic inhibitors (Fujino, 1967; Pemadasa and Jeyaseelen, 1976; Pemadasa and Koralege, 1977; see also review of light effects by Pemadasa, 1981).

The nature of the control on efflux remains unsolved. After transfer to the dark or addition of ABA, the efflux rises to a peak and then falls rapidly. In the experiments with Br$^-$, in which cytoplasmic and vacuolar contents can be estimated before the transfer to dark, it was possible to show by simulation of the $\bar{k}$ transient using trial values for the new fluxes that the efflux must fall more

TABLE 6.11. Effect of fusicoccin (FC) on ^{86}Rb$^+$ fluxes in isolated guard cells of *Commelina*

RbCl	pH	$\bar{k}_F/\bar{k}_C$	Flux	C	Time in FC (min)		
					15	30	45
30	3.9	0.65 ± 0.04 (12)	I	25.5	36.0	54.3	77.0
			E	22.8	14.9	15.7	17.1
			N	2.7	21.1	38.6	59.9
10	3.9	0.33 ± 0.03 (7)	I	6.0	13.0	28.5	36.3
			E	5.4	1.8	1.9	2.3
			N	0.6	11.2	26.6	34.0
30	6.7	0.057 ± 0.004 (8)	I	44.0	49.5	57.5	59.8
			E	43.3	2.7	2.7	3.0
			N	0.7	46.8	54.8	56.8
10	6.7	0.071 ± 0.006 (8)	I	13.5	25.0	48.8	61.0
			E	9.8	0.7	0.8	0.9
			N	3.7	24.3	48.0	60.1

NOTE: External concentration of RbCl in mmol l^{-1}. $\bar{k}_F/\bar{k}_C$, ratio of the apparent rate constant for efflux after the addition of FC to that before the addition; C, flux before the addition of 30 μmol l^{-1} FC (other figures give fluxes in successive periods after FC addition); I, influxes measured over 30-min periods (0–30, 15–45, and 30–60 min in FC), corresponding to the appropriate time period for efflux; E, efflux in successive 15-min periods after FC addition, calculated from $\bar{k}$ and adjusted tracer content in each period; N, net influx calculated from I and E. Fluxes in pmol mm^{-2} h^{-1}.

rapidly than the cytoplasmic content (MacRobbie, 1984). If the flux ϕ_{cv} (cytoplasm to vacuole) is reduced in the dark, and ϕ_{vc} (vacuole to cytoplasm) is increased, then the observed transient in $\bar{k}$ will be much less sharp than the true transient in plasmalemma efflux. Hence the falling ion content in the cytoplasm does not appear to be the primary factor in the fall in the peak efflux. Rather, some control factor may lead to the activation of a salt excretion process, followed by its deactivation. Two possible control factors are cytoplasmic pH and (perhaps more likely, since guard cells are so sensitive to calcium) cytoplasmic calcium concentration. The transient efflux may represent transient changes of one of these factors or of some other, unknown, cytoplasmic regulator.

Fluxes during FC-induced opening in Commelina. The flux transients already discussed concern the response to a closing signal and the control of the ability to stay open. It is also of interest to examine the flux response to fusicoccin (FC), which induces opening.

The effect of FC (30 μmol l^{-1}) on influx and efflux of ^{86}Rb$^+$ has been measured in 10 or 30 mmol l^{-1} RbCl at pH 3.9 or 6.7 in light (Clint and MacRobbie, 1984). The influx of ^{86}Rb$^+$ is markedly stimulated by the addition of FC (Table 6.11). The effect is seen within the first 15 minutes, with increases in influx and aperture continuing for at least 30 to 45 minutes, and sometimes for 135 minutes or more. The reverse effect is observed on efflux, which is strongly inhibited by addition of FC, best seen in the value of the ratios of the apparent rate constant after the addition of FC to that before the addition

($\bar{k}_F/\bar{k}_C$). Table 6.11 shows these ratios, together with the fluxes before and after FC addition. It may be noted that the effect of efflux develops fully within the first 15 minutes in FC, whereas the influx continues to rise for another 15 to 30 minutes. The net influx was also measured by loading tissue to a steady state in light and in the absence of FC, then adding FC to the labeling solution and following the increase in total ^{86}Rb$^+$ content in replicate samples; there was excellent agreement between the measured net flux and that calculated from the measured influxes and effluxes.

The osmotic effects of the increasing Rb-salt content in the first 20 minutes of the FC treatment are adequate to account for the observed aperture changes, but in the later stages of FC-induced opening this is no longer so (see Table 6.5). Thus the primary effect seems to be stimulation of the salt influx, but this is followed by other changes within the tissue, which also contribute to the increase in aperture.

Some measurements of the effect of FC on ^{82}Br$^-$ fluxes have also been made, but only at pH 3.9 (Clint, unpublished). At this pH the effect on the influx of ^{82}Br$^-$ is the opposite of that on ^{86}Rb$^+$, in that ^{82}Br$^-$ influx is inhibited by the addition of 30 μmol l^{-1} FC. Thus the influx was reduced to 75 percent of the control value in the first 30 minutes of FC treatment and to 61 to 71 percent of the control in the period 15 to 75 minutes in FC. But the effect on ^{82}Br$^-$ efflux is similar to that on ^{86}Rb$^+$ efflux, although less marked; and clear inhibition of efflux is observed. After 30 minutes in FC the efflux is only 52 percent of the control value, and after 90 minutes it is only 35 percent. Thus the results show that both influx and efflux of ^{82}Br$^-$ are inhibited by FC; whereas with ^{86}Rb$^+$, influx is stimulated, but efflux is inhibited. The combined effect would seem to produce a smaller net Br$^-$ influx than Rb$^+$ influx and hence to suggest that after FC treatment there may be significant malate synthesis, although in these conditions, that is, isolated strips floating on Rb(K)Cl(Br), malate synthesis in the control is insignificant.

It was argued (Clint and MacRobbie, 1984) that the inhibition of Rb$^+$ efflux by FC was much higher than would be expected as an effect of potential, given the size of the stimulation of influx. Clint's (unpublished) subsequent demonstration of inhibition also of Br$^-$ efflux gives further support to the argument that there is specific inhibition of the efflux by some other mechanism. The control of aperture is in part achieved by control of the efflux; the efflux can be stimulated by the closing signals, dark and ABA, but the FC results show that it can also be inhibited by changes in the internal state of the cell following FC treatment and consequent stimulation of the proton pump. It is interesting to note that the flux changes responsible for FC-induced opening differ in kind, and not just in direction, from those responsible for the dark-induced or ABA-induced closing. There is a marked stimulation of the ^{86}Rb$^+$ influx by FC, and this is largely responsible for the increased accumulation and consequent opening. In contrast, influx changes were not involved in the response to the two closing signals studied, for which control of efflux was responsible. It is therefore interesting that FC markedly inhibits the efflux of both Rb^+ and Br^- (at least at pH 3.9), the reverse of the effects on efflux seen in the closing responses.

Control of Ion Fluxes: Current Position and Unsolved Problems

The results discussed in this review suggest that two separate sets of processes are involved in the control of ion content and aperture in guard cells and that both sets of processes are subject to complex control by a number of different factors. First, there are the transport processes involved in ion uptake, likely to stem from the primary activity of the proton pump; and second, there are the processes involved in the metabolically mediated extrusion of ions in response to one of the closing signals. The ability to stay open seems to be regulated by the switching on or off of a specific process leading to salt extrusion rather than by control of influx. The control of the initiation of opening, however, seems to be governed by the switching on of the uptake processes in a state of low efflux.

Two factors that control ion fluxes may be considered. For both we have some information but lack a thorough understanding of the mechanisms involved. The first is light, which clearly controls more than one set of processes. In cells kept in darkness both Br^- influx and H^+ efflux are low; hence it appears that the uptake processes are suppressed. I suggested, however, that this is probably not a direct effect of darkness, since the uptake of ions in open guard cells is not immediately suppressed on transfer to the dark. It seems more likely that the reduction in ion content (probably cytoplasmic content), resulting from the enhanced efflux on transfer to the dark, produces changes that alter the level of activation of the uptake processes. The uptake can be switched on by FC in light or dark, which suggests that the limitation lies not in the energy supply, the availability of ATP, but rather in the control processes. The mechanism may be similar to that responsible for light effects on transport in other cells, where neither changes in ATP levels nor changes in cytoplasmic pH seem to be responsible for the light effect. On this point the direct evidence from perfused *Chara* cells is particularly important. No satisfactory explanation for the light effect in any system has yet emerged. In guard cells the problem is compounded by the fact that there are two separate light effects on aperture. Thus there is a blue-light effect, which saturates at low light intensities, and a second light effect, which has the action spectrum of photosynthesis, with peaks in both red and blue, and saturates at higher intensities (Hsiao, Allaway, and Evans, 1973; Ogawa et al., 1978; Zeiger, 1981, 1983). It is unclear which aspect of the ionic state is altered with each of these spectral sensitivities.

A second important factor in the control of ion accumulation in guard cells is the external ion concentration, and the evidence suggests that more than substrate availability is involved. Internal ion levels seem to be critical in regulating the ability of guard cells to retain high levels of net accumulation. The flux studies in *Commelina* show that at low external KCl the uptake is suppressed, that ϕ/C_o is low, giving evidence of internal control of some kind. Ion levels in the cytoplasm appear to regulate not only ion fluxes within the system but also non-ionic solute accumulation. Thus, increasing the level of KCl outside switches on ion accumulation, but the increase in osmotic pressure of guard cells is then greater than can be accounted for by the salt increase. The difference between the osmotic contribution of the salt increase and the osmotic requirement is particularly marked in the vacuole; we must therefore argue that the high cytoplasmic salt leads to increased transfer to the vacuole of both salt and some other solute. It looks as if the overall transport activity in the cell is very sensitive to ionic levels in the cytoplasm. An important aim for the future must be to test this hypothesis and to understand the mechanism whereby cytoplasmic levels of ions affect cellular transport.

The evidence considered in this review emphasizes the need to define the precise nature of any response we seek to understand. It is essential, in any situation where stomatal aperture changes, to establish exactly which fluxes change, of which ions, at which membrane, in which direction, and with what time course. It is not enough to observe the end results. We are still very much at the beginning of our attempts to understand the control of stomatal aperture and of solute contents of guard cells, and work continues to raise questions rather than provide answers.

REFERENCES

Allaway, W. G., and T. C. Hsiao. 1973. Preparation of rolled epidermis of *Vicia faba* L. so that stomata are the only visible cells: Analysis of guard cell potassium by flame photometry. *Australian Journal of Biological Sciences*, 26: 309–18.

Beilby, M. J., and N. A. Walker. 1981. Chloride transport in *Chara*, I. Kinetics and current-voltage curves for a probable proton symport. *Journal of Experimental Botany*, 32: 43–54.

Borle, A. B., T. Uchikawa, and J. H. Anderson. 1982. Computer simulation and interpretation of ^{45}Ca efflux profile patterns. *Journal of Membrane Biology*, 68: 37–46.

Bowman, B. J., S. E. Mainzer, K. E. Allen, and C. W. Slayman. 1978. Effects of inhibitors on the plasma membrane and mitochondrial adenosine triphosphatases of *Neurospora crassa*. *Biochimica biophysica acta*, 512: 13–28.

Bowman, B. J., and C. W. Slayman. 1979. The effects of vanadate on the plasma membrane ATPase of *Neurospora crassa*. *Journal of Biological Chemistry*, 254: 2928–34.

Campbell, N. A., and R. C. Garber. 1980. Vacuolar reorganisation in the motor cells of *Albizzia* during leaf movement. *Planta,* 148: 251–55.
Clint, G. M., and E. A. C. MacRobbie. 1984. Effects of fusicoccin in "isolated" guard cells of *Commelina communis* L. *Journal of Experimental Botany,* 35: 180–92.
Cocucci, M., A. Ballarin-Denti, and M. T. Marrè. 1980. Effects of orthovanadate on the H^+ secretion, K^+ uptake, electric potential differences, and membrane ATPase activities of higher plant tissues. *Plant Science Letters,* 17: 391–400.
Cram, W. J. 1976. Negative feedback regulation of transport in cells. In U. Lüttge and M. G. Pitman, eds., *Transport in plants* 2, Encyclopedia of Plant Physiology, n.s., 2A: 284–316. Berlin: Springer-Verlag.
———. 1980. Chloride accumulation as a homeostatic system: Negative feedback signals for concentration and turgor maintenance differ in a glycophyte and a halophyte. *Australian Journal of Plant Physiology,* 7: 237–49.
DuPont, F. M., L. L. Burke, and R. M. Spanswick. 1981. Characterization of a partially purified adenosine triphosphatase from a corn root plasma membrane fraction. *Plant Physiology,* 67: 59–63.
Fischer, R. A. 1968a. Stomatal opening: Role of potassium uptake by guard cells. *Science,* 160: 784–85.
———. 1968b. Stomatal opening in isolated epidermal strips of *Vicia faba,* I. Response to light and CO_2-free air. *Plant Physiology,* 43: 1947–52.
———. 1972. Aspects of potassium accumulation by stomata of *Vicia faba. Australian Journal of Biological Sciences,* 25: 1107–23.
Fischer, R. A., and T. C. Hsiao. 1968. Stomatal opening in isolated epidermal strips of *Vicia faba,* II. Response to KCl concentration and role of potassium absorption. *Plant Physiology,* 43: 1953–58.
Fujii, S., T. Shimmen, and M. Tazawa. 1979. Effect of intracellular pH on the light-induced potential change and electrogenic activity in tonoplast-free cells of *Chara australis. Plant Cell Physiology.* 20: 1315–28.
Fujino, M. 1967. Role of adenosinetriphosphate and adenosinetriphosphatase in stomatal movement. *Science Bulletin of the Faculty of Education, Nagasaki University,* 18: 1–47.
Gepstein, S., M. Jacobs, and L. Taiz. 1982. Inhibition of stomatal opening in *Vicia faba* epidermal tissue by vanadate and abscisic acid. *Plant Science Letters,* 28: 63–72.
Gutknecht, J., D. F. Hastings, and M. A. Bisson. 1978. Ion transport and turgor pressure regulation in giant algal cells. In G. Giebisch, D. C. Tosteson, and H. H. Ussing, eds., *Membrane transport in biology* 3: 125–74. Berlin: Springer-Verlag.
Hiatt, A. J. 1967. Relationship of cell sap pH to organic acid change during ion uptake. *Plant Physiology,* 42: 294–98.
Hiatt, A. J., and S. B. Hendricks. 1967. The role of CO_2 fixation in accumulation of ions by barley roots. *Zeitschrift für Pflanzenphysiologie,* 56: 220–32.
Hsiao, T. C. 1976. Stomatal ion transport. In U. Lüttge and M. G. Pitman, eds., *Transport in plants* 2, Encyclopedia of Plant Physiology, n.s., 2B: 195–221. Berlin: Springer-Verlag.
Hsiao, T. C., W. G. Allaway, and L. T. Evans. 1973. Action spectrum for guard cell Rb^+ uptake and stomatal opening in *Vicia faba. Plant Physiology,* 51: 82–88.
Humble, G. D., and T. C. Hsiao. 1969. Specific requirement for potassium for light-activated opening of stomata in epidermal strips. *Plant Physiology,* 44: 230–34.
Humble, G. D., and K. Raschke. 1971. Stomatal opening quantitatively related to potassium transport. Evidence from electron probe analysis. *Plant Physiology,* 48: 447–53.

Imamura, S.-I. 1943. Untersuchungen über den Mechanismus der Turgorschwankung der Spaltöffnungsschliesszellen. *Japanese Journal of Botany*, 12: 251–346.
Jacobs, M., and L. Taiz. 1980. Vanadate inhibition of auxin-enhanced H^+ secretion and elongation in pea epicotyls and oat coleoptiles. *Proceedings of the National Academy of Sciences*, 77: 7242–46.
Jarvis, R. G., and T. A. Mansfield. 1980. Reduced stomatal responses to light, carbon dioxide and abscisic acid in the presence of sodium ions. *Plant, Cell and Environment*, 3: 279–83.
Kauss, H. 1979. Osmotic regulation in algae. *Progress in Phytochemistry*, 5: 1–27.
Kikuyama, M., T. Hayama, S. Fujii, and M. Tazawa. 1979. Relationship between light-induced potential change and internal ATP concentration in tonoplast-free *Chara* cells. *Plant Cell Physiology*, 20: 993–1002.
Krause, G. H. 1973. The high energy state of the thylakoid system as indicated by chlorophyll fluorescence and chloroplast shrinkage. *Biochimica biophysica acta*, 292: 715–28.
Laffray, D., P. Louguet, and J. P. Garrec. 1982. Microanalytical studies of potassium and chloride fluxes and stomatal movements of two species: *Vicia faba* and *Pelargonium* × *hortorum*. *Journal of Experimental Botany*, 33: 771–82.
Macallum, A. B. 1905. On the distribution of potassium in animal and vegetable cells. *Journal of Physiology*, 32: 95–118.
MacRobbie, E. A. C. 1980. Osmotic measurements on stomatal cells of *Commelina communis* L. *Journal of Membrane Biology*, 53: 189–98.
———. 1981a. Ion fluxes in isolated guard cells of *Commelina communis* L. *Journal of Experimental Botany*, 32: 545–62.
———. 1981b. Effects of ABA in "isolated" guard cells of *Commelina communis* L. *Journal of Experimental Botany*, 32: 563–72.
———. 1983. Effects of light/dark on cation fluxes in guard cells of *Commelina communis* L. *Journal of Experimental Botany*, 34: 1695–1710.
———. 1984. Effects of light/dark on anion fluxes in isolated guard cells of *Commelina communis* L. *Journal of Experimental Botany*, 35: 707–26.
MacRobbie, E. A. C., and J. Lettau. 1980a. Ion content and aperture in "isolated" guard cells of *Commelina communis* L. *Journal of Membrane Biology*, 53: 199–205.
———. 1980b. Potassium content and aperture of "intact" stomatal and epidermal cells of *Commelina communis* L. *Journal of Membrane Biology*, 56: 249–56.
Maier-Maercker, U. 1979. Peristomal transpiration and stomatal movement: A controversial view, IV. Ion accumulation by peristomal transpiration. *Zeitschrift für Pflanzenphysiologie*, 91: 239–54.
———. 1983. A critical assessment of the role of potassium and osmolarity in stomatal opening. *Journal of Experimental Botany*, 34: 811–24.
Marrè, E. 1979. Fusicoccin: A tool in plant physiology. *Annual Review of Plant Physiology*, 30: 273–88.
Melis, A., and E. Zeiger. 1982. Chlorophyll *a* fluorescence transients in mesophyll and guard cells. *Plant Physiology*, 69: 642–47.
Moody, W., and E. Zeiger. 1978. Electrophysiological properties of onion guard cells. *Planta*, 139: 159–65.
Ogawa, T., D. Grantz, J. Boyer, and Govindjee. 1982. Effects of cations and abscisic acid on chlorophyll *a* fluorescence in guard cells of *Vicia faba*. *Plant Physiology*, 69: 1140–44.
Ogawa, T., H. Ishikawa, K. Shimada, and K. Shibata. 1978. Synergistic action of red and blue light and action spectra for malate formation in guard cells of *Vicia faba* L. *Planta*, 142: 61–65.

Outlaw, W. H. 1983. Current concepts on the role of potassium in stomatal movements. *Physiologia plantarum*, 59: 302–11.
Outlaw, W. H., and J. Kennedy. 1978. Enzymic and substrate basis for the anaplerotic step in guard cells. *Plant Physiology*, 62: 648–52.
Outlaw, W. H., and O. H. Lowry. 1977. Organic acid and potassium accumulation in guard cells during stomatal opening. *Proceedings of the National Academy of Sciences*, 74: 4434–38.
Outlaw, W. H., and J. Manchester. 1979. Guard cell starch concentration quantitatively related to stomatal aperture. *Plant Physiology*, 64: 79–82.
Outlaw, W. H., J. Manchester, C. A. Di Camelli, D. D. Randall, B. Rapp, and G. M. Veith. 1979. Photosynthetic carbon reduction pathway is absent in chloroplasts of *Vicia faba* guard cells. *Proceedings of the National Academy of Sciences*, 76: 6371–75.
Pallaghy, C. K., and R. A. Fischer. 1974. Metabolic aspects of stomatal opening and ion accumulation by guard cells in *Vicia faba*. *Zeitschrift für Pflanzenphysiologie*, 71: 332–44.
Pekarek, J. 1934. Über die Aziditätsverhältnisse in den Epidermis- und Schliesszellen bei *Rumex acetosa* im Licht und im Dunkeln. *Planta*, 21: 419–46.
Pemadasa, M. A. 1981. Photocontrol of stomatal movements. *Biological Reviews*, 56: 551–88.
Pemadasa, M. A., and K. Jeyaseelan. 1976. Some effects of sodium azide on stomatal closure in *Stachyarpheta indica*. *Annals of Botany*, 40: 655–58.
Pemadasa, M. A., and S. Koralege. 1977. Stomatal responses to 2,4-dinitrophenol. *New Phytologist*, 78: 573–78.
Penny, M. G., and D. J. F. Bowling. 1974. A study of potassium gradients in the epidermis of intact leaves of *Commelina communis* L. in relation to stomatal opening. *Planta*, 119: 17–25.
———. 1975. Direct determination of pH in the stomatal complex of *Commelina*. *Planta*, 122: 209–12.
Penth, B., and J. Weigl. 1971. Anionen-Influx, ATP-Spiegel und CO_2-Fixierung im *Limnophila gratioloides* und *Chara foetida*. *Planta*, 96: 212–23.
Pitman, M. G. 1969. Simulation of Cl^- uptake by low-salt barley roots as a test of models of salt uptake. *Plant Physiology*, 44: 1417–27.
Raschke, K. 1977. The stomatal turgor mechanism and its response to CO_2 and abscisic acid: Observations and a hypothesis. In E. Marrè and O. Ciferri, eds., *Regulation of cell membrane activities in plants*: 173–83. Amsterdam: North-Holland.
———. 1979. Movements of stomata. In W. Haupt and M. E. Feinleib, eds., *Physiology of movements*, Encyclopedia of Plant Physiology, n.s., 7: 383–441. Berlin: Springer-Verlag.
Rashcke, K., and G. D. Humble. 1973. No uptake of anions required by opening stomata of *Vicia faba*: Guard cells release hydrogen ions. *Planta*, 115: 47–57.
Roberts, J. K. M. 1984. Study of plant metabolism in vivo using NMR spectroscopy. *Annual Review of Plant Physiology*, 35: 375–86.
Saftner, R. A., and K. Raschke. 1981. Electric potentials in stomatal complexes. *Plant Physiology*, 67: 1124–32.
Sanders, D. 1980. The mechanism of Cl^- transport at the plasma membrane of *Chara corallina*, I. Co-transport with H^+. *Journal of Membrane Biology*, 53: 129–41.
———. 1984. Gradient-coupled chloride transport in plant cells. In G. A. Gerencser, ed., *Chloride transport coupling in cells and epithelia*: 63–119. Amsterdam: North-Holland.
Sanders, D., and U. P. Hansen. 1981. Mechanism of Cl^- transport at the plasma

membrane of *Chara corallina*, II. Transinhibition and the determination of H^+/Cl^- binding order from a reaction kinetic model. *Journal of Membrane Biology*, 58: 139–53.

Scarth, G. W. 1932. Mechanism of the action of light and other factors on stomatal movement. *Plant Physiology*, 7: 481–504.

Schwartz, A., and E. Zeiger. 1984. Metabolic energy for stomatal opening: Role of photophosphorylation and oxidative phosphorylation. *Planta*, 161: 129–36.

Small, J., and K. M. Maxwell. 1939. pH phenomena in relation to stomatal opening, I. *Coffea arabica* and some other species. *Protoplasma*, 32: 272–88.

Spanswick, R. M. 1980. Biophysical control of electrogenicity in the Characeae. In R. M. Spanswick, W. J. Lucas, and J. Dainty, eds., *Plant membrane transport: current conceptual issues*: 305–13. Amsterdam: North-Holland.

Spanswick, R. M., and A. G. Miller. 1977. The effect of CO_2 on the Cl^- influx and electrogenic pump in *Nitella translucens*. In H. Thellier, A. Monnier, M. Demarty, and J. Dainty, eds., *Transmembrane ionic exchanges in plants*: 239–45. Paris: CNRS.

Squire, G. R., and T. A. Mansfield. 1972. Studies on the mechanism of action of fusicoccin, the fungal toxin that induces wilting, and its interaction with abscisic acid. *Planta*, 105: 71–78.

———. 1974. The action of fusicoccin on stomatal guard cells and subsidiary cells. *New Phytologist*, 73: 433–40.

Stitt, M., R. M. Lilley, and H. W. Heldt. 1982. Adenine nucleotide levels in the cytosol, chloroplasts, and mitochondria of wheat leaf protoplasts. *Plant Physiology*, 70: 971–77.

Turner, N. C. 1973. Action of fusicoccin on the potassium balance of guard cells of *Phaseolus vulgaris*. *American Journal of Botany*, 60: 717–25.

Turner, N. C., and A. Graniti. 1969. Fusicoccin. A fungal toxin that opens stomata. *Nature*, 223: 1070–71.

Vaughn, K. C., and W. H. Outlaw. 1983. Cytochemical and cytofluorometric evidence for guard cell photosystems. *Plant Physiology*, 71: 420–24.

Weyers, J. D. B., and P. J. Fitzsimons. 1982. The non-osmotic volumes of *Commelina* guard cells. *Plant, Cell and Environment*, 5: 417–21.

Weyers, J. D. B., and J. R. Hillman. 1980. Effects of abscisic acid on $^{86}Rb^+$ fluxes in *Commelina communis* L. leaf epidermis. *Journal of Experimental Botany*, 31: 711–20.

Weyers, J. D. B., N. W. Paterson, P. J. Fitzsimons, and J. M. Dudley. 1982. Metabolic inhibitors block ABA-induced stomatal closure. *Journal of Experimental Botany*, 33: 1270–78.

Willmer, C. M., and T. A. Mansfield. 1969. A critical examination of the use of detached epidermis in studies of stomatal physiology. *New Phytologist*, 68: 363–75.

Yamashita, T. 1952. Influences of potassium supply upon various properties and movement of the guard cell. *Sieboldia acta biologica*, 1: 51–70. [In Japanese, with English summary.]

Zeiger, E. 1981. Novel approaches to the biology of stomatal guard cells: Protoplast and fluorescence studies. In P. G. Jarvis and T. A. Mansfield, eds., *Stomatal physiology*: 103–17. Cambridge: Cambridge University Press.

———1983. The biology of stomatal guard cells. *Annual Review of Plant Physiology*, 34: 441–75.

Zeiger, E., W. Moody, P. Hepler, and F. Varela. 1977. Light-sensitive membrane potentials in onion guard cells. *Nature*, 270: 270–71.

Zimmermann, U. 1977. Cell turgor pressure regulation and turgor pressure-mediated transport processes. In D. H. Jennings, ed., *Integration of activity in the higher plant*: 117–54. Symposia of the Society for Experimental Biology, 31. Cambridge: Cambridge University Press.

Zlotnikova, I. F., I. I. Gunar, and L. A. Panichkin. 1977. Light-induced changes of the membrane potential in *Tradescantia* leaf epidermal cells. *Izvestiya Timiryazevskoi selsko-khozyaistvenmoi akademii*, 3: 10–14.

7

Guard Cell Bioenergetics

Sarah M. Assmann and Eduardo Zeiger

Many biological processes involve the transduction of energy. Common examples of such processes are the capture of radiant energy in photosynthesis and the storage of energy in electrochemical gradients across cell membranes. It has become increasingly evident that energy transduction is also an important aspect of stomatal movements. In nature, stomata are often displaced from equilibrium conditions, and energy-requiring steps are involved in the modulation of stomatal aperture. A quantitative analysis of the energetics of stomatal opening has implications for our understanding of the biochemical and biophysical mechanisms underlying stomatal movement and the basis of their regulation. For instance, given available information on rates of proton extrusion and on ion concentrations in guard cells, we can evaluate the adequacy of proton pumping as a driving force for the ion fluxes underlying stomatal opening. We can also estimate the rate-limiting steps and predict the regulatory roles for energy supply and demand.

A need for an analysis of stomatal energetics is also evident from current trends in studies on the ecophysiology of plants and rates of gas exchange in leaves. Important advances in the evaluation of the energy budgets of plants underscore the relevance of the partitioning of limited resources among plant growth, development, and reproduction. A corresponding evaluation of the energy cost of stomatal movements has yet to be tackled.

In this paper we present an analysis of guard cell energetics. Our aim is to examine the primary energy-dependent processes in guard cells and to evaluate their energy balance. The analysis provides a framework for the examination of the bioenergetics of stomatal movements and emphasizes the information needed for further progress. We also assess current concepts of energy supply and demand in guard cells, in an attempt to achieve an integrated view of the bioenergetics of stomatal functioning.

For the purposes of this analysis, we have divided the topic of guard cell bioenergetics into three main sections. In the first two, we consider qualitatively the energy costs and potential energy sources in guard cells. In the third, we assess these costs and sources quantitatively, using data from experiments with *Vicia faba*.

Energy Costs

Maintenance. There are four major cellular processes that require continual energy expenditure and comprise the basic maintenance cost of the cell. The first is the use of energy for cell growth and reproduction. Mature guard cells, however, neither grow nor divide (see Sack, this volume); hence, this process does not require consideration here.

The second process is repair. Because many biological macromolecules have a limited lifetime, constant synthesis of new molecules is required in order to maintain cellular integrity. Guard cells may have unique capabilities in this area: guard cells in detached leaves of *Ginkgo* remain functional long after chlorophyll degradation and senescence have occurred in the surrounding tissue (Zeiger and Schwartz, 1982).

The third cellular process requiring energy is the maintenance of the electrochemical, or membrane, potential. In animal cells, the Na^+/K^+ ATPase, which maintains a negative membrane potential by a nonstoichiometric exchange of Na^+ and K^+, consumes approximately 30 percent of the ATP produced by a typical cell (Alberts et al., 1983). There is increasing evidence that, in plant cells, an analogous role is played by a plasmalemma ATPase, although it is not yet clear whether the ATPase is a primary proton pump or an H^+/K^+ antiporter (Poole, 1978; Briskin and Poole, 1983). The maintenance of an electrochemical potential results in transmembrane charge and concentration differences, which can drive the transport of other substances.

In the fourth cellular energy-requiring process, both charged and electroneutral substances may be selectively transported across the cell membrane via carriers or pumps.

Repair, maintenance of membrane potential, and active transport are the basic energy-requiring processes of the guard cell, and the energy they require could be considered the minimum amount of energy necessary to keep the cell alive.

Stomatal opening. For differentiated cells, in addition to maintenance costs there is the cost of the specialized function of the cell type. In the case of the guard cell, the specialized function is stomatal opening and closing. Although stomatal opening ultimately depends upon maintenance processes such as ion transport, it requires considerable expenditure of energy over that required by the guard cells when the stomata are closed. This additional component is considered here the energy cost for stomatal opening.

Stomata open when an influx of water into the guard cells causes them to swell. The influx of water is driven by an increase in cell solute concentration, which decreases the water potential of the cell. In most species, the ions that increase the concentration during stomatal opening are K^+, Cl^-, and $malate^{2-}$.

In theory, uptake of K^+, the largest ion flux associated with stomatal movements, could occur by active K^+ uniport, by H^+/K^+ antiport, or by a chemiosmotic mechanism relying on extrusion of protons. All of these processes would require energy, as confirmed by the many studies showing that metabolic inhibitors prevent stomatal opening (Walker and Zelitch, 1963; Humble and Hsiao, 1970; Raghavendra, 1981; Schwartz and Zeiger, 1984). Recent experiments (Assmann, Simoncini, and Schroeder, 1985) performed on guard cell protoplasts using the electrophysiological technique of whole-cell patch clamping (Hamill et al., 1981; Marty and Neher, 1983) support the chemiosmotic hypothesis. In this technique (Fig. 7.1a) a glass micropipette filled with an ionic solution is brought in contact with the cell membrane, and suction is applied to achieve a tight seal between the rim of the pipette tip and the plasmalemma. Further suction ruptures the enclosed membrane, resulting in diffusional equilibration between the cell contents and the pipette solution, and enabling experimental control of the ionic conditions inside the cell as well as in the bath solution. Any electrical signal detected can be directly attributed to ion movement across the plasmalemma. The experimenter can either clamp the membrane voltage to a constant value and observe current fluxes directly, or clamp the membrane current to a constant value and measure changes in membrane voltage.

Patch clamping of guard cell protoplasts of *Vicia faba* was used in conjunction with a dual-beam technique (Ogawa et al., 1978; Iino, Ogawa, and Zeiger, 1985) in which high fluence rates of background red light were used to saturate all guard cell photoresponses other than those of the blue-light photosystem. Subsequent application of a pulse of blue light caused a transient membrane hyperpolarization (Fig. 7.1b) analogous to that observed with conventional electrophysiological techniques following stimulation with white light (Gunar, Zlotnikova, and Panichkin, 1975; Zeiger et al., 1977; Moody and Zeiger, 1978; Ishikawa et al., 1983). This hyperpolarization is inconsistent with the functioning of a K^+ uniport, which would result in depolarization of the membrane potential. In addition, blue light stimulated hyperpolarization when K^+ was eliminated from both the pipette and external solutions (Fig. 7.1c), ruling out the operation of an obligate H^+/K^+ antiport. These and other patch-clamp experiments (Assmann, Simoncini, and Schroeder, 1985) demonstrated that the hyperpolarization results from active ion transport at the plasmalemma. The ion carrying the electrical charge was identified by measuring the pH of a suspension of *Vicia* guard cell protoplasts. A pulse of blue light caused a fluence-rate-dependent acidification that was clearly attributable to proton extrusion by the guard cell protoplasts (Shimazaki, Iino, and Zeiger, 1986; Zeiger et al., this volume). Taken together, the patch-clamp and acidification experiments demonstrate the operation of a proton pump at the guard cell plasmalemma, a conclusion supported by observations of proton

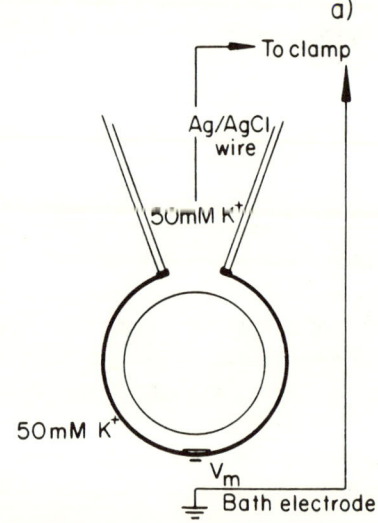
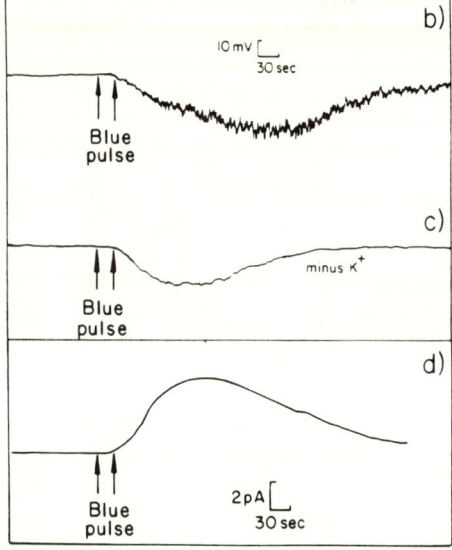

Fig. 7.1. *a*, Schematic of the whole-cell patch clamp configuration, which was used to obtain the data shown in Figures 1*b–d*. True ratio of pipette tip diameter to guard cell protoplast diameter is about 1:15. For more information on the patch clamp technique, see Sakmann and Neher (1983). *b*, Membrane-potential changes in a guard cell protoplast of *Vicia faba* held in current clamp mode upon application of a 30-s blue-light pulse (0.050 mmol m^{-2} s^{-1}) on a continuous background of 0.500 mmol m^{-2} s^{-1} red light. Both bath (pH 7.15, 500 mmol kg^{-1}) and pipette (pH 7.4, 450 mmol kg^{-1}) solutions contained 50 mM K$^+$ glutamate, 5 mM MgCl$_2$, 10 mM HEPES, and 3 mM NaOH; pipette solution contained in addition 9.2 mM (Na$^+$)$_2$ ATP, 20 mM TRIS, and 100 µM EGTA. Osmolarities were achieved by the addition of D-mannitol. *c*, Transient membrane hyperpolarization in response to the same illumination protocol as in Figure 7.1*b*. Solutions were the same as in Figure 7.1*b*, except that the bath solution contained 1 mM CaCl$_2$, and K$^+$ was replaced by the impermeant cation *n*-methylglucamine in both the bath and pipette solutions. *d*, Record of outward whole-cell membrane current in response to blue-light stimulation. Solutions and illumination protocol were the same as in Figure 7.1*b*. The cell was held in voltage clamp mode at $V_m = -4$ mV. (Figure 7.1*b–d* adapted from Assmann, Simoncini, and Schroeder, 1985.)

extrusion during stomatal opening (Raschke and Humble, 1973; Gepstein, Jacobs, and Taiz, 1982).

In chemiosmosis, energy is used to distribute an ionic species asymmetrically across a cell membrane, thereby generating an electrical gradient. Ions then redistribute passively, following the electrochemical driving force until equilibrium concentrations are attained. When protons are the actively transported species, as in guard cells, the free energy per mole of H^+ extruded is usually referred to as the "protonmotive force" (pmf). Its magnitude is given by the following equation:

$$\mu_{H^+_{out}} - \mu_{H^+_{in}} = RT \ln \frac{[H^+]_{out}}{[H^+]_{in}} - (z_{H^+}) FV_m \qquad (1)$$

where $[H^+]_{out}$ and $[H^+]_{in}$ refer respectively to the extracellular and intracellular concentrations of protons.

Equation 1 is more commonly rearranged as

$$\text{pmf} = \mu_{H^+_{in}} - \mu_{H^+_{out}} = (z_{H^+}) FV_m - RT \ln 10 \, \Delta pH. \qquad (2)$$

The derivation of this equation is given in Box 7.1. Note that the magnitude of the pmf is determined by two components: the electrical gradient across the membrane and the proton concentration gradient across the membrane.

Given a chemiosmotic mechanism, the ionic events underlying stomatal opening can be analyzed as follows. As a result of a hormonal or environmental signal, the rate of proton extrusion at the plasmalemma will increase. Under conditions in which the membrane is permeable to K^+ (Schroeder, Hedrich, and Fernandez, 1984), K^+ will start to enter the cell as soon as proton extrusion results in a membrane potential more negative than the Nernst potential for K^+. Potassium entry will tend to repolarize V_m toward its initial value; whether the membrane repolarizes or not, and the rate or extent of hyperpolarization changes, will depend on the relative rates of H^+ extrusion and K^+ influx, now occurring simultaneously. In guard cells, it seems that the membrane does not repolarize (Fig. 7.1b, c).

As H^+ extrusion continues, the ΔpH term in Equation 2 will increase to a value dependent on the net rate of proton pumping, the exchange properties of the cell wall, and the buffering capacity of the cytoplasm. With the exception of proton-pumping rates, these parameters are not well known for guard cells.

As the pmf increases, more energy will be required for net extrusion of a proton. The movement of H^+ and, therefore, of K^+ may eventually be limited by the amount of energy available to increase the pmf. Alternatively, although sufficient energy may be available for additional ion transport, specific regulatory mechanisms in the cell may stabilize guard cell ion content and volume below the highest possible values ensuing from purely energetic considerations. This hypothesis is attractive given the evidence (see "Quantitative Analysis of Guard Cell Bioenergetics," below) that energy in excess of that required for the amount of stomatal opening observed is often available to guard cells.

Regardless of the basis of regulation, a new steady-state stomatal aperture will result from the maintenance of a new transmembrane distribution of ions.

BOX 7.1. Chemical potential (after Nobel, 1974; Nicholls, 1982)

μ_j is defined as the chemical potential or free energy per mole of substance j. If μ_j differs between two areas in a system, that is, if

$$\mu_{j_1} \neq \mu_{j_2},$$

molecules of j will move passively toward the area of lowest potential. For example, if a permeant ion is present outside and inside the cell, and

$$\mu_{j_{out}} > \mu_{j_{in}},$$

ions will move into the cell.

The magnitude of μ_j is given by

$$\mu_j = \mu_j^* + RT\ell n a_j + \bar{V}_j P + z_j FE + mgh \qquad (1)$$

where μ_j^* is the free energy of j in some standard state
$RT\ell n a_j$ is a concentration term
$\bar{V}_j P$ is a pressure term
$z_j FE$ is an electrical term
mgh is a gravitational or potential energy term.

The free energy difference of a substance j across a cell membrane can therefore be written as

$$\Delta\mu_j = \mu_{j_{out}} - \mu_{j_{in}} = \mu_j^* - \mu_j^* + RT\ell n a_{out} - RT\ell n a_{in} + \bar{V}_j(P_{out} - P_{in}) +$$
$$z_j F(E_{out} - E_{in}) + mg(h_{out} - h_{in}). \qquad (2)$$

Across a cell membrane, Δh can be taken as 0, and the pressure term is negligible compared to the other terms (Nobel, 1974). Equation 2 then simplifies to

$$\Delta\mu_j = \mu_{j_{out}} - \mu_{j_{in}} = RT\ell n\left(\frac{a_{out}}{a_{in}}\right) - z_j F V_m \qquad (3)$$

where V_m is the membrane potential of the cell ($= E_{in} - E_{out}$).

Thus, given V_m and intracellular and extracellular activities of an ion (usually approximated by intracellular and extracellular concentrations), we can calculate whether

$$\mu_{j_{out}} > \mu_{j_{in}}$$

(a positive value of Eq. 3), in which case the ion would passively move into the cell, or

$$\mu_{j_{out}} < \mu_{j_{in}}$$

(a negative value of Eq. 3), in which case passive movement in the reverse direction would occur.

Note that in order to determine the condition under which no net movement of the ion would occur, we set $\Delta\mu = 0$. This formulation is often used to calculate the Nernst potential (E_{N_j}), the potential at which the observed intracellular and extracellular concentrations of the ion would be in equilibrium:

$$0 = RT\ell n\left(\frac{a_{j_{out}}}{a_{j_{in}}}\right) - z_j F V_m \qquad (4)$$

$$E_{N_j} = \frac{RT}{z_j F}\ell n\left(\frac{a_{j_{out}}}{a_{j_{in}}}\right). \qquad (5)$$

If the observed V_m does not equal E_{N_j}, the cell must be expending energy to maintain the $a_{j_{in}}$ observed.

TABLE 7.1. Guard cell ion concentrations in *Vicia faba*

Ion	Content/guard cell (pmol)[a]	Aperture (μm)[a]	Volume/guard cell (pℓ)[b]	Ion concentration (mM)	$\left(\frac{mmol}{kg}\right)$[c]
Potassium					
stomata open	2.12	12	5.0	424.0	446.1
stomata closed	0.10	2	2.6	38.5	38.8
difference	2.02	10	2.4	385.5	407.3
Chloride					
stomata open	0.110	12	5.0	22.0	23.2
stomata closed	0.008[d]	2	2.6	3.1	3.1
difference	0.102	10	2.4	18.9	20.1
Malate					
stomata open	2.33	12	5.0	464.6	488.8
stomata closed	0.125	2	2.6	48.1	48.9
difference	2.20	10	2.4	416.5	439.9

[a]Data from Humble and Raschke (1971), tables I and II; Van Kirk and Raschke (1978), fig. 3.
[b]Data from Sharpe and Wu (1978), fig. 8.
[c]Osmolalities were empirically determined for a solution of 424 mM K^+ + 464.6 mM malate^{2-} and a solution of 38.5 mM K^+ + 3.1 mM Cl^- + 48.1 mM malate^{2-} (unpublished observations).
[d]Table II of Humble and Raschke (1971) gives an average Cl^- content in guard cells of closed stomata as zero, owing to the uncertainty of the measuring technique. A zero Cl^- content seems unrealistic, as it would require constant active Cl^- extrusion against an infinitely steep Cl^- concentration gradient. We have chosen to calculate the Cl^- content of the closed stomata from table I of Humble and Raschke (1971), taking the average net counts for Cl^- as 3, and assuming a linear calibration curve; thus 41 counts/3 counts = 0.22 pmol Cl^-/x, and solving for x: x = 0.016 pmol Cl^-/stomata.

At steady state, net fluxes of H^+ and K^+ will once again be zero. This new steady state can be experimentally observed as an increase in external H^+ concentration, an increase in internal K^+ concentration, and a hyperpolarization of the membrane potential resulting from the changes in ion concentrations.

At steady-state stomatal apertures, the membrane potential should be close to the calculated Nernst potential of the major permeant ion, in this case K^+. According to chemiosmosis, K^+ should redistribute passively across the membrane until it is once more in electrochemical equilibrium at its Nernst potential. This prediction has been largely substantiated in experiments with *Tradescantia* and *Commelina* (Saftner and Raschke, 1981; Bowling and Edwards, 1984; Edwards and Bowling, 1984).

In the above analysis, the smaller flux of Cl^- that occurs during stomatal opening has been excluded for purposes of simplicity. However, since both concentration (Table 7.1) and electrical gradients across the cell membrane are in the wrong direction for passive Cl^- influx, active uptake of Cl^- must also be incorporated into the chemiosmotic model. Upon generation of the pmf, the driving force for passive movement of H^+ back into the cell is increased. Chloride uptake is hypothesized to occur via direct coupling with this H^+ backflux (Zeiger, Bloom, and Hepler, 1978).

Malate is another important anion participating in osmotic changes during stomatal opening. With the exception of the genus *Allium*, in which malate synthesis does not occur (Schnabl and Ziegler, 1977; Schnabl and Raschke, 1980), the net increase in anion content of the cytoplasm results partially from Cl^- uptake and partially from malate synthesis, with the relative contributions of the two ionic species being dependent on the external Cl^- concentration (Raschke and Schnabl, 1978; Van Kirk and Raschke, 1978). As in other cell types (O'Leary, 1982), malate probably serves both to balance charge and to regulate pH. Other organic anions, for example, citrate (Outlaw and Lowry, 1977), also appear to accumulate in the guard cells, but to a much lesser extent, and they are therefore ignored here.

Outlaw and Manchester (1979) demonstrated that the decrease in guard cell starch deposits during stomatal opening matched the increase they observed in malate synthesis. If we assume that malate synthesis results from starch breakdown, the direct cost of malate synthesis during stomatal opening can be calculated from the energetics of this process.

A calculation of the overall energetics of malate metabolism would require consideration of the cost of both malate synthesis and degradation. Recent immunocytochemical and physiological data may reopen the question of whether guard cells have the capacity to fix CO_2 photosynthetically (see "Quantitative Analysis of Guard Cell Bioenergetics," below). Biochemical data, however, suggest that ribulose-1,5-bisphosphate carboxylase (RuBPCase) is not active in guard cells (see Outlaw, 1982). If we assume an inactive Calvin cycle, carbon skeletons must be imported into the guard cells for synthesis of the starch from which malate is obtained. Preliminary data (Outlaw and Fisher, 1975; Outlaw, Fisher, and Christy, 1975) suggest that carbon could be imported in the form of sucrose. In phloem cells, sucrose import requires energy, proceeding via a chemiosmotic mechanism involving H^+-sucrose co-transport (Giaquinta, 1983). Sucrose uptake into the guard cell may proceed via a similar mechanism. The cost of import is considered here as part of the maintenance requirement of the guard cell since, unlike malate synthesis, it is not involved in opening.

Following the same reasoning, the cost of starch synthesis from sucrose is also included as a component of the maintenance requirement. Very little is known about the biochemistry of this process in the guard cell. In mesophyll cells, sucrose is first broken down to triose phosphates by glycolysis for entry into the chloroplasts via the triose phosphate–inorganic phosphate translocator (Heber and Heldt, 1981), and starch is then synthesized via gluconeogenesis. We can hypothesize that a similar process occurs in the guard cells. This pathway for starch synthesis is outlined in Figure 7.2.

If we again assume that the biochemistry of the mesophyll cells and that of the guard cells are similar, starch breakdown to malate probably begins by reverse gluconeogenesis of starch to 3-phosphoglyceric acid (3-PGA) (Kaiser and Bassham, 1979), which is translocated back into the cytoplasm and converted to phosphoenolpyruvate (PEP). Carboxylation of PEP to oxaloacetic

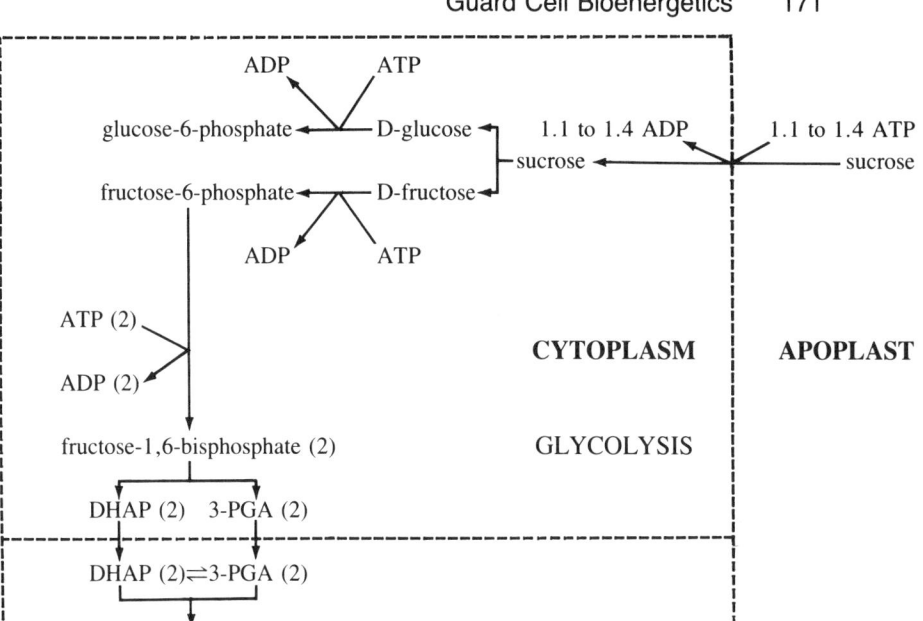

Fig. 7.2. Assumed pathway of carbon import and starch synthesis in the guard cell.

acid (OAA) followed by reduction would result in malate synthesis. Outlaw and Kennedy (1978) observed that PEP carboxylase (PEPCase) has an alkaline pH optimum and suggested that alkalinization of the cytoplasm during the generation of the pmf may activate PEPCase and result in malate synthesis during stomatal opening.

The fate of malate during stomatal closing is unclear. Malate may leave the guard cell (Van Kirk and Raschke, 1978), may be converted back to starch via gluconeogenesis, or may become a substrate for mitochondrial respiration. Current data (Outlaw, 1982) favor the last possibility. This complete pathway for starch breakdown is outlined in Figure 7.3. The net cost of malate synthesis could then be construed as the energy gain from malate respiration minus the costs of sucrose import and starch synthesis.

Maintenance energy costs in the guard cell are the costs of repair, of active transport, and of maintaining the membrane potential. The additional costs of stomatal opening are those of proton extrusion, necessary for K^+ and Cl^- uptake, and of malate synthesis.

Potential Energy Sources

The guard cell has two major energy sources: oxidative phosphorylation in the mitochondria and photophosphorylation in the chloroplasts. It has also been proposed that direct energy transduction via the blue-light photoreceptor may occur.

Mitochondria—oxidative phosphorylation. Guard cells have many mitochondria similar in size to those of the mesophyll but with better developed cristae (Allaway and Setterfield, 1972). Pearson and Milthorpe (1974) observed in *Vicia faba* a 3:1 ratio of mitochondria to chloroplasts in guard cells, as compared with a 1:1 ratio in mesophyll cells. *Vicia faba* guard cells have a dark respiration rate of 175 μmol O_2 (mg chlorophyll [chl])$^{-1}$ h^{-1} as compared with a rate for mesophyll cell protoplasts of 6 μmol O_2 (mg chl)$^{-1}$ h^{-1} (Shimazaki et al., 1983). On a per-cell basis, these values yield respiration rates of 0.350 pmol O_2 h^{-1} for guard cells and 0.948 pmol O_2 h^{-1} for mesophyll cells. Oxidative phosphorylation represents an important energy source in guard cells and is widely accepted as the energy source for stomatal opening in the dark.

Chloroplasts—photophosphorylation. Chloroplasts have been found in the guard cells of all species surveyed thus far except the orchids in the genus *Paphiopedilum* (Nelson and Mayo, 1975). On a per-cell basis, the chlorophyll contents of guard cells and of mesophyll cells of *Vicia* are 2 and 158 pg, respectively (Shimazaki et al., 1983). The classic work of Allaway and Setterfield (1972) showed that guard cells of *Vicia faba* have fewer chloroplasts than mesophyll cells and that the guard cell chloroplasts have smaller grana and more reduced thylakoids then their mesophyll counterparts. More recent work has shown that guard cell chloroplasts vary significantly across species in their morphology, in their number, and in the extent of their thylakoid membranes (D'Amelio and Zeiger, 1983; Sack, this volume).

Available evidence indicates that guard cell chloroplasts have most, if not all, of the functional capacity of mesophyll chloroplasts. Guard cell

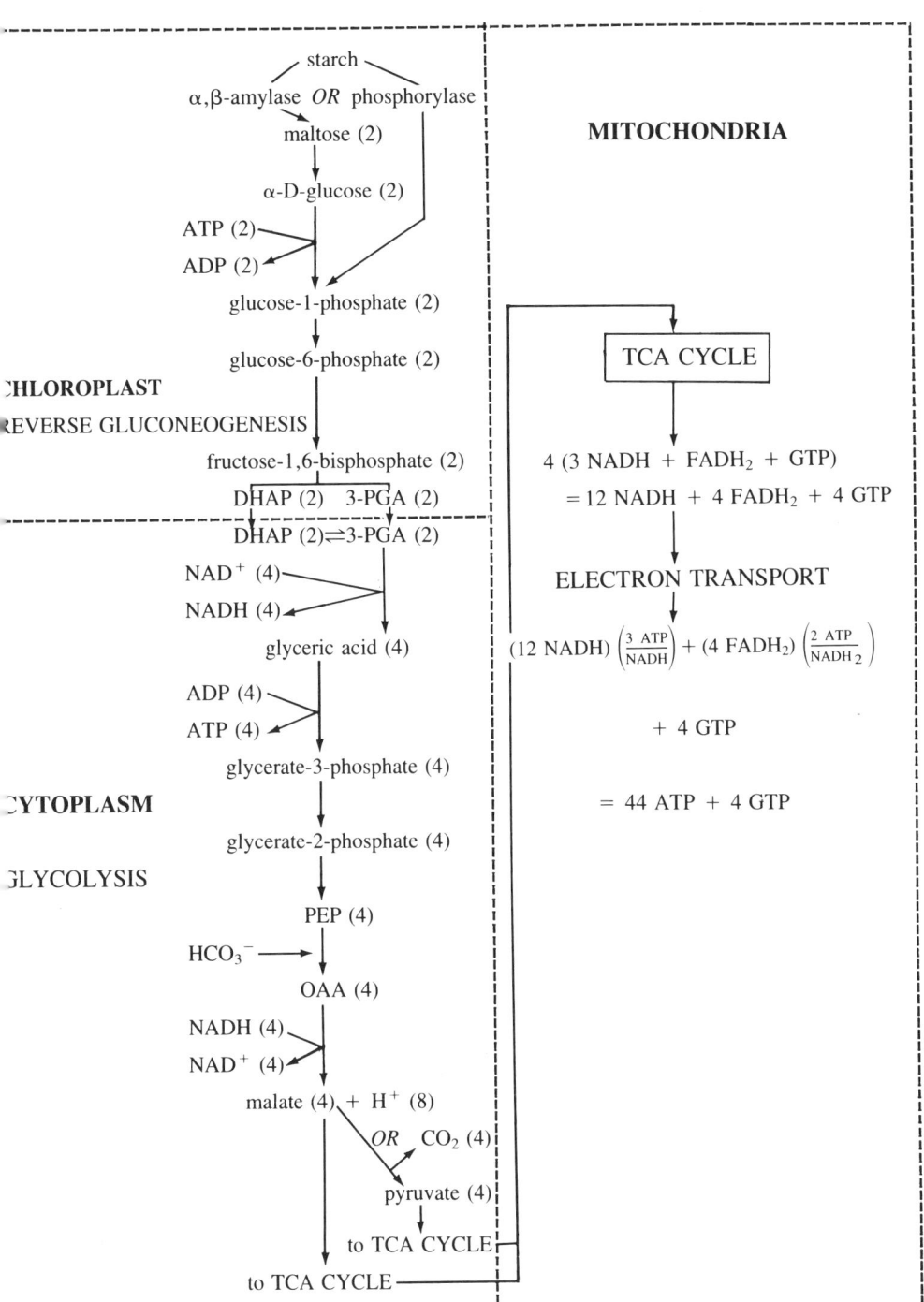

Fig. 7.3. Assumed pathway of malate formation and catabolism in the guard cell.

chloroplasts have been shown to have light-harvesting pigments of both photosystems (Outlaw et al., 1981; Zeiger, Armond, and Melis, 1981) and evolve oxygen at a rate of 150 μmol (mg chl)$^{-1}$ h^{-1} in white light (Shimazaki, Gotow, and Kondo, 1982; Shimazaki et al., 1983). Immunocytochemical evidence has confirmed the presence of light-harvesting proteins, the coupling factor of the chloroplast ATPase (Gepstein and Zeiger, 1984), and RuBPCase (Zemel and Gepstein, 1985). Studies of fluorescence transients (Melis and Zeiger, 1982; Ogawa et al., 1982; Mawson et al., 1984) and of the 518-nm electrochromic shift (Grantz, Graan, and Boyer, 1984) have provided indirect evidence for photophosphorylation, a feature directly confirmed in recent work with isolated guard cell chloroplasts of *Vicia faba* (Shimazaki and Zeiger, 1985). On a chlorophyll basis, rates of both cyclic and noncyclic photophosphorylation were found to be 50 to 80 percent of those observed with mesophyll chloroplasts. A role of photophosphorylation in guard cell chloroplasts in the supply of energy for stomatal opening is substantiated by physiological evidence showing that the PSII inhibitor 3-(3,4-dichlorophenyl)-1,1-dimethylurea (DCMU) prevents stomatal opening (Sharkey and Raschke, 1981; Schwartz and Zeiger, 1984), whereas the respiratory inhibitor potassium cyanide (KCN) has little effect on opening under red light (Schwartz and Zeiger, 1984).

Blue-light photosystem. Guard cell responses to light are mediated by a blue-light photosystem as well as by guard cell chloroplasts (see Sharkey and Ogawa, this volume). Despite the dramatic progress in our understanding of the properties of the blue-light photosystem in guard cells (see Zeiger et al., this volume), much remains to be learned about the specific phototransduction process resulting in stomatal movements and its role in the energy balance of guard cells.

Data obtained using the whole-cell patch-clamp technique (Fig. 7.1; Assmann, Simoncini, and Schroeder, 1985) provide some information on these questions. The observed hyperpolarization in response to a blue-light pulse does not begin immediately after onset of the light signal but shows a 25- to 35-second delay and does not reach a maximum for several minutes. These results suggest that blue light does not serve as a direct source of photochemical energy for H$^+$ extrusion but as an activator of a plasmalemma ATPase. Given that a role of the blue-light photosystem as a primary source of energy appears unlikely, photophosphorylation and oxidative phosphorylation remain the two obvious sources of energy for stomatal movements (see below).

A Quantitative Analysis of Guard Cell Bioenergetics

We now address the problem of guard cell bioenergetics quantitatively. We have made several assumptions in areas where our conceptual or experimental understanding of guard cell biology is incomplete. All such assumptions, however, have been clearly stated, to allow the reader to evaluate the analysis. Unless otherwise stated, all data are from experiments with *Vicia faba*, since this is the species for which the most extensive information on stomatal properties exists. When more than one value for a measurement was available from the literature, we chose the one that seemed to offer the highest reliability, based on the experimental method used. Whenever possible, we used data obtained under similar experimental conditions.

Energy Demands. The first calculation we need to make is that of a maintenance energy cost. No data are available on guard cell maintenance costs, but Penning de Vries (1975) has made some calculations at the whole-leaf level, assuming that the primary maintenance costs are replacement-molecule synthesis (equated with protein turnover) and ion transport. His calculations yielded a maintenance requirement of 13 to 23 mg of glucose per gram dry weight of leaf per day, assuming complete oxidation of glucose. From that value, we can estimate the maintenance cost of a guard cell, given a guard cell dry weight of 3 ng (Outlaw and Lowry, 1977) and an ATP yield of 36 mmol per millimole glucose:

$$\left(\frac{36 \text{ mmol ATP}}{\text{mmol glucose}}\right)\left(\frac{1 \text{ mmol glucose}}{180 \text{ mg glucose}}\right)\left(\frac{13 \text{ to } 23 \text{ mg glucose}}{\text{g dry weight} \cdot \text{day}}\right)$$

$$\left(\frac{3 \times 10^{-9} \text{ g dry weight}}{\text{guard cell}}\right)\left(\frac{1 \text{ day}}{24 \text{ h}}\right) = 0.325 \text{ to } 0.575 \frac{\text{pmol ATP}}{\text{h}}. \quad (3)$$

This basic maintenance requirement might be considered the rate of energy expenditure for a guard cell in darkness and normal air, conditions in which the stomata are closed. Let us now suppose that for a stomate of *Vicia faba*, conditions are changed such that the stomatal aperture increases from 2 to 12μm. Is it possible to evaluate an additional cost in moles of ATP for the observed stomatal opening and to estimate the contribution of each of the possible sources of energy?

Any increase in pore aperture results from a change in guard cell volume; however, this volume change occurs via the non–energy-requiring movement of water into the cell and is therefore peripheral to an estimate of the cost of stomatal opening. Sharpe, Wu, and Spence (this volume) discuss current models of the biomechanics of volume change in guard cells.

The influx of water into the guard cell results from a transient decrease in cell water potential during stomatal opening. This process can be understood if we consider a basic law of thermodynamics:

$$\psi = P - \pi \quad (4)$$

where P represents turgor pressure, π represents osmotic pressure, and ψ represents water potential. Use of this equation requires the assumption that the matrix potential, τ, is negligible. Initially the stomate is closed, $\psi_{out} = \psi_{in}$, and there is no net flux of water across the cell membrane. Signals for stomatal opening are transduced into a net increase in cell solute concentration, resulting in a greater π and therefore a decrease in ψ_{in}. With ψ_{out} greater than ψ_{in}, water flows down its potential gradient into the guard cell. The entry of water increases P_{in}, resulting in guard cell swelling and stomatal opening, until ψ_{out} once more equals ψ_{in} and equilibrium is reestablished.

The quantitative relationship between osmotic pressure and aperture is characterized by the parameter $\Delta\pi/\Delta$ *aperture*. MacRobbie (1980) explains how this parameter can be empirically obtained, and her derivation is restated in the appendix to this paper. Calculations of $\Delta\pi/\Delta$ *aperture* for isolated guard cells of *Vicia faba* give a value of 1.6 bars per micrometer of opening (Humble and Raschke, 1971). This number is an underestimate of $\Delta\pi/\Delta$ *aperture* in the intact leaf, where back pressure from the epidermal cells must also be overcome (MacRobbie, 1980, this volume).

The change in osmotic pressure is generated in the guard cells through energy-requiring accumulation of K^+, Cl^-, and malate^{2-}. Protons are also extruded from the guard cells, but their movement makes no significant contribution to $\Delta\pi$. We can write that

$$\left(\frac{\Delta\pi}{\Delta \text{ aperture}}\right)(\Delta \text{ aperture}) \approx RT \sum_j (\gamma_{j_{open}} c_{j_{open}} - \gamma_{j_{closed}} c_{j_{closed}}) \quad (5)$$

where j represents K^+, Cl^-, and malate^{2-} (for the derivation of Equation 5, see Box 7.2). Equation 5 requires measurements of the activities of these ions when the stomata are closed and open. The use of activity coefficients (γ_j) corrects for molecular interactions between solute molecules, such as those of electrostatics, which decrease the effective concentrations of solutes. As a first approximation, however, we can make the calculations assuming that the activity coefficients of the ions are equal to 1 and do not change in the course of stomatal opening. We can then use concentration values in Equation 5.

Humble and Raschke (1971) used electron-microprobe analysis to determine changes in guard cell K^+ and Cl^- content when stomata of *Vicia faba* opened from 2 to 12 μm. Van Kirk and Raschke (1978) measured changes in internal malate content over the same range of openings with an enzyme assay. Given the guard cell volumes cited in Sharpe and Wu (1978), we can calculate molar concentrations (Table 7.1). Determination of the osmolalities of solutions having the calculated osmolarities (Table 7.1) results in values that can be substituted into the right-hand side of Equation 5:

BOX 7.2. The Van't Hoff relation (after Nobel, 1974)

The activity of a substance is defined as

$$a_j = \gamma_j N_j \tag{1}$$

where γ_j is the activity coefficient of substance j, and N_j is its mole fraction, that is, the ratio of moles of j to total moles in the system. Thus, the activity of water is

$$a_w = \gamma_w N_w = \gamma_w \left(\frac{n_w}{n_w + \sum_{j \neq w} n_j} \right) = \gamma_w \left(1 - \frac{\sum_{j \neq w} n_j}{n_w + \sum_{j \neq w} n_j} \right). \tag{2}$$

For a pure substance, $\gamma_j = 1$. If the substance is almost pure, as in a very dilute solution of water, $(n_w >> \sum_{j \neq w} n_j)$ and γ_j is approximately 1. Then

$$a_w \approx 1 - \frac{\sum_{j \neq w} n_j}{n_w + \sum_{j \neq w} n_j} \tag{3}$$

and

$$\ell n\, a_w \approx \ell n \left(1 - \frac{\sum_{j \neq w} n_j}{n_w + \sum_{j \neq w} n_j} \right) \approx - \frac{\sum_{j \neq w} n_j}{n_w} \tag{4}$$

as determined by the series expansion of the logarithm $\ell n(1 - x)$.

The activity of water is also defined by the equation

$$RT \ell n\, a_w = - \bar{V}_w \pi \tag{5}$$

where $\bar{V}_w$ is the partial molal volume of water, and π is the osmotic pressure.

Substituting Equation 4 for $\ell n\, a_w$ in Equation 5, and solving for π, we obtain

$$\pi \approx RT \sum_j c_j. \tag{6}$$

The units of π are those of pressure or molality. Molality units require the expression of c_j as mol solute / kg solvent.

Equation 6 is valid for dilute solutions of ideal solutes and ideal solvents. The nonideality of solutes in real systems can be handled by replacing c_j with $\gamma_j c_j$ or a_j, giving

$$\pi \approx RT \sum_j a_j. \tag{7}$$

Equation 7 does not correct for the nonideality of real solvents and is therefore still an approximation.

Note that although π is usually referred to as osmotic pressure and is a positive term, an alternate convention often used in equations dealing with water potential is to define π as the osmotic potential, which is the negative of the osmotic pressure. Then

$$\psi = P + \pi \tag{8}$$

where π is a negative term.

$$\left(\frac{1.6 \text{ bars}}{\mu m}\right)(12 \ \mu m - 2 \ \mu m) = 16 \text{ bars} < \left(\frac{8.314 \times 10^{-5} \text{ kg bar}}{\text{mmol } °K}\right)$$

$$(298° \text{ K})\left\{\left[\left(446.1 \ \frac{\text{mmol}}{\text{kg}} \text{K}^+\right) - 38.8 \ \frac{\text{mmol}}{\text{kg}} \text{K}^+\right] + \right.$$

$$\left(23.2 \ \frac{\text{mmol}}{\text{kg}} \text{Cl}^- - 3.1 \ \frac{\text{mmol}}{\text{kg}} \text{Cl}^-\right) +$$

$$\left.\left(488.8 \ \frac{\text{mmol}}{\text{kg}} \text{ malate}^{2-} - 48.9 \ \frac{\text{mmol}}{\text{kg}} \text{ malate}^{2-}\right)\right\} = 21.5 \text{ bars.} \quad (6)$$

According to these calculations, the observed changes in solute concentration are more than adequate to cause the change in osmotic pressure required for the extent of stomatal opening observed, although it should be noted that values in the literature (Raschke, 1979) for $\Delta\pi/\Delta$ *aperture* range from 1.6 to 4.8 bars per micrometer.

We now need to calculate the energy cost for uptake of K^+ and Cl^-. Given a chemiosmotic mechanism, the cost of K^+ uptake is actually the cost of H^+ extrusion, which can be estimated using Equation 2. This cost will increase as the membrane hyperpolarizes and as ΔpH increases. We will calculate a cost range from the minimum that exists at the beginning of an opening response to the maximum that exists when a new steady-state aperture has been achieved and net H^+ efflux has ceased.

The choice of the values to use for V_m is not straightforward. Both V_m and internal ion concentrations will be affected by the concentration of ions in the external solution. Thus it would be best to use values of V_m measured in the same solutions as those used for the determination of the internal ion concentrations given in Table 7.1. Unfortunately, such data are not available for *Vicia faba*. The one report on the membrane potential of intact guard cells of *Vicia faba* (Ishikawa et al., 1983) gives an average white-light-induced hyperpolarization from -30 to -50 mV. These are the numbers we will use in subsequent calculations, keeping in mind that more negative membrane potentials have been observed for the guard cells of other species (Gunar, Zlotnikova, and Panichkin, 1975; Moody and Zeiger, 1978) and may better represent the situation of the intact leaf.

To approximate ΔpH, we can assume that the initial transmembrane proton gradient is zero (Bowling and Edwards, 1984) and use the vacuolar pH change of 0.5 units measured by Penny and Bowling (1975) in *Commelina communis*. Alternatively, we can assume that cytoplasmic pH remains constant, as has been reported by Bowling and Edwards (1984), and take the change in extracellular pH as ΔpH. This would yield a ΔpH value of approximately one unit (Raschke and Humble, 1973; Gepstein, Jacobs, and Taiz, 1982; Bowling and Edwards, 1984). To obtain an upper energy requirement, we will take the larger value, resulting in pmf ranges from

$$\text{pmf} = \left| (1)\left(\frac{96.487 \text{ J}}{\text{mol H}^+ \text{ mV}}\right)(-30 \text{ mV}) - \left(\frac{2{,}479 \text{ J}}{\text{mol}}\right)(\ell n \ 10)(0) \right| = \frac{2{,}895 \text{ J}}{\text{mol H}^+} \quad (7\text{a})$$

to

$$\text{pmf} = \left| (1)\left(\frac{96.487 \text{ J}}{\text{mol H}^+ \text{ mV}}\right)(-50 \text{ mV}) - \left(\frac{2{,}479 \text{ J}}{\text{mol}}\right)(\ell n \ 10)(1) \right| = \frac{10{,}532 \text{ J}}{\text{mol H}^+}. \quad (7\text{b})$$

In order to determine the amount of energy required to generate the pmf that actually develops across the guard cell membrane, we need an estimate of the magnitude of the H^+ efflux that occurs during stomatal opening. Accurate estimates of proton fluxes are difficult, because some of what is measured as proton "efflux" is actually ion exchange between the solution and the free space of the guard cell wall (Raschke and Humble, 1973) and because backfluxes result in a lower net proton extrusion. Raschke and Humble (1973), in experiments designed to minimize the free-space component, estimated a net flux of 0.1 to 0.55 pmol H^+ per guard cell per micrometer of opening. For 10 μm of opening, the proton flux would be

$$\left(\frac{0.1 \text{ to } 0.55 \text{ pmol H}^+}{\mu\text{m opening}}\right)(10 \ \mu\text{m}) = 1.0 \text{ to } 5.5 \text{ pmol H}^+. \quad (8)$$

The energy required for this flux of protons would be

$$(1.0 \text{ to } 5.5 \text{ pmol H}^+)\left(\frac{2{,}895 \text{ to } 10{,}532 \text{ pJ}}{\text{pmol H}^+}\right) = 2.90 \text{ to } 57.9 \text{ nJ}. \quad (9)$$

Because most calculations of cellular energetics are made on an ATP basis, we can convert these values to an ATP requirement. Energy gain per mole of ATP hydrolyzed depends on the energy charge of the cell. We will use a value typically assigned to ATP hydrolysis (Nicholls, 1982), 10 kcal per mole = 41.84 kJ per mole ATP. Thus

$$(2.90 \text{ to } 57.9 \text{ nJ})\left(\frac{1 \text{ pmol ATP}}{41.84 \text{ nJ}}\right) = 0.069 \text{ to } 1.38 \text{ pmol ATP}. \quad (10)$$

If we assume 100 percent efficiency of conversion between energy stored in the form of a pmf and the chemical-bond energy of ATP, the values in Equation 10 also represent the energy requirement for the observed amount of proton extrusion.

A different way of calculating the energy requirement for the establishment of a pmf would be to assume an ATP requirement per H^+ extruded. An accepted value for plant plasma-membrane proton pumps is $2 H^+$ extruded per ATP hydrolyzed (Poole, 1978). The average cost for extrusion of a proton will gradually increase as the electrochemical gradient against extrusion builds up; but if we take the above ratio as a rough estimate, we can calculate an energy requirement for guard cell proton extrusion of

$$\left(\frac{1.0 \text{ to } 5.5 \text{ pmol } H^+}{\text{guard cell}}\right)\left(\frac{1 \text{ pmol ATP}}{2 \text{ pmol } H^+}\right) = \frac{0.50 \text{ to } 2.75 \text{ pmol ATP}}{\text{guard cell}}. \quad (11)$$

The estimates in Equation 10 and 11 yield a range of values for the cost of H^+ extrusion. We will use Equation 11 rather than Equation 10 for subsequent calculations for three reasons: Equation 11 requires fewer assumptions; in Equation 11 the data on ATP costs of H^+ pumping are better substantiated; and Equation 11 is more likely to be an overestimate, and our practice has been to use maximum rather than minimum estimates.

The next step is the calculation of the energy requirement for Cl^- uptake. If we assume that Cl^- uptake is coupled with H^+ backflux, the cost of Cl^- uptake can be calculated as the cost of extruding those protons that later reenter the cell coupled with Cl^-. This cost was not included previously, because those calculations were based on net H^+ efflux, which is the parameter measured experimentally. We can, however, estimate the additional H^+ efflux required to drive Cl^- uptake by assuming a 1:1 coupling of Cl^- uptake and H^+ backflux.

Given the observed amount of Cl^- uptake (Table 7.1), and assuming the same energetics for H^+ extrusion as in Equation 11, we can calculate the cost for Cl^- uptake as

$$(0.102 \text{ pmol } Cl^-)\left(\frac{1 \text{ pmol } H^+ \text{ extruded and backfluxed}}{1 \text{ pmol } Cl^-}\right)$$

$$\left(\frac{1 \text{ pmol ATP}}{2 \text{ pmol } H^+ \text{ extruded}}\right) = 0.051 \text{ pmol ATP}. \quad (12)$$

TABLE 7.2. Malate metabolism: energy budget for one 12-carbon unit
(= 1 sucrose = 4 malate)

Energy supply[a]	
substrate level: − 2 ATP + 4 NADH + 4 ATP − 4 NADH	= 2 ATP
TCA cycle and	
electron transport	44 ATP + 4 GTP
net	46 ATP + 4 GTP
Energy cost	
sucrose import[b]	1.4 ATP
glycolysis	4 ATP
gluconeogenesis	2 ATP
net	7.4 ATP
Net energy gain (supply − cost)	38.6 ATP + 4 GTP

[a] Assuming starch breakdown by amylase.
[b] Assuming import via a chemiosmotic mechanism similar to that occurring in the phloem (Giaquinta, 1983).

Adding Equations 11 and 12, we obtain an energy requirement for ion uptake of

$$\left(\frac{0.50 \text{ pmol ATP}}{\text{total K}^+ \text{ uptake}} + \frac{0.051 \text{ pmol ATP}}{\text{total Cl}^- \text{ uptake}}\right) \text{ to}$$

$$\left(\frac{2.75 \text{ pmol ATP}}{\text{total K}^+ \text{ uptake}} + \frac{0.051 \text{ pmol ATP}}{\text{total Cl}^- \text{ uptake}}\right)$$

$$= 0.55 \text{ to } 2.80 \text{ pmol ATP}. \quad (13)$$

A complete calculation of the cost of stomatal opening also requires the computation of a cost of malate synthesis. For reasons outlined above (see p. 170) we will consider only the cost of the biochemical reactions involved in the breakdown of starch to malate. Assuming the pathway of malate synthesis given in Figure 7.3, we see that the synthesis of 4 mol malate actually results in a gain of 2 mol ATP (0.5 ATP per malate). Given (Table 7.1) an increase in guard cell malate content of 2.20 pmol, we can calculate that malate synthesis during stomatal opening results in an energy gain to the guard cell of

$$\left(\frac{2.20 \text{ pmol malate}}{\text{guard cell}}\right)\left(\frac{0.5 \text{ pmol ATP}}{\text{pmol malate}}\right) = 1.10 \text{ pmol ATP}. \quad (14)$$

Equation 14 shows that malate synthesis from starch is a potential source of energy of possible importance during phases of stomatal opening that impose a large energy demand on the guard cells.

For the sake of completeness, we have also calculated a total energy budget for malate (Table 7.2). The values indicate that, if guard cells are indeed heterotrophs and import sucrose as a carbon source, the whole process of sucrose import, starch synthesis, starch breakdown to malate, and malate respiration actually results in a net energy gain to the guard cell. This gain occurs because the import of sucrose represents not only import of carbon but also a net import of energy, as stored in the chemical bonds of the sucrose molecules. The considerable cost of reducing CO_2 to sucrose is not borne by the guard cells but by the mesophyll cells, where the sucrose is synthesized.

It is noteworthy, however, that some available evidence is consistent with photosynthetic CO_2 fixation in guard cell chloroplasts: first a CO_2 sensitivity (typically associated with carbon-fixation reactions) of fluorescence transients from guard cell chloroplasts of *Chlorophytum comosum* (Melis and Zeiger, 1982); second, the red-light-induced alkalinization of a suspension medium of *Vicia faba* guard cell protoplasts (Shimazaki, Iino, and Zeiger, 1986; Zeiger et al., this volume); and third, the immunocytochemical demonstration of RuBPCase in guard cell chloroplasts of *Vicia faba* (Zemel and Gepstein, 1985). Because the use of energy for CO_2 fixation would significantly alter the energy budget of the guard cell, it appears that more data are needed before a definitive analysis of the bioenergetics of carbon metabolism in guard cells can be made.

The total guard cell energy requirement during the opening period should also include the maintenance cost (calculated from Equation 3) during the three-hour opening process:

$$\left(\frac{0.55 \text{ to } 2.30 \text{ pmol ATP}}{3 \text{ h stomatal opening}}\right) + \left(\frac{0.32 \text{ to } 0.58 \text{ pmol ATP}}{1 \text{ h maintenance}}\right)(3 \text{ h})$$

$$= 1.51 \text{ to } 4.04 \text{ pmol ATP.} \quad (15)$$

Energy Supply. The analysis of energy supply for stomatal opening is even more difficult than that of energy cost. The relationship between ion fluxes and aperture is thought to be largely invariant, regardless of the conditions leading to the opening; whereas energy sources may vary with the nature of the opening signal. To simplify the analysis, we have considered energy supply for stomatal opening under only three sets of conditions, each set requiring some assumptions.

The first set of conditions is that of darkness and CO_2-free air, where it is assumed that the energy for stomatal opening is derived solely from mitochondrial respiration. This assumption is supported by the observation that dark opening is essentially nil in the presence of KCN (Schwartz and Zeiger, 1984) or sodium azide (Raghavendra, 1981). Energy supply can be calculated from measurements of O_2 uptake (Shimazaki et al., 1983). If we ignore substrate-level phosphorylation occurring during glycolysis and consider only the ATP formation coupled with electron flow, and if we assume an efficiency of 6 mol ATP produced per mole of O_2 consumed, we obtain an oxidative phosphorylation rate of

$$\left(\frac{0.175 \text{ pmol } O_2}{\text{pg chl} \cdot \text{hour}}\right)\left(\frac{2.0 \text{ pg chl}}{\text{guard cell}}\right)\left(\frac{6 \text{ pmol ATP}}{\text{pmol } O_2}\right) = \frac{2.10 \text{ pmol ATP}}{\text{guard cell} \cdot \text{hour}}. \quad (16)$$

We see that even without substrate-level phosphorylation the guard cell can produce in one hour almost half the maximum amount of ATP required for 10 μm of stomatal opening plus maintenance (Eq. 15). If we assume that the rate of fusicoccin-induced stomatal opening in the dark is the same as that of opening in the light (Zeiger, 1983), it would take three hours for the stomata to open 10 μm (Humble and Raschke, 1971; Gepstein, Jacobs, and Taiz, 1982). In that time, the guard cells would have produced approximately 1.4 times the ATP required for stomatal opening. The calculated excess warrants the conclusion that energy supply from oxidative phosphorylation is not limiting for stomatal opening in the dark.

A second set of conditions is red light of moderate intensity and normal air. We assume that the blue-light photoreceptor is not operative under these conditions and that respiration is not involved in energy supply. The first assumption seems logically justified; the second is supported by data of Schwartz and Zeiger (1984) that show no significant KCN inhibition of stomatal opening under red light. Based on evidence discussed previously (see pp. 172–74), we shall assume that photophosphorylation is the source of energy for stomatal opening in these conditions.

Shimazaki, Gotow, and Kondo (1982) have reported rates of O_2 evolution from guard cell protoplasts, measured under 0.788 mmol m^{-2} s^{-1} (saturating) white light. Assuming for simplicity of calculation that quantum yields of O_2 evolution under red and white light are the same, and assuming a noncyclic photophosphorylation stoichiometry of 3 mol ATP produced per mole O_2 evolved (Noble, 1983), then

$$\left(\frac{0.150 \text{ pmol } O_2}{\text{pg chl} \cdot \text{h}}\right)\left(\frac{2.0 \text{ pg chl}}{\text{guard cell}}\right)\left(\frac{3 \text{ pmol ATP}}{\text{pmol } O_2}\right)(3 \text{ h})$$

$$= \frac{2.70 \text{ pmol ATP}}{\text{guard cell}}. \quad (17)$$

This calculated value is within the range obtained in Equation 15 and is thus consistent with the notion that photophosphorylation capacity in guard cells is sufficient to supply the cost of stomatal opening and cell maintenance.

Alternative calculations for this set of conditions can be made with the rates of photophosphorylation directly measured in guard cell chloroplasts, which were 0.103 and 0.193 pmol ATP $[\text{pg chl}]^{-1}$ h^{-1} for noncyclic and cyclic photophosphorylation, respectively, under 0.610 mmol m^{-2} s^{-1} (saturating) orange light (Shimazaki and Zeiger, 1985). The ATP yield of photophosphorylation can be then calculated as

$$\left(\frac{0.103 \text{ pmol}}{\text{pg chl}}\right)\left(\frac{2.0 \text{ pg chl}}{\text{guard cell}}\right)(3 \text{ h}) = \frac{0.618 \text{ pmol ATP}}{\text{guard cell}}. \quad (18)$$

This value is lower than that yielded by Equation 17, a discrepancy most likely due to the lower photophosphorylation yields observed *in vitro* as compared with those prevailing *in vivo*.

The third set of conditions is that of low-intensity blue light and normal air. Earlier calculations for the bioenergetics of the stomatal blue-light response (Raven, 1981; Travis and Mansfield, 1981) have had to rely on extrapolations from other blue-light systems because of the paucity of data on the guard cell system. Recently, however, data on blue-light-stimulated stomatal opening (Schwartz and Zeiger, 1984), proton extrusion (Shimazaki, Iino, and Zeiger, 1986), and electrogenic ion pumping (Assmann, Simoncini, and Schroeder, 1985) have become available, allowing development of an analysis similar to that applied to white-light-induced stomatal opening.

Epidermal peels of *Vicia faba* irradiated with 0.025 mmol m^{-2} s^{-1} of blue light for three hours show a change in average stomatal aperture from 7.5 to 13 μm (Schwartz and Zeiger, 1984). According to Equation 8, the observed opening would require an H$^+$ flux of

$$\left(\frac{1.0 \text{ to } 5.5 \text{ pmol H}^+}{10 \text{ μm opening}}\right)(5.5 \text{ μm opening}) = 0.55 \text{ to } 3.02 \text{ pmol H}^+. \quad (19)$$

Data on blue-light-induced proton extrusion and membrane hyperpolarization measured with *Vicia* guard cell protoplasts allow us to compare this estimated requirement with actual guard cell responses. The data of Shimazaki, Iino, and Zeiger (1986) indicate that guard cell protoplasts stimulated with a 30-second pulse of 0.025 mmol m^{-2} s^{-1} blue light extrude protons at a maximum rate of about 5.0 pmol per square centimeter of plasmalemma per second, which, given an observed average guard cell protoplast diameter of 16 μm under their conditions, can be converted to

$$\left(\frac{5.0 \text{ pmol H}^+}{\text{cm}^2 \text{ s}}\right)\left(\frac{4\pi(8 \times 10^{-4})^2 \text{cm}^2 \text{ surface area}}{\text{guard cell}}\right)\left(\frac{60 \text{ sec}}{\text{min}}\right)$$
$$\left(\frac{60 \text{ min}}{\text{h}}\right)(3 \text{ h}) = \frac{0.43 \text{ pmol H}^+}{\text{guard cell}}. \quad (20)$$

The patch-clamp experiments measuring the electrical responses of guard cell protoplasts stimulated with blue light (Assmann, Simoncini, and Schroeder, 1985) provide an opportunity for an independent assessment of the energy requirement for proton extrusion. The measured membrane hyperpolarization resulted from current flowing out of the cell, with the data on proton extrusion (Shimazaki, Iino, and Zeiger, 1986) indicating that the ions carrying the charge were protons. It was observed that a 30-second pulse of 0.050 mmol m^{-2} s^{-1} blue light could result in an outward current from a single guard cell of 5.5 pA (Fig. 7.1d). Since current is flow of charge per unit time, the value of 5.5 pA can be converted to a rate of proton extrusion. Given that 1 ampere is 1 coulomb per second, and that 1 coulomb is 6×10^{18} units of charge (in this case, protons), we have

$$(5.5 \times 10^{-12} \text{A}) \left(\frac{1 \text{ coulomb s}^{-1}}{\text{A}} \right) \left(\frac{6 \times 10^{18} \text{ protons}}{\text{coulomb}} \right)$$

$$= 33 \times 10^6 \text{ protons s}^{-1}. \quad (21)$$

If we assume that this rate of proton extrusion is maintained during the opening process, total proton extrusion over a three-hour period would be

$$\left(\frac{33 \times 10^6 \text{ protons}}{\text{s}} \right) \left(\frac{1 \text{ mol protons}}{6.02 \times 10^{23} \text{ protons}} \right) \left(\frac{60 \text{ s}}{\text{min}} \right) \left(\frac{60 \text{ min}}{\text{h}} \right) (3 \text{ h})$$

$$= 0.59 \text{ pmol H}^+. \quad (22)$$

We see that the agreement between the estimate for proton extrusion obtained using Equation 8 and the values derived from actual measurements shown in Equations 20 and 22, each using data from a different experimental method, is reasonably good. This agreement lends confidence to the analytical methods that we have used to quantify the energy relations of guard cells.

If we take 0.59 pmol H$^+$ as the amount of protons extruded, and assume, as in Equation 11, that 2 H$^+$ are extruded per ATP hydrolysed and that the amount of Cl$^-$ taken up under blue light is the same as under white light (Eq. 12), the total energy required for the blue-light-induced stomatal opening would be

$$(0.59 \text{ pmol H}^+) \left(\frac{1 \text{ pmol ATP}}{2 \text{ pmol H}^+} \right) + 0.051 \text{ pmol ATP} +$$

$$\left(\frac{0.32 \text{ to } 0.58 \text{ pmol ATP}}{\text{h maintenance}} \right) (3 \text{ h}) = 1.31 \text{ to } 2.09 \text{ pmol ATP}. \quad (23)$$

We can now address the question of the possible energy sources supplying the ATP requirement. With the blue-light photosystem excluded as a direct source of energy, as discussed previously (p. 174), photophosphorylation and oxidative phosphorylation will be analyzed in turn.

The contribution from photophosphorylation can be estimated directly from the data of Shimazaki and Zeiger (1985), under the assumption of equal quantum yields in orange and blue light. At 0.025 mmol m^{-2} s^{-1} light, their data give a value of 0.015 pmol ATP formed per picogram of chlorophyll per hour. Converting to a per-guard-cell basis, the ATP produced over a three-hour period would be:

$$\left(\frac{0.015 \text{ pmol ATP}}{\text{pg chl} \cdot \text{h}}\right)\left(\frac{2.0 \text{ pg chl}}{\text{guard cell}}\right)(3 \text{ h}) = \frac{0.090 \text{ pmol ATP}}{\text{guard cell}}. \quad (24)$$

The calculation of Equation 24 suggests that blue-light-driven photophosphorylation alone does not produce enough energy for the blue-light response. The calculations also underscore the high apparent quantum efficiency of the blue-light photosystem. If, through a yet-unknown phototransduction process, the blue-light photosystem were to be the source of energy for the observed extrusion of protons, it would have to be over 14 times more efficient than the calculated production of ATP from guard cell photophosphorylation:

$$\frac{1.31 \text{ to } 2.09 \text{ pmol ATP required}}{0.090 \text{ pmol ATP from photophosphorylation}} = 14.6 \text{ to } 23.2. \quad (25)$$

Such a high phototransduction efficiency is unknown in photobiological processes in which white light is the primary source of energy.

A second potential source of energy for the blue-light response is oxidative phosphorylation. The observation of a blue-light response in guard cell protoplasts provided with an excess of ATP under patch-clamp conditions (Assmann, Simoncini, and Schroeder, 1985) makes it unlikely that the observed blue-light activation of the electrogenic pump is a result of a blue-light-dependent enhancement of respiration (Kowallik, 1983) that relieves an energy limitation on H$^+$ pumping. Rather, the results suggest that blue light causes utilization of the ATP that was available prior to the pulse. Calculations (see Eq. 16) indicate that oxidative phosphorylation could provide sufficient energy to drive blue-light-stimulated proton extrusion; furthermore, the blue-light response is inhibited by the respiratory poison KCN (Schwartz and Zeiger, 1984), a result that suggests an energetic role of oxidative phosphorylation in the response. However, secondary effects of KCN, for example, inhibition of an electron-transport chain involved in transduction of the blue-light signal (Zeiger, 1984), cannot be ruled out.

One final potential source of energy for the blue-light response is the breakdown of starch to malate. As illustrated in Figure 7.2, for each mole of malate produced, 0.5 mol ATP are formed. We can calculate that the amount of malate that would have to be produced in order to drive blue-light-induced stomatal opening would be

$$(1.31 \text{ to } 2.09 \text{ pmol ATP required}) \left(\frac{1 \text{ pmol malate}}{0.5 \text{ pmol ATP}} \right)$$

$$= 2.62 \text{ to } 4.18 \text{ pmol malate.} \quad (26)$$

Under 0.025 mmol m^{-2} s^{-1} blue light, a value of 3.5 nmol malate formed per milligram dry weight of epidermis per hour has been reported by Ogawa et al. (1978). Given values for dry weight per unit area of epidermis and stomatal density (Pearson, 1973), we can calculate the amount of malate formed in a single guard cell under these conditions:

$$\left(\frac{3.5 \times 10^{-9} \text{ mol malate}}{1 \times 10^{-3} \text{ g epidermis} \cdot \text{h}} \right) \left(\frac{4.6 \times 10^{-4} \text{ g epidermis}}{\text{cm}^2 \text{ epidermis}} \right)$$

$$\left(\frac{1 \text{ cm}^2 \text{ epidermis}}{12 \times 10^3 \text{ guard cells}} \right) (3 \text{ h}) = \frac{0.4 \text{ pmol malate}}{\text{guard cell}}. \quad (27)$$

According to these calculations, it is unlikely that malate provides more than about one-sixth of the energy required for blue-light-stimulated stomatal opening.

Current experimental data fail to pinpoint a specific energy source for the blue-light response. Further elucidation of the details of the phototransduction process associated with the blue-light-dependent modulation of stomatal movements should help in the understanding of this intriguing aspect of guard cell biology.

Our analysis of the qualitative and quantitative properties of guard cell bioenergetics points to the following conclusions. First, chemiosmosis appears to be the mechanism driving the ion fluxes. Second, based on estimates of guard-cell oxidative phosphorylation and photophosphorylation, the guard cells appear to be "overdimensioned" in terms of energy supply. The quantitative analysis of guard cell bioenergetics presented in this paper suggests that respiration or photophosphorylation alone can often supply sufficient energy for the proton extrusion that drives stomatal opening. Thus, under such conditions as white light and low intercellular CO_2 concentrations, where both these energy sources could be available, more than enough energy would be available to drive stomatal opening. Presumably there would be an integrated regulation of all energy sources, for maximum efficiency of energy production and use. These considerations underscore the importance of evaluating cellular bioenergetics when investigating physiological processes.

Finally, the calculations of this chapter point to three areas of guard cell bioenergetics for which information is notably lacking: intracellular ion activities, intracellular pH changes during stomatal opening, and the energetics of the blue-light response. Further research is certain to increase our understanding of these aspects of guard cell physiology and of their impact on the bioenergetics of stomatal movements.

APPENDIX: Determination of $d\pi/d$ *aperture* ($\Delta\pi/\Delta$ *aperture*) (after MacRobbie, 1980)

We start with the basic equation for equilibrium of water across a cell membrane:

$$\psi_{out} = \psi_{in} \tag{1}$$

or

$$P_{out} - \pi_{out} = P_{in} - \pi_{in}. \tag{2}$$

For the extracellular space, there is no turgor pressure, so $P_{out} = 0$; hence,

$$-\pi_{out} = P_{in} - \pi_{in} \tag{3}$$

or, using the definition of π derived in Box 2:

$$-\pi_{out} = P_{in} - RT\left(\frac{Q}{V}\right) \tag{4}$$

where Q is the total moles of solute, and V the total mass of solvent. Writing the complete differential of Equation 4 gives

$$d(-\pi_{out}) = dP_{in} - \frac{RT}{V} dQ + RT \frac{Q}{V^2} dV. \tag{5}$$

When we experimentally change π_{out} by adding solute to the external solution, we assume $dQ = 0$. Then

$$-d\pi_{out} = dP_{in} + RT \frac{Q}{V^2} dV. \tag{6}$$

During actual stomatal opening, there is no net change in the internal or external water potential, i.e. $-d\pi_{out} - d\psi_{in} - 0$. Equation 5 can therefore be set equal to 0:

$$0 = dP_{in} - \frac{RT}{V} dQ + RT \frac{Q}{V^2} dV \tag{7}$$

or

$$RT\frac{dQ}{V} = dP_{in} + RT\frac{Q}{V^2}dV. \tag{8}$$

That is, the change in solute concentration is balanced by changes in turgor pressures and volume. We see that the right-hand sides of Equations 6 and 8 are equal; therefore,

$$-d\pi_{out} = RT\frac{dQ}{V}. \tag{9}$$

If we determine changes in aperture with known changes in external solute concentration (a relatively straightforward procedure), we can use Equation 9 to estimate the dependence of the aperture change on the change in internal solute concentration of the guard cell that occurs during stomatal opening:

$$\frac{-d\pi_{out}}{d\ aperture} = \frac{RT\frac{dQ}{V}}{d\ aperture} \tag{10}$$

or

$$\frac{-d\pi_{out}}{d\ aperture} = \frac{RT\ dc}{d\ aperture}. \tag{11}$$

For nonideal solutes, the following equation is more correct:

$$\frac{-d\pi_{out}}{d\ aperture} = RT\frac{da}{d\ aperture}. \tag{12}$$

REFERENCES

Alberts, B., D. Bray, J. Lewis, M. Raff, K. Roberts, and J. D. Watson. 1983. *Molecular biology of the cell*: 290–94. New York: Garland.

Allaway, W. G., and G. Setterfield. 1972. Ultrastructural observations on guard cells of *Vicia faba* and *Allium porrum*. *Canadian Journal of Botany*, 50: 1405–13.

Assmann, S. M., L. Simoncini, and J. I. Schroeder. 1985. Blue light activates electrogenic ion pumping in guard cell protoplasts of *Vicia faba*. *Nature*, 318: 285–87.

Bowling, D. J. F., and A. Edwards. 1984. pH gradients in the stomatal complex of *Tradescantia virginiana*. *Journal of Experimental Botany*, 35: 1641–45.

Briskin, D. P., and R. J. Poole. 1983. The plasma membrane ATPase of higher plant cells. *What's New in Plant Physiology*, 14: 1–4.

D'Amelio, E., and E. Zeiger. 1983. Structural and functional divergence of guard cell plastids in the Orchidaceae. *Plant Physiology*, 72s: 101.

Edwards, A., and D. J. F. Bowling. 1984. An electrophysiological study of the stomatal complex of *Tradescantia virginiana*. *Journal of Experimental Botany*, 35: 562–67.

Gepstein, S., M. Jacobs, and L. Taiz. 1982. Inhibition of stomatal opening in *Vicia faba* epidermal tissue by vanadate and abscisic acid. *Plant Science Letters*, 28: 63–72.

Gepstein, S., and E. Zeiger. 1984. Immunocytochemistry of guard cell chloroplasts (GC-CH). *Plant Physiology*, 75s: 129.

Giaquinta, R. T. 1983. Phloem loading of sucrose. *Annual Review of Plant Physiology*, 34: 347–87.

Grantz, D. A., T. Graan, and J. S. Boyer. 1984. Chloroplast function in guard cells of *Vicia faba* L.: Measurement of the electrochromic absorbance change at 518 nm. *Plant Physiology*, 77: 956–62.

Gunar, I. I., I. F. Zlotnikova, and L. A. Panichkin. 1975. Electrophysiological investigation of cells of the stomate complex in spiderwort. *Soviet Plant Physiology*, 22: 704–7.

Hamill, O. P., A. Marty, E. Neher, B. Sakmann, and F. J. Sigworth. 1981. Improved patch-clamp techniques for high-resolution current recording from cells and cell-free membrane patches. *Pflügers Archiv*, 391: 85–100.

Heber, U., and H. W. Heldt. 1981. The chloroplast envelope: Structure, function, and role in leaf metabolism. *Annual Review of Plant Physiology*, 32: 139–68.

Humble, G. D., and T. C. Hsiao. 1970. Light-dependent influx and efflux of potassium of guard cells during stomatal opening and closing. *Plant Physiology*, 46: 483–87.

Humble, G. D., and K. Raschke. 1971. Stomatal opening quantitatively related to potassium transport. Evidence from electron probe analysis. *Plant Physiology*, 48: 447–53.

Iino, M., T. Ogawa, and E. Zeiger. 1985. Kinetic properties of the blue light response of stomata. *Proceedings of the National Academy of Sciences*, 82: 8019–23.

Ishikawa, H., H. Aizawa, H. Kishira, T. Ogawa, and M. Sakata. 1983. Light-induced changes of membrane potential in guard cells of *Vicia faba*. *Plant and Cell Physiology*, 24: 769–72.

Kaiser, W. M., and J. A. Bassham. 1979. Carbon metabolism of chloroplasts in the dark: Oxidative pentose phosphate cycle versus glycolytic pathway. *Planta*, 144: 193–200.

Kowallik, W. A. 1982. Blue light effects on respiration. *Annual Review of Plant Physiology*, 33: 51–72.

MacRobbie, E. A. C. 1980. Osmotic measurements on stomatal cells of *Commelina communis* L. *Journal of Membrane Biology*, 53: 189–98.

Marty, A., and E. Neher. 1983. Tight-seal whole-cell recording. In B. Sakmann and E. Neher, eds., *Single channel recording*: 107–22. New York: Plenum.

Mawson, B. T., A. Franklin, W. G. Filion, and W. R. Cummins. 1984. Comparative studies of fluorescence from mesophyll and guard cell chloroplasts in *Saxifraga cernua*. *Plant Physiology*, 74: 481–86.

Melis, A., and E. Zeiger. 1982. Chlorophyll *a* fluorescence transients in mesophyll and guard cells. Modulation of guard cell photophosphorylation by CO_2. *Plant Physiology*, 69: 642–47.

Moody, W., and E. Zeiger. 1978. Electrophysiological properties of onion guard cells. *Planta*, 139: 159–65.

Nelson, S. D., and J. M. Mayo. 1975. The occurrence of functional non-chlorophyllous guard cells in *Paphiopedilum* spp. *Canadian Journal of Botany*, 53: 1–7.

Nicholls, D. G. 1982. *Bioenergetics: An introduction to the chemiosmotic theory.* New York: Academic.

Nobel, P. S. 1974. *Biophysical plant physiology.* San Francisco: Freeman.

———. 1983. *Biophysical plant physiology and ecology.* San Francisco: Freeman.

Ogawa, T., D. Grantz, J. Boyer, and Govindjee. 1982. Effects of cations and abscisic acid on chlorophyll *a* fluorescence in guard cells of *Vicia faba*. *Plant Physiology*, 69: 1140–44.

Ogawa, T., H. Ishikawa, K. Shimada, and K. Shibata. 1978. Synergistic action of red and blue light and action spectra for malate formation in guard cells of *Vicia faba* L. *Planta*, 142: 61–65.

O'Leary, M. H. 1982. Phosphoenolpyruvate carboxylase: An enzymologist's view. *Annual Review of Plant Physiology*, 33: 297–315.

Outlaw, W. H., Jr. 1982. Carbon metabolism in guard cells. In L. L. Creasy and G. Hrazdina, eds., *Cellular and subcellular localization in plant metabolism:* 185–222. New York: Plenum.

Outlaw, W. H., Jr., and D. B. Fisher. 1975. Compartmentation in *Vicia faba* leaves, I. Kinetics of ^{14}C in the tissues following pulse labeling. *Plant Physiology*, 55: 699–703.

Outlaw, W. H., Jr., D. B. Fisher, and A. L. Christy. 1975. Compartmentation in *Vicia faba* leaves, II. Kinetics of ^{14}C-sucrose redistribution among individual tissues following pulse labeling. *Plant Physiology*, 55: 704–11.

Outlaw, W. H., Jr., and J. Kennedy. 1978. Enzymic and substrate basis for the anaplerotic step in guard cells. *Plant Physiology*, 62: 648–52.

Outlaw, W. H., Jr., and O. H. Lowry. 1977. Organic acid and potassium accumulation in guard cells during stomatal opening. *Proceedings of the National Academy of Sciences*, 74: 4434–38.

Outlaw, W. H., Jr., and J. Manchester. 1979. Guard cell starch concentration quantitatively related to stomatal aperture. *Plant Physiology*, 64: 79–82.

Outlaw, W. H., Jr., B. C. Mayne, V. E. Zenger, and J. Manchester. 1981. Presence of both photosystems in guard cells of *Vicia faba* L.: Implications for environmental signal processing. *Plant Physiology*, 67: 12–16.

Pearson, C. J. 1973. Daily changes in stomatal aperture and in carbohydrates and malate within epidermis and mesophyll of leaves of *Commelina cyanea* and *Vicia faba*. *Australian Journal of Biological Sciences*, 26: 1035–44.

Pearson, C. J., and F. L. Milthorpe. 1974. Structure, carbon dioxide fixation and metabolism of stomata. *Australian Journal of Plant Physiology*, 1: 221–36.

Penning de Vries, F. W. T. 1975. The cost of maintenance processes in plant cells. *Annals of Botany*, 39: 77–92.

Penny, M. G., and D. J. F. Bowling. 1975. Direct determination of pH in the stomatal complex of *Commelina*. *Planta*, 122: 209–12.

Poole, R. J. 1978. Energy coupling for membrane transport. *Annual Review of Plant Physiology*, 29: 437–60.

Raghavendra, A. S. 1981. Energy supply for stomatal opening in epidermal strips of *Commelina benghalensis*. *Plant Physiology*, 67: 385–87.

Raschke, K. 1979. Movements of stomata. In W. Haupt and M. E. Feinleib, eds.,

Physiology of movements, Encyclopedia of Plant Physiology, n.s., 7: 383–441. Berlin: Springer-Verlag.

Raschke, K., and G. D. Humble. 1973. No uptake of anions required by opening stomata of *Vicia faba:* Guard cells release hydrogen ions. *Planta,* 115: 47–57.

Raschke, K., and H. Schnabl. 1978. Availability of chloride affects the balance between potassium chloride and potassium malate in guard cells of *Vicia faba* L. *Plant Physiology,* 62: 84–87.

Raven, J. A. 1981. Light quality and solute transport. In H. Smith, ed., *Plants and the daylight spectrum*: 375–90. New York: Academic.

Saftner, R. A., and K. Raschke. 1981. Electrical potentials in stomatal complexes. *Plant Physiology,* 67: 1124–32.

Sakmann, B., and E. Neher, eds. 1983. *Single channel recording.* New York: Plenum.

Schnabl, H., and K. Raschke. 1980. Potassium chloride as stomatal osmoticum in *Allium cepa* L., a species devoid of starch in guard cells. *Plant Physiology,* 65: 88–93.

Schnabl, H., and H. Ziegler. 1977. The mechanism of stomatal movement in *Allium cepa* L. *Planta,* 136: 37–43.

Schroeder, J. I., R. Hedrich, and J. M. Fernandez. 1984. Potassium-selective single channels in guard cell protoplasts of *Vicia faba. Nature,* 312: 361–62.

Schwartz, A., and E. Zeiger. 1984. Metabolic energy for stomatal opening: Roles of photophosphorylation and oxidative phosphorylation. *Planta,* 161: 129–36.

Sharkey, T. D., and K. Raschke. 1981. Effect of light quality on stomatal opening in leaves of *Xanthium strumarium* L. *Plant Physiology,* 68: 1170–74.

Sharpe, P. J. H., and H. Wu. 1978. Stomatal mechanics, I. Volume changes during opening. *Plant, Cell and Environment,* 1: 259–68.

Shimazaki, K., K. Gotow, and N. Kondo. 1982. Photosynthetic properties of guard cell protoplasts from *Vicia faba* L. *Plant and Cell Physiology,* 23: 871–79.

Shimazaki, K., K. Gotow, T. Sakaki, and N. Kondo. 1983. High respiratory activity of guard cell protoplasts from *Vicia faba* L. *Plant and Cell Physiology,* 24: 1049–56.

Shimazaki, K., M. Iino, and E. Zeiger. 1986. Blue light–dependent proton extrusion by guard cell protoplasts of *Vicia faba. Nature,* 319: 324–26.

Shimazaki, K., and E. Zeiger. 1985. Cyclic and non-cyclic photophosphorylation in isolated guard cell chloroplasts from *Vicia faba* L. *Plant Physiology,* 78: 211–14.

Travis, A. J., and T. A. Mansfield. 1981. Light saturation of stomatal opening on the adaxial and abaxial epidermis of *Commelina communis. Journal of Experimental Botany,* 32: 1169–79.

Van Kirk, C. A., and K. Raschke. 1978. Presence of chloride reduces malate production in epidermis during stomatal opening. *Plant Physiology,* 61: 361–64.

Walker, D. A., and I. Zelitch. 1963. Some effects of metabolic inhibitors, temperature and anaerobic conditions on stomatal movement. *Plant Physiology,* 38: 390–96.

Zeiger, E. 1983. The biology of stomatal guard cells. *Annual Review of Plant Physiology,* 34: 441–75.

———1984. Blue light and stomatal function. In H. Senger, ed., *Blue light effects in biological systems*: 484–93. Berlin: Springer-Verlag.

Zeiger, E., P. Armond, and A. Melis. 1981. Fluorescence properties of guard cell chloroplasts. *Plant Physiology,* 67: 17–20.

Zeiger, E., A. J. Bloom, and P. K. Hepler. 1978. Ion transport in stomatal guard cells: A chemiosmotic hypothesis. *What's New in Plant Physiology,* 9: 29–31.

Zeiger, E., W. Moody, P. Hepler, and F. Varela. 1977. Light-sensitive membrane potentials in onion guard cells. *Nature,* 270: 270–71.

Zeiger, E., and A. Schwartz. 1982. Longevity of guard cell chloroplasts in falling leaves: Implication for stomatal function and cellular aging. *Science*, 218: 680–82.

Zemel, E., and S. Gepstein. 1985. Immunological evidence for the presence of ribulose biphosphate carboxylase in guard cell chloroplasts. *Plant Physiology*, 78: 586–90.

8

Stomatal Responses to Light

Thomas D. Sharkey and Teruo Ogawa

> A great deal has been written on the opening and closure of the stomata produced by light and darkness, but much remains to be done.
>
> —Francis Darwin, 1898

Direct stomatal responses to light were demonstrated 85 years ago by Francis Darwin (1898). He found that the surface of a leaf facing a bright window had open stomata but the surface away from the window had closed stomata. When the leaf was turned around, the stomata opened on the surface that had been dark but was now illuminated, and the stomata on the now-darkened side closed.

Stomata can respond to light absorbed by pigments within the guard cells. This is most often referred to as the direct response of stomata to light. In addition, guard cells can respond to CO_2, allowing an indirect response of stomata to light. As Scarth (1932) first proposed over 50 years ago, an increase in irradiance can cause an increase in photosynthesis, leading to a decreased level of CO_2 in the intercellular spaces. The stomata then open in response to the decrease in CO_2.

A third mechanism that has been suggested to account for some stomatal responses to light is the transmission of some agent from the mesophyll cells to the guard cells such that photosynthetic CO_2 assimilation in the mesophyll controls the degree of stomatal opening (Heath and Russell, 1954; Wong, Cowan, and Farquhar, 1979). This mechanism has been advanced to explain the observed correlation between photosynthetic rate and stomatal conductance. However, there is no direct evidence for the rapid response of stomata to a messenger from the mesophyll. Many investigators attribute the observed correlation between photosynthesis and conductance to parallel responses of guard cells and mesophyll cells to many environmental stimuli.

In this paper we will describe recent efforts to answer three questions. To what degree is the stomatal response to light in the intact leaf direct, as opposed to indirect (i.e. mediated by CO_2)? What are the photoreceptors involved in the direct stomatal response to light? What are the mechanisms by which light reception is transduced into stomatal opening?

CO_2 Mediation of the Stomatal Response to Light

Recent efforts at separating direct and indirect stomatal responses to light have been based on partial differentials of stomatal responses to light and CO_2

(Farquhar, Dubbe, and Raschke, 1978; Wong, Cowan, and Farquhar, 1978). This analysis was used to show that stomata of *Eucalyptus pauciflora* primarily respond directly to light (Wong, Cowan, and Farquhar, 1978). Sharkey and Raschke (1981a) obtained similar results for five herbaceous species, including *Zea mays*, and Morison and Jarvis (1981a, b) used similar arguments to show primarily direct stomatal responses to light in several other species.

Another way of investigating responses to light and CO_2 is Darwin's method, using the mesophyll as a shade. In many species that ordinarily have their leaves in a normal orientation to the sun's rays, the stomata on the surface facing the sun are much less sensitive to light than those on the shaded surface. Consequently, when the leaves are in their usual orientation with respect to a light source, the stomata on both surfaces open in concert as the light intensity increases. However, when a leaf is inverted, the stomata on the normally shaded side receive much more light than usual and so open at low photon-flux densities, and the stomata on the now-shaded surface do not open until exposed to very high photon-flux densities. Applying this method to leaves that had been poisoned with a PSII inhibitor to stabilize the intercellular CO_2 concentration (C_i) and to untreated control leaves, in which C_i varied with light, Sharkey and Raschke (1981b) found good agreement between their data and results predicted from the absorptance of the mesophyll of the leaf. But Raschke, Hanebuth, and Farquhar (1978) found no evidence of a direct stomatal response to light in corn grown in a growth chamber.

Observations of light responses in epidermal strips (Kuiper, 1964; Hsiao, Allaway, and Evans, 1973; Ogawa et al., 1978) and in isolated guard cell protoplasts (Zeiger and Hepler, 1977; and subsequent papers reviewed in Zeiger, 1983) provide further evidence that stomatal responses to light can be separate from those to changes in C_i, though the relative contribution of each response cannot be evaluated from these experiments. One instance in which experimental results from intact leaves and experimental results from epidermal strips differ qualitatively is the observation that stomata on leaves of *Paphiopedilum* species, whose guard cells lack chlorophyll, respond to red light, but stomata on epidermal strips from these species do not. In this instance the response observed in whole leaves is probably indirect and mediated by CO_2 (Zeiger, Assmann, and Meidner, 1983).

Additional evidence for direct stomatal responses to light is provided by experiments showing stomatal sensitivity to light in leaves poisoned with 3-(3,4-dichlorophenyl)-1,1-dimethylurea (DCMU) and other PSII inhibitors (Sharkey and Raschke, 1981a).

Light-quality experiments also indicate the importance of the direct stomatal responses to light. Sharkey and Raschke (1981a) showed that switching from red to blue light of equal quantum-flux densities caused assimilation to decrease and C_i to increase. If stomata responded indirectly to light, these responses might be expected to lead to stomatal closure; but the authors found that the stomata opened. Zeiger and Field (1982) found that the addition of low-intensity blue light to high-intensity red light caused both C_i and conductance to increase. These experiments indicate that the direct response to light overrides any concomitant CO_2 response under the conditions used.

In summary, several lines of evidence are consistent with the notion that stomatal responses to light are independent of changes in C_i. Most available data suggest that the direct responses to light can be very strong. However, it is reasonable to expect that variations in species and plant growth conditions will affect the degree to which stomatal responses to light are direct or indirect (via CO_2). For example, Raschke (1979) has shown that whereas *Zea mays* grown in a growth chamber primarily responds indirectly to light, field-grown *Zea mays* primarily responds directly.

Photoreceptors for the Direct Stomatal Response to Light

One method for determining the identity of a photoactive pigment is to compare the wavelength dependence of the light-induced action with the absorption spectrum of a pigment suspected to be involved. Some care must be taken in comparing a biological process (light-induced action) and a physical process (light absorption by a pigment). In this section we discuss some of the difficulties and limitations of action spectroscopy, experiments that indicate which pigments are photoactive in stomatal opening, and interactions between the photoactive pigments.

Action spectroscopy. In some photobiological and photochemical systems, for example, the bending of coleoptiles or the sensitization of silver salts to reduction, the action observed depends on the total dose of photons received, regardless of the period of time over which it was received. Steady-state conductance does not depend on light this way, although Iino, Ogawa, and Zeiger (unpublished) have found non-steady-state blue-light responses that exhibit reciprocity. If the steady-state stomatal conductance is considered the action (A) caused by light, then

$$A = I_0 \, \phi \, S \qquad (1)$$

where I_0 is the incident photon-flux areal density, ϕ is the photon-flux yield (A/I_a, where I_a is the absorbed light), and S is the ratio of absorbed to incident photons (I_a/I_0). S is also called the absorption cross section and is approximately equal to optical density when it is small.

Dependence of ϕ on the measure of action. It has long been recognized that the photon-flux density yield, ϕ, can be dependent on the measure of action chosen. The most obvious example is that at very high quantum-flux density, stomata will be maximally open and unable to respond to further increases in photon-flux density. At low light intensity, no increase in stomatal conductance is observed until some threshold light intensity is exceeded. Much has been

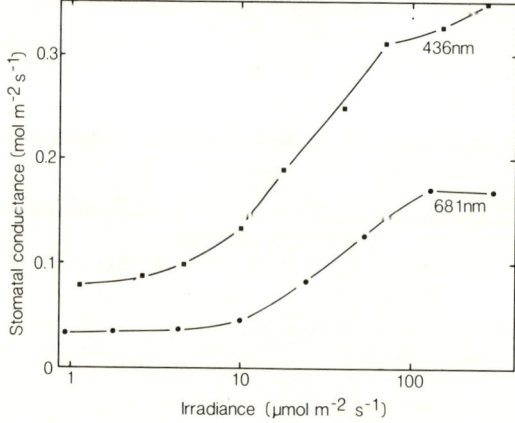

Fig. 8.1. Stomatal conductance in *Xanthium strumarium* as a function of irradiance in red light (681 nm; half-band width, 20 nm) and blue light (436 nm, half-band width, 20 nm). The CO_2 concentration was 250 µl l^{-1}. The light was shining on the abaxial surface, the surface most sensitive to light. (T. D. Sharkey, unpublished data.)

written about the threshold that can be readily observed in red light, but thresholds are also observed in blue light (Fig. 8.1) and in fact are observed in most photobiological systems. Because of the dependency of φ on the observed action, it is standard practice to keep A constant.

Dependence of φ on wavelength. In a photobiological system with a single active pigment, φ will be independent of wavelength, provided that no screening pigments are present and that only a small amount of the incident photon flux is absorbed (such that S approximately equals optical density). It is easy to determine whether this condition is met by plotting action versus log photon-flux density for the various wavelengths of interest. The resulting curves will be shifted by some amount proportional to S but should be of identical shape if φ is independent of wavelength. The reason why a semilog plot is appropriate is demonstrated by the following: suppose a photoreceptor pigment has an extinction coefficient ten times higher at 450 nm than at 650 nm. To get an equal action, ten times more red light than blue light is required, for example, 50 and 500 µmol photons m^{-2} s^{-1}, respectively. If we assume effective optical densities of 0.046 and 0.0044, respectively (i.e. 10 percent and 1 percent absorption), then 5 µmol m^{-2} s^{-1} are being absorbed in each case. To increase the action, for example, by increasing the blue light from 50 to 100 µmol m^{-2} s^{-1}, the red light would have to be increased from 500 to 1,000 µmol m^{-2} s^{-1}. In both cases I_a would increase from 5 to 10 µmol m^{-2} s^{-1}. Only by plotting action versus log photon-flux density will the changes from 50 to 100 and from 500 to 1,000 µmol m^{-2} s^{-1} appear the same.

Constructing an Action Spectrum. Equation 1 can be rearranged to

$$\frac{1}{I_0} = \frac{\phi}{A} S. \qquad (2)$$

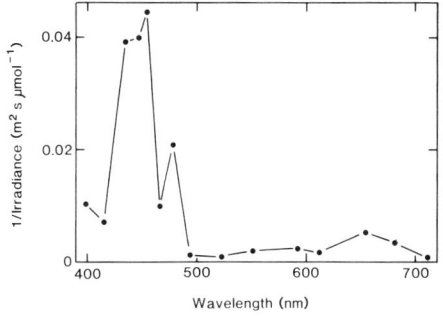

Fig. 8.2. Action spectrum of stomatal opening, based on the irradiance required to cause stomata to open to a conductance of 0.15 μmol m^{-2} s^{-1}. (Redrawn from Sharkey and Raschke, 1981b.)

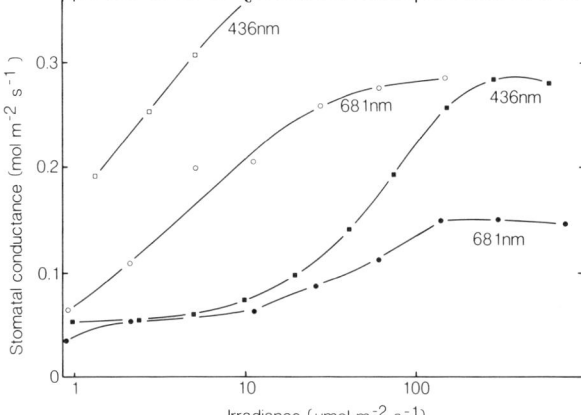

Fig. 8.3. Stomatal conductance as a function of irradiance for blue light and red light at 100 μl l^{-1} CO_2 (open circles) and 500 μl l^{-1} CO_2 (solid circles). (Redrawn from Sharkey and Raschke, 1981b.)

Provided that ϕ and A remain constant, a plot of $1/I_0$ versus wavelength will show how S varies with wavelength. Such a plot is called an action spectrum and should be directly comparable to the absorption spectrum of the photoactive pigment, provided that only one pigment is involved in the photoresponse and that there are no pigments shading the photoactive pigment. An action spectrum for stomatal opening is given in Figure 8.2. However, for stomatal opening it has been shown that one of the provisos, that of a single pigment being involved, does not hold.

Evidence for two photoreceptors in stomatal opening. In a single-pigment system the curves of action versus log photon-flux density at various wavelengths will be displaced depending on wavelength, but they should be parallel and of identical shape. Stomata open more in blue light than in red light (Figs. 8.1, 8.3; Schwartz and Zeiger, 1984, figs. 2, 3). In addition, Sharkey and Raschke (1981b) reported that on a semilog plot the average slope of the lines between 400 and 500 nm was 150 percent of the average slope of the lines between 600 and 700 nm.

The evidence for two photoreceptors can be summarized in five points:

1. The curves of A versus log photon-flux density at various wavelengths are not merely shifted but have a different slope as well (see above).

2. The maximum stomatal conductance in red light is substantially less than in blue light (see above).

3. DCMU and other inhibitors of PSII electron transport can eliminate the red-light response but have only a small effect on the blue-light response (Sharkey and Raschke, 1981a; Roth-Bejerano and Itai, 1981; Schwartz and Zeiger, 1984).

4. Background red illumination alters the blue-light response in a nonadditive fashion (Ogawa et al., 1978). If only one photoreceptor pigment were involved, the responses would be additive.

5. Blue light causes overshoots and oscillations in stomatal conductance in grasses, but an equivalent amount of red light (i.e. the amount that results in the same final stomatal aperture) does not (Brogårdh, 1975; Johnsson et al., 1976; Skaar and Johnsson, 1978).

One piece of evidence often taken to indicate two photoreceptors is the existence of an apparent threshold for red-light but not for blue-light responses. This apparent difference in thresholds results from plotting the responses to red and blue light on a linear scale appropriate to show both responses. Because stomata are more sensitive to blue than to red light, the threshold in blue light occurs at a lower light intensity. On a semilog plot (see *"Dependence of ϕ on wavelength,"* above) the thresholds in blue and red light often appear the same. On a linear scale the blue-light threshold will never be apparent. The appearance of a threshold intensity for blue light in semilog plots of stomatal conductance versus quantum flux depends on making the measurements at appropriate quantum-flux densities. As an example, Figure 8.3 shows red-light and blue-light responses at both high and low CO_2 concentrations. A threshold is apparent at high, but not at low, CO_2 concentration. It is apparent from this data that at low CO_2 concentration, measurements were not made at quantum-flux densities low enough to reveal the threshold. The presence of high CO_2 did not create the thresholds in Figure 8.3 but only shifted them into a region of quantum-flux density where observations were being made. Thus, thresholds cannot be invoked to prove the existence of two photoreceptors.

Red-light response of stomata. The involvement of chlorophyll in the stomatal response to light was hypothesized many years ago (Liebig, 1942). All available data are consistent with this hypothesis, and several experiments directly support the notion that chlorophyll is the pigment responsible for the red-light responses of stomata: DCMU and other PSII inhibitors close stomata in red light (Sharkey and Raschke, 1981a; Roth-Bejerano and Itai, 1981; Schwartz and Zeiger, 1984); and the light-quality response curves of stomatal conductance and photosynthesis between 600 and 700 nm are nearly identical (Sharkey and Raschke, 1981b).

Brogårdh (1975) argued that the red-light response of stomata is indirect and is mediated by changes in C_i. However, this is not the case, since the red-light response is observed in epidermal strips and in leaves in which the intercellular CO_2 pressure is kept constant. Further, by using the mesophyll as a shade, researchers have shown that the epidermis absorbs the red light that stimulates stomatal opening (Sharkey and Raschke, 1981a).

Since chlorophyll is accepted as the red photoactive pigment, induction kinetics of guard cell chlorophyll a have been studied recently and compared with results obtained from mesophyll chloroplasts. When mesophyll chloroplasts are first illuminated, fluorescence rises from a nonvariable level (O) to a peak value (P), then declines (S) and can increase to a second peak value (M), and finally decreases to a terminal value (T). Each transient has been assigned a cause, but in general the OPS section of the curve seems to reflect electron-transport phenomena directly, and the SMT section reflects the interaction between carbon reactions and electron transport, and especially the generation of a transmembrane potential necessary for ATP production. It is generally found that fluorescence transients of guard cells reach P faster than do those of mesophyll cells but lack a second peak (Zeiger, Armond, and Melis, 1981; Melis and Zeiger, 1982; Ogawa et al., 1982; Mawson et al., 1984). In addition Mawson et al. (1984) found that to saturate the rate of quenching (P-to-T transient) in guard cells more light was required than in mesophyll cells, though the total amount of quenching was independent of light intensity. However, they also found that less light was required in guard cells than in mesophyll cells to saturate the response to light.

Blue-light response of stomata. If chlorophyll were the photoactive pigment in stomatal opening, then stomatal sensitivity to blue and red light would be expected to match the absorption spectrum of chlorophyll. However, stomatal sensitivity to blue light is much greater than can be explained by chlorophyll alone. It is therefore generally accepted that a blue-light photoreceptor pigment is also involved in the stomatal response to light.

It is well established that in blue light, when both chlorophyll and the blue-light photoreceptor are stimulated, stomata open in very low photon-flux densities (1 µmol m^{-2} s^{-1} or less). This could result from either an extreme sensitivity of the blue-light receptor or from the synergistic action of the blue-light receptor and chlorophyll.

To distinguish between the responses of the two photoreceptors, Ogawa et al. (1978) determined a light-quality response curve for various wavelengths of blue light on a background of red light (Ogawa et al., 1978). In this environment the chlorophyll response should be constant at each wavelength, and the resulting spectrum should correspond to the blue-light photoreceptor. Ogawa et al. (1978) found that the blue-light photoreceptor's absorption spectrum resembled that of a flavin.

Experiments done with background red light in effect study the blue-light response when the chlorophyll response is saturated or very nearly so. To study the blue-light response in the absence of the chlorophyll response, Karlsson, Hoglund, and Klockare (1983) used wheat seedlings treated with SAN 9789, a compound that inhibits the synthesis of carotenoids. In the absence of carotenoids, chlorophyll is photobleached by low-intensity light. Stomata of chlorophyll-less wheat seedlings were able to respond to blue light, giving further support to the notion that the blue-light photoreceptor is a flavin rather than a carotenoid. However, plants that have been treated with toxins often do not behave normally. For example, although corn seedlings treated with SAN 9789 showed increased transpiration in blue light but not in red light, darkness did not cause the transpiration rate to fall to the original value (Vierstra and Sharkey, unpublished).

A more promising method for studying the blue-light response independent of the chlorophyll response uses orchids with guard cells that lack chlorophyll (Zeiger, Assmann, and Meidner, 1983). Stomata on epidermal strips from one such orchid respond to blue light but not to red light. The assertion of these authors that early reports of red-light responses in such orchid species must have resulted from an indirect light response mediated by CO_2 seems most plausible.

Interaction of the red-light and blue-light receptors. The surprising existence of two photosensory pigments in guard cells has been the subject of much conjecture. Sharkey and Raschke (1981b) noted that a one-pigment photoreceptor system is color-blind. Zeiger, Field, and Mooney (1981) proposed that the extreme sensitivity to blue light would stimulate stomatal opening before dawn, when the light is enriched in blue light.

It is now often accepted that the blue-light response of stomata is important at low light and the red-light (chlorophyll) response is important at high light intensity (Zeiger, 1983). This argument is often made because of the threshold for stomatal action in red light and the lack of a threshold for blue light. Other evidence that the blue-light photoreceptor is sensitive at lower quantum-flux densities than the chlorophyll response comes from experiments with *Paphiopedilum*, an orchid whose guard cells lack chlorophyll. Stomata of this plant respond to blue light at quantum-flux densities similar (10 μmol m^{-2} s^{-1}) to those observed in *Phragmipedium*, an orchid similar but with chlorophyll-containing guard cells. If the extreme sensitivity to blue light were simply the result of the interaction between the blue-light photoreceptor and chlorophyll, then DCMU treatment should eliminate the sensitivity to blue light of low quantum-flux density. Schwartz and Zeiger (1984) have shown that it does not; in their experiment DCMU did not eliminate sensitivity to quantum-flux densities below 10 μE m^{-2} s^{-1}.

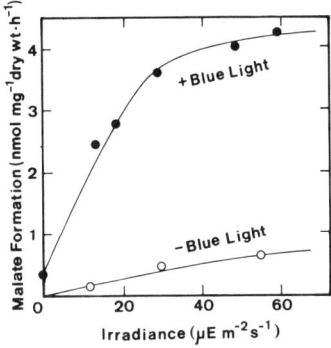

Fig. 8.4. Response of malate formation in epidermal strips to red light (675 nm) in the presence or absence of blue light (0.1 mW cm^{-2}, 360–580 nm). (Reproduced with permission from Ogawa, 1980.)

In monochromatic light, there is little doubt that blue light can cause almost maximal stomatal opening at quantum-flux densities that have little or no effect when given as red light. However, this evidence for the extreme sensitivity of the blue-light photoreceptor does not preclude the involvement of chlorophyll at low light intensity. Four examples illustrate this point. For the first, Figure 8.3 shows that the threshold for the red-light effect is around 10 μE m^{-2} s^{-1} in high CO_2. But when the CO_2 concentration was lower, the threshold was 1 μE m^{-2} s^{-1} or less. Hence, the threshold value for red-light action is not a function of red light alone but is a function of all stimuli to which stomata respond.

The second example is an experiment reported by Ogawa (1980), shown in Figure 8.4. A very small amount of background blue light, insufficient to cause substantial malate formation, significantly enhanced the effectiveness of red light in stimulating malate formation. This experiment demonstrates that the chlorophyll-mediated light responses can be important in the blue-light response at very low photon-flux densities (1 μmol m^{-2} s^{-1}), even though red light alone is not effective until much higher photon-flux densities (10 μmol m^{-2} s^{-1}) are given.

The third piece of evidence that chlorophyll could be involved at very low light intensity is that the slopes of lines in plots of A versus log I_0 are 50 percent higher in blue light than in red light. The higher slope in blue light is maintained to very low photon-flux densities and is not biphasic as would be expected if chlorophyll contributed to the stomatal response to light at high, but not at low, light.

The fourth piece of evidence comes from the experiments of Schwartz and Zeiger (1984) discussed earlier. In their experiment red light had no effect on stomatal opening at 10 μE m^{-2} s^{-1}, but in blue light DCMU reduced the effectiveness of light at 10 μE m^{-2} s^{-1} by one-third. Of course, it is always possible that DCMU had effects other than blocking PSII electron transport.

In summary, the blue-light photoreceptor appears to be sensitive to lower quantum-flux densities than chlorophyll, and it is required for maximum stomatal opening. It is not yet possible to know what contribution the chlorophyll response makes to stomatal opening in very low quantum-flux densities. However, there is good evidence that the chlorophyll response is important in the blue-light response below the threshold for action of monochromatic red light.

Mechanisms of Stomatal Opening in Response to Light

Chlorophyll-mediated response. Photosystems I and II are present in a wide variety of species (Outlaw et al., 1981; Zeiger, Armond, and Melis, 1981; Martin et al., 1984). A light-dependent, DCMU-sensitive O_2 evolution was demonstrated in *Commelina*, though the electron acceptor was unknown (Fitzsimons and Weyers, 1983). Fluorescence transients indicative of photophosphorylation have been observed (Melis and Zeiger, 1982); these fluorescence transients are affected by K^+ and abscisic acid (Ogawa et al., 1982) and by CO_2 (Melis and Zeiger, 1982). Because K^+, abscisic acid, and CO_2 can all affect stomatal aperture in darkness, the effects on fluorescence that have been observed are likely to be a secondary result of the guard cell response to these factors.

Although guard cell chloroplasts carry out light-dependent electron transport, they appear to be incapable of net CO_2 reduction. It has long been known that $^{14}CO_2$ is incorporated into the guard cells (Shaw and Maclachlan, 1954), but Raschke and Dittrich (1977) showed that CO_2 was fixed into malate and other dicarboxylic acids in *Commelina* guard cells. They found that $^{14}CO_2$ was not incorporated into sugar phosphates, probably because there was no ribulose 5-phosphate kinase and only low levels of ribulose-1,5-bisphosphate carboxylase (RuBPCase). Outlaw et al. (1979) found no RuBPCase in *Vicia* guard cells and only very little activity of ribulose 5-phosphate kinase and glyceraldehyde 3-phosphate dehydrogenase. Madhavan and Smith (1982) showed that stomata of C_3 plants generally have little and sometimes no RuBPCase. A more recent study by Zemel and Gepstein (1985) using improved techniques detected the presence of RuBPCase in guard cells of *Vicia faba*. It may be interesting to study the controls that regulate the levels of enzymes active in the photosynthetic carbon-reduction cycle. However, since a wide range of plants show little or no activity of one or more of these enzymes, it is most likely that stomatal responses to light do not involve photosynthetic carbon reduction.

Photophosphorylation. It is generally presumed that guard cells require ATP to translocate protons to promote stomatal opening. Since stomatal opening can occur in strips treated with potassium cyanide (KCN), it is presumed that photophosphorylation can produce the required ATP; and since opening can occur in DCMU-treated plant material, it is presumed that oxidative phosphorylation can also. Further, since guard cell protoplasts have a much higher respiratory activity than mesophyll protoplasts (Shimazaki et al., 1983), oxidative photophosphorylation may play a major role in ATP production in guard cells. If photophosphorylation were the only contribution of guard cell chloroplasts, treatment with DCMU would not be expected to result in stomatal closure in red light, since in theory cyclic photophosphorylation could occur. This is not a conclusive argument for an additional role of chloroplasts, however, since DCMU could upset the poising of electron transport and so inhibit cyclic photophosphorylation. It has been found that more light is required to saturate photophosphorylation in guard cells than in mesophyll cells (Shimazaki and Zeiger, 1985).

Generation of reductant. The many demonstrations of photosystem-II activity in guard cell chloroplasts suggest that photosynthetic electron transport can cause a net reduction in this activity: NADPH could reduce oxaloacetate to malate when guard cells make malate during stomatal opening.

Another interesting observation is that some enzymes in guard cells may be activated by reduction (Rao and Anderson, 1983). This is common for enzymes of the photosynthetic carbon-reduction cycle, and if the mechanism in guard cells is similar, whole-chain electron transport would be required.

Blue-light photoreceptor. The mechanism by which the blue-light photoreceptor causes stomatal opening is unknown. Two effects of blue light that could operate in guard cells are the blue-light stimulation of respiration (Kowallik, 1982) and the blue-light stimulation of phosphoenolpyruvate carboxylase activity (Ogasawara and Miyachi, 1970; Kamiya and Miyachi, 1975). Both of these responses have action spectra similar (Kowallik, 1967; Pickett and French, 1967; Kamiya and Miyachi, 1974) to that determined by Ogawa et al. (1978) for the blue-light response of stomata.

Stomata respond to blue light at very low photon-flux densities that may have insufficient energy to drive stomatal opening. One possibility is that the blue-light photoreceptor may function as a switch to start stomatal opening. Since the blue-light response is inhibited by KCN (Zeiger, Assmann, and Meidner, 1983) and by anoxia (Lurie, 1978), it may be that energy to drive stomatal opening in blue light can be provided by respiration.

The suggestion that blue light is absorbed by a flavin on the plasmalemma, causing protons to be pumped (Zeiger, 1983), is very plausible. Blue light may give a signal to switch on the pump, which is then driven by respiration or photophosphorylation.

REFERENCES

Brogårdh, T. 1975. Regulation of transpiration in *Avena*. Responses to red and blue light steps. *Physiologia plantarum*, 35: 303–9.

Darwin, F. 1898. Observations on stomata. *Philosophical Transactions of the Royal Society*, ser. B, 190: 531–621.

Farquhar, G. D., D. R. Dubbe, and K. Raschke. 1978. Gain of the feedback loop involving carbon dioxide and stomata: Theory and measurement. *Plant Physiology*, 62: 406–12.

Fitzsimons, P. J., and J. D. B. Weyers. 1983. Separation and purification of protoplast types from *Commelina communis* L. leaf epidermis. *Journal of Experimental Botany*, 34: 55–66.

Heath, O. V. S., and J. Russell. 1954. Studies in stomatal behaviour, VI. An investigation of the light responses of wheat stomata with the attempted elimination of control by the mesophyll, Part 2: Interactions with external carbon dioxide, and general discussion. *Journal of Experimental Botany*, 5: 269–92.

Hsiao, T. C., W. G. Allaway, and L. T. Evans. 1973. Action spectra for guard cell Rb^+ uptake and stomatal opening in *Vicia faba*. *Plant Physiology*, 51: 82–88.

Johnsson, M., S. Issaias, T. Brogårdh, and A. Johnsson. 1976. Rapid, blue-light-induced transpiration response restricted to plants with grass-like stomata. *Physiologia plantarum*, 36: 229–32.

Kamiya, A., and S. Miyachi. 1974. Effects of blue light on respiration and carbon dioxide fixation in colorless *Chlorella* mutant cells. *Plant and Cell Physiology*, 15: 927–37.

———. 1975. Blue light-induced formation of phosphoenolpyruvate carboxylase in colorless *Chlorella* mutant cells. *Plant and Cell Physiology*, 16: 729–36.

Karlsson, P. E., H.-O. Hoglund, and R. Klockare. 1983. Blue light induces stomatal transpiration in wheat seedlings with chlorophyll deficiency caused by SAN 9789. *Physiologia plantarum*, 57: 417–21.

Kowallik, W. 1967. Action spectrum for an enhancement of endogenous respiration by light in *Chlorella*. *Plant Physiology*, 42: 672–76.

———. 1982. Blue light effects on respiration. *Annual Review of Plant Physiology*, 33: 51–72.

Kuiper, P. J. C. 1964. Dependence upon wavelength of stomatal movement in epidermal tissue of *Senecio odoris*. *Plant Physiology*, 39: 952–55.

Liebig, M. 1942. Untersuchungen über die Abhängigkelt der Spaltweite der Stomata von Intensität und Qualität der Strahlung. *Planta*, 33: 206–57.

Lurie, S. 1978. The effect of wavelength of light on stomatal opening. *Planta*, 140: 245–49.

Madhavan, S., and B. N. Smith. 1982. Localization of ribulose bisphosphate carboxylase in the guard cells by an indirect, immunofluorescence technique. *Plant Physiology*, 69: 273–77.

Martin, G. E., W. H. Outlaw, L. C. Anderson, and S. G. Jackson. 1984. Photosynthetic electron transport in guard cells of diverse species. *Plant Physiology*, 75: 336–37.

Mawson, B. T., A. Franklin, W. G. Filion, and W. R. Cummins. 1984. Comparative studies of fluorescence from mesophyll and guard cell chloroplasts in *Saxifraga cernua*. Analysis of fluorescence kinetics as a function of excitation intensity. *Plant Physiology*, 74: 481–86.

Melis, A., and E. Zeiger. 1982. Modulation of guard cell photophosphorylation by CO_2. *Plant Physiology*, 69: 642–47.
Morison, J. I. L., and P. G. Jarvis. 1983a. Direct and indirect effects of light on stomata, I. In Scots pine and Sitka spruce. *Plant, Cell and Environment*, 6: 95–101.
———. 1983b. Direct and indirect effects of light on stomata, II. In *Commelina communis* L. *Plant, Cell and Environment*, 6: 103–9.
Ogasawara, N., and S. Miyachi. 1970. Regulation of CO_2-fixation in *Chlorella* by light and varied wavelengths and intensities. *Plant and Cell Physiology*, 11: 1–14.
Ogawa, T. 1980. Synergistic action of red and blue light on stomatal opening of *Vicia faba* leaves. In H. Senger, ed., *The blue light syndrome*: 622–28. Berlin: Springer-Verlag.
Ogawa, T., D. Grantz, J. Boyer, and Govindjee. 1982. Effects of cations and abscisic acid on chlorophyll *a* fluorescence in guard cells of *Vicia faba*. *Plant Physiology*, 69: 1140–44.
Ogawa, T., H. Ishikawa, K. Shimada, and K. Shibata. 1978. Synergistic action of red and blue light and action spectra for malate formation in guard cells of *Vicia faba* L. *Planta*, 142: 61–65.
Outlaw, W. H., J. Manchester, C. A. DiCamelli, D. D. Randall, G. Rapp, and G. M. Veith. 1979. Photosynthetic carbon reduction pathway is absent in chloroplasts of *Vicia faba* guard cells. *Proceedings of the National Academy of Sciences*, 76: 6371–75.
Outlaw, W. H., B. C. Mayne, V. E. Zenger, and J. Manchester. 1981. Presence of both photosystems in guard cells of *Vicia faba* L.: Implications for environmental signal processing. *Plant Physiology*, 67: 12–16.
Pickett, J. M., and C. S. French. 1967. The action spectrum for blue light stimulated oxygen uptake in *Chlorella*. *Proceedings of the National Academy of Sciences*, 57: 1587–93.
Rao, I. M., and L. E. Anderson. 1983. Light and stomatal metabolism, I. Possible involvement of light modulation of enzymes in stomatal movement. *Plant Physiology*, 71: 451–55.
Raschke, K. 1979. Movements of stomata. In W. Haupt and M. E. Feinleib, eds., *Physiology of movements*, Encyclopedia of Plant Physiology, n.s., 7: 383–441. Berlin: Springer-Verlag.
Raschke, K., and P. Dittrich. 1977. ^{14}C carbon-dioxide fixation by isolated leaf epidermes with stomata closed or open. *Planta*, 134: 69–75.
Raschke, K., W. F. Hanebuth, and G. D. Farquhar. 1978. Relationship between stomatal conductance and light intensity in leaves of *Zea mays* L., derived from experiments using the mesophyll as shade. *Planta*, 139: 73–77.
Roth-Bejerano, N., and C. Itai. 1981. Involvement of phytochrome in stomatal movement: Effect of blue and red light. *Physiologia plantarum*, 52: 201–6.
Scarth, G. W. 1932. Mechanism of the action of light and other factors on stomatal movement. *Plant Physiology*, 7: 481–504.
Schwartz, A., and E. Zeiger. 1984. The bioenergetics of stomatal guard cells: Energy supply for stomatal opening. *Planta*, 161: 129–36.
Sharkey, T. D., and K. Raschke. 1981a. Separation and measurement of direct and indirect effects of light on stomata. *Plant Physiology*, 68: 33–40.
———. 1981b. Effect of light quality on stomatal opening in *Xanthium strumarium* L. *Plant Physiology*, 68: 1170–74.
Shaw, M., and G. A. Maclachlan. 1954. The physiology of stomata, I. Carbon dioxide fixation in guard cells. *Canadian Journal of Botany*, 32: 784–94.
Shimazaki, K., K. Gotow, T. Sakaki, and N. Kondo. 1983. High respiratory activity of guard cell protoplasts from *Vicia faba* L. *Plant and Cell Physiology*, 24: 1049–56.

Shimazaki, K., and E. Zeiger. 1985. Cyclic and noncyclic photophosphorylation in isolated guard cell chloroplasts from *Vicia faba* L. *Plant Physiology*, 78: 211–14.

Skaar, H., and A. Johnsson. 1978. Rapid, blue-light induced transpiration in *Avena*. *Physiologia plantarum*, 43: 390–96.

Wong, S. C., I. R. Cowan, and G. D. Farquhar. 1978. Leaf conductance in relation to assimilation in *Eucalyptus pauciflora* Sieb. ex Spreng. Influence of irradiance and partial pressure of carbon dioxide. *Plant Physiology*, 62: 670–74.

―――. 1979. Stomatal conductance correlates with photosynthetic capacity. *Nature*, 282: 424–26.

Zeiger, E. 1983. The biology of stomatal guard cells. *Annual Review of Plant Physiology*, 34: 441–75.

Zeiger, E., P. Armond, and A. Melis. 1981. Fluorescence properties of guard cell chloroplasts. Evidence for linear electron transport and light-harvesting pigments of photosystems I and II. *Plant Physiology*, 67: 17–20.

Zeiger, E., S. M. Assmann, and H. Meidner. 1983. The photobiology of *Paphiopedilum* stomata: Opening under blue but not red light. *Photochemistry and Photobiology*, 38: 627–30.

Zeiger, E., and C. Field. 1982. Photocontrol of the functional coupling between photosynthesis and stomatal conductance in the intact leaf. Blue light and PAR-dependent photosystems in guard cells. *Plant Physiology*, 70: 370–75.

Zeiger, E., C. Field, and H. A. Mooney. 1981. Stomatal opening at dawn: Possible roles of the blue light response in nature. In H. Smith, ed., *Plants and the daylight spectrum*: 391–407. New York: Academic.

Zeiger, E., and P. K. Hepler. 1977. Light and stomatal function: Blue light stimulates swelling of guard cell protoplasts. *Science*, 196: 887–89.

Zemel, E., and S. Gepstein. 1985. Immunological evidence for the presence of ribulose biphosphate carboxylase in *Vicia* guard cell chloroplasts. *Plant Physiology*, 78: 586–90.

9

The Blue-Light Response of Stomata: Mechanism and Function

Eduardo Zeiger, Moritoshi Iino, Ken-Ichiro Shimazaki, and Teruo Ogawa

The importance of light in modulating stomatal function is clearly reflected in the extensive coverage of the subject in this volume: Sharkey and Ogawa review our current understanding of this light response; Tenhunen, Pearcy, and Lange document the dominant role of the stomatal response to light in the daily course of stomatal conductance under natural conditions; MacRobbie analyzes ion fluxes in guard cells under contrasting light and dark conditions; and Assmann and Zeiger discuss light as an energy source for stomatal movements. In this paper, we present recent findings on the blue-light (BL) response of stomata. These data have provided us with new insight on the mechanisms underlying the sensory transduction of BL and the modulation of guard cell metabolism and stomatal movements.

The increasing use of modern photobiological techniques since the late 1950's has provided the groundwork for the demonstration of a specific BL response of stomata separate from the response mediated by the absorption of BL in guard cell chloroplasts (Mouravieff, 1958, 1973; Mansfield and Meidner, 1966; Raschke, 1967; Meidner, 1968; Hsiao, Allaway, and Evans, 1973; see Pemadasa, 1981, for a review of this earlier literature). More recent work on the action spectra of the response and its metabolic correlates is presented by Sharkey and Ogawa (this volume; see also Zeiger, 1984; Karlsson, 1986a, b).

Despite the conclusive demonstration of the stomatal response to BL (Zeiger, 1984) and its general biological interest, that response has received little attention in the photobiological literature, mostly because of methodological limitations. The stomatal response to BL is difficult to separate experimentally from the responses of guard cell chloroplasts and from the effect that the light-dependent changes in photosynthetic activity in the mesophyll has on stomatal movements (Wong, Cowan, and Farquhar, 1979; Ramos and Hall,

The authors wish in gratitude to acknowledge research support from the National Science Foundation, the United States Department of Energy, and the United States Department of Agriculture.

1983). In addition, understanding of classical BL responses in plants, such as phototropism, has relied on extensive quantitation, which has been difficult to achieve in stomatal studies. Data from epidermal peels allow separation of the light response of stomata from that of the mesophyll but are usually obtained at discrete time intervals and show substantial variability between samples. Gas exchange experiments provide continuous measurements of the stomatal responses but are complicated by concomitant light responses of the mesophyll and by stomata-mesophyll interactions.

Stomatal Responses to Pulses of Blue Light in the Intact Leaf

In the research described in this paper, the above-mentioned difficulties were overcome by the combined use of two methodologies. Stomatal responses were measured using advanced gas exchange techniques, which provided real-time, continuous measurements of changes in stomatal conductance with high resolution (Iino, Ogawa, and Zeiger, 1985; Zeiger, Iino, and Ogawa, 1985). In addition, the specific BL response was isolated from guard-cell and mesophyll photosynthesis using short pulses of BL applied on a background of high fluence rates of red light, which maintained the photosynthetic responses at or near saturation.

Compared with continuous-light stimulation, the use of light pulses had proven effective in other photobiological studies (Iino and Schäfer, 1984), because of its better definition of the applied light fluence and the time course of the response. Blue-light pulses had been used to characterize a BL-dependent increase in transpiration in grasses (Skaar and Johnsson, 1978), but their potential for quantitation had not been fully exploited. Figure 9.1 shows a typical pulse response in *Commelina communis*. Two intact attached leaves, each from a different plant, were enclosed in the gas exchange cuvette and irradiated with 500 μmol m^{-2} s^{-1} of red light. A 30-second pulse of BL (fluence rate 250 μmol m^{-2} s^{-1}) was applied after the response to red irradiation had reached steady state (Zeiger, Iino, and Ogawa, 1985). An increase in stomatal conductance could be measured within 2 minutes of pulse application, with shorter response times being unresolvable because of intrinsic limitations of the gas exchange system arising from the temperature-control device of the cuvette. The conductance increase was maximal about 15 minutes after the pulse and then slowly returned to the values prevailing before pulse stimulation. This BL response was completed in 50 to 60 minutes and could be repeatedly induced with nearly identical time courses.

The high fluence rates of red light maintained during these experiments were intended to keep the photosynthetic response at or near saturation and to minimize thereby any intrinsic photosynthetic response to the BL pulses. The effectiveness of this dual-beam protocol is substantiated by a close examination of both the time course of net photosynthesis and calculated intercellular CO_2

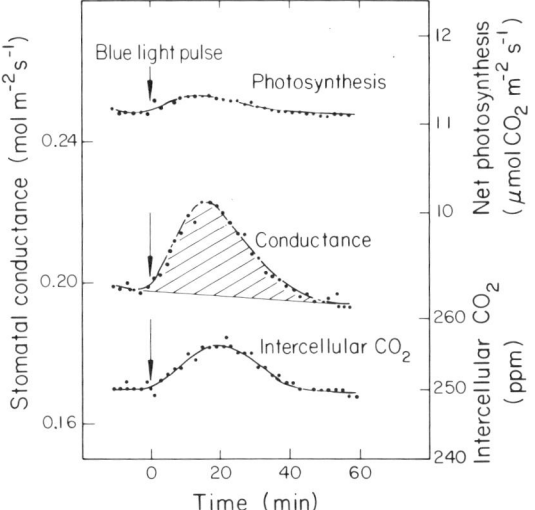

Fig. 9.1. Typical time courses of stomatal conductance, net photosynthesis, and calculated intercellular CO_2 concentrations in *Commelina communis* in response to a pulse of blue light (BL) applied in a background of high fluence rates of red light. Pulse duration: 30 s. Fluence rates: BL, 250; red light, 500 μmol m^{-2} s^{-1}. Measurement of the shaded area under the conductance curve allowed the quantitation of the response to BL pulses. For complete experimental details see Zeiger, Iino, and Ogawa (1985).

concentrations (Fig. 9.1). The absence of any detectable rapid change in photosynthetic rates in response to the BL pulse indicated that the conductance increases were not a consequence of a direct enhancement of photosynthesis. The slow increases observed in photosynthetic rates were most likely a response of the photosynthesizing mesophyll to the higher intercellular CO_2 concentrations resulting from the increases in stomatal conductances. Furthermore, red-light pulses (30 or 100 s; fluence rate 500 μmol m^{-2} s^{-1}) superimposed on the same red-light background did not cause conductance changes similar to those observed in response to BL. The ineffectiveness of red pulses strengthens the conclusion that the conductance increases in response to BL pulses were not photosynthetic responses but rather were specific stomatal responses to BL.

After establishing that the pulse-dependent increase in conductance was a true BL response, we investigated the relationship between light fluences and conductance changes by varying pulse duration at a constant fluence rate of 250 μmol m^{-2} s^{-1}. As shown in Figure 9.2A, the stomatal response to BL exhibits two typical photobiological features: it increases with light dose, and it saturates after a certain pulse duration (30 s in these experiments) is exceeded.

Further evaluation of the BL response required a means to quantify the observed changes in conductance. Since stomatal apertures were at steady state under the background red irradiation, we reasoned that the stomatal response to BL could be quantified by integrating the conductance increase in response to a BL pulse over the total response time (shaded area in Fig. 9.1). When quantified in this fashion and plotted relative to a saturating 50-second or 100-second response, the conductance increases approached saturation exponentially (Fig. 9.2B), with a half-time of about 9 seconds.

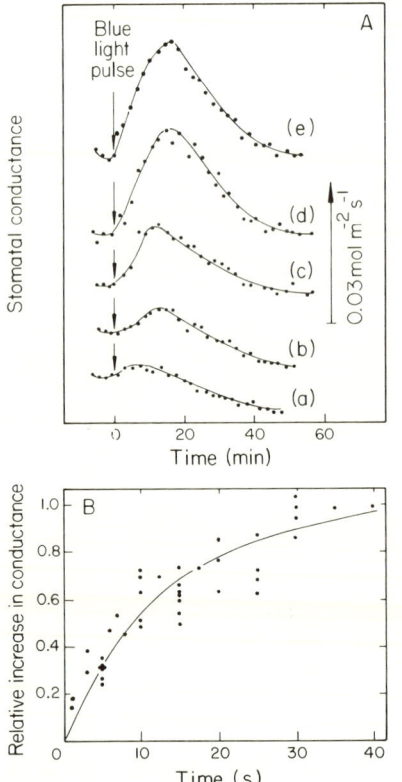

Fig. 9.2. Stomatal responses to blue light in the intact leaf. A, Time courses of changes in stomatal conductance of *Commelina communis* in response to 1-s (a), 3-s (b), 10-s (c), 30-s (d), and 100-s (e) pulses of BL, applied in a background of high fluence rates of red light. Fluence rates of red light and BL as in Figure 9.1. B, The relationship between conductance increases and pulse duration, relative to the response to a saturating 50-sec pulse. The responses were quantified as indicated in Figure 9.1. (Both figures reproduced with permission from Zeiger, Iino, and Ogawa, 1985.)

The photobiological property of reciprocity was also tested. Reciprocity is demonstrated when a photobiological response can be shown to depend on total applied light fluence, irrespective of fluence rate or exposure time. Thus, if a photoresponse obeys reciprocity, a series of pulses of varying duration of fluence rate but yielding identical fluences would give identical responses. We tested reciprocity at two different fluences, 1.2 and 2.3 × 10^3 μmol m^2, applied for 5 and 48 seconds and for 10 and 96 seconds, respectively (Zeiger, Iino, and Ogawa, 1985). In both cases, the conductance increases were a function of fluence, indicating that the BL response of stomata does obey reciprocity. A more thorough demonstration of reciprocity was recently provided by a study of the stomatal response to BL in wheat (Karlsson, 1986b). The stomatal response to BL pulses therefore indicates that guard cells primarily respond to the number of applied blue quanta and that BL perception is transduced into changes in stomatal conductance of a magnitude dependent on the incident fluence. It is also clear that the BL response is saturable and continues at high rates long after the application of the pulse.

The observation that a sufficiently long pulse saturated the BL response but

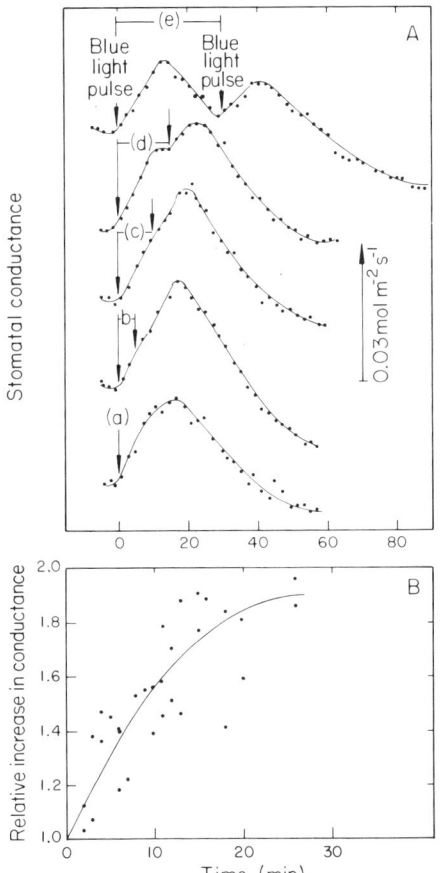

Fig. 9.3. Stomatal responses to two blue-light pulses in the intact leaf. *A*, Time courses of the changes in stomatal conductance in leaves of *Commelina communis* in response to a single pulse (*a*) and two consecutive 30-s BL pulses (arrows), applied at intervals of 5 min (*b*), 10 min (*c*), 15 min (*d*), and 30 min (*e*) under high fluence rates of red light. Fluence rates as in Figure 9.1. (Reproduced with permission from Zeiger, Iino, and Ogawa, 1985.) *B*, The relationship between total conductance increases and time between pulses, relative to the response to two 50-sec pulses, 50 min apart. The responses were quantified as indicated in Figure 9.1. (Reproduced with permission from Iino, Ogawa, and Zeiger, 1985.)

that a second pulse, applied after the completion of the initial response, could induce a new response indicated that the stomatal capacity to respond to pulses was slowly restored. The time course of that recovery was characterized in experiments testing the response to two saturating pulses separated by increasingly longer intervals (Zeiger, Iino, and Ogawa, 1985). Figure 9.3A depicts typical conductance profiles obtained in these experiments and shows that the combined stomatal response to two pulses increases with the time between them. The relationship between the increases in conductance in response to two pulses and the time between the pulses (Fig. 9.3B) is very similar to the fluence-response curve observed with single pulses (Fig. 9.2B) but shows a response half-time of about 9 minutes.

The results of the single- and double-pulse experiments led to the formulation of a kinetic model of the BL response (Iino, Ogawa, and Zeiger, 1985). The model postulates that one component of the phototransduction process can exist in two interconvertible forms, A and B: A is an inactive form that is photoconverted to an active form B by BL; B is converted back to A in a thermal, or dark, reaction. The model assumes that the BL-dependent increases in stomatal conductance are proportional to the concentration or activity of B. Since the half-time of the dark reaction is much slower than that of the light reaction (Figs. 9.2B, 9.3B), the response to a single pulse reaches apparent saturation when the BL fluence is sufficient to convert all available A into B. Immediately after the pulse, the thermal reaction starts to regenerate A from B; hence, the longer the interval between two saturating pulses, the higher the amount of regenerated A available for photoconversion, and the greater the response to a second pulse.

The model also predicts that, under continuous BL irradiation, prevailing amounts of B will depend on BL fluence and on the rates of the light and thermal reactions. Rate constants for these reactions were calculated from the data of the single- and double-pulse experiments, allowing the estimation of theoretical relative concentrations of B under different fluence rates of BL (Iino, Ogawa, and Zeiger, 1985). The calculated concentrations closely matched the observed BL-dependent increases in stomatal conductance under different fluence rates of continuous BL irradiation and thus supported the model (Iino, Ogawa, and Zeiger, 1985).

Detailed kinetic measurements and determinations of rate constants for the light and thermal reactions have so far been obtained only for *Commelina*, but we found that leaves of *Vicia faba*, soybean, sugarcane, *Phaseolus vulgaris*, and the orchid *Paphiopedilum harrissianum* exposed to short pulses of BL on a background of high fluence rates of red light all show similar transient increases in stomatal conductance (Assmann, Simoncini, and Schroeder, 1985; Grantz, Larqué-Saavedra, and Zeiger, unpublished). These observations and those previously reported with grasses (Skaar and Johnsson, 1978; Karlsson, 1986a, b) suggest that the photobiological properties of the BL response expressed in pulse-induced increases in conductance are common to both monocots and dicots. It will be of interest to establish whether the rate constants for the light and dark reactions are generally invariant, hence reflecting basic properties of the BL response, or they vary with species, age, or environmental history.

Recent results with two Mexican varieties of *Phaseolus vulgaris* point to some inherent variability in the stomatal response to BL. Cacahuate 72 is a bush bean variety cultivated as a single crop in full sunlight; Negro 150 is a climbing vine-type variety grown with maize and thus shaded during most of the growth season (Rodríguez and Larqué-Saavedra, 1983). When leaves of plants from the two varieties, grown under identical conditions in a greenhouse at Stanford University, Stanford, California, were compared in their conductance response to saturating BL pulses under high fluence rates of background red light (Fig. 9.4), Cacahuate 72 consistently exhibited greater responses than Negro 150 (Larqué-Saavedra and Zeiger, unpublished). Further analysis of the BL

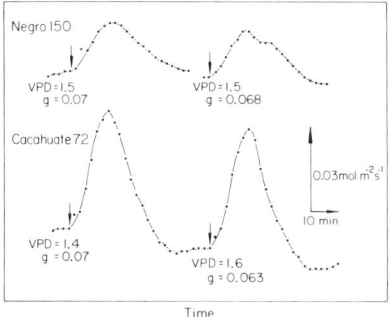

Fig. 9.4. A comparison of the responses to BL pulses applied in a background of high fluence rates of red light in two Mexican cultivars of *Phaseolus vulgaris* L., Cacahuate 72 and Negro 150. For these experiments, plants from the two cultivars were grown in a greenhouse at Stanford University, Stanford, Calif., under identical conditions. Blue-light pulses (arrows) were 100 s long, and their fluence rate was 250 μmol m^{-2} s^{-1}. The fluence dependency of the pulse response was not established, but measurements of responses to pulse durations longer than 100 s showed that a 100-s pulse saturated the response. Fluence rate of red light: 500 μmol m^{-2} s^{-1}. Traces shown are typical responses to two consecutive pulses. Differences between the cultivars in the height of the response peak and the area under the conductance curves were significant ($n = 19$; $p > 0.0001$ and 0.001, respectively). As indicated by the values in the figure, overall conductance levels (g, mol^{-2}s^{-1}) and prevailing water-vapor pressure differences (VPD, kPa) in the two cultivars were nearly identical. (Data from A. Larqué-Saavedra and E. Zeiger, unpublished.)

responses of these varieties and of other species with contrasting growth habits should be valuable for understanding the mechanistic implications of the BL response and its role in the ecophysiology of leaves.

Responses of Isolated Guard Cell Protoplasts to Light

Guard cell protoplasts extrude protons in response to pulses of blue light. The capacity of leaves to exhibit discrete increases in stomatal conductance in response to BL pulses led us to investigate earlier metabolic events mediating the transduction of BL absorption into conductance changes. Previous experiments showing that guard cell protoplasts of onion swell in response to continuous BL irradiation had suggested the involvement of a proton pump (Zeiger and Hepler, 1977; Zeiger, Bloom, and Hepler, 1978; see also Assmann and Zeiger, this volume); we tested that hypothesis in a study of the effects of BL pulses on proton extrusion by guard cell protoplasts (Shimazaki, Iino, and Zeiger, 1986).

Guard cell protoplasts are a valuable experimental tool for studying the properties of guard cells without interference from other cell types present in the leaf. Since the first report on a technique to isolate guard cell protoplasts (Zeiger and Hepler, 1976), methods developed for their large-scale purification (e.g. Shimazaki, Gotow, and Kondo, 1982) have facilitated many physiological and biochemical studies (for review see Weyers et al., 1983). Concerns about mesophyll contamination as an intrinsic limitation in preparations of guard-cell protoplasts (Outlaw, 1982) have been dispelled (Shimazaki and Zeiger, 1985).

Guard cell protoplasts have distinct advantages for studies of proton extrusion (Shimazaki, Iino, and Zeiger, 1986). Earlier work has demonstrated light- and fusicoccin(FC)-dependent proton extrusion using guard cells in epidermal peels (Raschke and Humble, 1973; Gepstein, Jacob, and Taiz, 1982). Resolution, however, was limited by the thick guard cell walls, which have a large ion-exchange capacity (Raschke and Humble, 1973), and by the slow, continuous lysis of epidermal cells in the preparation, which results in considerable unspecific acidification (Shimazaki and Zeiger, unpublished). Both disadvantages can be overcome using guard cell protoplasts.

We analyzed BL-dependent proton extrusion by guard cell protoplasts by measuring the pH of *Vicia* protoplast preparations kept in 0.35 M mannitol, 1 mM $CaCl_2$, 10 mM KCl, and 0.5 mM 2-[N-morpholino]ethanesulfonic acid (Mes)-NaOH, pH 6.2. The suspension, containing about 10^6 protoplasts in 1.2 ml of medium, was stirred in an open vessel at 25° C, and the pH of the medium was continuously monitored. Protoplasts were irradiated with a high fluence rate of red light (800 μmol m^{-2} s^{-1}) known to saturate photosynthesis (Shimazaki and Zeiger, 1985). Blue-light pulses, applied after the pH of the suspension had stabilized under red light, consistently induced acidification of the medium. The acidification response was observed with *Commelina* protoplasts (Shimazaki and Zeiger, unpublished) and was therefore correlated with the BL-dependent increases in conductance described in the preceding section. Detailed studies of the response, however, were conducted with guard cell protoplasts of *Vicia*, which were more amenable to consistent, reproducible isolation (Shimazaki, Gotow, and Kondo, 1982). Pulse-dependent increases in stomatal conductance in *Vicia* leaves have also been observed (Assmann, Simoncini, and Schroeder, 1985).

Proton extrusion from guard cell protoplasts of *Vicia* was detectable within 30 seconds of pulse application (Fig. 9.5) and exhibited maximum rates within 2 minutes (Shimazaki, Iino, and Zeiger, 1986). Acidification continued for about 9 minutes, followed by a stabilization of medium pH or, often, by a small pH increase. In this series of experiments, fluence dependence was studied at a constant pulse duration of 30 seconds by varying the fluence rate. As shown in Figure 9.6, the response saturated at about 50 μmol m^{-2} s^{-1}, with half-saturation observed at around 180 μmol m^{-2}.

Unambiguous interpretation of the observed response as proton extrusion by the guard cell protoplasts demanded discrimination between acidification specifically caused by protons and that caused by CO_2. In solution, CO_2 equilibrates with HCO_3^- in a reaction: $CO_2 + H_2O \rightleftharpoons HCO_3^- + H^+$. This reaction has a pK_a of 6.3; thus at pH 6.3, the CO_2 and HCO_3^- are in equal

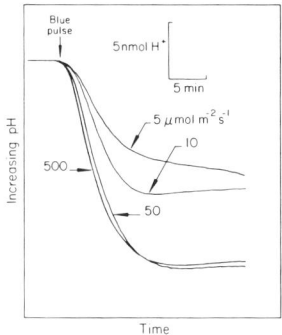

Fig. 9.5. Time courses of acidification in a suspension of *Vicia* guard cell protoplasts in response to BL pulses applied under high fluence rates (800 μmol m^{-2} s^{-1}) of red light. Fluence rates of the 30-s BL pulses are indicated. For the protoplast isolation procedure, see Shimazaki, Gotow, and Kondo (1982). Protoplasts (about 10^6) were suspended in 1.2 ml 0.35 M mannitol, 10 mM KCl, 1mM CaCl$_2$, and 0.5 mM MES-NaOH buffer, pH 6.2. (Reproduced with permission from Shimazaki, Iino, and Zeiger, 1986.)

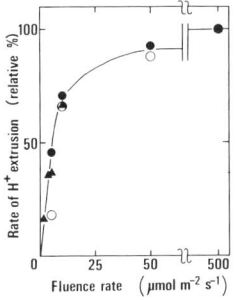

Fig. 9.6. Fluence-rate dependence of the blue-light-induced medium acidification rate by *Vicia* guard cell protoplasts, relative to the magnitude of acidification observed in response to a fluence rate of 500 μmol m^{-2} s^{-1}. Pulse duration: 30 s; background red light: 800 μmol m^{-2} s^{-1}. Different symbols correspond to experiments with different protoplast isolates. (Reproduced with permission from Shimazaki, Iino, and Zeiger, 1986.)

concentrations. Because of this equilibrium, addition of CO_2 to a solution will displace the chemical equilibrium toward HCO_3^- and generate protons, which will acidify the medium. Conversely, removal of CO_2 will result in the opposite reaction and cause alkalinization. Note that another species in this equilibrium reaction, CO_3^{2-}, is ignored, because its concentration would be negligible at the pH of our experiments.

Blue light has been reported to enhance respiration in several experimental systems (Kowallik, 1982); a similar response of the protoplast preparation could increase CO_2 evolution in the suspension medium and cause acidification. We studied that possibility by measuring the extent of acidification in response to BL pulses at different pH's. As shown in Figure 9.7, the BL-dependent acidification was essentially independent of medium pH in the range of pH 5 to 7. The dotted line in Figure 9.7 shows the calculated relative concentrations of HCO_3^-; in a CO_2-induced acidification, the amount of H^+ released would be expected to change with the dotted line. We therefore concluded that the observed acidification was not caused by CO_2 evolution but primarily by proton extrusion from the guard cell protoplasts. It is noteworthy that the pH dependence of the BL-induced acidification is very similar to that reported for the *Neurospora* proton pump (Sanders, Ballarin-Denti, and Slayman, 1984).

Like the pulse-dependent increases in conductance in the intact leaf, proton

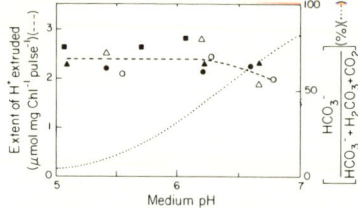

Fig. 9.7. Blue-light-dependent acidification by *Vicia* guard cell protoplasts as a function of medium pH. Pulse duration and fluence rates of red and blue light as in Figure 9.6. The dotted line shows relative HCO_3^- concentrations as percentage of total concentrations of CO_2, HCO_3^-, and H_2CO_3 in water, calculated from the Henderson-Hasselbalch equation (pK_1 = 6.35, 25° C). In a CO_2-dependent acidification, the amount of protons released as a function of medium pH would increase with the dotted line. (Reproduced with permission from Shimazaki, Iino, and Zeiger, 1986.)

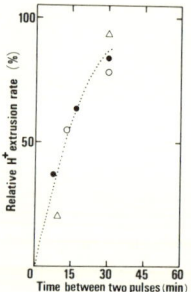

Fig. 9.8. Relationship between proton extrusion and the time between two saturating BL pulses. In all cases, pulse duration was 30 s. Fluence rates were: BL, 50 μmol m^{-2} s^{-1}; red light, 800 μmol m^{-2} s^{-1}. In each experiment, the rate of H^+ extrusion in response to the first pulse was taken as 100 percent. Symbols indicate different protoplast isolates. (Reproduced with permission from Shimazaki, Iino, and Zeiger, 1986.)

extrusion from guard cell protoplasts in response to a second pulse increased in rate with the time between the two pulses (Fig. 9.8). That correlation further indicates that the BL-dependent proton extrusion is a component of the sensory transduction process preceding the observed increases in stomatal conductance.

The mechanism underlying the extrusion of protons was investigated with metabolic inhibitors (Shimazaki, Iino, and Zeiger, 1986). Carbonyl cyanide m-chlorophenyl hydrazone (CCCP), a proton translocator that dissipates proton gradients across membranes, completely abolished the BL-dependent acidification at 10 μM, supporting the notion of an involvement of proton fluxes. Diethylstilbestrol (DES), a putative inhibitor of the plasmalemma ATPase (Balke and Hodges, 1977), completely inhibited the response at 50 μM. Unexpectedly, vanadate, another inhibitor of the plasmalemma ATPase, was ineffective, in contrast with its previously reported inhibition of FC- and white-light-dependent proton extrusion from guard cells in epidermal peels (Gepstein, Jacobs, and Taiz, 1982). It remains to be established whether the discrepancy between these results ensues from an intrinsic difference between the responses to blue and white light or from an ineffectiveness of vanadate with guard cell protoplasts. It is noteworthy that in our experiments FC induced high rates of proton extrusion from guard cell protoplasts without an inhibitory effect by vanadate (Shimazaki, Iino, and Zeiger, 1986).

The effect of abscisic acid (ABA; see Raschke, this volume, on the role of ABA in stomatal function and water stress in plants) on the BL-dependent acidification was also tested. ABA at 5 μM inhibited the acidification response to a saturating BL pulse by about 40 percent (Fig. 9.9). Higher concentrations of ABA did not increase the level of inhibition.

Red-light-dependent, medium alkalinization by guard-cell protoplasts. The red light irradiation required for the measurements of BL-dependent acidification provided us with an opportunity to study the effect of red light on the pH of protoplast suspensions. Unexpectedly, we found that at the onset of red-light irradiation, dark-adapted preparations of *Vicia* guard cell protoplasts markedly alkalinized their suspension medium (Fig. 9.10; Shimazaki and Zeiger, unpublished). Similar time courses of the red-light-dependent alkalinization were consistently observed in many experiments. This red-light effect on guard cells has not been previously reported.

The response was further analyzed by simultaneous measurements of pH and O_2 tension in the medium, obtained with a Clark-type, O_2 microelectrode

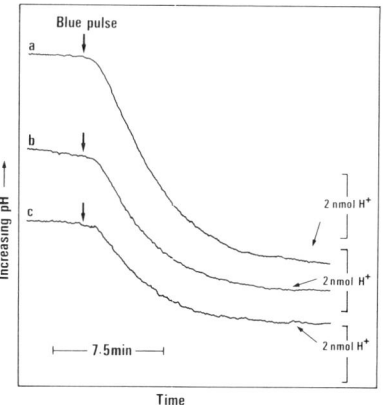

Fig. 9.9. Effect of ABA on the BL-dependent proton extrusion by *Vicia* guard cell protoplasts. ABA was added to the suspension medium 20 min before the application of the BL pulses (30 s with 50 μmol m^{-2} s^{-1}). Addition of ABA caused a slight acidification of the medium. The magnitude of the BL-dependent acidification (trace *a*) was reduced by about 40 percent by the addition of 5 μM ABA (trace *b*). Addition of 50 μM ABA (trace *c*) increased the inhibition only slightly. Background red light: 800 μmol m^{-2} s^{-1}. (Reproduced with permission from Shimazaki, Iino, and Zeiger, 1986.)

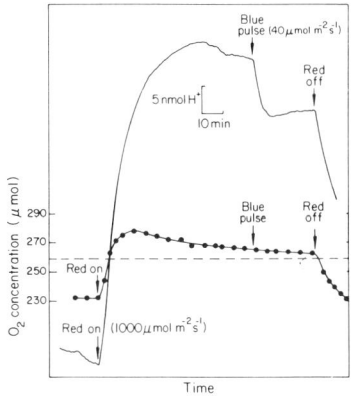

Fig. 9.10. Red-light-induced alkalinization in a suspension of *Vicia* guard cell protoplasts. Experimental conditions as in Figure 9.5. The pH profile also shows the relationship between the alkalinization at the onset of red-light irradiation and the acidification in response to a BL pulse. Closed circles show the O_2 concentration in the solution, measured with a Clark-type oxygen electrode. The plotted O_2 concentrations were calculated using standard O_2 solubility data at 25° C. The broken line indicates the measured O_2 tension just below the surface of the suspension, which was in equilibrium with the atmosphere. (Data from K. Shimazaki and E. Zeiger, unpublished.)

(Revsbech, 1983). At the onset of red irradiation, medium alkalinization and oxygen evolution were observed; when the red light was turned off, medium acidification and oxygen uptake ensued (Fig. 9.10). Both oxygen evolution and medium alkalinization were inhibited by addition of the photosynthetic inhibitor 3-(3,4-dichlorophenyl)-1,1-dimethylurea (DCMU) at 5 µM (Shimazaki and Zeiger, unpublished).

The close correlation between the red-light-dependent alkalinization and oxygen evolution, and the DCMU sensitivity of both phenomena, indicate that the alkalinization is coupled to photosynthetic activity in the guard cell protoplasts. In suspensions of blue-green algae (Miller and Colman, 1980) and of mesophyll cells or protoplasts, light-dependent alkalinization is commonly attributed to photosynthetic CO_2 fixation. Guard cell chloroplasts have been shown to have competence for electron transport (Zeiger, Armond, and Melis, 1981; Ogawa et al., 1982; Shimazaki, Gotow, and Kondo, 1982) and for cyclic and noncyclic photophosphorylation (Shimazaki and Zeiger, 1985), but they are widely thought to be devoid of Calvin-cycle activity and therefore unable to fix CO_2 photosynthetically (see Outlaw, this volume; Sharkey and Ogawa, this volume).

Several observations, however, suggest that guard cell chloroplasts may carry out photosynthetic CO_2 fixation. In an analysis of fluorescence transients, Melis and Zeiger (1982) found the same CO_2 sensitivity in guard cell chloroplasts as in mesophyll chloroplasts. Sensitivity to CO_2 in mesophyll chloroplasts has been shown to be a result of photosynthetic CO_2 fixation. In addition, stomata of the orchid *Paphiopedilum harrisianum*, which are completely devoid of chlorophyll, show a markedly reduced CO_2 sensitivity (Assmann and Zeiger, 1985), pointing to a correlation between guard cell chloroplast activity and the well-known stomatal response to CO_2 (Morison, this volume). Furthermore, Zemel and Gepstein (1985) recently reported the detection of ribulose-1,5-bisphosphate carboxylase (RuBPCase), the CO_2-fixing enzyme in chloroplasts, in *Vicia* guard cell chloroplasts by two different immunological methods. Zemel and Gepstein interpreted the difference between their findings and the negative results of an immunocytochemical study with *Vicia* (Madhavan and Smith, 1982) as a consequence of antibody penetration problems. Pretreatment of guard cells with cellulolytic enzymes made it possible for the antibodies to cross the thick guard cell walls (Zemel and Gepstein, 1985). Of further interest is the evidence for marked proteolytic activity in guard cell homogenates (Zemel and Gepstein, 1985), which could account for the lack of RuBPCase activity in *in vitro* assays (Outlaw et al., 1979).

Taken together, the immunological demonstration of RuBPCase in *Vicia* guard cells and the red-light-induced alkalinization by guard cell protoplasts strongly suggest photosynthetic CO_2 fixation in guard cells. Additional evidence for this hypothesis has been recently reported (Zeiger et al., 1986).

Pulses of blue light activate an electrogenic proton pump at the guard cell plasmalemma. An important question emerging from the studies of BL-dependent acidification discussed in the previous section is the cellular basis of proton extrusion. In plant cells, including guard cells, proton pumping is widely thought to depend on the activity of a plasmalemma ATPase (Poole, 1978; Spanswick, 1981), and chemiosmotic proton pumping has been proposed as the basis of active ion uptake during stomatal movements (Zeiger, 1983; Assmann and Zeiger, this volume). In an effort to characterize the mechanism underlying BL-dependent proton extrusion from guard cell protoplasts, Assmann, Simoncini, and Schroeder (1985) used patch-clamping techniques to measure the electrical changes in *Vicia* guard cell protoplasts stimulated with BL pulses. Methodological details of patch clamping are given in Assmann and Zeiger (this volume). The patch-clamp study showed that protoplasts irradiated with the same dual-beam protocol used for the intact leaf and acidification studies responded to a BL pulse with a transient membrane hyperpolarization of up to 45 mV (see Fig. 7.1). The onset of the hyperpolarization showed a delay of 25 to 30 seconds, with the response peaking several minutes after the pulse. Pulse-dependent proton extrusion and membrane hyperpolarization exhibited similar delays (compare Figs. 7.1 and 9.5), with the hyperpolarization, acidification, and conductance increase in the intact leaf showing increasingly longer total response times (Figs. 7.1, 9.2A, 9.5). This temporal relationship is consistent with a hypothesis that the phenomena are consecutive events in the sensory transduction of BL.

A comparison of the patch-clamping results with the previously reported light-dependent hyperpolarization in *Allium* and *Vicia* guard cells (Moody and Zeiger, 1978; Ishikawa et al., 1983) obtained with conventional electophysiological techniques underscores the resolving power of patch clamping. Light-dependent hyperpolarizations measured with a single electrode in a guard cell (Zeiger et al., 1977) are consistent with proton pumping, but alternative explanations, such as increases in membrane potentials ensuing from K^+ uptake, could not be excluded (Moody and Zeiger, 1978). Furthermore, in conventional electrophysiological studies, the tip of the electrode is usually located inside the vacuole; that positioning results in electrical recordings of the plasmalemma and the tonoplast in series. Under these conditions it is difficult to determine the localization of the underlying ion fluxes. In contrast, membrane hyperpolarizations measured under patch-clamp conditions and concomitant observations of outward currents of up to 5.5 pA, could be unambiguously interpreted as evidence for BL-dependent, active ion pumping at the guard cell plasmalemma (Assmann, Simoncini, and Schroeder, 1985). Measurements of proton extrusion under identical light conditions (Shimazaki, Iino, and Zeiger, 1986) indicate that the pumped ions carrying the electrical charges are protons. Patch-clamping also allowed the critical testing of a K^+ requirement for pump activity. Elimination of K^+ from the pipette and medium solution did not prevent the hyperpolarization (Assmann, Simoncini, and Schroeder, 1985). The substantial reduction of the voltage noise of the response as compared with that observed in the presence of the cation (Fig. 7.1, traces b and c) suggested that the noise observed in the presence of K^+ represented passive flux of K^+ through membrane channels (Schroeder, Hedrich, and Fernandez, 1984).

Available information therefore indicates that the BL-dependent increases in stomatal conductance observed in the intact leaf are the result of a specific activation of an electrogenic proton pump at the guard cell plasmalemma. The observed fluence dependency of proton extrusion by guard cell protoplasts also indicates that the magnitude of the electrochemical gradient generated by the proton pump depends on prevailing fluences of BL.

A conclusive demonstration of an electrogenic proton pump at the guard cell plasmalemma provides direct evidence for a chemiosmotic mechanism in stomatal movements (Zeiger, 1983). A substantiation of chemiosmosis and available data on the BL response indicate that the fluence-dependent proton gradient drives the passive uptake of K^+ and Cl^- and results in quantitative changes in guard cell osmotic potential and turgor, and concomitant modulations of pore apertures and stomatal conductances.

Conclusions and Perspectives

Properties of the blue-light photoreceptor. The identity of the BL photoreceptor or photoreceptors in plant cells remains to be established (Schmidt, 1984). Muñoz and Butler (1975) have proposed that flavins could absorb BL and reduce a cytochorome in a membrane-bound electron transport system at the onset of sensory transduction. Blue-light-dependent redox reactions in membrane preparations have been studied by several groups (e.g. Borgeson and Bowman, 1985), but the *in vivo* physiology of these photochemical reactions remains unknown. Spectroscopic data of several BL responses, including that in stomata (Ogawa et al., 1978; Sharkey and Raschke, 1981; Karlsson, 1986b), give evidence of the involvement of flavins in BL absorption, but definitive data remain to be obtained (Schmidt, 1984).

The BL response of stomata could prove valuable for the identification of the BL photoreceptor. The response is localized in a single cell type, the guard cell, and sensory transduction is specifically associated with reversible changes in stomatal apertures and concomitant modulation of ion fluxes. The recently postulated kinetic model of the BL response (Iino, Ogawa, and Zeiger, 1985) offers specific temporal predictions of light-dependent, relative activities of the hypothetical interconvertible forms *A* and *B*, which could be tested biochemically. It will also be interesting to establish whether *A* and *B* are interconvertible forms of the photoreceptor itself, or of a downstream molecular component of the sensory transduction process.

The blue-light response, proton pumping, and ion transport in stomatal movements. The mechanism whereby BL activates proton pumping at the guard cell plasmalemma has central implications for the understanding of BL sensory transduction. The properties of the BL response to pulses, which proceeds at high rates long after pulse stimulation, indicates that continuous BL irradiation is not necessary for the response; therefore BL is ruled out as a primary source of metabolic energy for proton pumping. That conclusion is strengthened by the patch-clamp experiments in which high concentrations of ATP were made available within the cell without any detectable stimulation of proton pumping prior to the BL irradiation (Assmann, Simoncini, and Schroeder, 1985). Thus the BL effect appears to be regulatory, rather than a relief of an energy limitation for proton pumping. The ATP requirement for the BL-dependent membrane hyperpolarizations (Assmann, Simoncini, and Schroeder, 1985) and the inhibition of proton extrusion by the ATPase inhibitor diethylstilbestrol (DES; Shimazaki, Iino, and Zeiger, 1986) provide indirect evidence for a BL stimulation of a plasmalemma ATPase. If this is the case, the regulation could be on the ATPase proper or mediated by redox reactions (Zeiger, 1984). Of further interest is the quantitative relationship between the extent of BL-stimulated proton pumping and the saturated rates of proton pumping in guard cells. In intact leaves, the increases in stomatal conductance in response to saturating fluences of BL did not exceed 20 percent of the conductance levels attainable under background high fluence rates of red irradiation (Zeiger, Iino, and Ogawa, 1985). Assuming that stomatal opening in response to red light is also dependent on proton pumping, the specific BL-stimulated component can be envisioned as a separate population of BL-sensitive ATPases, or a further stimulation of a single ATPase population, or the specific proton-pumping activity of a membrane-bound redox system. Studies of the BL response in membrane vesicles from guard cell protoplasts could provide further information on these questions.

The emerging properties of the stomatal response to BL have further implications for our understanding of the regulation of stomatal apertures by the proton pump. Although proton pumping is widely recognized as the driving force for stomatal movements, some investigators (e.g. MacRobbie, 1984) have argued that the regulation of stomatal apertures depends solely on the control of ion efflux and that proton pumping *per se* has no regulatory function (see also MacRobbie, this volume). The findings reported in the present paper rule out an exclusive role of ion efflux in the regulation of stomatal apertures and show that a fluence-dependent modulation of proton pumping provides a specific mechanism for the control of stomatal apertures. Stomatal movements therefore appear to be regulated by both precise rates of proton pumping that drives stomatal opening and by ion efflux that causes closing in response to such stimuli as ABA (Raschke, this volume; MacRobbie, this volume), calcium (Schwartz, 1985), and high water-vapor pressure deficits (see Tenhunen, Pearcy, and Lange, this volume; Schulze et al., this volume).

The function of the blue-light response of stomata and its interaction with the response of guard cell chloroplasts. The existence of two distinct photoreceptor systems in guard cells—the guard cell chloroplast and a specific BL-dependent photosystem—remains intriguing. The widespread occurrence of the orchid genus *Paphiopedilum*, with achlorophyllous guard cells, demonstrates that stomata devoid of chloroplasts can be functional, albeit limited in their maximal rates of stomatal conductance. In contrast, all naturally occurring stomatal systems yet studied exhibit the BL response, indicating that the response has an adaptive value. Isolation of mutants deficient in or lacking entirely the BL photosystem should help characterize the functional properties of the stomatal response to BL.

Much remains to be learned about BL-dependent proton pumping under continuous BL irradiation and its modulation. The low saturation levels of the response, about 10 mmol m^{-2} (Shimazaki, Iino, and Zeiger, 1986; Zeiger, Iino, and Ogawa, 1985), imply that it is light-saturated throughout most of the day. Hence, the BL response could function as a "light on" signal during daylight, with its metabolic correlates, including proton pumping (Assmann, Simoncini, and Schroeder, 1985; Shimazaki, Iino, and Zeiger, 1986) and stimulation of malate biosynthesis (Ogawa et al., 1978), gearing stomatal responses for the functional requirements of the daylight activity of the leaf. In addition, the BL response has been shown to drive stomatal opening at dawn (Zeiger, Field, and Mooney, 1981) and could be important in the modulation of stomatal movements under sun flecks, a main source of light energy for leaves in the understory of dense forests or within the canopy of many agricultural crops (Zeiger, Iino, and Ogawa, 1985). Further characterization of the metabolic and ecophysiological roles of the BL response should enrich our understanding of stomatal function.

REFERENCES

Assmann, S. M., L. Simoncini, and J. I. Schroeder. 1985. Blue light activates electrogenic ion pumping in guard cell protoplasts of *Vicia faba*. *Nature*, 318: 285–87.

Assmann, S. M., and E. Zeiger. 1985. Stomatal responses to CO_2 in *Paphiopedilum* and *Phragmipedium*: Role of the guard cell chloroplast. *Plant Physiology*, 77: 461–64.

Balke, N. E., and T. K. Hodges. 1977. Inhibition of ion absorption in oat roots: Comparison of diethylstilbestrol and oligomycin. *Plant Science Letters*, 10: 319–25.

Borgeson, C. E., and B. J. Bowman. 1985. Blue light-reducible cytochromes in membrane fractions of *Neurospora crassa*. *Plant Physiology*, 78: 433–37.

Gepstein, S., M. Jacobs, and L. Taiz. 1982. Inhibition of stomatal opening in *Vicia faba* epidermal tissue by vanadate and abscisic acid. *Plant Science Letters*, 28: 63–72.

Hsiao, T. C., W. G. Allaway, and L. T. Evans. 1973. Action spectra for guard cell Rb^+ uptake and stomatal opening in *Vicia faba*. *Plant Physiology*, 51: 82–88.

Iino, M., T. Ogawa, and E. Zeiger. 1985. Kinetic properties of the blue-light response of stomata. *Proceedings of the National Academy of Sciences*, 82: 8019–23.

Iino, M., and E. Schäfer. 1984. Phototropic response of the Stage I *Phycomyces* sporangiophore to a pulse of blue light. *Proceedings of the National Academy of Sciences*, 81: 7103–7.

Ishikawa, H., H. Aizawa, H. Kishira, T. Ogawa, and M. Sakata. 1983. Light-induced changes of membrane potential in guard cells of *Vicia faba*. *Plant and Cell Physiology*, 24: 769–72.

Karlsson, P. E. 1986a. Blue light regulation of stomata in wheat seedlings, I. Influence of red background illumination and initial conductance level. *Physiologia plantarum*, 66: 202–6.

———. 1986b. Blue light regulation of stomata in wheat seedlings, II. Action spectrum and search for action dichroism. *Physiologia plantarum*, 66: 207–10.

Kowallik, W. 1982. Blue light effects on respiration. *Annual Review of Plant Physiology*, 33: 51–72.

MacRobbie, E. A. C. 1984. Light effects on ion fluxes in guard cells. In W. J. Cram, K. Janáček, R. Rybová, and K. Sigler, eds., *Membrane transport in plants*: 129–34. New York: Wiley.

Madhavan, S., and B. N. Smith. 1982. Localization of ribulose biphosphate carboxylase in the guard cells by an indirect, immunofluorescence technique. *Plant Physiology*, 69: 273–77.

Mansfield, T. A., and H. Meidner. 1966. Stomatal opening in light of different wavelengths: Effect of blue light independent of carbon dioxide concentrations. *Journal of Experimental Botany*, 17: 510–21.

Meidner, H. 1968. The comparative effects of blue and red light on the stomata of *Allium cepa* L. and *Xanthium pennsylvanicum*. *Journal of Experimental Botany*, 19: 146–51.

Melis, A., and E. Zeiger. 1982. Chlorophyll *a* fluorescence transients in mesophyll and guard cells. Modulation of guard cell photophosphorylation by CO_2. *Plant Physiology*, 69: 642–47.

Miller, A. G., and B. Colman. 1980. Evidence for HCO_3^- transport by the blue-green alga (Cyanobacterium) *Coccochloris peniocystis*. *Plant Physiology*, 65: 397–402.

Moody, W., and E. Zeiger. 1978. Electrophysiological properties of onion guard cells. *Planta*, 139: 159–65.

Mouravieff, I. 1958. Action de la lumière sur le cellule végétale. *Bulletin de la Société botanique de France*, 105: 467–75.

———. 1973. Microphotométrie des fluctuations de la teneur en amidon des stomates éclairés par la lumière de 436 nm et 665 nm en absence ou en présence de gaz carbonique. *Annales des sciences naturelles (botanique)*, 14: 377–83.

Muñoz, V., and W. L. Butler. 1975. Photoreceptor pigment for blue light in *Neurospora crassa*. *Plant Physiology*, 55: 421–26.

Ogawa, T., D. Grantz, J. Boyer, and Govindjee. 1982. Effects of cations and abscisic acid on chlorophyll *a* fluorescence in guard cells of *Vicia faba*. *Plant Physiology*, 69: 1140–44.

Ogawa, T., H. Ishikawa, K. Shimada, and K. Shibata. 1978. Synergistic action of red and blue light and action spectra for malate formation in guard cells of *Vicia faba* L. *Planta*, 142: 61–65.

Outlaw, W. H., Jr. 1982. Carbon metabolism in guard cells. In L. L. Creasy and G. Hrazdina, eds., *Cellular and subcellular localization in plant metabolism*: 185–222. New York: Plenum.
Outlaw, W. H., Jr., J. Manchester, C. A. DiCamelli, D. D. Randall, B. Rapp, and G. M. Veith. 1979. Photosynthetic carbon reduction pathway is absent in chloroplasts of *Vicia faba* guard cells. *Proceedings of the National Academy of Sciences*, 76: 6371–75.
Pemadasa, M. A. 1981. Photocontrol of stomatal movements. *Biological Reviews*, 56: 551–88.
Poole, R. J. 1978. Energy coupling for membrane transport. *Annual Review of Plant Physiology*, 29: 437–60.
Ramos, C., and A. E. Hall. 1983. Effects of photon fluence rate and intercellular CO_2 partial pressure on leaf conductance and CO_2 uptake rate in *Capsicum* and *Amaranthus*. *Photosynthetica*, 17: 34–42.
Raschke, K. 1967. Der Einfluss von Rot- und Blaulicht auf die Öffnungs- und Schliessgeschwindigkeit der stomata von *Zea mays*. *Naturwissenchaften*, 54: 72–73.
Raschke, K., and G. D. Humble. 1973. No uptake of anions required by opening stomata of *Vicia faba*: Guard cells release hydrogen ions. *Planta*, 115: 47–57.
Revsbech, N. P. 1983. In situ measurement of oxygen profiles of sediments by use of oxygen microelectrodes. In E. Gnaiger and H. Forstner, eds., *Polarographic oxygen sensors*: 265–73. Heidelberg: Springer-Verlag.
Rodríguez, M. T., and A. Larqué-Saavedra. 1983. Abscisic acid accumulation in *Phaseolus vulgaris* L. with different growth habits. *Journal of Plant Growth Regulation*, 2: 225–28.
Sanders, D., A. Ballarin-Denti, and C. L. Slayman. 1984. The role of transport in regulation of cytoplasmic pH. In W. J. Cram, K. Janáček, R. Rybová, and K. Sigler, eds., *Membrane transport in plants*: 303–8. New York: Wiley.
Schmidt, W. 1984. Bluelight physiology. *Bioscience*, 34: 698–704.
Schroeder, J. I., R. Hedrich, and J. M. Fernandez. 1984. Potassium-selective single channels in guard cell protoplasts of *Vicia faba*. *Nature*, 312: 361–62.
Schwartz, A. 1985. Role of Ca^{2+} and EGTA on stomatal movements in epidermal peels of *Commelina communis* L. *Plant Physiology*, 79: 1003–5.
Sharkey, T. D., and K. Raschke. 1981. Effects of light quality on stomatal opening in leaves of *Xanthium strumarium* L. *Plant Physiology*, 68: 170–74.
Shimazaki, K., K. Gotow, and N. Kondo. 1982. Photosynthetic properties of guard cell protoplasts of *Vicia faba*. *Plant and Cell Physiology*, 23: 871–79.
Shimazaki, K., M. Iino, and E. Zeiger. 1986. Blue light–dependent proton extrusion by guard cell protoplasts of *Vicia faba*. *Nature*, 319: 324–26.
Shimazaki, K., and E. Zeiger. 1985. Cyclic and noncyclic photophosphorylation in isolated guard cell chloroplasts from *Vicia faba*. L. *Plant Physiology*, 78: 211–14.
Skaar, H., and A. Johnsson. 1978. Rapid, blue-light induced transpiration in *Avena*. *Physiologia plantarum*, 43: 390–96.
Spanswick, R. M. 1981. Electrogenic ion pumps, 1981. *Annual Review of Plant Physiology*, 32: 267–89.
Weyers, J. D. B., P. J. Fitzsimons, G. M. Mansey, and E. S. Martin. 1983. Guard cell protoplasts—Aspects of work with an important new research tool. *Physiologia plantarum*, 58: 331–39.
Wong, S. C., I. R. Cowan, and G. D. Farquhar. 1979. Stomatal conductance correlates with photosynthetic capacity. *Nature*, 282: 424–26.
Zeiger, E. 1983. The biology of stomatal guard cells. *Annual Review of Plant Physiology*, 34: 441–75.

_____. 1984. Blue light and stomatal function. In H. Senger, ed., *Blue light effects in biological systems*: 484–94. Berlin: Springer-Verlag.

Zeiger, E., P. Armond, and A. Melis. 1981. Fluorescence properties of guard cell chloroplasts. *Plant Physiology*, 67: 17–20.

Zeiger, E., A. J. Bloom, and P. K. Hepler. 1978. Ion transport in stomatal guard cells: A chemiosmotic hypothesis. *What's New in Plant Physiology*, 9: 29–32.

Zeiger, E., C. Field, and H. A. Mooney. 1981. Stomatal opening at dawn: Possible roles of the blue light response in nature. In H. Smith, ed., *Plants and the daylight spectrum*: 391–407. New York: Academic.

Zeiger E., K. Gotow, B. Mawson, and S. Taylor. 1986. The guard cell chloroplast: Properties and function. In J. Biggins, ed., *Proceedings of the 7th International congress on photosynthesis*. The Hague: Junk. [In press.]

Zeiger, E., and P. K. Hepler. 1976. Production of guard cell protoplasts from onion and tobacco. *Plant Physiology*, 58: 492–98.

_____. 1977. Light and stomatal function: Blue light stimulates swelling of guard cell protoplasts. *Science*, 196: 887–89.

Zeiger, E., M. Iino, and T. Ogawa. 1985. The blue light response of stomata: Pulse kinetics and some mechanistic implications. *Photochemistry and Photobiology*, 42: 759–63.

Zeiger, E., W. Moody, P. Hepler, and F. Varela. 1977. Light-sensitive membrane potentials in onion guard cells. *Nature*, 270: 270–71.

Zemel, E., and S. Gepstein. 1985. Immunological evidence for the presence of ribulose biphosphate carboxylase in guard cell chloroplasts. *Plant Physiology*, 78: 586–90.

10

Intercellular CO_2 Concentration and Stomatal Response to CO_2

James I. L. Morison

In most plant species investigated to date the degree of stomatal opening decreases with increases in the CO_2 concentration either around or inside the leaf. Scarth (1932) first suggested that the cause of light-induced stomatal opening was the photosynthetic removal of intercellular CO_2, and early evidence supporting this hypothesis was provided by Freudenberger (1940), who studied stomatal movement in *Canna* leaves in response to a wide range of CO_2 concentrations. Freudenberger (1940) observed that decreasing the CO_2 concentration stimulated opening and that increasing the CO_2 concentration initiated closure. This pattern of stomatal response to CO_2 was also found in extensive studies by Heath and co-workers in the 1950's. Several reviewers since then have discussed those and later results (Heath, 1959; Meidner and Mansfield, 1968; Raschke, 1975a, 1979). However, the role of CO_2 in stomatal movement is still controversial, and there is no unifying hypothesis on its action.

Stomatal Response to CO_2 Concentration

Stomata have been shown to respond to changes in CO_2 concentration in over 50 species, including angiosperms and gymnosperms, dicots and monocots, and species with C_3, C_4, and CAM photosynthetic pathways (Table 10.1). We can therefore assume that stomatal response to CO_2 is a general phenomenon. However, it should be noted that most experimental work has been carried out on fewer than ten species. Stomata can respond to CO_2 when isolated, when in epidermal peels, in leaf discs, in intact leaves, or in whole plants. Reports of stomatal CO_2 responses range from complete insensitivity to CO_2 concentrations between 0 and 2,000 μmol mol^{-1} in cotton (Bierhuizen and Slatyer, 1964) and Scotch pine (Ng, 1978) to a tenfold decrease in conductance in response to an increase in CO_2 from 300 to 500 μmol mol^{-1} in apple leaves (Warritt, Landsberg, and Thorpe, 1980). Such differences are not restricted to interspecific contrasts but occur also within the same species under different environmental conditions, owing to interactions between CO_2 and other factors that control stomatal aperture.

TABLE 10.1. Responses of stomata to CO_2

Species	Parameter	Type	CO_2 concentration $\mu mol\ mol^{-1}$	Reference
C_3 species				
Allium cepa	vP		0, atm	Heath & Meidner, 1957
Avena sativa	E	C_a	0, atm	Brogårdh & Johnsson, 1975
Bacopa caroliniana	μm	C_a	0, atm	Mouravieff, 1958
Brassica pusna	E	C_a	0–1,340	Gaastra, 1959
Canna indica	μm	C_a	30–25,000	Freudenberger, 1940
Capsicum annuum		C_i	50–300	Ramos & Hall, 1983
Cassia obtusifolia	g_s	C_a	350, 675	Patterson & Flint, 1982
Commelina communis	r_s	C_i	100–400	Morison & Jarvis, 1983b
Crotalaria spectabilis	g_s	C_a	350, 675	Patterson & Flint, 1982
Desmodium paniculatum	g_s	C_a	350–1,000	Wulff & Strain, 1982
Eucalyptus pauciflora	g_s	C_i	100–300	Wong, Cowan & Farquhar, 1978
Fagus sylvatica	g_s	C_i	100–800	Koch, 1969
Glycine max	g_s	C_a	330–1,000	Rogers et al., 1980
Gossypium hirsutum	g_s	C_i	100–500	Radin & Ackerson, 1981
Helianthus annuus	r_s	C_i	0–500	Whiteman & Koller, 1967a
Hordeum vulgare	E	C_a	190–450	Louwerse, 1980
Kochia indica	E	C_a	0, atm	Whiteman & Koller, 1967b
Liquidambar styraciflua	g_s	C_a	330–1,000	Rogers et al., 1980
Lycopersicon esculentum	g_s	C_i	50–600	Bradford, Sharkey & Farquhar, 1983
Malus pumila	g_s	C_i	100–300	Warrit, Landsberg & Thorpe, 1980
Oryza sativa	g_s	C_i	100–700	Morison & Gifford, 1983
Paphiopedilum leeanum	r_s	C_a	300, 1,150	Nelson & Mayo, 1975
Pelargonium zonale	vP		0, atm	Heath, 1948
Perilla frutescens	g_s	C_i	150–330	Sharkey & Raschke, 1981a
Phalaris aquatica	g_s	C_i	100–700	Morison & Gifford, 1983
Phaseolus vulgaris	g_s	C_i	150–330	Sharkey & Raschke, 1981a
Phyllitus scolopendrium	μm	C_a	0, 880, 10,000	Mansfield & Willmer, 1969
Picea sitchensis	g_s	C_i	50–500	Morison & Jarvis, 1983a
Pinus sylvestris	g_s	C_i	50–500	Morison & Jarvis, 1983a
Pisum sativum	g_s	C_a	350, 675	Paez, Hellmers & Strain, 1983

Species	Method	Range	Reference
Polypodium vulgare	μm	0–400	Lösch, 1976
Populus deltoides	g_s	50–1,000	Regehr, Bazzaz & Boggess, 1975
Rumex conglomeratus	vP	0, atm	Allaway & Mansfield, 1967
Sesamum indicum	r_s	50–300	Hall & Kaufmann, 1975
Spinacia oleracea	r_s	20–500	Harris et al., 1983
Taraxacum officinale	E	0–800	Akita & Moss, 1972
Triticum aestivum	C_a	0–840	Heath & Russell, 1954
Vicia faba	vP	0, atm	Fischer, 1968
Vigna luteola	μm		Ludlow & Wilson, 1971
Vitis vinifera	r_s	0–1,200	Kriedemann, Sward & Downton, 1976
Xanthium strumarium	E	atm, 1,200	Sharkey & Raschke, 1981b
	g_s	0–400	
C_4 species			
Amaranthus hypochondriacus	C_i	50–300	Ramos & Hall, 1983
A. powelli	g_s	0–450	Dubbe, Farquhar & Raschke, 1978
A. retroflexus	E	0–800	Akita & Moss, 1972
Atriplex halimus	E	0, atm	Whiteman & Koller, 1967b
Bothriochloa caucasia	C_a	0–500	Coyne & Bradford, 1984
Hilaria rigida	g_s	0–700	Nobel, 1980
Paspalum plicatulum	C_a	100–700	Morison & Gifford, 1983
Pennisetum typhoides	g_s	100–500	McPherson & Slatyer, 1973
Setaria viridis	C_a	0–800	Akita & Moss, 1972
Sorghum almum	E	0–800	Ludlow & Wilson, 1971
S. bicolor	C_a	100–1,800	van Bavel, 1974
S. sudanense	r_s	0–550	Downes, 1971
Spartina townsendii	r_s	0–1,000	Long & Woolhouse, 1978
Zea mays	C_a	0–400	Raschke, Hanebuth & Farquhar, 1978
	g_s		
CAM plants			
Agave americana	E	200–1,000	Neales, 1970
A. deseri	C_a	atm, 2%	Cockburn, Ting & Sternberg, 1979
Opuntia ficus-indica	C_a	atm, 2%	Cockburn, Ting & Sternberg, 1979

NOTE: g_s, stomatal conductance; r_s, stomatal resistance; vP, viscous-flow porometer; μm, aperture; E, transpiration measurement. C_i, C_a, and atm are intercellular, ambient, and normal atmospheric CO_2 concentrations, respectively, in μmol mol^{-1}; str, experiments on epidermal strips. Only single references for each species have been given as examples.

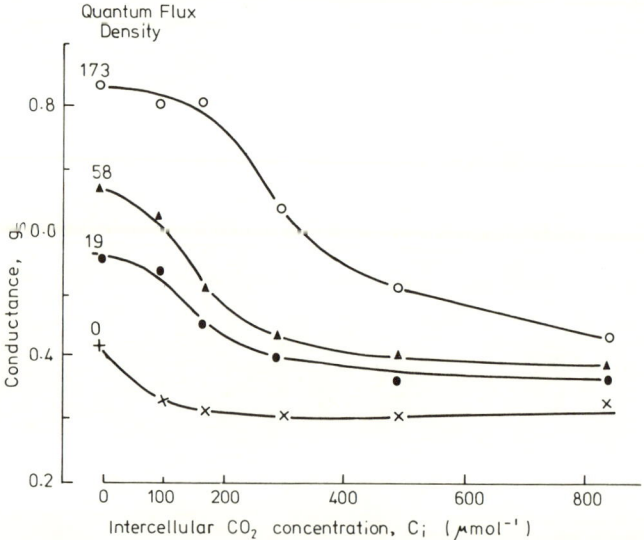

Fig. 10.1. Response of conductance, g_s, calculated as the reciprocal of viscous flow resistance in Gregory-Pearse units to CO_2 concentration in wheat leaves at four light intensities. (Adapted from Morison and Jarvis, 1983b: 107; original data from Heath and Russell, 1954.)

The response of stomatal conductance to CO_2 concentration is generally curvilinear, showing larger changes in response to concentrations below 300 μmol mol^{-1} than in response to higher concentrations (Fig. 10.1). In some reports the maximum stomatal opening, or conductance, has occurred at 100 μmol mol^{-1} ambient CO_2 concentration (Raschke, 1976; Dubbe, Farquhar, and Raschke, 1978); in most others, at 0 μmol mol^{-1}. The response can be very rapid (e.g. < 1 min in maize leaves, using a viscous flow porometer: Raschke, 1972), or it can take several hours (e.g. 3 h in Scotch pine: Morison, 1980).

Stomata respond to CO_2 in darkness or in light. Normally, the opening in response to low CO_2 concentrations is more pronounced in the light than in the dark (Fig. 10.1). At high light intensities, normal atmospheric CO_2 concentrations may not reduce stomatal aperture below the maximum (Downes, 1971; Long and Woolhouse, 1978; Sharkey and Raschke, 1981a), and high CO_2 concentrations can prevent the opening effect of increasing light intensity (Wong, Cowan, and Farquhar, 1978; Morison and Jarvis, 1983b).

The action of CO_2 on stomata varies with other factors, such as growth conditions and leaf age (Fig. 10.2). In particular, stomatal physiologists over several years have been concerned with the effects of water stress on stomatal response to CO_2. Many such effects have been attributed to the action of abscisic acid (ABA; see Raschke, this volume). Raschke (1975b) originally suggested that ABA "sensitized" stomata to CO_2 and that ABA was required

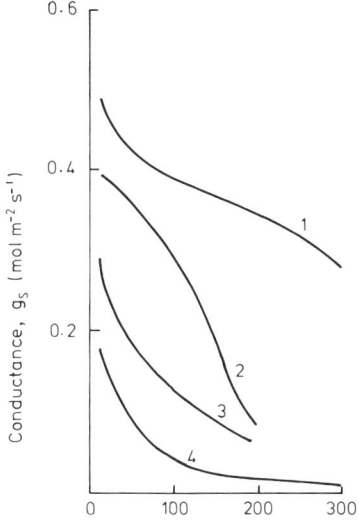

Fig. 10.2. Response of conductance to intercellular CO_2 concentration in maize leaves. Curve *1* is from a leaf from a field-grown plant; curves *2–4*, from plants grown in growth chambers. Curves *2* and *3* are from comparable leaves, but leaf *3* has received 10^{-5} M ABA. Curve *4* is from a different batch of plants. (Redrawn from Raschke, 1979: 428.)

for stomata to respond to CO_2 concentration, and vice versa. However, Mansfield (1976), working with the same species as Raschke (*Xanthium strumarium*), argued that the effects of CO_2 and ABA were separate and additive, not interactive. Since then it has become clear from the detailed work of Mansfield, Davies, and colleagues at Lancaster with epidermal strips and from work on intact leaves (Radin and Ackerson, 1981; Radin, Parker, and Guinn, 1982; Kubik and Antoszewski, 1983) that the effect of ABA varies with experimental conditions, particularly pretreatment, plant material (age, nutrition, abaxial or adaxial strips), and the cytokinin and auxin content of the tissue, which presumably varies with the growth conditions (see Mansfield and Davies, this volume). Indeed, it has been suggested that it is difficult to define accurately the ABA sensitivity of stomata, because of the other factors involved (Snaith and Mansfield, 1982). By the same argument it may also not be possible to define the sensitivity of stomata to CO_2, a point we shall return to.

Stomata and Intercellular CO_2 Concentration

The original suggestion was that stomata open in response to the reduction of the intercellular CO_2 concentration caused by mesophyll photosynthesis. Stomata apparently do not respond directly to the CO_2 concentration around the leaf. The CO_2 sensor for stomatal action is located in the epidermis and is presumably in the guard cells (Mouravieff, 1956; Pallaghy, 1971), the inner lateral walls of which are permeable to CO_2 (Meidner and Mansfield, 1965).

Therefore most workers have assumed that the CO_2 concentration in the guard cells is dependent on the concentration in the intercellular spaces and not on the external (ambient) CO_2 concentration. The intercellular CO_2 concentration (C_i) depends on the flux of CO_2 through the stomatal pore and is determined by the ambient CO_2 concentration (C_a), the net assimilation rate (A), and the stomatal conductance (g_s). Intercellular CO_2 concentration is not normally measured independently but is calculated from measurements of A, C_a, and g_s.

Moss and Rawlins (1963) first published estimates of C_i from the exchange of CO_2 and water vapor in single leaves. This calculation has become routine, though the exact calculation procedures have recently been improved (see Ball, this volume). Improvements in equipment for measuring leaf gas exchange and the apparent importance of estimates of C_i linking mesophyll assimilation activity and stomatal conductance have resulted in many calculations of intercellular CO_2 concentrations. However, direct determinations of C_i had been made only in large leaves of CAM plants (Cockburn, Ting, and Sternberg, 1979; Spalding et al., 1979), leaving some doubt over the accuracy of these estimates for other species. In particular, the calculation assumes that the diffusion pathway for CO_2 is identical to that for water vapor, and the calculated C_i is therefore the average CO_2 concentration at the sites of evaporation. As most of the evaporation through the stomatal pore occurs at nearby cell surfaces (Meidner, 1975; Cowan, 1977; Tyree and Yianoulis, 1980), it has been suggested that the C_i so calculated is the concentration near the guard cells, which may not be the appropriate concentration for the mesophyll air spaces. However, researchers have validated the calculations by using an independent direct-measurement technique (Sharkey et al., 1982), the results of which suggest that the concentration gradient of CO_2 inside the leaf is only 5 to 10 μmol mol^{-1}.

Is Intercellular CO_2 Concentration a Controlling Signal?

Two central questions that have emerged over the last decade are whether C_i is the important signal for light-induced stomatal opening and what influence C_i exerts on stomatal regulation (e.g. Jarvis and Morison, 1981). Certainly, a general correlation exists between light intensity, mesophyll assimilation, and conductance. Since C_i declines as assimilation increases, and since conductance in many cases increases with decreasing C_i, it has been supposed that assimilation controls conductance by effecting changes in C_i (e.g. Raschke, 1976). For example, Figure 10.3 shows typical measurements of the response of assimilation and conductance to increasing quantum flux density in leaves of *Commelina communis*. Other examples are taken up by Tenhunen, Pearcy, and Lange (this volume). In most species examined, as light intensity increases, assimilation and conductance increase, and C_i decreases, sharply at first, then

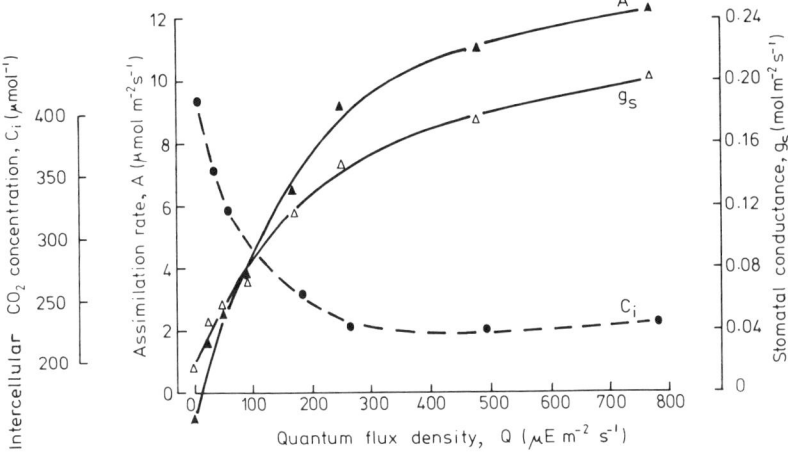

Fig. 10.3. Relationship between stomatal conductance (g_s: △), CO_2 assimilation rate (A: ▲), intercellular CO_2 concentration (C_i: ●), and increasing quantum flux density (Q) in a shoot of *Commelina communis*. (Redrawn from Morison and Jarvis, 1983b: 108.)

to a fairly constant value for light intensities above 10 percent of full sunlight. At a leaf-to-air vapor pressure difference (VPD) of 2 kPa, C_i values of about 100 and 230 μmol mol^{-1} are typical for C_4 and C_3 species, respectively, when measured in normal atmospheric CO_2 concentrations. This plateau value of C_i has suggested to several workers (e.g. Raschke, 1975a) that stomatal movement is controlled so as to maintain a nearly constant C_i in changing conditions. In addition, in several cases when plants have been gradually exposed to water stress C_i has also remained constant (e.g. *Zea mays*, Fig. 10.4; *Amaranthus palmeri*, Ehleringer, 1983) though not in all (e.g. *Notholaena parryi*, Nobel, 1978; *Vigna unguiculata*, Hall and Schulze, 1980, and Nagarajah and Schulze, 1983).

Since a constant C_i with varying environmental conditions results from the linear correlation of A with g_s, only experiments that manipulate C_i permit firm conclusions on the response of stomata to CO_2 concentration. Schulze and Hall (1982) have recently examined the environmental and internal factors that influence stomata and have described many linear correlations between CO_2 assimilation rates and conductance in natural conditions. They concluded that, since conductance was not correlated with maximum assimilation in all cases, the linear correlations were due not to a close physiological coupling of A and g_s but rather to the long-term matching of g_s and A, and that C_i is thus a conservative value. Similarly, Jarvis and Morison (1981) argued that, because the correlation between g_s and A can be broken so readily in experiments (e.g. in CO_2-free air) and in fluctuating natural conditions, stomatal conductance is largely independent of the rate of carbon fixation in the leaf over short periods

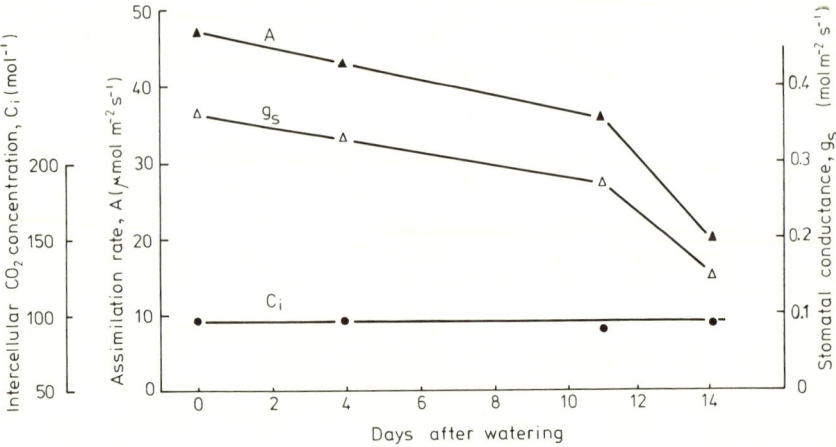

Fig. 10.4. Relationship between stomatal conductance (g_s: △), net assimilation rate (A: ▲), intercellular CO_2 concentration (C_i: ●), and increasing water stress (days after watering) in maize leaves. (Data from Wong, 1979a.)

of time. However, over longer periods the epidermis is dependent on the mesophyll for carbon supply, and substantial transport of sugars and amino acids to the epidermis has been measured (Willmer, 1981; Thorpe and Milthorpe, 1984). Though the role of the movement of such compounds (and others, such as hormones) and their rates and amounts *in vivo* has yet to be clarified, we suggest that this movement affects the long-term, perhaps daily, matching of conductance and assimilation. Any short-term (tens of minutes) matching that occurs is caused by the parallel responses of assimilation and conductance to light, which result in a nearly constant C_i. Indeed, the examples given in Jarvis and Morison (1981) and those detailed below indicate that C_i is not constant, though it is conservative, because it is both the result and an effector of stomatal aperture.

Several studies have examined the CO_2 response of stomatal conductance in intact leaves and the effect of other variables on the response and on the relationship between conductance and assimilation. Figure 10.5 shows the response of conductance to CO_2 in leaves of *Eucalyptus pauciflora* at four quantum flux densities (Wong, Cowan, and Farquhar, 1978). Stomatal conductance increased with decreasing C_i, but the response was dependent on quantum flux density. While C_i was constant at any one value of C_a across a range of light intensities, C_i changed when C_a was changed, as for example in Figure 10.6, showing data for *C. communis*. Intercellular CO_2 concentration remained a constant proportion of C_a, independent of light intensity, in this and many other species (see references in Jarvis and Morison, 1981; and in Bell, 1982). The only examples in which C_i was maintained approximately constant with varying ambient CO_2 concentrations have been in maize (Raschke, 1965;

Dubbe, Farquhar, and Raschke, 1978; Louwerse, 1980), though there have been reports of a constant C_i/C_a for this species (Wong, Cowan, and Farquhar, 1979; Sharkey and Raschke, 1981a). A constant C_i/C_a suggests an interdependence of conductance with mesophyll assimilation and has suggested a mechanistic coupling (e.g. Wong, Cowan, and Farquhar, 1979), though this has yet to be demonstrated (see Jarvis and Morison, 1981).

In C_4 species the typical value of C_i/C_a is lower than that in C_3 species; and as Ramos and Hall (1982, 1983) and Schulze and Hall (1982) have argued, it may be that plants of the C_4 photosynthetic pathway show a larger sensitivity of stomatal aperture to C_i. However, at high humidities and moderate light intensities, stomata of two C_4 and two C_3 grass species showed the same quantitative response to CO_2 (Morison and Gifford, 1983), indicating that there is little inherent difference in the CO_2 sensitivity of stomata of the two groups of plants.

The response of stomata to air humidity, independent of bulk leaf water status, provides a second example of the slight effect that the small changes of C_i observed in natural conditions have on stomata. Since small changes of humidity can affect stomata without directly affecting assimilation rate in many species (see Schulze et al., this volume), C_i varies with humidity, as shown for maize, rice, *Paspalum*, and *Phalaris* in Figure 10.7. Similar results have been found in widely differing species (*Nicotiana glauca*, Farquhar, Schulze, and Küppers, 1980; *Betula pendula* and *Gmelina arborea*, Osunobi and Davies, 1980a; *Geraea canescens* and *Perityle emoryii*, Ball and Berry, 1982; *Amaranthus tricolor*, Lin and Ehleringer, 1983; *Diplacus aurantiacus*, Mooney and Chu, 1983; *Amaranthus palmeri*, Ehleringer, 1983).

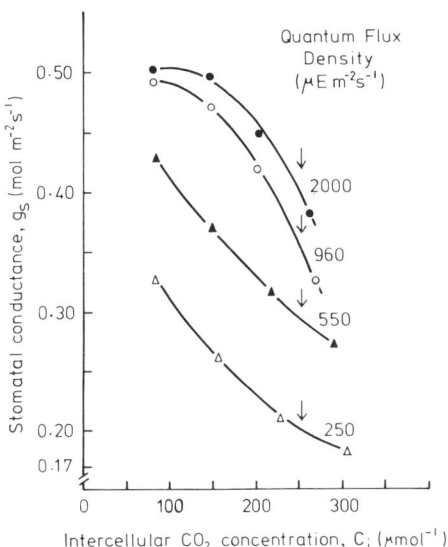

Fig. 10.5. Response of stomatal conductance (g_s) to intercellular CO_2 concentration (C_i) at four quantum flux densities for *Eucalyptus pauciflora*. Arrows indicate g_s and C_i at each flux density for C_a = 330 μmol mol^{-1}. (Redrawn from Wong, Cowan, and Farquhar, 1978: 672.)

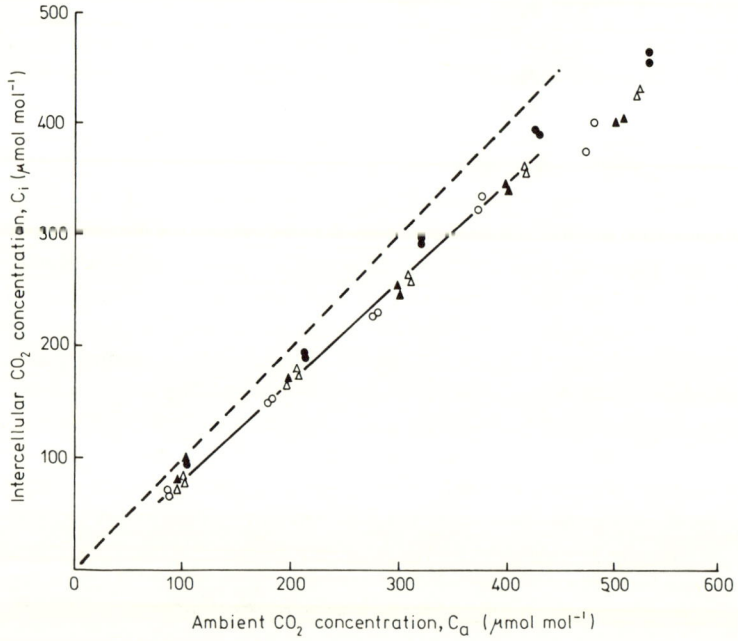

Fig. 10.6. Relationship between intercellular (C_i) and ambient (C_a) CO_2 concentration at four quantum flux densities for *Commelina communis* Quantum flux densities: 130 (●), 240 (△), 490 (▲), and 950 (○) $\mu E\ m^{-2}\ s^{-1}$. The dashed line is the line of unit slope; the solid line is a linear regression through all points where $C_a < 450\ \mu mol\ mol^{-1}$. (Redrawn from Morison and Jarvis, 1983b: 105.)

Hällgren, Sundbom, and Strand (1982), examining the temperature response of two *Betula* species, attributed declining C_i to increased temperature, but in this experiment the vapor pressure deficit increased from 0.5 to 2.5 kPa. The effect of temperature alone on the relationship between conductance and assimilation and therefore on C_i is little known. In C_3 species, which have a broad temperature optimum for assimilation, the few reported measurements at constant leaf-to-air VPD (Hall and Kaufmann, 1975; Ferrar, 1980; Osonubi and Davies, 1980b) suggest that conductance and assimilation do not respond in parallel to temperature, so that C_i varies substantially. The response of stomata to temperature in C_4 species, which have a narrower temperature optimum for assimilation, has not been studied. Some measurements of the effect of VPD on the CO_2 response of stomata have suggested no interaction (Hall and Kaufmann, 1975; Lösch, 1977; Hall and Schulze, 1980; Morison and Gifford, 1983), but work with tomato indicates that stomata do not respond to changes in C_i if VPD is small (Bradford, Sharkey, and Farquhar, 1983).

Several studies indicate that leaf conductance and assimilation follow different time courses during leaf ontogenesis (see review of Solárová and

Pospíšilová, 1983). In some studies C_i increased with increasing leaf age (e.g. Čatský, Tichá, and Solárová, 1976; Davis and McCree, 1978); in others, C_i decreased (Lin and Ehleringer, 1982; Prange, Ormrod, and Proctor, 1983). Stomatal closure is correlated with the decline of assimilation, the loss of chlorophyll, and other processes occurring during leaf senescence. This might suggest again that mesophyll assimilation is controlling stomatal aperture through changes in C_i, or possibly that stomatal closure controls foliar senescence (Thimann and Satler, 1979a, b). However, Wardle and Short (1983) have shown that the ability of stomata in strips from senescent *Vicia faba* leaves to respond to stimuli decreases in parallel with the other changes in the mesophyll. Therefore, although C_i is generally conservative, it is not precisely regulated.

However, it is not sufficient to argue that C_i is not truly constant; we need to analyze the CO_2 sensitivity of stomata. The contribution that the C_i response makes to stomatal opening with increasing light intensity has been evaluated in several recent studies. For the C_3 species examined, workers have agreed that in well-watered conditions the contribution of the CO_2 response is slight (Farquhar, Dubbe, and Raschke, 1978; Wong, Cowan, and Farquhar, 1978; Sharkey and Raschke, 1981a; Morison and Jarvis, 1983b; Ramos and Hall, 1983). The only exceptions seem to be data from the C_4 species *Amaranthus*

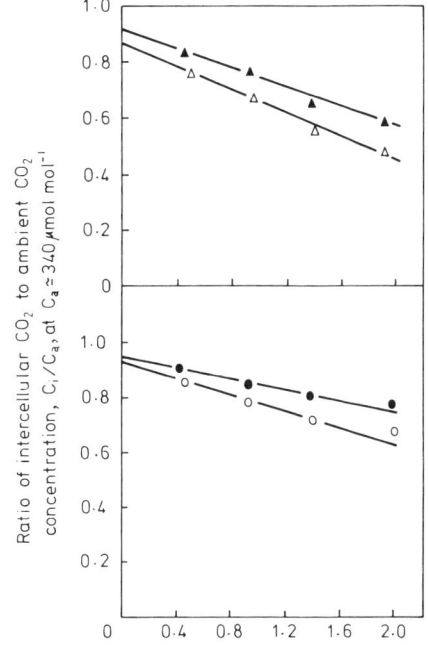

Fig. 10.7. Change of the ratio of intercellular (C_i) to ambient (C_a) CO_2 concentration with changing leaf-to-air vapor pressure difference. Data for (a) two C_4 species, maize (△) and *Paspalum* (▲), and (b) two C_3 species, rice (●) and *Phalaris* (○). Redrawn from Morison and Gifford, 1983: 795.)

hypochondriacus (Ramos and Hall, 1983) and *Zea mays* (Dubbe, Farquhar, and Raschke, 1978; Sharkey and Raschke, 1981a). In simple qualitative terms, stomatal conductance in *Eucalyptus*, *Commelina*, and wheat leaves was least sensitive to C_i at low quantum flux densities (0–250 µE m^{-2} s^{-1}), where the largest changes in C_i normally occur (e.g. as in Fig. 10.1). At higher quantum flux densities (250–2,000 µE m^{-2} s^{-1}), conductance was more sensitive to CO_2, but in these ranges C_i did not decrease further with increasing light in normal atmospheric CO_2 concentrations. Quantitatively, the analyses of Farquhar, Dubbe, and Raschke (1978), Wong, Cowan, and Farquhar (1978), Sharkey and Raschke (1981a), and Ramos and Hall (1983) indicate that the response of stomata to the changes in CO_2 is indeed slight.

However, owing to the complexity of the interactions between variables such as light intensity, temperature, and VPD, we have to be cautious in stating that the CO_2 response of stomata is not important. For example, in natural conditions low light is correlated with low temperatures and low VPD. The CO_2 response may be more important in these conditions in stomatal opening in the early morning or closure in the evening. In addition, the possible significance of endogenous rhythms should not be overlooked; and the many internal factors governing the CO_2 sensitivity of stomata may mean that just as we cannot readily define the ABA, cytokinin, or IAA sensitivity of stomata, neither can we define their sensitivity to CO_2, light, or humidity.

Stomatal Conductance and Water-Use Efficiency in High CO_2 Concentrations

In spite of the uncertainty over the exact role of changes in C_i in the daily behavior of stomata, we know that high CO_2 concentrations cause reductions in stomatal aperture and conductance. As the mean atmospheric CO_2 concentration is increasing (Clark, 1982), stomatal conductance will decrease, with a consequent increase in the plant's water-use efficiency, a subject that has been much debated recently. Current predictions are that the atmospheric CO_2 concentration will have doubled from the preindustrial concentration by the end of next century, from about 300 to about 600 µmol mol^{-1}. Many researchers have measured stomatal responses to high CO_2 concentration, using a variety of techniques, methods, and plant materials and examining interactions with light intensity, VPD, nutrient supply, and water stress. From the data in 23 such reports we have derived, either directly or by curvilinear interpolation, the response of g_s (calculated by measurements of water-vapor diffusion) to a doubling of ambient CO_2 concentration from the present 330 µmol mol^{-1}. Only those studies that included measurements at or sufficiently close to 330 and 660 µmol mol^{-1} were used. Details of the sources of the data are given in Table 10.2; 80 observations were compiled from 23 reports on 25 species (9 C_4 species and 16 C_3 species). Most of the measurements were made using whole

TABLE 10.2. Reports of the effect of doubling ambient CO_2 concentration on stomatal conductance

Species	Pretreatment	Variables	Reference
C_4 species			
Bothriochloa caucasia	atm	temperature	Coyne & Bradford, 1984
Hilaria rigida	atm		Nobel, 1980
Paspalum plicatulum	atm	humidity	Morison & Gifford, 1983
P. plicatulum	gr	water stress	Gifford & Morison, 1985
Pennisetum typhoides	atm	light	McPherson & Slatyer, 1973
Sorghum almum	atm		Ludlow & Wilson, 1971
S. bicolor	atm		van Bavel, 1974
S. sudanense	atm	light	Downes, 1971
Spartina townsendii	atm		Long & Woolhouse, 1978
Zea mays	atm	humidity	Morison & Gifford, 1983
Z. mays	gr	nutrients	Wong, 1979a
Z. mays	gr	nutrients	Wong, 1979b
Z. mays	atm	field study	Rogers et al., 1980
C_3 species			
Cassia obtusifolia	gr	nutrients	Patterson & Flint, 1982
Commelina communis	gr	light	Morison & Jarvis, 1983b
Crotalaria spectabilis	gr	nutrients	Patterson & Flint, 1982
Glycine max	gr	nutrients	Patterson & Flint, 1982
G. max	gr	field studies	Rogers et al., 1980, 1984
G. max	gr	field studies	Sionit et al., 1984
Gossypium hirsutum	atm	nutrients	Wong, 1979a
G. hirsutum	gr	nutrients	Wong, 1979b
Hordeum vulgare	atm	field study	Louwerse, 1980
Liquidambar styraciflua	gr	field study	Rogers et al., 1980
Lycopersicon esculentum	atm		Bradford, Sharkey & Farquhar, 1983
Malus pumila	atm	leaf age	Warrit, Landsberg & Thorpe, 1980
Nerium oleander	atm		Downton, Björkman & Pike, 1980
Oryza sativa	atm	humidity	Morison & Gifford, 1983
Phalaris aquatica	atm	humidity	Morison & Gifford, 1983
Pisum sativum	gr	water stress	Paez, Hellmers & Strain, 1983
Populus deltoides	atm		Regehr, Bazzaz & Boggess, 1975
Spinacia oleracea	atm		Harris, Cheesbrough & Walker, 1983
Vigna luteola	atm		Ludlow & Wilson, 1971

NOTE: atm, plants grown in normal atmospheric CO_2 prior to measurement; gr, plants grown in high CO_2 concentrations prior to measurement.

leaves, but some were from leaf discs, and others from plants in the field. In 12 reports (9 species) the plants had been grown in high CO_2 concentrations prior to measurement (indicated "gr" in Table 10.2). Only two reports have intentionally not been included: data for *Picea sitchensis* (Beadle, Jarvis, and Neilson, 1979), which showed no effect on g_s of ambient CO_2 concentrations between 20 and 600 μmol mol^{-1} over a range of plant water potentials, and for *Gossypium hirsutum* (Bierhuizen and Slatyer, 1964), which also showed no effect of CO_2 concentrations between 200 and 2,000 μmol mol^{-1}.

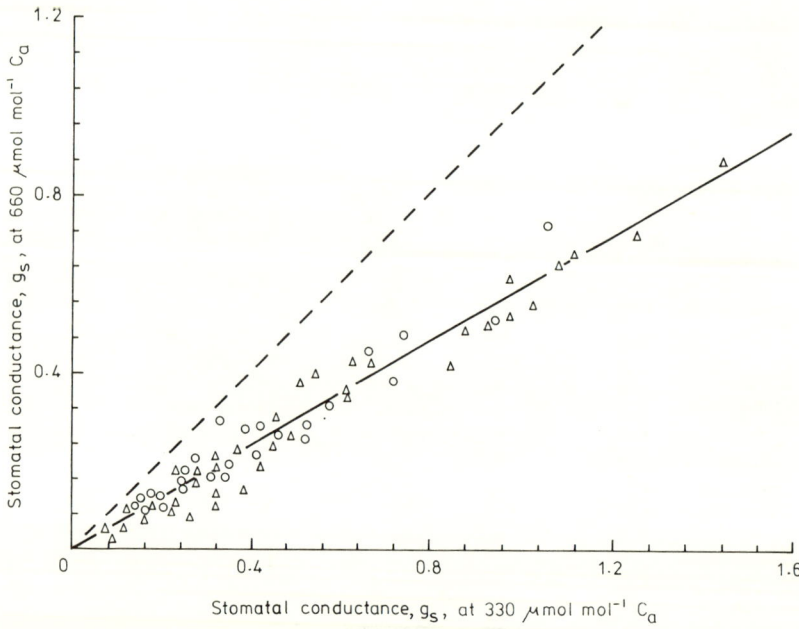

Fig. 10.8. Relationship between stomatal conductance (mol m^{-2} s^{-1}) at present and at twice-present atmospheric CO_2 concentrations. Data interpolated from the references in Table 10.2. The line of unit slope is shown (broken line). Separate regressions for the data for C_3 (o) and C_4 (△) species were not significantly different, so a single regression was fitted to the pooled data, y = 0.595, intercept not significantly different from zero. 95% confidence limits = 0.042; n = 80. Overlapping symbols have been omitted for clarity.

All 80 of the observations are plotted in Figure 10.8 as g_s in high CO_2 concentration (660 μmol mol^{-1}) against g_s in normal atmospheric CO_2 concentration. Though there is appreciable scatter of the data, g_s at 660 μmol mol^{-1} CO_2 is apparently linearly related to g_s at 330 μmol mol^{-1}. The relationships for C_3 and C_4 species are indistinguishable, and a simple linear relationship suffices as a starting point for prediction of stomatal conductance in future atmospheres: for a doubling of CO_2 concentration, conductance is reduced by 40 percent (± 5%; 95% confidence limits). Because of the many feedbacks that may occur to reduce or negate the impact of high CO_2, however, caution should be exercised in extrapolating from this relationship to stomatal responses and, in particular, to plant growth and water use in future CO_2 atmospheres. For example, several workers have reported that assimilation rate adjusts over periods of days when plants are transferred from normal to high CO_2 concentrations (e.g. Kriedemann and Wong, 1984). Further, plants of some species grown at high CO_2 concentrations have lower assimilation rates at those concentrations than plants grown at normal concentrations but measured at high concentrations (e.g. Wong, 1979b; Wulff and Strain, 1982), though this is not always the case (Gifford, 1977). However, only one report (Imai and Murata, 1978) so far has indicated that conductance adjusted similarly. In addition, stomatal insensitivity to high CO_2 concentrations has been noted (see above). Thus, the relationship shown in Figure 10.8 should be treated as an approximation only.

Further, it is difficult to predict the effect of stomatal closure on transpiration rate in plants in natural conditions, because the reduced stomatal aperture results in reduced transpiration, which in turn changes the energy balance of leaves. This negative feedback results in smaller reductions of transpiration than of g_s, but the magnitude of the effect depends on local conditions and the boundary layer conductance of leaves (see Morison and Gifford, 1984a). An additional negative feedback may reduce the effect of stomatal closure in high CO_2 concentration on transpiration but only on the larger scale of expanses of vegetation. The vapor pressure deficit over and within vegetation is partly dependent on evaporation (see McNaughton and Jarvis, 1983); hence, transpiration may be still less affected by reduced stomatal conductance in response to high CO_2 concentrations.

Nevertheless, it is likely that for the same climatic conditions some improvements in water-use efficiency will occur, caused in part by reduced transpiration and in part by increased assimilation. From a compilation of 770 experimental observations on 56 species, Kimball (1983) concluded that plant yield will increase 32 percent ($\pm$ 6%; 95% confidence limits) with a doubling of atmospheric CO_2 concentration. Taken together with a 30 to 40 percent decrease in transpiration, this suggests that water-use efficiency may increase 80 to 100 percent (Kimball and Idso, 1983). Morison and Gifford (1984b) compared 18 species, grown as young plants in normal and twice-present atmospheric CO_2 concentrations, and found a 67 percent increase in water-use efficiency on average. However, the range they reported, from 40 to 80 percent, suggests that there is considerable variation between species, which warrants further investigation before generalizations can be made.

Mechanisms of CO_2 Effects on Stomatal Opening

Our knowledge of stomatal responses to CO_2 and the interrelationships between CO_2 concentration and other variables has expanded over recent years, but our understanding of the mechanisms by which CO_2 affects stomata has not. It remains a central dilemma that the best-characterized process in which CO_2 is involved in guard cells, the formation of organic anions, is positively correlated with stomatal aperture, yet CO_2 closes stomata. Carbon metabolism of guard cells is discussed in detail by Outlaw elsewhere in this volume and by other recent reviewers (e.g. Raschke, 1979; Mansfield, Travis, and Jarvis,

1981; Jarvis and Morison, 1981; Zeiger, 1983). It should be pointed out that CO_2 affects stomata in either blue or red light (Sharkey and Raschke, 1981b; Zeiger and Field, 1982; Morison and Jarvis, 1983a), which act in different ways to cause opening, presumably through the same basic ion-accumulation mechanism (see MacRobbie, this volume; Sharkey and Ogawa, this volume). Therefore, it seems likely that CO_2 also acts at some point in the ion-accumulation mechanism, as Mansfield, Travis, and Jarvis (1981) and Zeiger (1983) have argued.

Conclusions

In summary we are left with a paradox. Stomata respond markedly to experimentally manipulated CO_2 concentrations in whole leaves and in epidermal strips. It appears that in whole leaves stomata respond to C_i, though we cannot yet point to the cellular mechanisms involved. However, in intact leaves in changing environmental conditions and in many different species (e.g. Werk et al., 1983), C_i is generally conservative, owing to the correlation of mesophyll assimilation rate and stomatal conductance. This conservative C_i does not appear to be a result of high sensitivity of the stomata to C_i, as many examples, including changing C_a, have demonstrated. In addition, we should note that the ratio C_i/C_a is also conservative but, again, over the short term it is not sufficiently constant to indicate a mechanistic link operating over periods of minutes. It appears that in the limited set of conditions examined so far, CO_2 is not the major signal for stomata but one of a large and varying set of signals. The absolute C_a, however, does set a particular range of conductances as other environmental variables change. Indeed, this influence of CO_2 on stomatal conductance will have profound effects on plant growth in future atmospheric CO_2 concentrations through reductions in water loss per unit leaf area. Furthermore, the observation that C_i is conservative at about 100 and 230 μmol mol^{-1} in C_4 and C_3 species, respectively, should deter stomatal physiologists from using C_a and CO_2-free air as "plus" and "minus" CO_2 treatments.

A concern of stomatal physiologists has always been to determine the link that we believe should exist between the mesophyll and the guard cells. A link through CO_2 flux or concentration seemed both reasonable and attractive. It appears, however, that we have to suggest either that evolution has resulted in parallel responses of assimilation and conductance to environmental stimuli or that other links between the mesophyll and stomata are present and active. The interactions between the effects of environmental variables, mesophyll assimilation, and conductance and ABA, cytokinins, and auxins that are now being unraveled (see Incoll and Jewer, this volume; Davies and Mansfield, this volume) seem likely to provide such links.

REFERENCES

Akita, S., and D. N. Moss. 1972. Differential stomatal response between C_3 and C_4 species to atmospheric CO_2 concentration and light. *Crop Science*, 12: 789–93.
Allaway, W. G., and T. A. Mansfield. 1967. Stomatal responses to changes in carbon dioxide concentration in leaves treated with 3-(4-chlorophenyl)-1,1-dimethylurea. *New Phytologist*, 66: 57–63.
Ball, J. T., and J. A. Berry. 1982. The C_i/C_s ratio: A basis for predicting stomatal control of photosynthesis. *Carnegie Institute of Washington Year Book, 1981*: 88–92.
van Bavel, C. H. M. 1974. Antitranspirant action of carbon dioxide on intact sorghum plants. *Crop Science*, 14: 208–12.
Beadle, C. L., P. G. Jarvis, and R. E. Neilson. 1979. Leaf conductance as related to xylem water potential and carbon dioxide concentrations in Sitka spruce. *Physiologia plantarum*, 45: 158–66.
Bell, C. J. 1982. A model of stomatal control. *Photosynthetica*, 16: 486–95.
Bierhuizen, J. F., and R. O. Slatyer. 1964. Photosynthesis of cotton leaves under a range of environmental conditions in relation to internal and external diffusive resistances. *Australian Journal of Biological Sciences*, 17: 348–59.
Bradford, K. J., T. D. Sharkey, and G. D. Farquhar. 1983. Gas exchange, stomatal behaviour, and $\delta^{13}C$ values of the *flacca* tomato mutant in relation to abscisic acid. *Plant Physiology*, 72: 245–50.
Brogårdh, T., and A. Johnsson. 1975. Regulation of transpiration in *Avena*: Responses to white light steps. *Physiologia plantarum*, 35: 115–25.
Čatský, J., I. Tichá, and J. Solárová. 1976. Ontogenetic changes in the internal limitations to bean-leaf photosynthesis, I. Carbon dioxide exchange and conductances for carbon dioxide transfer. *Photosynthetica*, 10: 349–402.
Clark, W. C. 1982. *Carbon dixoide review, 1982*. Oxford: Clarendon.
Cockburn, W., I. P. Ting, and L. O. Sternberg. 1979. Relationships between stomatal behaviour and internal carbon dioxide concentration in crassulacean acid metabolism plants. *Plant Physiology*, 63: 1029–32.
Cowan, I. R. 1977. Stomatal behaviour and environment. *Advances in Botanical Research*, 4: 117–228.
Coyne, P. I., and J. A. Bradford. 1984. Leaf gas exchange in "Caucasian" Bluestem in relation to light, temperature, humidity, and CO_2. *Agronomy Journal*, 76: 107–13.
Davis, S. D., and K. J. McCree. 1978. Photosynthetic rate and diffusion conductance as a function of age in leaves of bean plants. *Crop Science*, 18: 280–82.
Downes, R. W. 1971. Adaptation of sorghum plants to light intensity: Its effects on gas exchange in response to changes in light, temperature and CO_2. In M. D. Hatch, C. B. Osmond, and R. O. Slatyer, eds., *Photosynthesis and photorespiration*: 57–62. New York: Wiley.
Downton, W. J. S., O. Björkman, and C. S. Pike. 1980. Consequences of increased atmospheric concentrations of carbon dioxide for growth and photosynthesis of higher plants. In G. I. Pearman, ed., *Carbon dioxide and climate: Australian research*: 143–52. Canberra: Australian Academy of Sciences.
Dubbe, D. R., G. D. Farquhar, and K. Raschke. 1978. Effect of abscisic acid on the gain of the feedback loop involving carbon dioxide and stomata. *Plant Physiology*, 62: 413–17.
Ehleringer, J. 1983. Ecophysiology of *Amaranthus palmeri*, a Sonoran Desert summer annual. *Oecologia*, 57: 107–12.

Farquhar, G. D., D. R. Dubbe, and K. Raschke. 1978. Gain of the feedback loop involving carbon dioxide and stomata: Theory and measurement. *Plant Physiology*, 62: 406–12.

Farquhar, G. D., E.-D. Schulze, and M. Küppers. 1980. Responses to humidity by stomata of *Nicotiana glauca* L. and *Carylus avellana* L. are consistent with the optimisation of carbon dioxide uptake with respect to water loss. *Australian Journal of Plant Physiology*, 7: 315–27.

Ferrar, P. J. 1980. Environmental control of gas exchange in some savanna woody species. *Oecologia*, 47: 204–12.

Fischer, R. A. 1968. Stomatal opening: Role of potassium uptake by guard cells. *Science*, 160: 784–85.

Freudenberger, H. 1940. Die Reaktion der Schliesszellen auf Kohlensäure and Sauerstoffentzug. *Protoplasma*, 35: 15–54.

Gaastra, P. 1959. Photosynthesis of crop plants as influenced by light, carbon dioxide, temperature and stomatal diffusion resistance. *Mededelingen van de Landbouwhogeschool*, Wageningen, 59: 1–68.

Gifford, R. M. 1977. Growth pattern, carbon dioxide exchange and dry weight distribution in wheat growing under differing photosynthetic environments. *Australian Journal of Plant Physiology*, 4: 99–110.

Gifford, R. M., and J. I. L. Morison. 1985. Photosynthesis, water use and growth of a C_4 grass stand at high CO_2 concentration. *Photosynthesis Research*, 7: 69–76.

Hall, A. E., and M. R. Kaufmann. 1975. Stomatal response to environment with *Sesamum indicum* L. *Plant Physiology*, 55: 455–59.

Hall, A. E., and E.-D. Schulze. 1980. Stomatal response to environment and a possible interrelation between stomatal effects on transpiration and CO_2 assimilation. *Plant, Cell and Environment*, 3: 467–74.

Hällgren, J. E., E. Sundbom, and M. Strand. 1982. Photosynthetic responses to temperature in *Betula pubescens* and *Betula tortuosa*. *Physiologia plantarum*, 54: 275–82.

Harris, G. G., J. K. Cheesbrough, and D. A. Walker. 1983. Measurement of CO_2 and H_2O vapor exchange in spinach leaf discs: Effects of orthophosphate. *Plant Physiology*, 71: 102–7.

Heath, O. V. S. 1948. Control of stomatal movement by a reduction in the normal carbon dioxide content of the air. *Nature*, 161: 179–81.

———. 1959. Light and carbon dioxide in stomatal movements. In W. Ruhland, ed., *Physiology of movements*, Handbuch der Pflanzenphysiologie, 17: 415–64. Berlin: Springer-Verlag.

Heath, O. V. S., and H. Meidner. 1957. Effects of carbon dioxide and temperature on stomata of *Allium cepa* L. *Nature*, 180: 181–82.

Heath, O. V. S., and J. Russell. 1954. Studies in stomatal behaviour, VI. An investigation of the light response of wheat stomata with the attempted elimination of control by the mesophyll. Part 2: Interactions with external carbon dioxide and general discussion. *Journal of Experimental Botany*, 5: 269–92.

Imai, K., and Y. Murata. 1978. Effect of carbon dioxide concentration on growth and dry matter production of crop plants, V. Analysis of after-effect of carbon dioxide–treatment on apparent photosynthesis. *Japanese Journal of Crop Science*, 47: 587–95.

Jarvis, P. G., and J. I. L. Morison. 1981. The control of transpiration and photosynthesis by the stomata. In P. G. Jarvis and T. A. Mansfield, eds., *Stomatal physiology*: 247–79. Cambridge: Cambridge University Press.

Kimball, B. A. 1983. *Carbon dioxide and agricultural yield: An assemblage and analysis of 770 prior observations*. USDA, Agricultural Research Service, Water Conservation Laboratory Report 14. Phoenix, Ariz.

Kimball, B. A., and S. B. Idso. 1983. Increasing atmospheric CO_2: Effects on crop yield, water use and climate. *Agricultural Water Management*, 7: 55–72.

Koch, W. 1969. Untersuchungen über die Wirkung von CO_2 auf die Photosynthese einiger Holzegewächse unter Laboratoriumsbedingungen. *Flora*, Abt. B, 158: 402–28.

Kriedemann, P. E., R. J. Sward, and W. J. S. Downton. 1976. Vine response to carbon dioxide enrichment during heat therapy. *Australian Journal of Plant Physiology*, 3: 605–18.

Kriedemann, P. E., and S. C. Wong. 1984. Growth response and photosynthetic adaption to carbon dioxide: Comparative behaviour in some C_3 species. *Advances in Photosynthesis Research*, 4(3): 209–12.

Kubik, M., and R. Antoszewski. 1983. Altering the sensitivity of strawberry plants to exogenous abscisic acid. *Physiologia plantarum*, 57: 505–8.

Lin, Z. F., and J. Ehleringer. 1982. Effects of leaf age on photosynthesis and water use efficiency of papaya. *Photosynthetica*, 16: 514–19.

———. 1983. Photosynthetic characteristics of *Amaranthus tricolor*, a C_4 tropical, leafy vegetable. *Photosynthesis Research*, 4: 171–78.

Long, S. P., and H. W. Woolhouse. 1978. The responses of net photosynthesis to vapour pressure deficit and CO_2 concentration in *Spartina townsendii* (*sensu lato*), a C_4 species from a cool temperate climate. *Journal of Experimental Botany*, 29: 567–77.

Lösch, R. 1976. Die Reaktion der Stomata isolierter Epidermen auf Luftfeuchtigkeit, Temperatur, CO_2-Gehalt der Luft und Wasseranspannung. (Untersuchungen an *Polypodium vulgare*). Dissertation, University of Würzburg.

———. 1977. Response of stomata to environmental factors: Experiments with isolated epidermal strips of *Polypodium vulgare*, I. Temperature and humidity. *Oecologia*, 29: 85–87.

Louwerse, W. 1980. Effects of CO_2-concentration and irradiance on the stomatal behaviour of maize, barley and sunflower plants in the field. *Plant, Cell and Environment*, 3: 391–98.

Ludlow, M. M., and G. L. Wilson. 1971. Photosynthesis of tropical pasture plants, I. Illuminance, carbon dioxide concentration, leaf temperature and leaf-air vapor pressure difference. *Australian Journal of Biological Science*, 124: 449–70.

McNaughton, K. G., and P. G. Jarvis. 1983. Predicting the effects of vegetation changes on transpiration and evaporation. In T. T. Kozlowski, ed., *Water deficits and plant growth* 7: 1–47. New York: Academic.

McPherson, H. G., and R. O. Slatyer. 1973. Mechanisms regulating photosynthesis in *Pennisetum typhoides*. *Australian Journal of Biological Science*, 26: 329–39.

Mansfield, T. A. 1976. Delay in the response of stomata to abscisic acid in CO_2-free air. *Journal of Experimental Botany*, 27: 559–69.

Mansfield, T. A., A. J. Travis, and R. G. Jarvis. 1981. Responses to light and carbon dioxide. In P. G. Jarvis and T. A. Mansfield, eds., *Stomatal physiology*: 119–35. Cambridge: Cambridge University Press.

Mansfield, T. A., and C. M. Willmer. 1969. Stomatal responses to light and carbon dioxide in the Hart's-Tongue Fern, *Phyllitis scolopendrium* Newm. *New Phytologist*, 68: 63–66.

Meidner, H. 1975. Water supply, evaporation and vapour diffusion in leaves. *Journal of Experimental Botany*, 26: 666–73.

Meidner, H., and T. A. Mansfield. 1965. Stomatal responses to illumination. *Biological Reviews*, 40: 483–509.

———. 1968. *Physiology of stomata*. London: McGraw-Hill.

Mooney, H. A., and C. Chu. 1983. Stomatal response to humidity of coastal and interior populations of a Californian shrub. *Oecologia*, 57: 148–50.

Morison, J. I. L. 1980. Light and CO_2 effects on stomata. Dissertation, University of Edinburgh.

Morison, J. I. L., and R. M. Gifford. 1983. Stomatal sensitivity to carbon dioxide and humidity: A comparison of two C_3 and two C_4 grass species. *Plant Physiology*, 71: 789–96.

———. 1984a. Plant growth and water use with limited water supply in high CO_2 concentrations, I. Leaf area, water use and transpiration. *Australian Journal of Plant Physiology*, 11: 361–74.

———. 1984b. Plant growth and water use with limited water supply in high CO_2 concentrations, II. Plant dry weight, partitioning and water use efficiency. *Australian Journal of Plant Physiology*, 11: 375–84.

Morison, J. I. L., and P. G. Jarvis. 1983a. Direct and indirect effects of light on stomata, I. In Scots pine and Sitka spruce. *Plant, Cell and Environment*, 6: 95–101.

———. 1983b. Direct and indirect effects of light on stomata, II. In *Commelina communis* L. *Plant, Cell and Environment*, 6: 103–9.

Moss, D. N., and S. L. Rawlins. 1963. Concentration of carbon dioxide inside leaves. *Nature*, 197: 1320–21.

Mouravieff, I. 1956. Action du CO_2 et de la lumière sur l'appareil stomatique separé de mésophylle. *Le botaniste*, 40: 195–212.

———. 1958. Action de la lumière sur la cellule végétale. *Bulletin de la Société botanique de France*, 105: 467–75.

Nagarajah, S., and E.-D. Schulze. 1983. Responses of *Vigna unguiculata* (L.) Walp. to atmospheric and soil drought. *Australian Journal of Plant Physiology*, 10: 385–94.

Neales, T. F. 1970. Effect of ambient carbon dioxide concentration on the rate of transpiration of *Agave americana* in the dark. *Nature*, 228: 880–82.

Nelson, S. D., and J. M. Mayo. 1975. The occurrence of functional, non-chlorophyllous guard cells in *Paphiopedilum* sp. *Canadian Journal of Botany*, 53: 1–7.

Ng, P. A. P. 1978. Response of stomata to environmental variables in *Pinus sylvestris* L. Dissertation, University of Edinburgh.

Nobel, P. S. 1978. Microhabitat, water relations and photosynthesis of a desert fern, *Notholaena parryi*. *Oecologia*, 31: 293–309.

———. 1980. Water vapor conductance and CO_2 uptake for leaves of a C_4 desert grass, *Hilaria rigida*. *Ecology*, 61: 252–58.

Osonubi, O., and W. J. Davies. 1980a. The influence of plant water stress on stomatal control of gas exchange at different levels of atmospheric humidity. *Oecologia*, 46: 1–6.

———. 1980b. The influence of water stress on photosynthetic performance and stomatal behaviour of tree seedlings subjected to variations in temperature and irradiance. *Oecologia*, 45: 3–10.

Paez, A., H. Hellmers, and B. R. Strain. 1983. CO_2 enrichment, drought stress and growth of Alaska pea plants (*Pisum sativum*). *Physiologia plantarum*, 58: 161–65.

Pallaghy, C. K. 1971. Stomatal movement and potassium transport in epidermal strips of *Zea mays*: The effect of CO_2. *Planta*, 101: 287–95.

Patterson, D. T., and E. P. Flint. 1982. Interacting effects of CO_2 and nutrient concentration. *Weed Science*, 30: 389–94.

Prange, R. K., D. P. Ormrod, and J. T. A. Proctor. 1983. Effect of water stress on gas exchange in fronds of the ostrich fern (*Matteuccia struthiopteris* [L.] Todaro). *Journal of Experimental Botany*, 34: 1108–16.
Radin, J. W., and R. C. Ackerson. 1981. Water relations of cotton plants under nitrogen deficiency, III. Stomatal conductance, photosynthesis and abscisic acid accumulation during drought. *Plant Physiology*, 67: 115–19.
Radin, J. W., L. L. Parker, and G. Guinn. 1982. Water relations of cotton plants under nitrogen deficiency, V. Environmental control of abscisic acid accumulation and stomatal sensitivity to abscisic acid. *Plant Physiology*, 70: 1066–70.
Ramos, C., and A. E. Hall. 1982. Relationships between leaf conductance, intercellular CO_2 partial pressure and CO_2 uptake rate in two C_3 and two C_4 plant species. *Photosynthetica*, 16: 343–55.
———. 1983. Effects of photon fluence rate and intercellular CO_2 partial pressure on leaf conductance and CO_2 uptake rate in *Capsicum* and *Amaranthus*. *Photosynthetica*, 17: 34–42.
Raschke, K. 1965. Die Stomata als Glieder eines schwindungs-fähigen CO_2-Regelsystems. Experimenteller Nachweis an *Zea mays* L. *Zeitschrift für Naturforschung*, 20b: 1261–70.
———. 1972. Saturation kinetics of the velocity of stomatal closing in response to CO_2. *Plant Physiology*, 49: 229–34.
———. 1975a. Stomatal action. *Annual Reviews of Plant Physiology*, 26: 309–40.
———. 1975b. Simultaneous requirement of carbon dioxide and abscisic acid for stomatal closing in *Xanthium strumarium* L. *Planta*, 125: 243–59.
———. 1976. How stomata resolve the dilemma of opposing priorities. *Philosophical Transactions of the Royal Society*, ser. B, 273: 551–60.
———. 1979. Movements of stomata. In W. Haupt and M. E. Feinleib, eds., *Physiology of movements*, Encyclopedia of Plant Physiology, n.s., 7: 381–441. Berlin: Springer-Verlag.
Raschke, K., W. F. Hanebuth, and G. D. Farquhar. 1978. Relationship between stomatal conductance and light intensity in leaves of *Zea mays* L., derived from experiments using the mesophyll as shade. *Planta*, 139: 73–77.
Regehr, D. L., F. A. Bazzaz, and W. R. Boggess. 1975. Photosynthesis, transpiration and leaf conductance of *Populus deltoides* in relation to flooding and drought. *Photosynthetica*, 9: 52–61.
Rogers, H. H., G. E. Bingham, J. D. Cure, W. W. Heck, A. S. Heagle, D. W. Israel, J. M. Smith, K. A. Surand, and J. P. Thomas. 1980. *Response of vegetation to carbon dioxide, no. 001. Field studies of plant responses to elevated carbon dioxide levels*. Washington, D. C.: U.S. Department of Energy.
Rogers, H. H., N. Sionit, J. D. Cure, J. M. Smith, and G. E. Bingham. 1984. Influence of elevated carbon dioxide on water relations of soybeans. *Plant Physiology*, 74: 233–38.
Scarth, G. W. 1932. Mechanism of the action of light and other factors on stomatal movement. *Plant Physiology*, 7: 481–504.
Schulze, E.-D., and A. E. Hall. 1982. Stomatal responses to water loss and CO_2 assimilation rates of plants in contrasting environments. In O. L. Lange, P. S. Nobel, C. B. Osmond, and H. Ziegler, eds., *Physiological plant ecology 2, Water relations and carbon assimilation*, Encyclopedia of Plant Physiology, n.s., 12B: 181—330. Berlin: Springer-Verlag.
Sharkey, T. D., K. Imai, G. D. Farquhar, and I. R. Cowan. 1982. A direct confirmation of the standard method of estimating intercellular partial pressure of CO_2. *Plant Physiology*, 69: 657–60.

Sharkey, T. D., and K. Raschke. 1981a. Separation and measurement of direct and indirect effects of light on stomata. *Plant Physiology*, 68: 33–40.

———. 1981b. Effect of light quality on stomatal opening in leaves of *Xanthium strumarium* L. *Plant Physiology*, 68: 1170–74.

Sionit, N., H. H. Rogers, G. E. Bingham, and B. R. Strain. 1984. Photosynthesis of container- and field-grown soybeans and stomatal conductance with CO_2 enrichment on leaf photosynthesis and stomatal conductance under field conditions. *Agronomy Journal*, 76: 447–51.

Snaith, P. J., and T. A. Mansfield. 1982. Stomatal sensitivity to abscisic acid: Can it be defined? *Plant, Cell and Environment*, 5: 309–11.

Solárová, J., and J. Pospíšilová. 1983. Photosynthetic characteristics during ontogenesis of leaves, VIII. Stomatal diffusive conductance and stomatal reactivity. *Photosynthetica*, 17: 101–51.

Spalding, M. H., D. K. Stumpf, M. S. B. Ku, R. H. Burns, and G. E. Edwards. 1979. Crassulacean acid metabolism and diurnal variations of internal CO_2 and O_2 concentrations in *Sedum praealtum*, D.C. *Australian Journal of Plant Physiology*, 6: 557–67.

Thimann, K. V., and S. O. Satler. 1979a. Relation between leaf senescence and stomatal closure: Senescence in darkness. *Proceedings of the National Academy of Sciences*, 76: 2270–73.

———. 1979b. Relation between leaf senescence and stomatal closure: Senescence in the light. *Proceedings of the National Academy of Sciences*, 76: 2295–98.

Thorpe, N., and F. L. Milthorpe. 1984. Transport of metabolites between the mesophyll and epidermis of *Commelina cyanea* R. Br. *Australian Journal of Plant Physiology*, 11: 59–68.

Tyree, M. T., and P. Yianoulis. 1980. The site of water evaporation from sub-stomatal cavities, liquid path resistances and hydroactive stomatal closure. *Annals of Botany*, 46: 175–93.

Wardle, K., and K. C. Short. 1983. Stomatal responses and the senescence of leaves. *Annals of Botany*, 52: 411–12.

Warrit, B., J. J. Landsberg, and M. R. Thorpe. 1980. Responses of apple leaf stomata to environmental factors. *Plant, Cell and Environment*, 3: 13–22.

Werk, K. S., J. Ehleringer, I. N. Forseth, and C. S. Cook. 1983. Photosynthetic characteristics of Sonoran Desert winter annuals. *Oecologia*, 59: 101–5.

Whiteman, P. C., and D. Koller. 1967a. Interactions of carbon dioxide concentration, light intensity and temperature on plant resistances to water vapour and carbon dioxide diffusion. *New Phytologist*, 66: 463–73.

———. 1967b. Species characteristics in whole plant resistances to water vapour and CO_2 diffusion. *Journal of Applied Ecology*, 4: 363–77.

Willmer, C. M. 1981. Guard cell metabolism. In P. G. Jones and T. A. Mansfield, eds., *Stomatal physiology*: 87–102. Cambridge: Cambridge University Press.

Wong, S. C. 1979a. Stomatal behaviour in relation to photosynthesis. Dissertation, Australian National University.

———. 1979b. Elevated atmospheric partial pressure of CO_2 and plant growth, I. Interactions of nitrogen nutrition and photosynthetic capacity in C_3 and C_4 plants. *Oecologia*, 44: 68–74.

Wong, S. C., I. R. Cowan, and G. D. Farquhar. 1978. Leaf conductance in relation to assimilation in *Eucalyptus pauciflora* Sieb. ex Spreng. *Plant Physiology*, 62: 670–74.

———. 1979. Stomatal conductance correlates with photosynthetic capacity. *Nature*, 282: 424–26.

Wulff, R. D., and B. R. Strain. 1982. Effects of CO_2 enrichment on growth and photosynthesis in *Desmodium paniculatum* (Leguminosae). *Canadian Journal of Botany*, 60: 1084–91.

Zeiger, E. 1983. The biology of stomatal guard cells. *Annual Review of Plant Physiology*, 34: 441–75.

Zeiger, E., and C. Field. 1982. Photocontrol of the functional coupling between photosynthesis and stomatal conductance in the intact leaf. *Plant Physiology*, 70: 370–75.

11

Action of Abscisic Acid on Guard Cells

Klaus Raschke

Requirement of Signals for Stomatal Closing and Prevention of Opening

Plant cells possess a mechanism to build up turgor. It enables the plant to grow, to maintain the rigidity of its thin-walled organs, and, in herbaceous species, to maintain the posture of the whole plant. In pulvinar tissue and guard cells, this mechanism has been put to other uses. In pulvini, changes in turgor cause nyctinastic and thigmonastic leaf movements. In guard cells, variations in turgor produce changes in cell volume and shape, leading to stomatal movement. If we continue to consider stomatal movement only, we recognize that, if the water relations of a leaf are strained, reins have to be put on the guard cells' osmoregulatory mechanism. A signal must be present that opposes the tendency, inherent in all plant cells, to inflate, and that opposes effects of signals calling for stomatal opening, for instance, in response to light or to an absence of CO_2. Increased water loss will ensue, and water stress can be the consequence. A remedy is to reduce stomatal aperture. A brief analysis of stomatal sensitivity to changes in the water potential of the tissue will demonstrate that stomatal closure can be initiated only if a closing signal triggers loss of solutes from the guard cells.

Stomatal aperture is a function of guard cell volume (see Chap. 4); guard cell volume depends on turgor, p, which in turn is determined by the relationship between the water potential of the cell (in equilibrium with the surrounding tissue), Ψ, and the osmotic pressure, π, in the guard cell: $\Psi = p - \pi$. Both, turgor and osmotic pressure, are related to cell volume, V. We wish to evaluate by how much the volume of a cell will change when the water potential, Ψ, of the tissue changes. The magnitude of a change in volume relative to initial volume, dV/V, is determined by the magnitude of the pressure change, dp, and the stiffness of the cell walls, represented by their modulus of elasticity, ϵ. If we assume, for the sake of simplicity, that the walls are ideally

---

The author wishes to thank U. Brill and U. Lohmann for providing the data for Figures 11.1, 11.3, 11.4, and 11.5; R. Behl, E. Fischer, R. Hedrich, and W. Lahr for discussions and critique; and U. Brill, A. Schön, and R. Seibert for help during the preparation of the manuscript.

elastic, we can write

$$\frac{dV}{V} = \frac{dp}{\epsilon}$$

or

$$\frac{dp}{dV} = \frac{\epsilon}{V}.$$

The osmotic pressure of a cell is determined in first approximation by the concentration of its solutes, s: $\pi = \frac{RTs}{V}$. By differentiation we obtain

$$\frac{d\pi}{dV} = -\frac{RTs}{V^2} = -\frac{\pi}{V}.$$

Because $\frac{d\Psi}{dV} = \frac{dp}{dV} - \frac{d\pi}{dV}$, we can write

$$\frac{d\Psi}{dV} = \frac{\epsilon}{V} + \frac{\pi}{V}$$

and the volume change, dV/V, that occurs in response to a change in water potential, $d\Psi$, is, by rearrangement of the preceding equation,

$$dV/V = \frac{d\Psi}{\epsilon + \pi}.$$

This means that volume changes will be small if the cell walls are stiff and the solute contents are high. If, for instance, $\epsilon = 60$ bar and $\pi = 40$ bar, a reduction in the water potential of the tissue by 10 bar would cause a reduction of the cell volume by a factor of $10/(60 + 40) = 0.1$; that is, cell volume would be reduced by 10 percent of its initial value. Because stomatal apertures are roughly linearly related to guard cell volumes, at least in *Vicia faba*, this 10-bar reduction in water potential would cause a 10 percent reduction in pore width (Raschke, 1979). This change in aperture would in many cases be far too small to correct a decline in the water content of the tissue.

Quite clearly, the plant must possess a mechanism to reduce the solute content of the guard cells, a mechanism to convert a change in the water potential in the tissue to a change in the osmotic pressure in the guard cells and to prevent a buildup of high osmotic pressures if a lack of water is impending when stomata are opening. It is possible that such a mechanism is present in the guard cells themselves. Although there is some evidence for it (Raschke, 1979), it is difficult to imagine how it could be very sensitive to changes in Ψ, because π is high in guard cells of open stomata, and a high π reduces the magnitude of the volume change guard cells experience upon a change in Ψ, as was shown in the last equation presented above. Positive feedback and a high amplification would be required to make such a system operate effectively. Another possibility would be to use cells of low ϵ and π as sensors. They would respond to changes in Ψ with larger changes than guard cells in relative volume, as explained above, and in turgor:

$$dp = \frac{\epsilon}{\epsilon + \pi} d\Psi.$$

The last equation says that changes in water potential cause changes in turgor that are "buffered" by the osmotica present. Cells low in π, or in both ϵ and π, would have to send a signal to the guard cells that announced a reduction in Ψ and triggered a loss of solutes from the guard cells or prevented an accumulation of osmotica in them. In the following sections, a brief historical account will be given of the discovery that abscisic acid (ABA) may be both the postulated messenger and the substance that prevents stomatal opening. Evidence will follow that supports the hypothesis that this phytohormone plays a key role in the stomatal control of water loss.

An Indication of a Metabolic Response to Wilting, and the Discovery of ABA as an Agent Causing Stomatal Closure

Stålfelt reported in 1929 that stomata in leaves of *Vicia faba* did not close immediately after deprivation of water; rather it took 13 minutes of withholding water before stomata began to respond. Once the closing movement had begun, resupplying the leaves with water could not stop it. Stålfelt suspected that a metabolic response was involved in initiating stomatal closure and preventing opening. He called this mechanism the "hydroactive system." It was not until 1968 that Little and Eidt discovered that applying ABA to woody species reduced transpiration. Mittelheuser and van Steveninck (1969) showed that this effect was due to stomatal closure. The observation that wilting caused the synthesis of ABA in leaves gave a boost to the hypothesis that ABA is a messenger between the mesophyll and the epidermis. Wright and Hiron (1969) reported that ABA formation was correlated with wilting and that the ensuing stomatal closure led to a recovery of turgor (Hiron and Wright, 1973).

The response can occur within a few minutes after addition of ABA to the transpiration stream (Mittelheuser and van Steveninck, 1969; Cummins, Kende, and Raschke, 1971; Kriedemann et al., 1972) and can be quickly reversed if the ABA solution feeding a leaf is replaced by water (Cummins, Kende, and Raschke, 1971). Experiments involving measurements of the gas exchange of leaves (Cummins, Kende, and Raschke, 1971) and experiments with epidermal strips proved that ABA acted on guard cells: not only did ABA initiate closure, but it also prevented opening. Prevention of opening was observed when ABA was sprayed on leaves of *Xanthium pennsylvanicum* and *Nicotiana tabacum* (Jones and Mansfield, 1970) or when epidermal strips of *Commelina communis* (Tucker and Mansfield, 1971), *Kalanchoe daigremontiana* (Jewer, Incoll, and Howarth, 1981), *Pisum sativum* var. Argenteum (Jewer, Incoll, and Shaw, 1982), and *Valerianella locusta* (Hartung, 1983) were incubated on solutions of ABA. In general, a minimum concentration of 10^{-9} to 10^{-8} mol l^{-1} (10^{-3} to 10^{-2} μM) (±)-ABA was required to produce an inhibition; the inhibition was half maximal at about 10^{-6} mol l^{-1} (1 μM), and maximal suppression of opening occurred at 10^{-4} mol l^{-1} (100 μM). However, the CAM plant *K. daigremontiana* presented a great surprise; for full inhibition its stomata required only 1/10,000 of the ABA concentration that produced a maximal response in other species: 10^{-8} mol l^{-1} (±)-ABA was sufficient for maximal suppression. A concentration as low as 1.5×10^{-11} mol l^{-1} inhibited opening in 50 percent of the stomata.

Stomata in epidermal peels may close within three minutes of exposure to 10^{-5} mol l^{-1} ABA, as Raschke, Firn, and Pierce (1975) observed on strips from *C. communis*. Weyers and Hillman (1979a) saw stomata in epidermal strips of the same species close within 10 minutes after placement on 10^{-4} mol l^{-1} ABA. It can be concluded that ABA, applied at low concentrations, has a dual effect on guard cells. When stomata are closed, it prevents solute accumulation; when they are open, it causes a release of solutes from guard cells and thereby stomatal closure. Both processes, prevention of stomatal opening and initiation of stomatal closure, if determined on leaves of the same species and comparable history, show a similar dependence on ABA concentration (Fig. 11.1). It is possible, but not yet proven, that both effects have a common response mechanism.

The next six sections will summarize our present knowledge of the action of ABA on guard cells. In the last four sections of this paper, we shall return to questions concerning the synthesis and metabolism of ABA, the distribution of ABA under stress, the "direct" effect of ABA on the photosynthetic apparatus, and the evidence for the operation in leaves of a feedback loop that involves ABA and the stomata.

Observations on Specificity and Site of Action of ABA

Abscisic acid is a sesquiterpenoid that occurs in two enantiomers:

(+)-(S)-ABA   (−)-(R)-ABA

Action of Abscisic Acid on Guard Cells 257

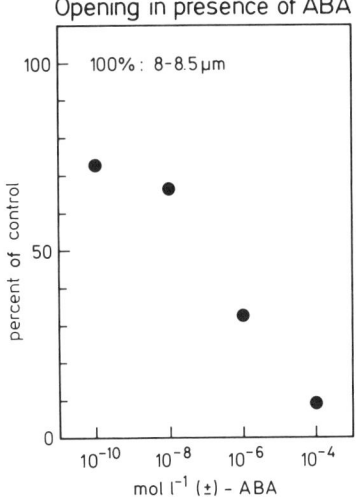
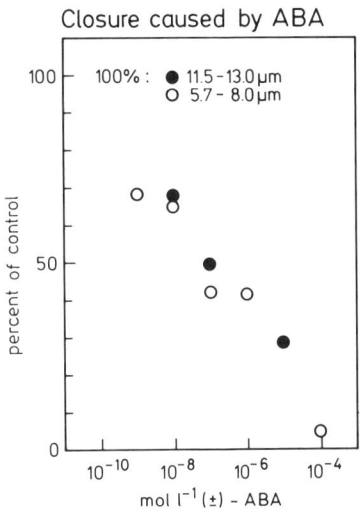

Fig. 11.1. Prevention of stomatal opening and initiation of stomatal closure by (±)-ABA in epidermal peels from leaves of *Commelina communis* show similar dependences on ABA concentration. Samples from the lower epidermis were taken from leaves of comparable crops, darkened for 2 hours for the opening experiment, or illuminated for 3 hours for the closing experiment. The epidermal peels were incubated on 100 mmol l^{-1} KCl, pH 7, plus ABA additions, in the light and in CO_2-free air at 30° C. Stomatal apertures were determined after an exposure for 3 hours (opening) or 1 hour (closure). Each data point is the average of measurements on 80 to 100 stomata.

The path of biosynthesis in higher plants leads to (+)-ABA only (see p. 267). Commercially available ABA is usually a 1:1 mixture of (+)- and (−)-ABA (referred to in this paper as (±)-ABA), although enriched (+)-ABA can now be purchased. The enantiomers can be separated by immunoaffinity chromatography (Mertens, Stüning, and Weiler, 1982) or by high-performance liquid chromatography on a chiral stationary phase (Vaughan and Milborrow, 1984). Cummins and Sondheimer (1973) found that stomata responded more to (+)-ABA than to (−)-ABA. Apparently, the receptor sites for ABA on or in the guard cells possess a higher affinity for (+)-ABA than for (−)-ABA. In epidermal strips of *Commelina communis* and *Pisum sativum* var. Argenteum, the concentration of (−)-ABA required to produce a 50 percent reduction in stomatal aperture varied between 3×10^{-7} and 2×10^{-5} mol l^{-1}, whereas that of (+)-ABA varied between 0.8×10^{-7} mol l^{-1} and 0.6×10^{-5} mol l^{-1} (Raschke, unpublished). This indicates that, on the average, (+)-ABA is three to four times as effective in producing stomatal closure as (−)-ABA.

Hornberg and Weiler (1984) in their recent binding studies found no action at all of (−)-ABA on stomata of *Vicia faba* and no competition with ^{3}H-(+)-ABA for presumed receptor proteins, as determined by photoaffinity labeling. Analogs of ABA like *trans*-ABA and all-*trans* farnesol were ineffective, and analogs that produced some stomatal closing caused a roughly proportional displacement of the radiolabeled (+)-ABA from the binding sites. The question remains, whether the (−)-enantiomer has some effect on guard cells, as the bioassay indicates, or none at all, as must be concluded from the binding studies of Hornberg and Weiler (1984). Further experiments will have to be conducted with extremely pure samples of the two forms. That the receptor sites on the guard cells are not absolutely selective for ABA can be concluded from experiments in which the metabolite of ABA, phaseic acid (PA; see p. 270), also caused narrowing of the stomatal pores (Sharkey and Raschke, 1980). In unpublished work of Schwinning and Raschke, the concentration of PA that inhibited stomatal opening in epidermal strips of *C. communis* by 50 percent was, on the average, 300 times higher than that of (±)-ABA. An alternative explanation for the low effectiveness of PA in closing stomata would be the presence of a separate receptor for this compound in guard cells.

Experiments with epidermal peels all yielded a noteworthy result, independent of the species investigated: the widths of the final apertures attained depended on the concentration of ABA applied and, within reasonable limits, not on the duration of exposure. A steady state appears to exist between the rate of binding and the rate of removal of ABA from the receptor sites in guard cells. In this context, results of an experiment of Hartung (1983) are important. He found that acid-treated epidermis of *Valerianella locusta* accumulated a significant amount of ABA if incubated at pH 5, but hardly any at pH 8. Nevertheless, ABA was as effective in causing stomatal closure at pH 8, when virtually no ABA had entered the guard cells, as it was at pH 5, when ABA could be detected in the cells. Therefore the receptor sites for ABA must be at the surface of the guard cells.

The amounts of ABA required to cause a stomatal response are of the order of a few fmol mm^{-2} of epidermal tissue (as reviewed by Raschke, 1982). Hornberg and Weiler (1984) concluded from their experiments involving photoaffinity labeling that guard cell protoplasts of *Vicia faba* have more than 2,000 ABA-specific binding sites per μm^2 of plasmalemma. This number is large in relation to the number of potassium-selective channels in guard cell protoplasts of the same species. Schroeder, Hedrich, and Fernandez (1984), using patch-clamp measurements, estimated the channel density to be 1 per 15 μm^2 of cell surface. If the plasmalemma of a guard cell of *V. faba* has a surface of approximately 4,000 μm^2, the number of the potassium channels would be on the order of 300 per guard cell.

The evidence provided by Hartung (1983) and by Hornberg and Weiler (1984) points to the plasmalemma as the site of action of ABA on guard cells. Particularly during ABA-initiated stomatal closure, however, responses occur that require the transmission of a signal into the cytoplasm and to the tonoplast. Osmotica can leave the vacuole of guard cells only if there are channels in the

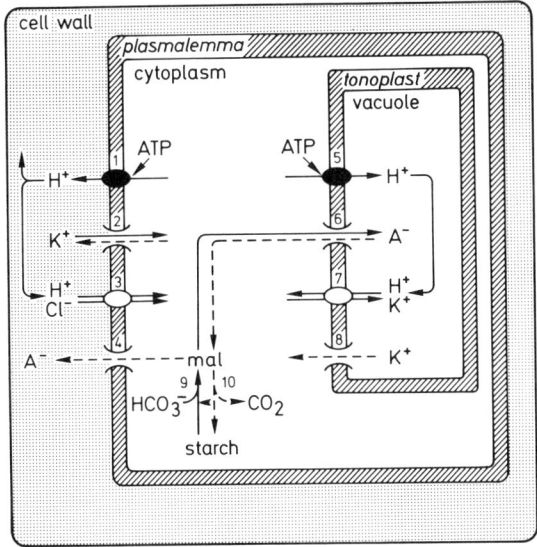

Fig. 11.2. A simplified model of the stomatal mechanism as related to the action of ABA on guard cells. *Solid lines*, opening; *dashed lines*, closing; *1, 5*, proton pumps; *2*, K$^+$ channels; *3*, H$^+$−Cl$^-$ cotransport mechanism; *4*, anion channels, closed during opening; *6*, anion channels, may be closed; *7*, H$^+$−K$^+$ exchange mechanism; *8*, K$^+$ channels, closed during opening; *9*, malate production from starch, including carboxylation of phosphoenolpyruvate and glycolysis; *10*, malate removal by oxidation and gluconeogenesis. The action of the pumps at the plasmalemma (*1*) can be accelerated by application of FC (Marrè, 1979). Closure occurs if the proton pumps are inhibited and ion channels (*4, 8*) are open, which eliminates the Donnan properties of the cytoplasm and the vacuole and allows salt gradients to dissipate.

tonoplast. Before we explore what may occur in a guard cell after it has received ABA, we shall examine the mechanisms that could be involved.

Stomatal Metabolism and Ion Transport in Relation to ABA Action

The model in Figure 11.2 shows how ion transport and carbon metabolism of guard cells may be related in stomatal reactions to ABA. This model is a highly speculative simplification. For models emphasizing other elements of the stomatal mechanism than ion transport and acid metabolism or favoring different views from those expressed here, see Chapters 6–10.

Stomatal opening results from an accumulation of potassium chloride and potassium malate in the vacuoles of the guard cells. Closure occurs when the potassium salts leave the vacuoles and are released from the guard cells or are metabolized, as is the case with malate. During opening also, ions will cross the plasmalemma and tonoplast simultaneously, and these movements will be coupled to metabolic processes in the cytoplasm, chloroplasts, and mitochondria. These processes supply (Fig. 11.2, *9*) or remove (Fig. 11.2, *10*) malate and provide the energy (ATP) to drive the system. During stomatal opening, guard cells excrete H^+ (Raschke and Humble, 1973) through the operation of a proton pump in the plasmalemma (Fig. 11.2, *1*). If all anion channels (Fig. 11.2, *4*) are shut, the cytoplasm becomes electrically negative with respect to the cell wall. Potassium ions follow the potential difference and enter the cell through K^+-selective channels (Fig. 11.2, *2*; Schroeder, Hedrich, and Fernandez, 1984). Although electroneutrality can thus be maintained, the H^+ concentration in the cytoplasm would quickly decrease through H^+–K^+ exchange.

The alkalinization that would follow is prevented by two mechanisms that supply anions to the cytoplasm. One imports Cl^-, if it is available in the wall, possibly through a H^+–Cl^- symport (Fig. 11.2, *3*), as in *Chara* (Sanders, 1984); the other produces malic acid (Fig. 11.2, *9*; Allaway, 1973) from carbon skeletons derived from starch (Dittrich and Raschke, 1977b; Outlaw, this volume). This latter process requires HCO_3^- for the carboxylation of phosphoenolpyruvate. Guard cells of *Allium cepa* do not contain starch and therefore rely solely on Cl^- import during stomatal opening (Schnabl and Ziegler, 1977; Schnabl and Raschke, 1980). Import of K^+ and Cl^- and production of malate would quickly come to a halt if these ions were not loaded from the cytoplasm into the vacuole.

We know even less about the transport processes occurring in the tonoplast than about those in the plasmalemma. If one extrapolates to guard cells the findings in CAM plants (Lüttge, Smith, and Marigo, 1982), there should be a proton pump in the tonoplast also (Fig. 11.2, *5*); this pump, in combination with anion channels (Fig. 11.2, *6*) and with a K^+–H^+ exchange mechanism (Fig. 11.2, *7*), could account for the observed accumulation of K^+ salts, as long as the postulated K^+ channels (Fig. 11.2, *8*) remain closed. The possibility exists that a nonselective channel for both, anions and cations, exists in the tonoplast combining the functions of channels *6* and *8* in Figure 11.2. The increase in pH in guard cell vacuoles during stomatal opening (Penny and Bowling, 1975) could have its cause in a stimulation of the K^+–H^+ exchange mechanism in the tonoplast by an increase in the K^+ content of the cytoplasm. The pH of the vacuoles is already below that of the cytoplasm when the stomata are closed.

When the proton pumps (Fig. 11.2, *1*, *5*) stop and the anion and K^+ channels (Fig. 11.2, *4*, *8*) remain closed, guard cells will stay inflated. The cells will collapse, and stomata will close as soon as these channels (Fig. 11.2, *4*, *8*) open and vacuole and cytoplasm are no longer Donnan spaces. In this model, stomatal closure would require a release of K^+, Cl^-, and malate from the guard cells. Simultaneously, malate would be used up in the citric-acid cycle, and a third fraction of malate would return to starch via decarboxylation (Fig. 11.2, *10*; Dittrich and Raschke, 1977a).

It is conceivable that some anion channels (Fig. 11.2, *4*) in the plasmalemma are part of the H^+-Cl^- symport mechanism (Fig. 11.2, 3) and the K^+ channels (Fig. 11.2, 8) in the tonoplast are part of the H^+-K^+ antiporter (Fig. 11.2, 7), their functions being controlled by the ion concentrations in the compartments they connect. (See Sanders, 1984, for possible properties of the H^+-Cl^- symport mechanism.) There must also be the possibility of an H^+–cation exchange at the plasmalemma, because guard cell protoplasts can alkalinize their medium when they lose solutes (see Fig. 11.4). Stomatal opening can be prevented and closure caused by interference at various points with the mechanism shown in Figure 11.2. In the next section, we return to reported effects of ABA that might indicate sites of action of ABA in guard cells.

The analysis of the stomatal mechanism has been facilitated by the use of the fungal toxin fusicoccin (FC), which induces stomatal opening (Turner and Graniti, 1969) and stimulates proton expulsion from plant cells (Marrè, 1979), including guard cells (Raschke, 1977; Gepstein, Jacobs, and Taiz, 1982). If the proton pump at the plasmalemma (Fig. 11.2, 1) is accelerated, all other processes involved in stomatal opening should also run at higher than normal rates.

First Measurable Changes after Application of ABA

Papers published between 1969 and 1972 report that stomata in detached leaves and in epidermal peels begin to close within minutes after application of ABA and are closed within 15 minutes, although closure often takes longer. Thus, ABA triggers a rapid release of osmotica from the vacuole through tonoplast, cytoplasm, and plasmalemma. Starch that had disappeared from the guard cell chloroplasts during opening reappears. For instance, Mansfield and Jones (1971) determined that when stomata closed in epidermal samples of *Commelina communis* in response to 10^{-4} mol l^{-1} ABA, the osmotic pressure and the K^+ content of the guard cells decreased, and the starch content of the chloroplasts increased. In experiments of Dittrich and Raschke (1977a) and Van Kirk and Raschke (1978), malate leaked from epidermal strips of *C. communis* and *Vicia faba* when they were placed on solutions of ABA. MacRobbie (this volume) presents evidence for a transient loss of K^+ ($^{86}Rb^+$) and Cl^- ($^{82}Br^-$) from guard cells of *C. communis* that had been treated with 2×10^{-5} mol l^{-1} ABA. When guard cell protoplasts of *V. faba* and *Allium cepa* were suspended in solutions of ABA, they rapidly shrank (Schnabl, 1978; Schnabl, Bornman, and Ziegler, 1978); this response was confirmed with protoplasts of *V. faba* (Gotow, Kondo, and Syōno, 1982) and with guard cell protoplasts prepared from the Argenteum mutant of *Pisum sativum* (Jewer, Incoll, and Shaw, 1982). Obviously, the presence of functional epidermal cells is not required for a stomatal response to ABA, contrary to the claim of Itai and

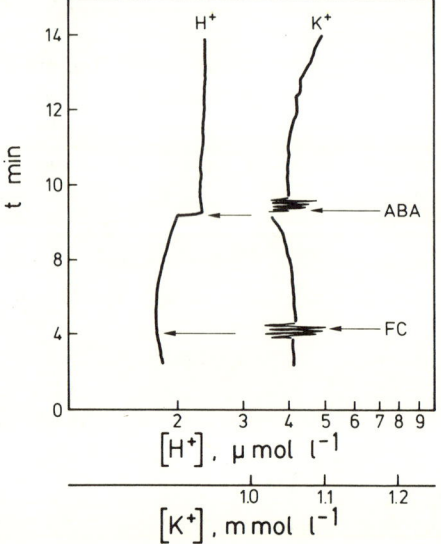

Fig. 11.3. Inhibition by (±)-ABA (36 μmol l^{-1}) of H$^+$ excretion and initiation of K$^+$ efflux from guard cell protoplasts of *Vicia faba* whose ion exchange had been activated by FC (1 μmol l^{-1}) detected by miniature electrodes submersed in a suspension of 2 × 10^5 protoplasts in 100 μl. Displacements of the traces at the times of additions to the suspensions were caused by the acidity of the added ABA.

Meidner (1978). But frequently stomata do respond more to ABA when whole leaves are treated than when epidermal strips are floated on ABA solutions.

The fastest guard cell responses to ABA recorded so far are changes in pH and K$^+$ content of the medium in which guard cell protoplasts are suspended (Fig. 11.3). Addition of FC to the ABA-free medium stimulated proton excretion and K$^+$ uptake within about two minutes. Both processes stopped within a minute after injection of ABA into the medium. About two minutes after ABA was added, K$^+$ began to leak out of the cells. Occasionally, application of ABA caused a reversal of the proton flow: the pH of the suspension began to increase. In early experiments with epidermal strips, proton excretion from guard cells was stimulated by increased K$^+$ concentrations in the medium (Raschke and Humble, 1973). Similarly, FC-induced proton excretion by guard cell protoplasts proceeded at a higher rate in a solution of 20 mEq l^{-1} K$^+$ than in a solution of 1 mEq l^{-1} (Fig. 11.4, right panel). Addition of ABA stopped proton excretion when the external K$^+$ concentration was low; at higher external K$^+$ concentration ABA reduced the rate of proton export. The recorded phenomena may imply an inhibition of the proton pump at the plasmalemma in combination with an opening of ion channels. Inhibition of a presumed H$^+$-exporting ATPase in the plasmalemma by vanadate (e.g. O'Neill and Spanswick, 1984) also stopped proton excretion by guard cell protoplasts and caused alkalinization of the medium when its pH was near 5 and thus below the presumed pH of the protoplast contents.

High external concentrations of KCl prevent stomatal closure in response to ABA, as has been observed in epidermal strips of *C. communis* (Wilson,

Ogunkanmi, and Mansfield, 1978; Weyers and Hillman, 1979b) and of *V. faba* (Wardle and Short, 1981). Experiments conducted with KCl, K-iminodiacetate, and bis-tris-propane chloride showed that the reduced response to ABA is the combined effect of high external concentrations of K^+ and Cl^- (Fig. 11.5). K^+ and Cl^-, each given with a nonpenetrating counterion, were effective in opposing stomatal responses to ABA.

Control by ABA of Solute Accumulation and Loss

Abscisic acid appears to bind to proteins in the plasmalemma of guard cells (Hornberg and Weiler, 1984) and causes a rapid reduction, even stoppage, of proton excretion (Figs. 11.3, 11.4). The simplest explanation for this effect would be that ABA inhibits the proton pump at the plasmalemma. Because only net flows of protons can be recorded in the media by monitoring pH (or by titration, for that matter), it is also possible that ABA activates a proton pump directed into the cell. There is some evidence to support this view. First, the opposing effects of FC and ABA appear to be additive. But a counterargument is that FC may set in motion an H^+ pump in addition to the H^+ pump ordinarily working in the plasmalemma. Second, ABA not only inhibits transport processes, it also enhances a few. One of them is sucrose accumula-

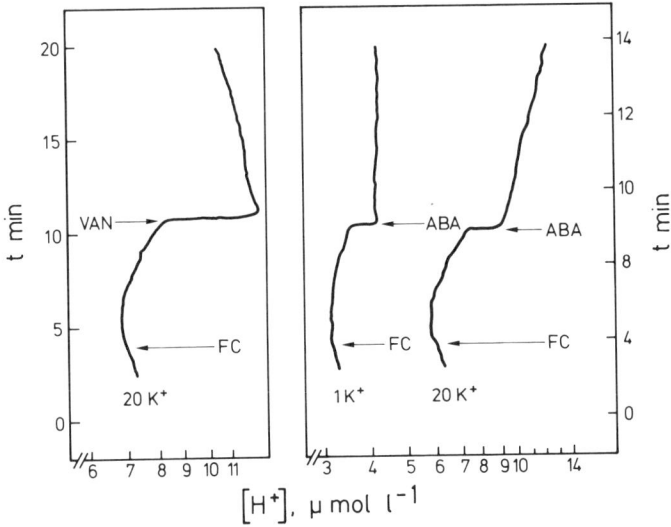

Fig. 11.4. Inhibition of the FC (1 μmol l^{-1})-induced proton expulsion from guard cell protoplasts by vanadate (VAN, 120 μmol l^{-1}) or (±)-ABA (36 μmol l^{-1}), detected by a miniature pH electrode immersed in suspensions of 2×10^5 protoplasts from *Vicia faba* in 100 μl of solutions of KCl or K_2SO_4 containing 1 or 20 milliequivalents of K^+ per liter, as indicated. Displacements of the traces at the times of additions were caused by the acidity of the added vanadate or ABA.

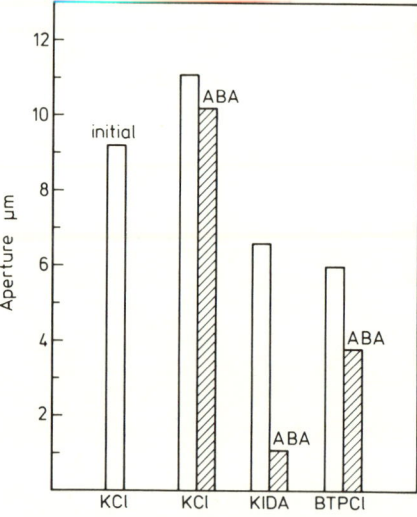

Fig. 11.5. Prevention of ABA-induced stomatal closure by high concentrations of K^+ and Cl^- in epidermal peels from leaves of *Commelina communis*. Epidermal peels were floated on 200 mmol l^{-1} of KCl, K-iminodiacetate (KIDA), or bis-tris-propane Cl (BTP Cl), without (open columns) or with (hatched columns) 10^{-5} M (±)-ABA for 1 hour in the light at 30° C.

tion in beet roots, although this has been described as resulting from K^+-sucrose co-transport (Saftner and Wyse, 1984). Third, ABA-initiated stomatal closure can be stopped by inhibiting respiration (Weyers et al., 1982). I found that uncouplers (and sulfhydryl blockers) will do the same. Apparently, energy is needed for ABA to have an effect on guard cells; but loss of response to ABA in the presence of energy-supply inhibitors may be related to greatly decreased effluxes from "nonenergized" cells (Komor, Weber, and Tanner, 1979). If we apply Ockham's razor and make only a minimum of assumptions we should state that ABA inhibits proton expulsion. That would explain the suppression of stomatal opening.

Inhibition of the proton pump, however, will not alone explain stomatal closure. As was already discussed (see p. 260), channels must be opened in the plasmalemma and in the tonoplast to release electrolytes from the vacuole and from the cytoplasm. Because the permeability of the guard cell membranes for cations is high (Saftner and Raschke, 1981), we should expect variations in anion permeability (and anion gradients) to have a strong effect on the leakage of osmotica from guard cells. The fact that the closing effect of ABA on stomata can be reduced or even abolished if the external concentrations of K^+ and Cl^- are high (> 100 mmol l^{-1}, versus 100 to 450 mmol l^{-1} inside) demonstrates how important the control of leakage is in the guard cells' response to ABA.

Rapid onset of efflux of solutes from guard cells and rapid stomatal closing require the transmission of a signal from the plasmalemma to the tonoplast. Of course, ABA itself could be the messenger. The ABA concentration required to produce a response at the tonoplast would probably be on the order of 10^{-7} mol l^{-1}, or even less. This concentration could quickly be established when the external ABA concentration is being raised; the necessary amounts of imported ABA would be extremely small. Other candidates as messengers are Ca^{2+} and H^+. Calcium ions indeed appear to be involved in stomatal responses to ABA: the addition of Ca^{2+} to ABA solutions reinforces responses to ABA. Exposure of epidermal strips to chelators of Ca^{2+} (e.g. EGTA, EDTA, citric acid, ATP) reduces the magnitude of the closing response (W. Pagel, unpublished). These observations, however, are not sufficient evidence for a role of Ca^{2+} as a second messenger in guard cells, because Ca^{2+} could equally well modify the ABA receptors or the channel proteins in the plasmalemma. If we turn to H^+, we recognize that a change in the activity of the proton pumps could cause a change in the pH of the cytoplasm. It is easy to see how such a change could affect properties of plasmalemma and tonoplast simultaneously and guarantee synchronous passage of salts through both membranes. A change in cytoplasmic pH could also couple synthesis and metabolism of malate to ion transport. By shifting the equilibrium of the reaction $CO_2 + H_2O \rightleftharpoons HCO_3^- + H^+$ to the left side, a decline in pH would reduce the HCO_3^- available as substrate for the production of malate and would move the pH out of the optimal range for phosphoenolpyruvate carboxylase into the range favoring decarboxylation of malate by malic enzyme (Willmer, Pallas, and Black, 1973).

A hypothesis involving the cytoplasmic pH in the mechanism of response to ABA could explain why the magnitude of the closing response varies so greatly among plants and leaves, because intracellular pH (and the cells' success in regulating it) varies. Such a hypothesis could also explain the interaction between stomatal responses to ABA and those to CO_2 and indoleacetic acid (IAA), which will be discussed in the next section.

To sum up: it seems as if ABA prevents stomatal opening by inhibiting the proton pump at the plasmalemma or by activating a proton pump directed inward. Stomatal closure requires the additional opening of ion channels in the plasmalemma and the tonoplast. Opening of the channels could be triggered by a direct action of ABA on these channels or could be mediated by a change in the cytoplasmic concentration of an ion, possibly H^+. Conceivably, this change in the cytoplasmic ion concentration is linked to the inhibition of net proton expulsion. No part of this hypothesis has been unequivocally proven.

Interactions among Stomatal Responses to ABA, CO_2, and IAA

In leaves of many species, stomata can be in a state unresponsive to CO_2. Administration of ABA will correct this situation, sensitizing guard cells to CO_2 (Raschke, 1975; Dubbe, Farquhar, and Raschke, 1978). It is a general experience that wilting will also sensitize stomata to CO_2, and Raschke, Pierce, and Popiela (1976) reported sensitization of stomata to CO_2 by chilling. Conversely, the presence of CO_2 enhances stomatal response to ABA; half-saturation of this enhancement occurred in leaves of *Xanthium strumarium* at a CO_2 concentration of 200 µl l^{-1} (Raschke, 1975). Sensitization of stomata to CO_2 could have ecological implications: if drought is impending, CO_2 sensitivity would enable stomata to follow changes in assimilation with increased accuracy and thus save water through feedback of the intercellular CO_2 concentration. This feedback would be superimposed on stomatal responses to light and would attune stomatal conductance to variations in the environment.

Amplitudes of stomatal responses to CO_2 and ABA vary greatly, and interactions among stomatal responses to CO_2 and ABA are not always found (Mansfield and Wilson, 1981). Snaith and Mansfield (1982) offered an explanation for the conflicting observations. These authors tested the phytohormone IAA as an additional factor in the control of stomatal aperture and showed that at high levels of IAA (10^{-5} to 10^{-4} mol l^{-1}) in the incubation medium, stomata in epidermal peels of *Commelina communis* were less sensitive to ABA, and particularly to CO_2, than at low levels (10^{-8} to 10^{-7} mol l^{-1}). Increasing concentrations of IAA reduced the stomatal response to CO_2, and ABA restored it. Apparently IAA can stimulate proton expulsion not only in growing tissue but also in guard cells, and thus it can antagonize the effects of ABA and CO_2 on the proton exchange of guard cells. Abscisic acid and CO_2 possibly reinforce each other in causing acidification of the cytoplasm by reducing net expulsion of protons and by providing a substrate for the production of malate (Raschke, 1975). In chilled plants of *Phaseolus vulgaris*, however, the interplay between responses to ABA, CO_2, and IAA appears to deviate from that just described (Eamus and Wilson, 1984).

The effects of other plant hormones on responses to ABA require further studies. We can expect that knowledge of these interactions may clarify the perplexing variability of stomatal response to the environment. At the same time, we may obtain some insight into the complicated mechanism by which plants adjust gas exchange in accordance with their history, with their developmental state, and with the prevailing environmental conditions.

A Long-Term Effect of ABA on Stomata

Stomata in epidermal peels taken from nonstressed *Vicia faba* plants do not close entirely, not in darkness, and not even when guard cells have been plasmolyzed. The mechanical properties of the walls apparently do not allow complete closure when the peel has been separated from the leaf. In other plants, stomata on the leaf may stay open in darkness. In these cases, the pressure of the ordinary epidermal cells on the guard cells does not suffice to force the stomatal apertures closed. This situation changes after plants have experienced drought. It seems that under stress the properties of guard cell walls can change, despite the differentiation of these cells. Microautoradiography of epidermal peels of tulip exposed to a solution of ^{14}C glucose

1-phosphate (Dittrich and Raschke, 1977b) indicated that metabolism continued in the walls of mature guard cells.

Changes in the mechanics of stomatal movement also occur under the influence of ABA. The tomato mutants *sitiens* and *flacca* wilt because their stomata are unable to close, even after guard cells have been plasmolyzed (Tal, 1966). Similar observations were made on "droopy," a wilty mutant of potato (Quarrie, 1982). These three mutants are characterized by low ABA content (Tal and Imber, 1972; Quarrie, 1982). Spraying *flacca* plants with solutions of ABA will cure them of their proneness to wilting; their stomata acquire the ability to shut (Tal and Imber, 1972; Bradford, Sharkey, and Farquhar, 1983). Repeated sprays over several days are required, although some narrowing of the residual apertures, determined after plasmolysis of the guard cells, occurred in material that had been treated with ABA already for only 3.5 hours (Tal, Imber, and Gardi, 1974). Apparently, ABA can produce a second effect in guard cells, affecting the mechanical properties of the cell walls. This effect develops more slowly than the reduction of proton excretion and could be related to other morphogenetic responses to ABA, such as formation of adventitious roots, which takes five days to develop (Hartung, Ohl, and Kummer, 1980).

Biosynthesis and Metabolism of ABA in Leaves

Abscisic acid appears to be a signal in a feedback loop between the stomata and other components of the hydraulic system of the plant. If stomata are effectors in regulating water loss, they must rely on the timing of arrival and disappearance of ABA. Synthesis, storage, transfer, and removal of ABA by metabolism and sequestration will largely determine the quality of stomatal regulation of water loss. This section will deal with the turnover of ABA; the next section, with possible transfer and compartmentation.

Abscisic acid is a sesquiterpenoid very likely derived from mevalonic acid. Two pathways could lead to ABA (for details, see Milborrow, 1983; Neill, Horgan, and Walton, 1984). The direct (C_{15}) route would lead to ABA through the precursor farnesyl pyrophosphate; on the alternative (C_{40}) route, ABA could be derived from xanthoxin (Fig. 11.6), which is a natural cleavage product of the xanthophyll violaxanthin (Taylor, 1968; Taylor and Burden, 1970). Plants can convert xanthoxin into ABA (Taylor and Burden, 1973), and stomata close in leaves when xanthoxin has been added to the transpiration stream (Raschke, Firn, and Pierce, 1975). However, it appears unlikely that the C_{40} path would have the metabolic capacity to produce ABA rapidly enough. Neill, Horgan, and Walton (1984) believe that the evidence indicates that "most ABA arises via a direct C_{15} pathway," but new studies by Creelman and Zeevaart (1984a) on the incorporation of ^{18}O into ABA in wilted leaves of

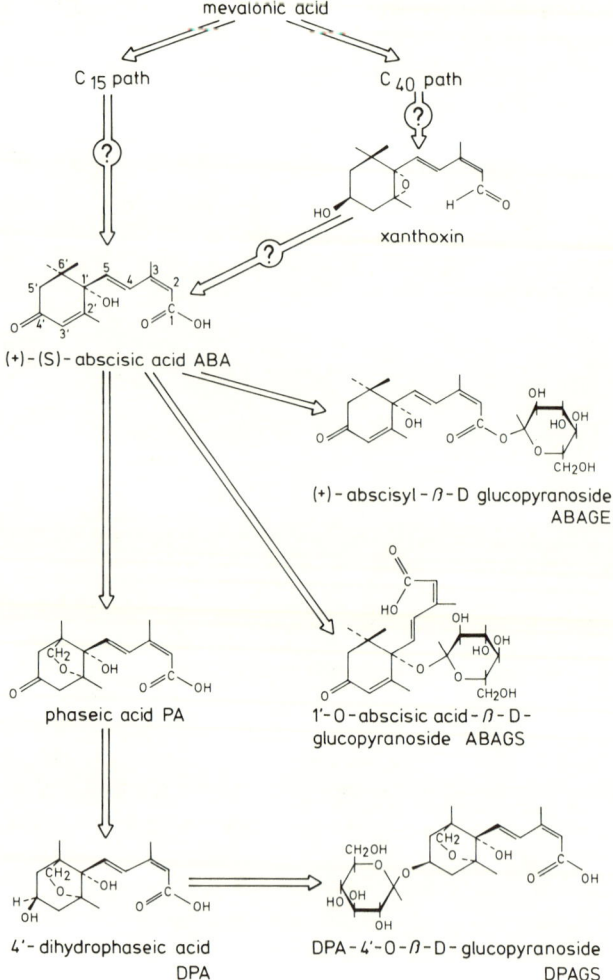

Fig. 11.6. Formulas of xanthoxin, abscisic acid, and some of the metabolites of abscisic acid. The following metabolites also occur in plants but were not entered: *epi*-DPA and the conjugates of PA, DPA, and *epi*-DPA (Zeevaart and Milborrow, 1976).

Xanthium strumarium revive the possibility that ABA formation follows the indirect pathway. We do not know yet how ABA is made in plants.

In general, except in the epidermis (Loveys, 1977; Dörffling et al., 1980) leaf tissue can synthesize ABA. Epidermal cells are also unable to convert xanthoxin to ABA (Raschke, Firn, and Pierce, 1975). Seen teleologically, the inability of epidermal cells to make ABA would enable them to recognize ABA as a messenger of stress when it arrives from the mesophyll. The increased levels of ABA found in osmotically stressed guard cell protoplasts (Weiler,

Schnabl, and Hornberg, 1982) would not accord with this view. It is therefore important that the ability of guard cells to produce ABA be reinvestigated, particularly since Weiler, Schnabl, and Hornberg reported that guard cell protoplasts accumulated ABA at 0° C, a temperature at which metabolism is believed to proceed slowly or not at all.

Turning to the sites of ABA synthesis within the cells, we note that Milborrow (1974) concluded that ABA was produced within the chloroplasts. Hartung, Heilmann, and Gimmler (1981), however, determined that spinach chloroplasts were unable to convert radioactive mevalonic acid into ABA, although they were able to take up this precursor. Rather, synthesis of ABA proceeded in the cytoplasmic fraction of the protoplasts. Additional evidence will be needed to establish that the cytoplasm is the site of production for ABA.

The presumed role of ABA as a phytohormone requires its biosynthesis to be strictly controlled, but all plant tissues contain ABA all the time. There is ABA in leaves that are not stressed, and it seems that ABA is produced and metabolized continuously. Nobody has determined the rate of this basal turnover, although it is important to know it in order to assess how variation in the rate of synthesis or metabolism may affect the ABA level in the leaf. There must be an endogenous component of control that is involved in the ontogeny of the plant and that may produce a diurnal rhythm in ABA content. Stress appears to trigger a production of ABA that is superimposed on the basal rate. Of the various insults to plants (drought, chilling, salinization, etc.) only drought can unequivocally cause a dramatic rise in the level of ABA. During early investigations it was recognized that the water potential of the investigated tissue had to fall below a threshold value before ABA would begin to accumulate. An analysis of the relationships between ABA synthesis and water potential, osmotic pressure, and turgor in individual leaves by Pierce and Raschke (1980) showed that loss of turgor was the event that initiated ABA accumulation. When the solute content of leaves increased during the course of osmoregulation in response to water stress, the water potential at which turgor was lost and likewise the threshold for the acceleration of ABA production decreased. Using an experimental approach different from that of Pierce and Raschke (1980), Creelman and Zeevaart (1984b) confirmed that turgor loss was responsible for increased ABA production.

How a loss of turgor initiates a biochemical event that leads to ABA synthesis remains unknown. A change in turgor is not equivalent to a change in absolute pressure. Turgor is a pressure difference across the plasmalemma and the wall, stretching the plasma membrane and causing a stress in its plane that in turn results in membrane compression. A change in absolute pressure causes only membrane compression. If a leaf is put into a pressure bomb and the pressure is raised, it will matter very much to the leaf whether the cut end of the petiole is allowed to protrude into the external air and is at atmospheric pressure or whether it is also enclosed in the bomb and experiences the bomb pressure. It should not surprise us that leaves subjected to changes in absolute pressure respond with ABA productions at rates different from those caused by turgor changes, as Pierce (1981) and Ackerson and Radin (1983) observed.

Osmotic pressures vary among the cells of a leaf, even more so among leaves, and these varying osmotic pressures introduce variations in the water potential at which turgor is lost: remembering that $\Psi = p - \pi$ (see page 253), we obtain $\Psi = -\pi$ for $p = 0$. Thus, when Ψ is being lowered, turgor will be lost in all cells at the same Ψ only if all cells had been identical in their water-relations parameters. If in a population of cells π is distributed over a range, the sharpness of turgor loss will disappear, and so will the sudden increase in the rise in ABA production. Likewise, it should be obvious that a causal relationship between a plant's ABA synthesis and one of the plant's water-status parameters cannot clearly emerge if ABA production is related to the average turgor of a population of leaves.

Synthesis of ABA in the leaves of a crop appears to depend not only on the distribution of π and Ψ within and among plants, and on the adjustment of π to changing values of Ψ, but also on an additional, unknown mechanism that makes ABA accumulation dependent on the rate at which water stress develops in the leaves (Henson, 1985).

After the onset of wilting, ABA synthesis does not proceed so quickly as was once thought. In the experiments of Pierce and Raschke (1980) it took an hour or more after turgor loss before a measurable increase in ABA content could be detected. It required six to eight hours before steady-state levels were reached. Because wilt-induced accumulation of ABA proceeds so slowly, doubts may arise about the function of ABA as a hormone involved in controlling water loss, even if the ABA concentrations increased from about 0.1 μmol l^{-1} in unstressed leaves to 1.5 or even 5 μmol l^{-1} in stressed ones. It may be possible to dispel the doubts later, in the following section.

Accumulation of ABA in wilting leaves results from synthesis (and possibly import). Milborrow (1978), Zeevaart (1980), Pierce and Raschke (1981), and Neill, Horgan, and Heald (1983) determined that ABA is not derived from the hydrolysis of conjugates stored in the tissue. Recently, Cornish and Zeevaart (1985) discovered that excised roots of *Xanthium strumarium* and *Lycopersicon esculentum* accumulated considerable amounts of ABA after they had lost water to an extent of 60 to 70 percent of their fresh weight. Conceivably, some of the ABA found in leaves may have been imported from roots exposed to dry soil.

Plants metabolize ABA by either of two pathways (for details, see Loveys and Milborrow, 1984). The more important one involves an oxidation catalyzed by an oxygenase (Creelman and Zeevaart, 1984a), followed by a rearrangement to phaseic acid (PA; Fig. 11.6). Phaseic acid is then reduced to 4'-dihydrophaseic acid (DPA), and to a smaller extent to *epi*-DPA (Zeevaart and Milborrow, 1976). DPA accumulates in the tissue but apparently is not the ultimate deactivation product of ABA. The other path of ABA deactivation involves esterification with glucose to form (+)-abscisyl-β-D-glucopyranoside (ABAGE; Fig. 11.6). In a number of plant species, a second conjugate with glucose appears as a major metabolite: 1'-O-abscisic acid-β-D-glucopyranoside (ABAGS; Fig. 11.6). Phaseic acid and dihydrophaseic acid can also be converted to conjugates. Conjugate formation appears to be an unspecific mechanism. When Mertens, Stüning, and Weiler (1982) fed the unnatural (−)-enantiomer of ABA to leaf discs of *Vicia faba*, they found that virtually all the (−)-ABA was converted into conjugates. Vaughan and Milborrow (1984) confirmed this fate for (−)-ABA in experiments on tomato shoots, which converted (−)-ABA mostly to ABAGE and ABAGS. The (+)-ABA ultimately yielded predominantly DPA-4'-O-β-D-glucopyranoside (DPAGS; Fig. 11.6), indicating that the natural enantiomer had been metabolized following the phaseic-acid route and that the DPA had then been conjugated.

Conversion of ABA to PA follows mass action; in addition, it is turgor dependent. Pierce and Raschke (1981) found that when a wilted bean leaf recovered turgor during rehydration, its rate of ABA synthesis dropped to zero within three hours, but the rate of conversion of ABA to PA accelerated until it was sufficient to convert almost half the ABA initially present in the tissue to PA within one hour.

Metabolism of ABA proceeds faster in darkness than in the light, perhaps because ethylene is involved (Zeevaart, 1983). Alternatively, metabolism of ABA may be slow in the light because, under this condition, much of it is trapped in the chloroplasts (Heilmann, Hartung, and Gimmler, 1980) and protected from metabolism; the enzymes involved reside in the cytoplasm (Hartung, Ohl, and Kummer, 1980).

It takes a wilted leaf about three to four hours to raise its ABA content to half the maximum; after rehydration, it takes about one hour to remove half the ABA by metabolism. Again the question arises whether these rates are fast enough for the process to be directly involved in correcting excessive water loss and in speedy recovery. Apparently they are not. Rather they seem to allow integrating the experiences of wilting within a sliding window in time, to prepare the plant for situations in which water stress could get worse. One should keep in mind that only a fraction of the cells capable of ABA production may need to lose turgor for an anticipatory rise in the level of the hormone; wilting may often be just incipient.

Compartmentation and Redistribution of ABA

In this section we close the circle. We began with the requirement of ABA as a signal. Now we return by asking how ABA can be a signal in the stomatal regulation of water loss; ABA is synthesized too slowly during wilting and not metabolized fast enough to keep pace with rehydration. In some cases, stomata stayed closed even if the level of ABA in the leaves had returned to normal (Beardsell and Cohen, 1975; Ludlow, Ng, and Ford, 1980; Henson, 1981). Very likely guard cells do not gauge the bulk ABA content of a leaf but only sense the ABA delivered to them through the apoplast by the transpiration stream. Perhaps the following scheme applies:

$$\begin{array}{c} \text{Guard cells} \\ \uparrow \\ \text{Apoplast} \\ \updownarrow \\ \text{Synthesis} \rightarrow \text{ABA in the cytoplasm} \rightarrow \text{Metabolism} \\ \text{of mesophyll cells} \\ \updownarrow \\ \text{Chloroplasts} \end{array}$$

We have no indication whether the path to the guard cells ends there or not, nor whether ABA is sequestered or metabolized in the epidermis to a degree that makes the transfer virtually unidirectional. The evidence and theoretical considerations that can be called upon to support this scheme are summarized below.

First, when ABA was supplied to the apoplast of a detached leaf of *Xanthium strumarium* through the water supply, an increase as small as 2 percent of the total amount of ABA present in a leaf was sufficient to cause stomatal closing (Raschke, 1975). The ABA in the transpiration stream evidently bypasses the ABA sequestered in the mesophyll.

Second, Cowan et al. (1982) suggested that ABA is distributed over the tissue by free permeation and by equilibration of the undissociated acid across the membranes. Because ABA is a weak acid (pK 4.8), it will dissociate in varying degrees in various compartments, depending on their pH values. Alkaline compartments will act as traps for the ABA anion. Once their pH decreases, they will release ABA. In the light, chloroplasts should be effective traps, because of the alkalinization of the stroma; when darkened, they should be a source. Cowan et al. (1982) propose that ABA uptake and release may even be involved in stomatal responses to light. They estimate that upon a change in illumination, the half-time of the redistribution of ABA between apoplast and the protoplasts of the leaf cells would be 19 minutes; this is in agreement with observations. These calculations do not prove that ABA is essential for stomatal movement in response to changes in light, but they demonstrate that the participation of chloroplasts in ABA storage and release is a viable hypothesis.

Third, if the scheme contains some truth, stressed tissue should release ABA. Hartung, Kaiser, and Burschka (1983) measured increased loss of ABA from the apoplasts of leaf slices when the slices were suspended in hypertonic solutions. Cornish and Zeevaart (1984) showed convincingly that a leaf can give off ABA before accelerated ABA synthesis has begun. They placed a leaf of *Xanthium strumarium* in a pressure bomb and analyzed the xylem exudate (apoplastic ABA) and the leaf tissue (bulk ABA). Following leaf wilting, the apoplastic ABA rose before the bulk ABA. The authors concluded that "the first response to water stress is a movement of ABA into the apoplast. This may be relevant for rapid stomatal closure in stressed leaves."

It is not yet possible to explain satisfactorily the delayed stomatal opening after periods of reduced water supply. There are three possible aftereffects. First, ABA lingers longer in the epidermis than in the rest of the leaf. This has not yet been shown. Second, the metabolite of ABA, phaseic acid, remains in the epidermis and inhibits stomatal opening. Although phaseic acid does have an effect on stomata, its concentration in the tissue will not be high enough to produce a strong inhibition (see page 258). Third, proteolysis has occurred. Full stomatal activity requires resynthesis of enzymes and pump and carrier proteins. This possibility has not yet been tested.

Looking back, it seems possible that the amounts of ABA normally turning over in the tissue are sufficient for a communication between mesophyll and the stomata. When additional ABA is synthesized in response to wilting, it is to reinforce the messenger pool and amplify the message.

Effect of ABA on Photosynthesis in Whole Leaves

Until quite recently, it was thought that stomatal closure in response to ABA was the only way this phytohormone could affect the rate of CO_2 assimilation. A series of experiments conducted over a number of years, however, proved beyond any doubt that ABA can affect the photosynthetic apparatus independent of its effect on the stomatal system (Cornic and Miginiac, 1983; Raschke, 1982; Raschke and Hedrich, 1985). Stomatal and photosynthetic responses are often proportional to each other, with the result that in many cases the intercellular CO_2 concentration remains constant. However, there is a wide variation, from no response at all to ABA, to a decrease of CO_2 assimilation, to a small fraction of the initial rate. We do not know whether ABA is involved in throttling photosynthesis also under natural conditions. Abscisic acid probably cannot attack the photosynthetic apparatus directly but rather produces a change in pH, or in the Mg^{2+} concentration that affects enzyme activities and the efficiency of energy conversion. Fischer, Raschke, and Stitt (1986) analyzed the pattern of Calvin-cycle intermediates in rapidly killed samples from whole leaves treated with ABA; they determined that *in vivo* activity of ribulose-1,5-bisphosphate carboxylase decreased after application of ABA. Ogawa et al. (1982) reported that ABA suppressed the K^+-induced decline in chlorophyll fluorescence in guard cells. Again, it is unlikely that this effect was a direct one of ABA on the photosynthetic apparatus. Perhaps the primary mechanism of action of ABA is on the plasmalemma and is identical in guard cells and in the photosynthetic tissue.

The sequence

ABA → stomatal closure → reduction in CO_2 supply →
  reduction in CO_2 assimilation

may have to be replaced by

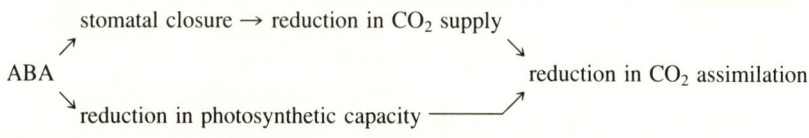

(Arrows are to be read as meaning "causes.")

The Presumed Role of ABA as Stress Hormone Needs Further Evidence

Abscisic acid is the presumed messenger substance the plant needs to deflate guard cells and shut stomata when water stress sets in. The ironclad proof for this function of ABA does not yet exist. So far, only the wilty and "droopy" mutants of tomato and potato demonstrate that plants need ABA to stay turgid. This is insufficient proof for the key role of ABA, because only the importance of the morphogenetic effect of ABA has thus been shown, and not yet the presumed involvement of the hormone in the rapid reduction of water loss. Nevertheless, under experimental conditions, stomatal closure in response to ABA can be so strikingly fast and large in amplitude that it is difficult to ward off thoughts that the response may have a function in the whole plant in the real world. New experiments, for instance, of the kind performed by Cornish and Zeevaart (1984), will have to demonstrate that strain on the water relations of a plant indeed causes a rapid release of ABA from storage sites into the epidermis and thus causes stomatal closure, which in turn will result in an improvement of the water status of the plant and in reduced release of ABA. Further, it would have to be shown that, after the water status had improved, ABA was removed from the receptor sites on the guard cells by diffusion, sequestration, or metabolism in the epidermis. Stomata should then reopen. In short, the operation of the entire presumed feedback loop would have to be demonstrated on one and the same system. Similarly, it will be necessary to show that the specific receptor proteins discovered by Hornberg and Weiler (1984), if occupied by ABA, indeed check proton pumping and, directly or indirectly, open ion channels. The role of ABA in the effector mechanism of the guard cells will have to be elucidated beyond any doubt. Experimental approaches to these problems exist.

REFERENCES

Ackerson, R. C., and J. W. Radin. 1983. Abscisic acid accumulation in cotton leaves in response to dehydration at high pressure. *Plant Physiology*, 71: 432–33.
Allaway, W. G. 1973. Accumulation of malate in guard cells of *Vicia faba* during stomatal opening. *Planta*, 110: 63–70.
Beardsell, M. F., and D. Cohen. 1975. Relationships between leaf water status, abscisic acid levels, and stomatal resistance in maize and sorghum. *Plant Physiology*, 56: 207–12.
Bradford, K. J., T. D. Sharkey, and G. D. Farquhar. 1983. Gas exchange, stomatal behaviour, and $\delta^{13}C$ values of the *flacca* tomato mutant in relation to abscisic acid. *Plant Physiology*, 72: 245–50.
Cornic, G., and E. Miginiac. 1983. Nonstomatal inhibition of net CO_2 uptake by ($\pm$) abscisic acid in *Pharbitis nil*. *Plant Physiology*, 73: 529–33.
Cornish, K., and J. A. D. Zeevaart. 1984. The effect of water stress on the movement of abscisic acid into the apoplast. *Plant Physiology*, 75 (supplement): 137.
———. 1985. Abscisic acid accumulation by roots of *Xanthium strumarium* L. and *Lycopersicon esculentum* Mill. in relation to water stress. *Plant Physiology*, 79: 653–58.
Cowan, I. R., J. A. Raven, W. Hartung, and G. D. Farquhar. 1982. A possible role for abscisic acid in coupling stomatal conductance and photosynthetic carbon metabolism in leaves. *Australian Journal of Plant Physiology*, 9: 489–98.
Creelman, R. A., and J. A. D. Zeevaart. 1984a. Incorporation of oxygen into abscisic acid and phaseic acid from molecular oxygen. *Plant Physiology*, 75: 166–69.
———. 1984b. Abscisic acid accumulation in spinach leaf slices in the presence of penetrating and non-penetrating solutes. *Plant Physiology*, 75 (supplement): 102.
Cummins, W. R., H. Kende, and K. Raschke. 1971. Specificity and reversibility of the rapid stomatal response to abscisic acid. *Planta*, 99: 347–51.
Cummins, W. R., and E. Sondheimer. 1973. Activity of the asymmetric isomers of abscisic acid in a rapid bioassay. *Planta*, 111: 365–69.
Dittrich, P., and K. Raschke. 1977a. Malate metabolism in isolated epidermis of *Commelina communis* L. in relation to stomatal functioning. *Planta*, 134: 77–81.
———. 1977b. Uptake and metabolism of carbohydrates by epidermal tissue. *Planta*, 134: 83–90.
Dörffling, K., D. Tietz, J. Streich, and M. Ludewig. 1980. Studies on the role of abscisic acid in stomatal movements. In F. Skoog, ed., *Plant growth substances, 1979*: 274–85. Berlin: Springer-Verlag.
Dubbe, D. R., G. D. Farquhar, and K. Raschke. 1978. Effect of abscisic acid on the gain of the feedback loop involving carbon dioxide and stomata. *Plant Physiology*, 62: 406–17.
Eamus, D., and J. M. Wilson. 1984. A model for the interaction of low temperature, ABA, IAA, and CO_2 in the control of stomatal behaviour. *Journal of Experimental Botany*, 35: 91–98.
Fischer, E., K. Raschke, and M. Stitt. 1986. Effects of abscisic acid on photosynthesis in whole leaves: Changes in CO_2 assimilation, levels of carbon reduction cycle intermediates, and activity of ribulose 1,5-bisphosphate carboxylase. *Planta*, 169, in press.
Gepstein, S., M. Jacobs, and L. Taiz. 1982. Inhibition of stomatal opening in *Vicia faba* epidermal tissue by vanadate and abscisic acid. *Plant Science Letters*, 28: 63–72.

Gotow, K., N. Kondo, and K. Syono. 1982. Effect of CO_2 on volume change of guard cell protoplast from *Vicia faba* L. *Plant and Cell Physiology*, 23: 1063–70.

Hartung, W. 1983. The site of action of abscisic acid at the guard cell plasmalemma of *Valerianella locusta*. *Plant, Cell and Environment*, 6: 427–28.

Hartung, W., B. Heilmann, and H. Gimmler. 1981. Do chloroplasts play a role in abscisic acid synthesis? *Plant Science Letters*, 22: 235–42.

Hartung, W., W. M. Kaiser, and C. Burschka. 1983. Release of abscisic acid from leaf strips under osmotic stress. *Zeitschrift für Pflanzenphysiologie*, 112: 131–38.

Hartung, W., B. Ohl, and V. Kummer. 1980. Abscisic acid and the rooting of runner bean cuttings. *Zeitschrift für Pflanzenphysiologie*, 98: 95–103.

Heilmann, B., W. Hartung, and H. Gimmler. 1980. The distribution of abscisic acid between chloroplasts and cytoplasm of leaf cells and the permeability of the chloroplast envelope for abscisic acid. *Zeitschrift für Pflanzenphysiologie*, 97: 67–78.

Henson, I. E. 1981. Changes in abscisic acid content during stomatal closure in pearl millet (*Pennisetum americanum* [L.] Leeke). *Plant Science Letters*, 21: 121–27.

―――. 1985. Dependence of abscisic acid accumulation in leaves of pearl millet (*Pennisetum americanum* [L.] Leeke) on rate of development of water stress. *Journal of Experimental Botany*, 36: 1232–39.

Hiron, R. W. P., and S. T. C. Wright. 1973. The role of endogenous abscisic acid in the response of plants to stress. *Journal of Experimental Botany*, 24: 769–81.

Hornberg, C., and E. W. Weiler. 1984. High-affinity binding sites for abscisic acid on the plasmalemma of *Vicia faba* guard cells. *Nature*, 310: 321–24.

Itai, C., and H. Meidner. 1978. Functional epidermal cells are necessary for abscisic acid effects on guard cells. *Journal of Experimental Botany*, 29: 765–70.

Jewer, P. C., L. D. Incoll, and G. L. Howarth. 1981. Stomatal responses in isolated epidermis of the crassulacean acid metabolism plant *Kalanchoe daigremontiana* Hamet et Perr. *Planta*, 153: 238–45.

Jewer, P. C., L. D. Incoll, and J. Shaw. 1982. Stomatal responses of *Argenteum*—A mutant of *Pisum sativum* L. with readily detachable leaf epidermis. *Planta*, 155: 146–53.

Jones, R. J., and T. A. Mansfield. 1970. Suppression of stomatal opening in leaves treated with abscisic acid. *Journal of Experimental Botany*, 21: 714–19.

Komor, E., H. Weber, and W. Tanner. 1979. Greatly decreased susceptibility of nonmetabolizing cells towards detergents. *Proceedings of National Academy of Sciences*, 76: 1814–18.

Kriedemann, P. E., B. R. Loveys, G. L. Fuller, and A. C. Leopold. 1972. Abscisic acid and stomatal regulation. *Plant Physiology*, 49: 842–47.

Little, C. H. A., and D. C. Eidt. 1968. Effect of abscisic acid on budbreak and transpiration in woody species. *Nature*, 220: 498–99.

Loveys, B. R. 1977. The intracellular location of abscisic acid in stressed and non-stressed leaf tissue. *Physiologia plantarum*, 40: 6–10.

Loveys, B. R., and B. V. Milborrow. 1984. Metabolism of abscisic acid. In A. Crozier and J. R. Hillman, eds., *The biosynthesis and metabolism of plant hormones*: 71–104. Society for Experimental Biology, Seminar Series, 23. Cambridge: Cambridge University Press.

Ludlow, M. M., T. T. Ng, and C. W. Ford. 1980. Recovery after water stress of leaf gas exchange in *Panicum maximum* var. *trichoglume*. *Australian Journal of Plant Physiology*, 7: 299–313.

Lüttge, U., J. A. C. Smith, and G. Marigo. 1982. Membrane transport, osmoregulation, and the control of CAM. In I. P. Ting and M. Gibbs, eds., *Crassulacean acid metabolism*: 69–91. Rockville, Md.: American Society of Plant Physiologists.

Mansfield, T. A., and R. J. Jones. 1971. Effects of abscisic acid on potassium uptake and starch content of stomatal guard cells. *Planta*, 101: 147–58.

Mansfield, T. A., and J. A. Wilson. 1981. Regulation of gas exchange in water stressed plants. In C. B. Johnson, ed., *Physiological processes limiting plant productivity*: 237–52. London: Butterworths.

Marrè, E. 1979. Fusicoccin: A tool in plant physiology. *Annual Review of Plant Physiology*, 30: 273–88.

Mertens, R., M. Stüning, and E. W. Weiler. 1982. Metabolism of tritiated enantiomers of abscisic acid prepared by immunoaffinity chromatography. *Naturwissenschaften*, 69: 595.

Milborrow, B. V. 1974. Biosynthesis of abscisic acid by a cell-free system. *Phytochemistry*, 13: 131–36.

———. 1978. The stability of conjugated abscisic acid during wilting. *Journal of Experimental Botany*, 29: 1059–66.

———. 1983. Pathways to and from abscisic acid. In F. T. Addicott, ed., *Abscisic acid*: 79–111. New York: Praeger.

Mittelheuser, C. J., and R. F. M. van Steveninck. 1969. Stomatal closure and inhibition of transpiration induced by (RS)-abscisic acid. *Nature*, 221: 281–82.

Neill, S. J., R. Horgan, and J. K. Heald. 1983. Determination of the levels of abscisic acid–glucose ester in plants. *Planta*, 157: 371–75.

Neill, S. J., R. Horgan, and D. C. Walton. 1984. Biosynthesis of abscisic acid. In A. Crozier and J. R. Hillman, eds., *The biosynthesis and metabolism of plant hormones*: 43–70. Society for Experimental Biology, Seminar Series, 23. Cambridge: Cambridge University Press.

Ogawa, T., D. Grantz, J. Boyer, and Govindjee. 1982. Effects of cations and abscisic acid on chlorophyll *a* fluorescence in guard cells of *Vicia faba*. *Plant Physiology*, 69: 1140–44.

O'Neill, S. D., and R. M. Spanswick. 1984. Effects of vanadate on the plasma membrane ATPase of red beet and corn. *Plant Physiology*, 75: 586–91.

Penny, M. G., and D. J. F. Bowling. 1975. Direct determination of pH in the stomatal complex of *Commelina*. *Planta*, 122: 209–12.

Pierce, M. L. 1981. Turgor dependence of biosynthesis and metabolism of abscisic acid. Ph.D. dissertation, Michigan State University, East Lansing.

Pierce, M. L., and K. Raschke. 1980. Correlation between loss of turgor and accumulation of abscisic acid in detached leaves. *Planta*, 148: 174–82.

———. 1981. Synthesis and metabolism of abscisic acid in detached leaves of *Phaseolus vulgaris* L. after loss and recovery of turgor. *Planta*, 153: 156–65.

Quarrie, S. A. 1982. Droopy: A wilty mutant of potato deficient in abscisic acid. *Plant, Cell and Environment*, 5: 23–26.

Raschke, K. 1975. Simultaneous requirement of carbon dioxide and abscisic acid for stomatal closing in *Xanthium strumarium* L. *Planta*, 125: 243–59.

———. 1977. The stomatal turgor mechanism and its responses to CO_2 and abscisic acid: Observations and a hypothesis. In E. Marrè and O. Ciferri, eds., *Regulation of cell membrane activities in plants*: 173–83. Amsterdam: Elsevier.

———. 1979. Movements of stomata. In W. Haupt and M. E. Feinleib, eds., *Physiology of movements*, Encyclopedia of Plant Physiology, n.s., 7: 381–441. Berlin: Springer-Verlag.

———. 1982. Involvement of abscisic acid in the regulation of gas exchange: Evidence

and inconsistencies. In P. F. Wareing, ed., *Plant growth substances, 1982*: 581–90. London: Academic.
Raschke, K., R. D. Firn, and M. L. Pierce. 1975. Stomatal closure in response to xanthoxin and abscisic acid. *Planta*, 125: 149–60.
Raschke, K., and R. Hedrich. 1985. Simultaneous and independent effects of abscisic acid on stomata and the photosynthetic apparatus in whole leaves. *Planta*, 163: 105–18.
Raschke, K., and G. D. Humble. 1973. No uptake of anions required by opening stomata of *Vicia faba*: Guard cells release hydrogen ions. *Planta*, 115: 47–57.
Raschke, K., M. Pierce, and C. C. Popiela. 1976. Abscisic acid content and stomatal sensitivity to CO_2 in leaves of *Xanthium strumarium* L. after pretreatments in warm and cold growth chambers. *Plant Physiology*, 57: 115–21.
Saftner, R. A., and K. Raschke. 1981. Electrical potentials in stomatal complexes. *Plant Physiology*, 67: 1124–32.
Saftner, R. A., and R. E. Wyse. 1984. Effect of plant hormones on sucrose uptake by sugarbeet root tissue discs. *Plant Physiology*, 74: 951–55.
Sanders, D. 1984. Gradient-coupled chloride transport in plant cells. In G. A. Gerencser, ed., *Chloride transport coupling in biological membranes and epithelia*: 63–120. Amsterdam: Elsevier.
Schnabl, H. 1978. The effect of Cl^- upon the sensitivity of starch-containing and starch-deficient stomata and guard cell protoplasts towards potassium ions, fusicoccin and abscisic acid. *Planta*, 144: 95–100.
Schnabl, H., C. H. Bornman, and H. Ziegler. 1978. Studies on isolated starch-containing (*Vicia faba*) and starch-deficient (*Allium cepa*) guard cell protoplasts. *Planta*, 143: 33–39.
Schnabl, H., and K. Raschke. 1980. Potassium chloride as stomatal osmoticum in *Allium cepa* L., a species devoid of starch in guard cells. *Plant Physiology*, 65: 88–93.
Schnabl, H., and H. Ziegler. 1977. The mechanism of stomatal movement in *Allium cepa* L. *Planta*, 136: 37–43.
Schroeder, J. I., R. Hedrich, and J. M. Fernandez. 1984. Potassium-selective single channels in guard cell protoplasts of *Vicia faba*. *Nature*, 312: 361–62.
Sharkey, T. D., and K. Raschke. 1980. Effects of phaseic acid and dihydrophaseic acid on stomata and the photosynthetic apparatus. *Plant Physiology*, 65: 291–97.
Snaith, P. J., and T. A. Mansfield. 1982. Control of the CO_2 responses of stomata by indol-3-ylacetic acid and abscisic acid. *Journal of Experimental Botany*, 33: 360–65.
Stålfelt, M. G. 1929. Die Abhängigkeit der Spaltöffnungsreaktionen von der Wasserbilanz. *Planta*, 8: 287–340.
Tal, M. 1966. Abnormal stomatal behaviour in wilty mutants of tomato. *Plant Physiology*, 41: 1387–91.
Tal, M., and D. Imber. 1972. The effect of abscisic acid on stomatal behaviour in *flacca*, a wilty mutant of tomato, in darkness. *New Phytologist*, 71: 21–84.
Tal, M., D. Imber, and I. Gardi. 1974. Abnormal stomatal behaviour and hormonal imbalance in *flacca*, a wilty mutant of tomato. Effect of abscisic acid and auxin on stomatal behaviour and peroxidase activity. *Journal of Experimental Botany*, 25: 51–60.
Taylor, H. F. 1968. Carotenoids as possible precursors of abscisic acid in plants. In *Plant growth regulators*: 22–33. Society of Chemical Industry Monographs, 31. London.
Taylor, H. F., and R. S. Burden. 1970. Identification of plant growth inhibitors produced by photolysis of violaxanthin. *Phytochemistry*, 9: 2217–23.

_____. 1973. Preparation and metabolism of 2-^{14}C-*cis,trans*-xanthoxin. *Journal of Experimental Botany*, 24: 873–80.
Tucker, D. J., and T. A. Mansfield. 1971. A simple bioassay for detecting "antitranspirant" activity of naturally occurring compounds such as abscisic acid. *Planta*, 98: 157–63.
Turner, N. C., and A. Graniti. 1969. Fusicoccin: A fungal toxin that opens stomata. *Nature*, 223: 1070–71.
Van Kirk, C. A., and K. Raschke. 1978. Release of malate from epidermal strips during stomatal closure. *Plant Physiology*, 61: 474–75.
Vaughan, G. T., and B. V. Milborrow. 1984. The resolution by HPLC of RS-2-^{14}C-Me-1′,4′-cis-diol of abscisic acid and the metabolism of ($-$)-R- and ($+$)-S-abscisic acid. *Journal of Experimental Botany*, 35: 110–20.
Wardle, K., and K. Short. 1981. Responses of stomata in epidermal strips of *Vicia faba* to carbon dioxide and growth hormones when incubated on potassium chloride and potassium iminodiacetate. *Journal of Experimental Botany*, 32: 303–9.
Weiler, E. W., H. Schnabl, and C. Hornberg. 1982. Stress-related levels of abscisic acid in guard cell protoplasts of *Vicia faba* L. *Planta*, 154: 24–28.
Weyers, J. D. B., and J. R. Hillman. 1979a. Uptake and distribution of abscisic acid in *Commelina* leaf epidermis. *Planta*, 144: 167–72.
_____. 1979b. Sensitivity of *Commelina* stomata to abscisic acid. *Planta*, 146: 623–28.
Weyers, J. D. B., N. W. Paterson, P. J. Fitzsimons, and J. M. Dudley. 1982. Metabolic inhibitors block ABA-induced stomatal closure. *Journal of Experimental Botany*, 33: 1270–78.
Willmer, C. M., J. E. Pallas, Jr., and C. C. Black, Jr. 1973. Carbon dioxide metabolism in leaf epidermal tissue. *Plant Physiology*, 52: 448–52.
Wilson, J. A., A. B. Ogunkanmi, and T. A. Mansfield. 1978. Effects of external potassium supply on stomatal closure induced by abscisic acid. *Plant, Cell and Environment*, 1: 199–201.
Wright, S. T. C., and R. W. P. Hiron. 1969. ($+$)-Abscisic acid, the growth inhibitor induced in detached wheat leaves by a period of wilting. *Nature*, 224: 719–20.
Zeevaart, J. A. D. 1980. Changes in the levels of abscisic acid and its metabolites in excised leaf blades of *Xanthium strumarium* during and after water stress. *Plant Physiology*, 66: 672–78.
_____. 1983. Metabolism of abscisic acid and its regulation in *Xanthium* leaves during and after water stress. *Plant Physiology*, 71: 477–81.
Zeevaart, J. A. D., and B. V. Milborrow. 1976. Metabolism of abscisic acid and the occurrence of *epi*-dihydrophaseic acid in *Phaseolus vulgaris*. *Phytochemistry*, 15: 493–500.

12

Cytokinins and Stomata

L. D. Incoll and P. C. Jewer

Cytokinins are plant growth regulators that have been detected in a diversity of plant material, including roots, leaves, fruits, seeds, phloem sap, and xylem sap (see table 1 of Van Staden and Davey, 1979). The concentration of endogenous cytokinins in a particular tissue, however, does not necessarily correlate with the capacity of that tissue to synthesize these regulators. Circumstantial evidence, largely from bioassays, suggests that the leaves of vegetative plants probably depend on cytokinins that have been synthesized in roots (Van Staden and Davey, 1979; Sembdner et al., 1980). Bioassays have also shown cytokinin-like activity in xylem exudates of many species (Van Staden and Davey, 1979). Because stomata provide the penultimate resistance to the movement of water through the plant to the atmosphere, the transpiration stream may represent the major supply of cytokinins to stomata. Accordingly, those conditions that favor growth or maintenance of healthy roots, such as good nutrition, and aeration, optimum temperature, and high water potential in the soil, may favor synthesis of cytokinin and thereby raise its concentration in the apoplastic solution bathing the plasmalemma of the guard cell. Conversely, water stress, osmotic stress, anaerobiosis, nutrient deficiency, and chilling may reduce the titer of cytokinin. The concentration of endogenous cytokinins in leaves, as measured by bioassay, does fluctuate with environmental conditions around roots, and also with the stage of development of the plant, and with season (see refs. in Van Staden and Davey, 1979). Thus cytokinin may be a signal that enables the plant to adjust the exchange of gases between leaves and the atmosphere in response to perturbations in the edaphic environment. An extensive literature has concentrated on the role of abscisic acid (ABA) in the alleviation of stress in plants. Mizrahi (1980) and Bradford (1982) have suggested, however, that the plant's response to stress may be better understood as the result of opposing effects of cytokinin and ABA.

There have been many experiments on stressed and unstressed plants in which it could be inferred that the application of cytokinins or the imposition of treatments that alter the levels of endogenous cytokinins had influenced the

The authors wish to thank K. J. Bradford and W. J. Davies for preprints of their papers, and J. M. Cheeseman both for his advice and for allowing us to use his electrophysiological equipment.

TABLE 12.1. Effect of cytokinins on transpiration of excised leaves in normal air and in the light

Reference	Species	Cytokinin[a]
Respond by increased transpiration		
Livnè & Vaadia, 1965	Hordeum vulgare	Kn
Mittelheuser & van Steveninck, 1969	Triticum aestivum	Kn
Cooper, Digby & Cooper, 1972	H. vulgare	Kn
Kuraishi, 1976	Brassica campestris	Kn
Incoll & Whitelam, 1977	H. vulgare	Kn
Incoll & Whitelam, 1977	Anthephora pubescens	Kn
Kuraishi & Ishikawa, 1977	B. campestris	BA
Biddington & Thomas, 1978	Avena sativa	Z, BA
Bengtson, Falk & Larsson, 1979	T. aestivum	Kn
Respond by increased aperture or decreased stomatal resistance		
Livnè & Vaadia, 1965	H. vulgare	Kn
Meidner, 1967	H. vulgare	Kn
Kuraishi, Hashimoto & Shiraishi, 1981	Helianthus annuus	Z

NOTE: Data arranged chronologically. All experiments were completed within 8 h of start of treatment.
[a]BA, N^6-benzyladenine; Kn, kinetin; Z, zeatin.

response of leaves to stress via effects on stomatal physiology. These experiments, however, often extended over periods of days and undoubtedly gave rise to complex responses that should not be attributed simply to a direct action of cytokinins on stomata. For example, if senescence in the mesophyll has been retarded or prevented by cytokinins, then consequent differences in photosynthetic rate between control and treated leaves may cause different internal CO_2 concentrations and thereby different stomatal apertures.

We will therefore confine our discussion to experiments lasting eight hours or less, many of them designed specifically to examine the role of cytokinins in stomatal movement. We have assumed that eight hours is more than enough time for exogenous cytokinins to be transported to stomata, to reach an effective concentration at the guard cell, and to allow changes in transpiration or aperture to be measured. In fact many of the experiments to which we will refer were completed within three or four hours. Consequently we have excluded from discussion many of the papers cited in the reviews of Livnè and Vaadia (1972) and Vaadia (1976) and some of the work of Kuraishi and co-workers in Japan and of Thimann and co-workers in California.

Effects of Exogenous Cytokinins on Excised Leaves, Epidermal Strips, and Guard Cell Protoplasts

Effects in the presence of the opening stimuli: light and CO_2-free air. The first report that a cytokinin could increase the rate of transpiration from leaves and stomatal aperture (Table 12.1; Livnè and Vaadia, 1965) followed soon after the discovery that xylem exudate contained compounds with kinetin-like activity (Kende, 1965). Many of the subsequent reports that kinetin (Kn) causes excised leaves to transpire rapidly in light and normal air have been for species of grass

(Table 12.1). The effect, however, is not confined to grasses; the dicotyledonous species *Brassica campestris* and *Helianthus annuus* respond in the same way (Table 12.1).

Direct effects on stomata are best tested on isolated epidermal strips or on guard cell protoplasts, because indirect effects via the mesophyll are thus eliminated. It was the discovery of an easily stripped grass, *Anthephora pubescens*, a C_4 species, that provided the first evidence that stimulation of transpiration by Kn was the result of a direct effect of Kn on stomata (Incoll and Whitelam, 1977). Subsequent work in our laboratory showed that stomata of this species responded to a number of natural and synthetic cytokinins (Table 12.2) mostly with an optimum concentration of 10 mmol m^{-3} (10 μM) but of 0.1 mmol m^{-3} for N^6-[Δ^2-isopentenyl] adenine (I^6Ade) and N^6-[Δ^2-isopentenyl] adenosine (I^6Ado). Furthermore the stomata of the dicotyledonous CAM plant *Kalanchoe daigremontiana* also opened in response to Kn and zeatin (Z), again at 10 mmol m^{-3} (Jewer and Incoll, 1981). These reports of enhanced opening stand alone with that of Das, Rao, and Raghavendra (1976) for stomata of *Commelina benghalensis* and *Tridax procumbens* (Table 12.2). Further inspection of Table 12.2 shows that some species respond to all cytokinins, some respond to none, and two species respond to N^6-benzyladenine (BA) but not to Kn. Within one genus, *Commelina*, we find reports of no response, of response to BA but not to Kn, and, surprisingly, of closure with increasing concentration of Kn between 1 and 100 mmol m^{-3} (Blackman and Davies, 1983).

TABLE 12.2. Effect of cytokinins on stomatal aperture in isolated epidermis in CO_2-free air and in the light

Reference	Species	Cytokinin[a] and Response[b]
Horton, 1971	*Vicia faba*	Kn, 0; BA, 0
Tucker & Mansfield, 1971	*Commelina communis*	Kn, 0
Ogunkanmi, Tucker & Mansfield, 1973	*V. faba*	Kn, 0
Das, Rao & Raghavendra, 1976	*Commelina benghalensis*	Kn, 0; BA, +
Das, Rao & Raghavendra, 1976	*Tridax procumbens*	Kn, 0; BA, +
Incoll & Whitelam, 1977	*Anthephora pubescens*	Kn, +
Cooper & Cockburn, 1979	*V. faba*	BA, 0
Jewer & Incoll, 1980	*A. pubescens*	Kn, +; BA, +; Z, +[c]
Jewer & Incoll, 1981	*Kalanchoe daigremontiana*	Kn, +; Z, +
Wardle & Short, 1981	*V. faba*	Kn, 0
Jewer, Incoll & Shaw, 1982	*Pisum sativum*	Kn, 0
Blackman & Davies, 1983	*C. communis*	Kn, −
Blackman & Davies, 1983	*Zea mays*[d]	Kn, 0; Z, 0

NOTE: Data arranged chronologically.

[a]Abbreviations as in Table 12.1.

[b]Response: +, aperture > control; 0, no difference; −, aperture < control. All experiments were completed within 5 h of start of treatment.

[c]Also kinetin riboside (KR), N^6-benzyladenosine (BO), zeatin riboside (ZR), N^6-[Δ^2-isopentenyl] adenine (I^6Ade), N^6-[Δ^2-isopentenyl] adenosine (I^6Ado), dihydrozeatin (DHZ), dihydrozeatin riboside (DHZR): all +.

[d]Epidermis stripped from pieces of leaf after incubation.

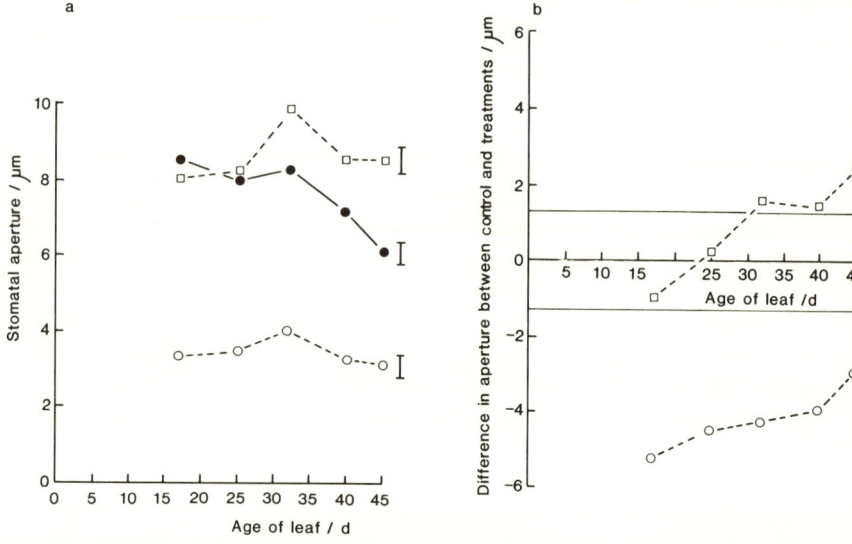

Fig. 12.1. The effect of age of plant on the response of stomata in isolated adaxial epidermis from leaf 4 of 18-d-old plants of the *Arg* mutant of *Pisum sativum* to exogenous ABA and Kn. Leaf 4 was fully expanded on 18-d-old plants; for other details of technique see Jewer, Incoll, and Shaw (1982). *a*, Apertures measured after 4 h at 25° C in the light and CO_2-free air on 10 mol m^{-3} MES buffer + 25 mol m^{-3} KCl (control, ●), plus 1 mmol m^{-3} ABA (○), and plus 10 mmol m^{-3} Kn (□). *b*, Differences between stomatal apertures of the control and either ABA (○) or Kn (□) after 4 h. Each value is the mean of 300 measurements. Vertical bars and horizontal lines represent least significant differences at $P = 0.05$.

It is not always clear whether the stomata that did not respond could have opened further, that is, whether they were not fully turgid. If the epidermis already contains optimal levels of endogenous cytokinins, then stomata in light and CO_2-free air may be unable to respond to exogenously added cytokinins. In our opinion the age of material in many of the experiments in Table 12.2 is not adequately defined. Variation in the physiology of leaves with differing ontogeny is well known, so that guard cells in "youngest fully expanded" leaves could experience different levels of endogenous growth regulators, depending on their position on the plant axis. We have found that although stomata in epidermis from just fully expanded leaves of *Pisum sativum* did not respond to Kn (Table 12.2), they did respond increasingly as the leaf aged (Fig. 12.1), the greatest response being in visibly senesced 45-day-old leaves. Therefore the stage of growth of the leaf within its own life cycle needs to be exactly defined.

The validity of some of the experimental methods must be questioned. All experiments before 1977 used non-zwitterionic buffers, the disadvantages of which have been described by Good and Izawa (1972). Ogunkanmi, Tucker, and Mansfield (1973) also used one zwitterionic buffer, but with a high concentration of Na^+ (100 mol m^{-3} or 100 mM) and no K^+, as in the method of Tucker and Mansfield (1971). Jarvis and Mansfield (1980) have subsequently shown that high concentrations of Na^+ (50 and 100 mol m^{-3}) reduce the sensitivity of stomata to various stimuli, including the plant growth regulator ABA; response to Kn may have been similarly affected. The other problem with epidermal strips is choosing the correct potassium concentration in the medium for incubation. Different K^+ concentrations can drastically alter equilibrium apertures (Zeiger, 1983). For this reason experiments with *A. pubescens* are more credible, because this species does not require an external source of K^+. *A. pubescens*, however, in common with other grasses, has disadvantages: its maximum apertures are small (< 6 µm), and its guard cell protoplasts cannot be isolated.

Despite the amount of work being done on guard cell protoplasts, cytokinins have been tested on the protoplasts of only one species, *P. sativum* (Jewer, Incoll, and Shaw, 1982). This experiment confirmed the result for the epidermal strips of the same age; that is, the guard cell protoplasts did not increase in volume when incubated in 1 mmol m^{-3} Kn.

Although all these experiments last for at least three hours, the stomatal response to cytokinins is not slow. Kuraishi, Hashimoto, and Shiraishi (1981) have measured significantly larger pore areas two minutes after spraying sunflower leaves with aqueous Z solution (50 mmol m^{-3}). We have measured the membrane potential difference (MPD) of the subsidiary cell in isolated epidermis of *A. pubescens*. The MPD increased within 1.5 minutes of the addition of Kn to the bathing medium (Fig. 12.2); the rapidity of this response makes it unlikely that the electrical changes were Nernstian potentials resulting from diffusion. Our data show, however, that subsidiary cells hyperpolarized on addition of Kn to the medium (Fig. 12.2); a mechanism actively pumping K^+ out should depolarize subsidiary cells. The electrical potential difference of guard cells and subsidiary cells have been compared in open and closed stomata, but the results are inconsistent. Gunar, Zlotnikova, and Panichkin (1975) showed that changes in potential difference in guard cells during alternation of light and darkness were of opposite sign to those in cells of the epidermis of *Tradescantia albiflora*. The potential differences of guard cells and subsidiary cells, however, remained similar in open and closed stomata of *T. nova* and *C. communis* (Lyalin et al., 1979; Saftner and Raschke, 1981). These measurements were made on nongraminaceous species, in which K^+ for stomatal movement is derived from all cells of the epidermis. Differences between our data and previous measurements might be expected, because in graminaceous species during opening and closing, K^+ is shuttled only between subsidiary cells and guard cells (Raschke and Fellows, 1971).

Effects in the presence of the closing stimuli: darkness, abscisic acid, and CO_2. Although all the experiments listed in Tables 12.1 and 12.2 were performed

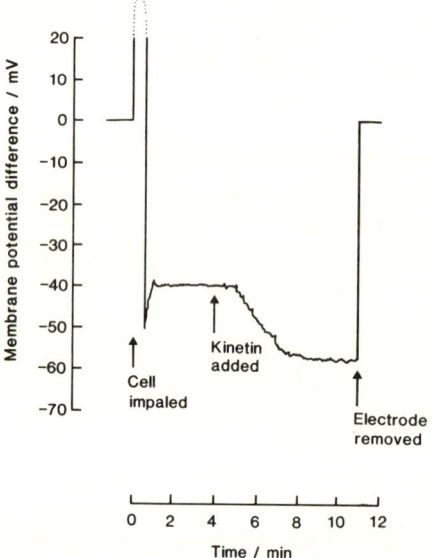

Fig. 12.2. The effect of kinetin on the membrane potential difference of a subsidiary cell of an epidermal strip of *Anthephora pubescens*. This response, typical of 10 replicate strips, was detected with a micropipette electrode. The strip was bathed in 10 mol m^{-3} MOPS buffer (pH 7.2) to which 10 mmol m^{-3} Kn was added after the potential difference had stabilized. Aperture in MOPS alone was 2.31 μm; in MOPS + Kn it was 3.45 μm (n = 10). For details of growth of plants and preparation and incubation of strips, see Jewer and Incoll (1980).

in the light, light is not necessary for response to cytokinins. Jewer and Incoll (1981) found that the response of *K. daigremontiana* stomata to Kn and Z did not require light. We have found also that *A. pubescens* stomata respond to Kn in the dark (unpublished). However, *Vicia faba* stomata preincubated with BA in the dark did not open (Cooper and Cockburn, 1979).

When we come to examine the reported effects of cytokinins in the presence of ABA the only generalization we can make is that ABA alone, that is, the control treatment, always reduces apertures. In experiments with transpiring wheat leaves, the effect of 1.9 mmol m^{-3} ABA in reducing transpiration was partly countered by 25 mmol m^{-3} Kn (Mittelheuser and van Steveninck, 1969). Cooper, Digby, and Cooper (1972) reported a statistical interaction between the effects of Kn alone and Kn in the presence of ABA. Their data were not analyzed to test the effect of various concentrations of Kn in the presence of one concentration of ABA, that is, the comparison of ABA alone versus ABA plus Kn. An important test, it seems to us, is whether the addition of Kn to treatments with ABA increases transpiration. Factorial experimental designs are not ideal, because they allow statistical tests of the interactions of factors without the interactions being biologically meaningful. In the absence of the appropriate tests, it appears that increasing Kn from 0.1 to 10 mmol m^{-3} did not reduce the closing effect of ABA at any of the three concentrations of ABA (0.1, 1, and 10 mmol m^{-3}) that were tested by Cooper, Digby, and Cooper (1972).

With epidermal strips, interactions of ABA with Kn and BA were not detected in systems based on non-zwitterionic buffers or on high concentrations of Na^+ (Horton, 1971; Tucker and Mansfield, 1971; Ogunkanmi, Tucker, and Mansfield, 1973), except for the report that 50 mmol m^{-3} BA partly overcame the closing effect of ABA (Das, Rao, and Raghavendra, 1976; concentration of ABA not specified). More recently Blackman and Davies (1983) have used the same basic design as Cooper, Digby, and Cooper (1972) for experiments on pieces of maize leaf and strips and pieces from *Commelina* leaves. They also found statistically significant interactions between treatments with cytokinins and with cytokinins plus ABA. In the presence of a single high concentration of ABA (100 mmol m^{-3}), cytokinins (Kn and Z) at concentrations of up to 100 mmol m^{-3} did not overcome the ABA-induced closure of *Commelina* stomata. With increasing concentration, however, both cytokinins progressively reversed the closure of maize stomata, which, as noted before, did not respond to these cytokinins in the absence of ABA.

A possible interpretation of the lack of response of *Commelina* is that saturating levels of ABA masked any effect of cytokinins. If cytokinins compete with ABA for sites in membranes associated with the transport of ions, then the reversal of the effect of ABA may be more apparent at lower concentrations of ABA. The lowest concentration of ABA tested so far in combination with cytokinins is 0.1 mmol m^{-3} (Cooper, Digby, and Cooper, 1972), whereas ABA affects apertures at concentrations as low as 10^{-6} mmol m^{-3} (Jewer, Incoll, and Howarth, 1981).

There are few reports of the effects of cytokinins on the CO_2-induced closure of stomata. Blackman and Davies (1984) have shown for pieces of maize leaf that Kn and Z increasingly overcome CO_2-induced closure as their concentrations are increased from 1 to 100 mmol m^{-3}. Closure induced by CO_2 and ABA together, however, could not be reversed by cytokinins. Wardle and Short (1981, 1983) have shown that Kn reverses the closing effect of CO_2 on stomata in epidermal strips from *V. faba* and from *Chrysanthemum* plantlets (Kn at concentrations of 100 and 200 mmol m^{-3}, respectively). For *V. faba*, final aperture in the presence of CO_2 and Kn (11.6 μm) was greater than that in the presence of Kn alone (10.5 μm), leading the authors to propose that CO_2 was required for maximal opening in the presence of Kn. There are no comparable data for the effect of cytokinins on CO_2-induced closure for species with stomata that open with cytokinins alone; neither *Zea mays* nor *V. faba* show a response to cytokinins alone, and Wardle and Short (1983) did not report on the response of *Chrysanthemum* to Kn alone.

Conclusions. The responses of stomata to endogenous cytokinins in the presence of other opening and closing stimuli have now been reported in species from two monocotyledonous families (Gramineae and Commelinaceae) and four dicotyledonous families (Crassulaceae, Compositae, Cruciferae, and

Leguminosae). From these limited published results we conclude that cytokinins can act directly on stomata to increase their aperture and that this effect is rapid and independent of light. We predict that stomata of all species will respond to cytokinins, but that the responses in some species may be masked by the presence of optimal concentrations of endogenous cytokinins. In others the response to cytokinins may be apparent only on interaction with closing stimuli, such as CO_2 and ABA. Further, the response may be affected by aging and stress, both of which may alter the balance of endogenous cytokinins and ABA. The extent of these interactions may be resolved when levels of endogenous cytokinins and ABA in the epidermis are assayed with modern analytical methods. Clearly, data for more species are needed from experiments in which experimental protocols are carefully chosen and the physiological state of the leaves or the stomata are well defined.

Mode of Action of Cytokinins on Stomata: A Hypothesis

There are no published hypotheses for the mechanism by which cytokinins affect stomatal movements. We therefore present a hypothesis that we hope will stimulate critical discussion of how cytokinins act on stomata. Our hypothesis, for which evidence is presented below, is highly speculative and assumes that cytokinins have a primary action on the fluxes of ions through guard cell membranes. The hypothesis states that cytokinins initially induce the release of Ca^{2+} from guard cells. The released Ca^{2+} is subsequently bound by the calcium-dependent regulator protein calmodulin, forming a Ca-calmodulin complex. The complex binds to and stimulates a membrane ATPase that consequently induces ion flux and stomatal movement.

Recently Zeiger (1983) has emphasized that the transport of ions in guard cells is a major locus of control of stomatal movement and that such dissimilar factors as temperature, CO_2, and ABA could all modulate ionic fluxes. This view is consistent with the unifying theory that growth regulators act on membranes and thereby alter cytoplasmic ion balance, which causes changes in the growth and development of the plant (Trewavas, 1976, 1979).

Many of the best-characterized effects of cytokinins are on membranes, for example, the regulation of synthesis of betacyanin in cotyledons of *Amaranthus tricolor* and of anthocyanin in seedlings of red cabbage. N^6-benzyladenine (BA) increases the extrusion of protons from *Amaranthus* cotyledons and protoplasts (Batchelor and Elliott, 1981) and from leaf pieces and cotyledons of other plants (Marrè, 1977). The accumulation of betacyanin is stimulated by K^+ in the presence of BA in the external medium (Elliott, 1979a), and it is increased synergistically by BA and fusicoccin (a phytotoxin that enhances H^+ extrusion and K^+ uptake) and by BA and red light (Elliott, 1979b). The conclusion that cytokinins affect ion fluxes, particularly of K^+, is also supported by the inhibition of the response to BA by uncouplers that destroy proton fluxes and by inhibitors of ATPases in membranes (Elliott, 1979c). Our measurements of MPDs in subsidiary cells of *Anthephora pubescens* in the absence and presence of Kn (Fig. 12.2) also support that conclusion. Our data remain the only evidence for an effect of cytokinins on stomata via changes in electrical charge.

It is likely that cytokinins exert at least part of their effect on fluxes of ions across plant membranes by modulating intracellular Ca^{2+} concentration. Cytokinins have been shown to enhance the release of bound Ca^{2+} from hyphae of the fungus *Achlya* (Le John and Cameron, 1973). The cytokinin analogue 1-methyl adenine (1-MeAde) controls meiosis in starfish oocytes by interacting with stereospecific receptors localized on the plasma membrane of the oocyte (Dorée and Guerrier, 1975); transduction of the hormonal message involves the release of Ca^{2+} in the cortex of the oocyte (Dorée and Kishimoto, 1981). Furthermore, although cytokinin stimulates mitosis in protonemata of the moss *Funaria hygrometrica*, the initial division can be induced in the absence of cytokinin by a Ca^{2+} ionophore in a medium containing Ca^{2+} (Saunders and Hepler, 1982). Thus regulation of mitosis by cytokinin may be due, at least in part, to the modulation of intracellular Ca^{2+} concentration. Clearly there are interrelationships between calcium and cytokinins that involve the functioning of membranes.

It is possible that Ca^{2+} acts as an intracellular secondary messenger in plants (Marmé, 1982). Much recent evidence indicates that it is impossible to consider a role for Ca^{2+} as a messenger without invoking the involvement of the calcium-dependent regulator protein calmodulin (Marmé, 1982). Calmodulin, which is itself inactive, binds to Ca^{2+} to form a biologically active Ca-calmodulin complex (Cheung, 1980). Calmodulin is an intracellular acceptor of Ca^{2+}, acting when the concentration of Ca^{2+} increases in response to a stimulus. The Ca-calmodulin binding purportedly induces a distinct change in the shape of the calmodulin molecule, and the resulting complex can bind to and regulate several enzymes (Marx, 1980), including ATPases (Cheung, 1980). Dorée and Kishimoto (1981) have suggested that a short-lived Ca^{2+} complex, possibly Ca-calmodulin, plays a key role in the reinitiation of meiosis in the oocytes of starfish. Although the total amount of calmodulin does not change in starfish oocytes following the addition of 1-MeAde (Dorée, 1980), a transient change in the amount complexed with Ca^{2+} might provide a signal for the stimulation of phosphorylation of protein.

Tests of this hypothesis are clearly required. The effects of Ca^{2+} on stomatal movements are not well defined, and it is not known whether Ca^{2+} affects the response of stomata to cytokinins. Moreover, a role for calmodulin in stomatal movement has not been investigated.

REFERENCES

Batchelor, S. M., and D. C. Elliott. 1981. The isolation of protoplasts of *Amaranthus tricolor* and their use in proton extrusion studies. *Proceedings of the Australian Biochemical Society*, 14: 117.

Bengtson, C., S. T. Falk, and S. Larsson. 1979. Effects of kinetin on transpiration rate and abscisic acid content of water-stressed young wheat plants. *Physiologia plantarum*, 45: 183–88.

Biddington, N. L., and T. H. Thomas. 1978. Influence of different cytokinins on the transpiration and senescence of excised oat leaves. *Physiologia plantarum*, 42: 369–74.

Blackman, P. G., and W. J. Davies. 1983. The effects of cytokinins and ABA on stomatal behaviour of maize and *Commelina*. *Journal of Experimental Botany*, 34: 1619–26.

———. 1984. Modification of the CO_2 responses of maize stomata by abscisic acid and by naturally-occurring and synthetic cytokinins. *Journal of Experimental Botany*, 35: 174–79.

Bradford, K. J. 1982. Regulation of shoot responses to root stress by ethylene, abscisic acid and cytokinin. In P. F. Wareing, ed., *Plant growth substances, 1982*: 599–608. London: Academic.

Cheung, W. Y., ed. 1980. *Calcium and Cell Function*. Vol. 1, *Calmodulin*. New York: Academic.

Cooper, M. J., J. Digby, and P. J. Cooper. 1972. Effects of plant hormones on the stomata of barley: A study of the interaction between abscisic acid and kinetin. *Planta*, 105: 43–49.

Cooper, S. D., and W. Cockburn. 1979. Osmotically induced water stress, potassium uptake, and stomatal aperture in epidermal strips of *Vicia faba* L. *Journal of Experimental Botany*, 30: 913–18.

Das, V. S. R., I. M. Rao, and A. S. Raghavendra. 1976. Reversal of abscisic acid–induced stomatal closure by benzyl adenine. *New Phytologist*, 76: 449–52.

Dorée, M. 1980. Calmodulin content does not change following hormone-induced meiosis reinitiation in starfish oocytes. *Experientia*, 36: 932–33.

Dorée, M., and P. Guerrier. 1975. Site of action of 1-methyladenine in inducing oocyte maturation in starfishes: Kinetic evidence for receptors localised on the cell membrane. *Experimental Cell Research*, 91: 296–300.

Dorée, M., and T. Kishimoto. 1981. Calcium-mediated transduction of the hormonal message in 1-methyladenine-induced meiosis reinitiation of starfish oocytes. In J. Guern and C. Peaud-Lenoel, eds., *Metabolism and molecular activities of cytokinins*: 338–48. Berlin: Springer-Verlag.

Elliott, D. C. 1979a. Ionic regulation for cytokinin-dependent betacyanin synthesis in *Amaranthus* seedlings. *Plant Physiology*, 63: 264–68.

———. 1979b. Temperature-sensitive responses of red light–dependent betacyanin synthesis. *Plant Physiology*, 64: 521–24.

———. 1979c. Differential effects on fusicoccin- and cytokinin-stimulated betacyanin synthesis by ATPase inhibitors and uncouplers. *Plant Science Letters*, 15: 251–64.

Good, N. E., and S. Izawa. 1972. Hydrogen ion buffers. In A. San Pietro, ed., *Methods in enzymology 25, Photosynthesis and nitrogen fixation*: 53–68. New York: Academic.

Gunar, I. I., I. F. Zlotnikova, and L. A. Panichkin. 1975. Electrophysiological

investigation of cells of the stomatal complex in spiderwort. *Soviet Plant Physiology*, 22: 704–7.
Horton, R. F. 1971. Stomatal opening: The role of abscisic acid. *Canadian Journal of Botany*, 49: 583–85.
Incoll, L. D., and G. C. Whitelam. 1977. The effect of kinetin on stomata of the grass *Anthephora pubescens* Nees. *Planta*, 137: 243–45.
Jarvis, R. G., and T. A. Mansfield. 1980. Reduced stomatal responses to light, carbon dioxide and abscisic acid in the presence of sodium ions. *Plant, Cell and Environment*, 3: 278–83.
Jewer, P. C., and L. D. Incoll. 1980. Promotion of stomatal opening in the grass *Anthephora pubescens* Nees by a range of natural and synthetic cytokinins. *Planta*, 150: 218–21.
———. 1981. Promotion of stomatal opening in detached epidermis of *Kalanchoe daigremontiana* Hamet et Perr. by natural and synthetic cytokinins. *Planta*, 153: 317–18.
Jewer, P. C., L. D. Incoll, and G. L. Howarth. 1981. Stomatal responses in isolated epidermis of the crassulacean acid metabolism plant *Kalanchoe daigremontiana* Hamet et Perr. *Planta*, 153: 238–45.
Jewer, P. C., L. D. Incoll, and J. Shaw. 1982. Stomatal responses of *Argenteum*—A mutant of *Pisum sativum* L. with readily detachable epidermis. *Planta*, 155: 146–53.
Kende, H. 1965. Kinetinlike factors in the root exudate of sunflowers. *Proceedings of the National Academy of Sciences*, 53: 1302–7.
Kuraishi, S. 1976. Ineffectiveness of cytokinin-induced chorophyll retention in hypostomatous leaf discs. *Plant and Cell Physiology*, 17: 875–85.
Kuraishi, S., Y. Hashimoto, and M. Shiraishi. 1981. Latent periods of cytokinin-induced stomatal opening in the sunflower leaf. *Plant and Cell Physiology*, 22: 911–16.
Kuraishi, S., and F. Ishikawa. 1977. Relationship between transpiration and amino acid accumulation in *Brassica* leaf discs treated with cytokinins and fusicoccin. *Plant and Cell Physiology*, 18: 1273–79.
Le John, H. B., and L. E. Cameron. 1973. Cytokinins regulate calcium binding to a glycoprotein from fungal cells. *Biochemical and Biophysical Research Communications*, 54: 1053–60.
Livnè, A., and Y. Vaadia. 1965. Stimulation of transpiration rate in barley leaves by kinetin and gibberellic acid. *Physiologia plantarum*, 18: 658–64.
———. 1972. Water deficits and hormone relations. In T. T. Kozlowski, ed., *Water deficits and plant growth* 3, *Plant responses and control of water balance*: 255–75. New York: Academic.
Lyalin, O. O., I. N. Ktitorova, A. A. Kazaryan, and I. S. Akhmedov. 1979. Electrical contacts between guard cells and accessory cells. *Soviet Plant Physiology*, 26: 548–51.
Marmé, D. 1982. The role of Ca^{2+} in signal transduction of higher plants. In P. F. Wareing, ed., *Plant growth substances, 1982*: 419–26. London: Academic.
Marrè, E. 1977. Physiologic implications of the hormonal control of ion transport in plants. In P. E. Pilet, ed., *Plant growth regulation*: 54–66. Berlin: Springer-Verlag.
Marx, J. L. 1980. Calmodulin: A protein for all seasons. *Science*, 208: 274–76.
Meidner, H. 1967. The effect of kinetin on stomatal opening and rate of intake of carbon dioxide in mature primary leaves of barley. *Journal of Experimental Botany*, 18: 556–61.
Mittelheuser, C. J., and R. F. M. van Steveninck. 1969. Stomatal closure and inhibition of transpiration induced by (RS)-abscisic acid. *Nature*, 221: 281–82.

Mizrahi, Y. 1980. The role of plant hormones in plant adaptation to stress conditions. In *Physiological aspects of crop productivity: 15th colloquium of the International Potash Institute*: 75–86. Worblaufen-Bern: International Potash Institute.

Ogunkanmi, A. B., D. J. Tucker, and T. A. Mansfield. 1973. An improved bioassay for abscisic acid and other antitranspirants. *New Phytologist*, 72: 277–82.

Raschke, K., and M. P. Fellows. 1971. Stomatal movement in *Zea mays*: Shuttle of potassium and chloride between guard cells and subsidiary cells. *Planta*, 101: 296–316.

Saftner, R. A., and K. Raschke. 1981. Electrical potentials in stomatal complexes. *Plant Physiology*, 67: 1124–32.

Saunders, M. J., and P. K. Hepler. 1982. Calcium ionophore A23187 stimulates cytokinin-like mitosis in *Funaria*. *Science*, 217: 943–45.

Sembdner, G., D. Gross, H.-W. Liebisch, and G. Schneider. 1980. Biosynthesis and metabolism of plant hormones. In J. MacMillan, ed., *Hormonal regulation of development* 1: 281–444. Berlin: Springer-Verlag.

Trewavas, A. 1976. Plant growth substances. In J. A. Bryant, ed., *Molecular aspects of gene expression in plants*: 249–98. London: Academic.

―――. 1979. What is the molecular basis of plant hormone action? *Trends in Biochemical Science*, 4: N199–N202.

Tucker, D. J., and T. A. Mansfield. 1971. A simple bioassay for detecting "antitranspirant" activity of naturally occurring compounds such as abscisic acid. *Planta*, 98: 157–63.

Vaadia, Y. 1976. Plant hormones and water stress. *Philosophical Transactions of the Royal Society*, ser. B, 273: 513–22.

Van Staden, J., and J. E. Davey. 1979. The synthesis, transport and metabolism of endogenous cytokinins. *Plant, Cell and Environment*, 2: 93–106.

Wardle, K., and K. C. Short. 1981. Responses of stomata in epidermal strips of *Vicia faba* to carbon dioxide and growth hormones when incubated on potassium chloride and potassium iminodiacetate. *Journal of Experimental Botany*, 32: 303–9.

―――. 1983. Stomatal responses of in vitro cultured plantlets, I. Responses in epidermal strips of *Chrysanthemum* to environmental factors and growth regulators. *Biochemie und Physiologie der Pflanzen*, 178: 619–24.

Zeiger, E. 1983. The biology of stomatal guard cells. *Annual Review of Plant Physiology*, 34: 441–75.

Note added in proof. Since the completion of our literature review in 1984 there have been reports on the action of Ca^{2+} on stomata and on the interaction of ABA, Ca^{2+}, and calmodulin; see *Journal of Plant Physiology*, 118 (1985): 177–87, and *New Phytologist*, 100 (1985): 473–82, and 101 (1985): 555–63.

13

Auxins and Stomata

W. J. Davies and T. A. Mansfield

Studies of the effects of auxins on stomata have a long history. The subject did not assume prominence in the minds of stomatal physiologists until recently, however, because a role for endogenous auxin could not be unambiguously demonstrated. Only in the past two years has the possibility of a major involvement of auxins in determining stomatal behavior seemed tenable. Thus, our discussion will be largely restricted to new data, and the significance of some of the observations must be a matter of speculation.

Early Studies

Boysen-Jensen (1936) supplied excised leaves with indol-3-ylacetic acid (IAA) via their petioles but found no effects. Some later studies with synthetic auxins, however, showed that they could induce stomatal closure. Brown (1946) found a reduction in the transpiration of plants sprayed with 2,4-dichlorophenoxyacetic acid (2,4-D), and Ferri and Lex (1948) and Ferri and Rachid (1949) confirmed these observations and showed that the same response occurred with naphth-2-yloxyacetic acid (NOXA) in both dicotyledons and monocotyledons. Bradbury and Ennis (1952) studied the effect of the ammonium salt of 2,4-D on *Phaseolus vulgaris* and suggested that the closure it induced was a direct effect on the stomata and was not the result of disturbed water relations of the plants. Further studies by Zelitch (1963), Mansfield (1967), Pemadasa (1979), Wardle and Short (1981), and others confirmed that synthetic auxins cause stomata to close and showed that naphth-1-ylacetic acid (NAA) was at least as effective as NOXA.

Whether the effects on the stomata were direct or indirect was not established in most of these early experiments. Transpiration was usually used to indicate stomatal opening, and the effects of the auxins on photosynthesis or respiration, or both, could have altered the CO_2 concentration in the leaf to bring about a change in stomatal aperture. Synthetic auxins can stimulate respiration, and both Mansfield (1967) and Pemadasa (1979) found that the effects of NAA and NOXA were reduced in CO_2-free air. Part of their action could thus be attributed to CO_2 accumulation in the leaves, but Pemadasa concluded that there was also a direct action on K^+ accumulation and starch metabolism in the guard cells.

The Discovery of the Effects of IAA

A notable absentee from most of these studies was IAA. We do not know how many researchers since Boysen-Jensen have looked for an effect of the natural auxin but have not reported their results because they were negative. During the evaluation of the use of epidermal strips for bioassays, IAA was incorporated into the media, but no effects were seen (Tucker and Mansfield, 1971; Ogunkanmi, Tucker, and Mansfield, 1973). It will be evident from the discussion that follows, however, that the choice of conditions for these bioassays would have minimized the effects of IAA. Thus the view became firmly established that if there was an auxin effect on stomata, it was to produce stomatal closure, not opening. The findings of Pemadasa (1982a) were thus unexpected, for he found that IAA caused stomatal opening. Pemadasa's long-standing interest has been the disparity of opening capacity in adaxial and abaxial stomata. In the majority of amphistomatous leaves there are fewer stomata on the adaxial surface, and it is not surprising that the diffusive conductance of this surface is sometimes much lower than that of the abaxial surface, as for example in tobacco (Turner, 1970). When allowance is made for the difference in stomatal density, however, the characteristics of the stomatal responses in the two epidermes are found to be different. In particular, the sensitivity to light is greater in the normally shaded abaxial stomata (Turner, 1970; Pemadasa, 1979), and the maximum opening of the pore under optimum conditions is greater in the abaxial than in the adaxial epidermis. The differences in actual stomatal opening thus reinforce those of stomatal density to lower the diffusive conductance of the adaxial epidermis.

Pemadasa (1981) treated adaxial stomata with fusicoccin, and this was the first agent discovered to cause them to open as widely as the abaxial stomata (Fig. 13.1). A further study showed that IAA was also highly effective in overcoming the reluctance of adaxial stomata to open (Pemadasa, 1982a). Foliar application of IAA led to virtually identical openings on the two epidermes for two or three days after treatment (Fig. 13.2). It was also found that IAA reduced the closing effects of darkness, CO_2, and abscisic acid (ABA) on both adaxial and abaxial stomata, though the effects on the abaxial stomata were small under some of the experimental conditions. But when stomatal opening in abaxial epidermis was limited by the supply of K^+ in the incubation medium ($0-25$ mol m^{-3} or $0-25$ mM K^+), IAA invariably promoted opening (Pemadasa, 1982a). It seems likely that the failure of early studies to reveal the effects of IAA probably resulted from a choice of experimental conditions that obscured its action. For example, in the bioassay system using epidermal strips, the main objective in the choice of environmental conditions was to promote stomatal opening, so that the inhibitory effects of compounds like ABA could

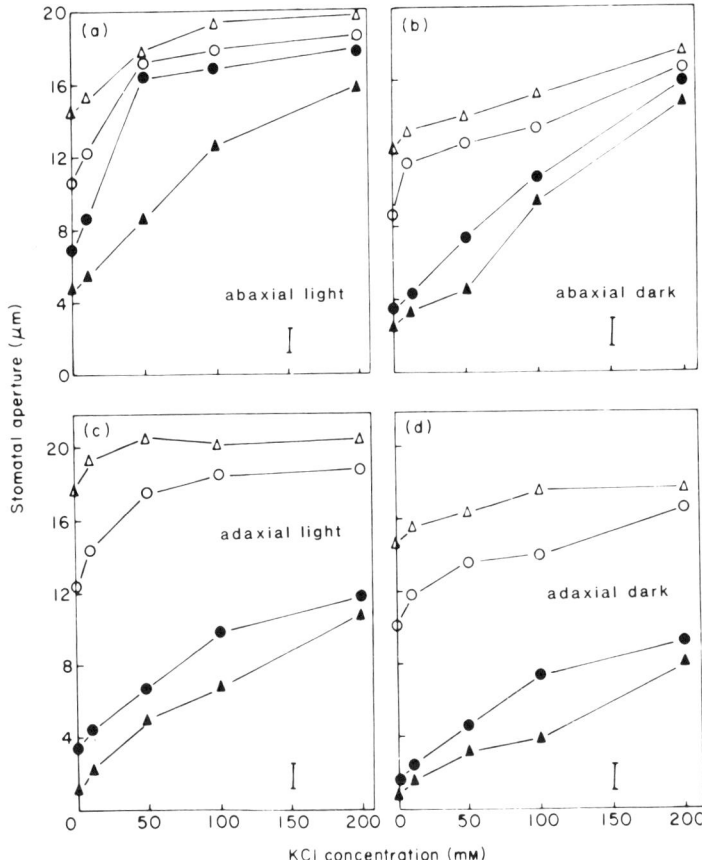

Fig. 13.1. The effect of fusicoccin (FC) on abaxial (*a*, *b*) and adaxial (*c*, *d*) stomata on epidermal strips of *Commelina communis*. The strips were incubated in buffered media with different KCl concentrations in light (*a*, *c*) or darkness (*b*, *d*) in the presence (△, ▲) or absence (○, ●) of CO_2. △, ambient air plus FC; ▲, ambient air minus FC; ○, CO_2-free air plus FC; ●, CO_2-free air minus FC. The vertical bars are the least significant differences at $P = 0.05$. (Redrawn from Pemadasa, 1981: 378.)

be seen. Because substituting Na^+ for K^+ had been found to give more reliable opening in the controls, Na^+ became a standard ingredient of the incubation media. Later work showed that this altered the sensitivity to ABA (Jarvis and Mansfield, 1980) and that it probably also disguised the effects of IAA (Pemadasa, 1983).

Pemadasa (1982a) suggested three possible explanations for the contribution that auxins might make toward the normal disparity in the ability of adaxial and abaxial stomata to open: the endogenous auxin content is lower in the adaxial than in the abaxial epidermis; the sensitivity to auxin is lower in adaxial than in abaxial guard cells; and both auxin content and sensitivity are lower in the

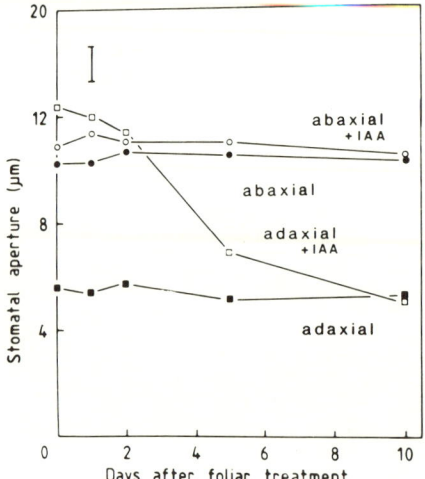

Fig. 13.2. The effect of application of 1 mM (1 mol m^{-3}) IAA on abaxial (○, ●) and adaxial (□, ■) stomatal opening in *Commelina communis*. IAA was applied to the surface of the intact leaves on day 0, and epidermis was removed at the times shown and incubated for 3 h under conditions favorable for stomatal opening. ○, □, plus IAA; ●, ■, untreated controls. (Redrawn from Pemadasa, 1982a: 216.)

adaxial epidermis. There is still no way of deciding which, if any, of these is correct. The possibility of a differential synthesis or transport of IAA that establishes a higher concentration in the abaxial than in the adaxial epidermis is worthy of investigation, and more studies of the factors controlling sensitivity to IAA (e.g. K$^+$ concentration, ABA) are needed.

When conditions are such that IAA has an effect on stomata, it induces opening, but the synthetic auxins invariably induce closure. Pemadasa (1982b) found that phenylacetic acid (PAA), which is thought to be an endogenous auxin of moderate activity (Schneider and Wightman, 1978), induced closure. However, PAA binds covalently to irradiated flavins *in vitro* and could therefore inhibit photoresponses that are activated by blue light, one of which is stomatal opening. Thus PAA might conceivably have an effect on guard cells that is unrelated to the effects of IAA and the synthetic auxins. It is important, however, to note that low concentrations of PAA act in the same way as IAA and can stimulate the opening of adaxial stomata, though higher concentrations are inhibitory (Pemadasa, 1982c). The inhibitory action of compounds like NAA and NOXA may be the result of their ability to bind to membrane sites and block the action of IAA. It is surprising that NAA, in particular, does not behave like IAA, because it does so in most systems in which endogenous IAA is thought to be important (Letham, Goodwin, and Higgins, 1978). Further investigation of the action of these synthetic auxins on stomata will be of much interest.

Interaction between IAA and Other Agents

The possibility that the magnitude of the response of stomata to other agents might be affected by IAA was further investigated by Snaith and Mansfield

(1982a, b), who found that the dose-response curves for ABA-induced closure of stomata in epidermal strips were markedly affected by the inclusion of IAA in the incubation medium (Fig. 13.3); indeed, there was virtually no inhibitory effect of ABA in the presence of 10^{-1} mol m^{-3} (100 μM) IAA. When epidermis was incubated in a CO_2-free environment in the light, IAA had no effect on stomatal aperture; but in the presence of CO_2, opening became strongly dependent on IAA concentration (Fig. 13.4a). A high concentration of IAA suppressed the inhibitory effect of CO_2 on the stomata, but this effect was counteracted by the inclusion of ABA in the incubation medium (Fig. 13.4b). A complex interplay between CO_2, ABA, and IAA was thus revealed.

Regulation by CO_2 and interactions with ABA and auxins. Stomata often close in elevated CO_2 concentrations, but they do not always do so, and recent reviewers (e.g. Farquhar and Sharkey, 1982) have drawn attention to the variability of the response. With respect to the restriction of assimilation, it is possible to assess the advantages of a stomatal reaction to CO_2. Farquhar and

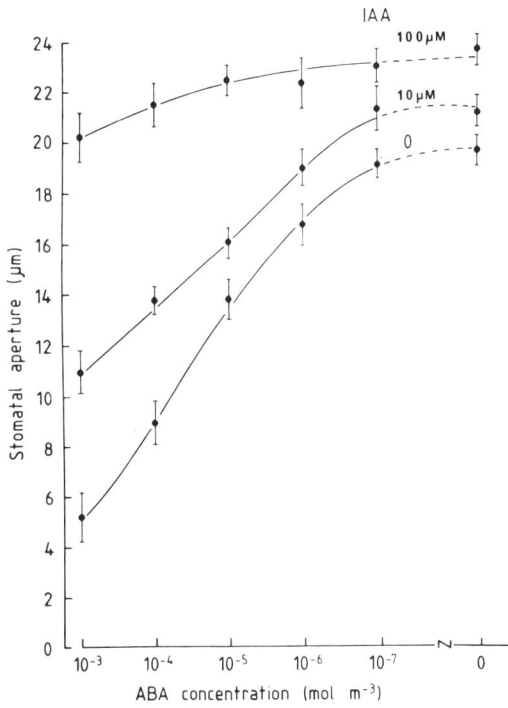

Fig. 13.3. The influence of IAA on the response of stomata in the abaxial epidermis of *Commelina communis* to ABA. A favorable concentration of KCl (100 mol m^{-3}) was present in all treatments; 95% confidence limits are shown. (Redrawn from Snaith and Mansfield, 1982a: 310.)

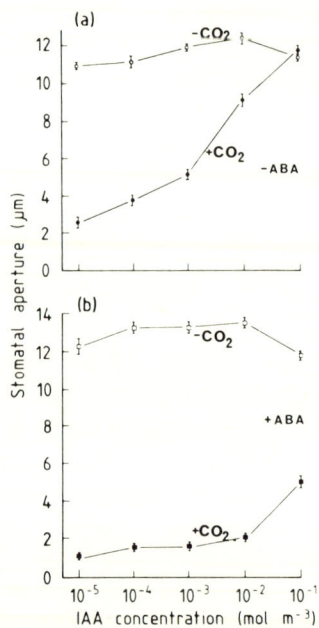

Fig. 13.4. *a*, Stomatal opening on detached abaxial epidermis of *Commelina communis* after incubation for 3 h in light in different IAA concentrations. ○, zero CO_2; ●, 700 mol mol^{-1} CO_2. The standard errors of the means are shown. *b*, As *a*, but with 10^{-5} mol m^{-3} ABA in the incubation medium. □, zero CO_2; ■, 700 mol mol^{-1} CO_2. (Redrawn from Snaith and Mansfield, 1982b: 362.)

Sharkey have pointed out that the stomata of C_4 plants should ideally be insensitive to intercellular CO_2 concentration (C_i) when it is below the saturation level for photosynthesis, but they should become very sensitive to C_i when it is above this level, because their closure will reduce transpiration without affecting assimilation. The same arguments apply for C_3 plants, except that there is no level of C_i at which stomatal closure will curtail transpiration without some sacrifice of assimilation rate.

Recognition of this situation in C_3 plants means that ideally there should be a mechanism for altering the CO_2 sensitivity of the stomata according to the water relations of the plant. When the water supply is abundant, the plant can bear the cost of having a small stomatal response to C_i. In this case, the external conditions that change C_i can do so without a reaction of the stomata, and the implications for the economy of water use can be disregarded. For a leaf carrying on photosynthesis at a constant rate in light, the main external factor determining C_i will be wind speed. As wind speed and boundary layer conductance increase, C_i must also increase, because the conductance of the pathway from the external atmosphere to the intercellular spaces will rise. Grace (1977) has shown that, particularly at low levels of available energy, increases in wind speed can markedly increase transpiration if the stomata remain open. If the plant's water supply is limited, it will therefore be important for the stomata to close partially as wind speed rises, so that evaporation is reduced. A closing response to an increase in C_i provides a likely mechanism for producing the necessary fall in transpiration. Morison and Jarvis (1983) have noted that stomatal responses to CO_2 by some plants are particularly marked at low irradiances.

The ecophysiological constraints point to a need for a variable stomatal response, ranging from insensitivity to C_i (or even a promotion of opening by CO_2) to closing in response to a small increase in C_i. Do the curves in Figure 13.4, which were obtained for epidermal strips, provide evidence of such a mechanism in the intact leaf? This is a question that can only be answered satisfactorily by future experiments, but evidence in the literature suggests they do.

Raschke (1975a, b) presented evidence that the stomata of some species are insensitive to CO_2 when the plants are grown under conditions that impose little water stress on the tissues. He found that this was the case for glasshouse-grown plants of *Xanthium strumarium*, whose stomata could be sensitized to CO_2 by raising the level of ABA in the leaf. Similarly, the stomatal responses to ABA were enhanced by CO_2.

Attempts to show statistically an interaction between CO_2 and ABA in *X. strumarium* (Mansfield, 1976) were not successful, however, even though there was evidence of some interdependence between the two regulators. Further studies with *Commelina communis* by Mansfield and Wilson (1981) and Wilson (1981) also failed to show an interaction between CO_2 and ABA; indeed, the dose-response curves for ABA were virtually parallel at two CO_2 concentrations over the range of maximum sensitivity to ABA (between 10^{-6} and 10^{-5} mol m^{-3}; Fig. 13.5). Wilson did, however, find that dose-response curves assumed a different character in epidermis from plants that had been water-stressed or had been pretreated with ABA one day prior to use (Fig. 13.6). The essential difference between these dose responses has not been explained, but we can suggest that after water stress or pretreatment with ABA a third agent becomes involved. The effect of this third agent would be to enhance the response of the stomata to CO_2, as can be seen if Wilson's data in Figures 13.5 and 13.6 are plotted as the difference in aperture between the $+CO_2$ and $-CO_2$ treatments (Fig. 13.7). Proposing IAA as the third agent is a realistic hypothesis. Recently, Blackman and Davies (1984a) have reported an interplay between the effects of CO_2, ABA, and naturally occurring or synthetic cytokinins on stomata that is remarkably similar to that shown in Figure 13.4 for CO_2, ABA, and IAA. It seems clear that future research on the hormone relations of leaves under water stress should consider changes in endogenous IAA and cytokinins as well as changes in ABA.

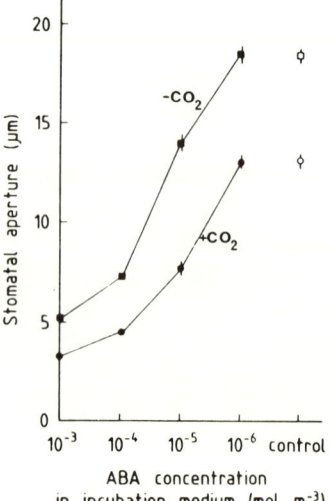

Fig. 13.5. The effects of CO_2 and ABA on stomata and epidermal strips taken from well-watered plants of *Commelina communis*. Air containing 3 (■) or 330 (●) mol mol^{-1} CO_2 was bubbled through the incubation medium. The open symbols are the controls incubated in the absence of ABA, and standard error values are shown. (Redrawn from Wilson, 1981: 264.)

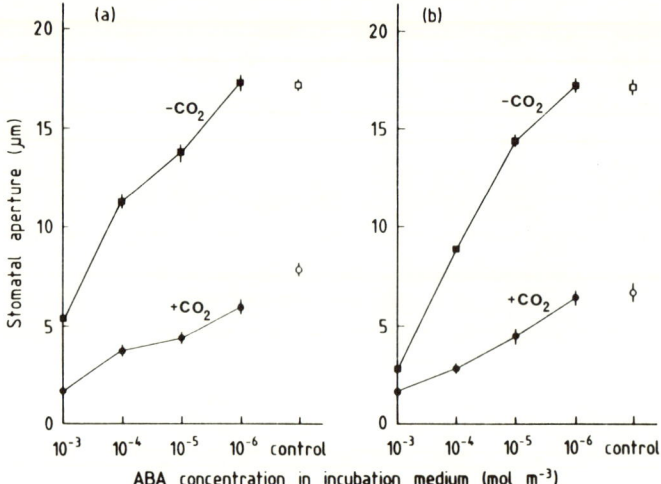

Fig. 13.6. *a*, The effects of CO_2 and ABA on stomata on epidermal strips taken from water-stressed plants of *Commelina communis*. Air containing 3 (■) or 330 (●) mol mol^{-1} CO_2 was bubbled through the incubation medium. The open symbols are the controls incubated in the absence of ABA, and standard error values are shown. *b*, As *a*, except that the epidermis was taken from well-watered plants that had been sprayed with 10^{-1} mol m^{-3} ABA one day prior to use. (Redrawn from Wilson, 1981: 266–67.)

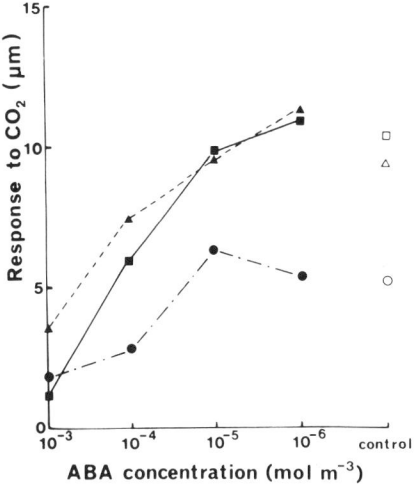

Fig. 13.7. The absolute response to CO_2 (change in aperture in μm) when different concentrations of ABA were supplied in the incubation medium to epidermal strips of *Commelina communis*. Prior to removal of the epidermis, the plants had been well watered (●), water-stressed (▲), or sprayed with 10^{-1} mol m^{-3} ABA (■). The open symbols represent the responses in epidermis from well-watered plants without the addition of ABA to the incubation medium. (Redrawn from Wilson, 1981: 268.)

Guard cell activities controlled by IAA, ABA, and CO_2. Studies by Weyers and Hillman (1980), MacRobbie (1981), and others have suggested that the action of ABA on guard cells results from a stimulation of the efflux of K^+ rather than from a reduction of its influx. The much-reduced sensitivity to ABA when the guard cells are bathed in Na^+ instead of K^+ also suggests that the response is regulated via K^+ fluxes rather than through the activity of the proton pump (Jarvis and Mansfield, 1980). We know very little about the mechanisms involved in the CO_2 responses of stomata, and recent hypotheses have tended to emphasize the importance of organic acid metabolism rather than the control of ionic fluxes (Raschke, 1975b). However, Zeiger (1983) has argued that modifications in membrane permeability by CO_2 are more feasible; he also discussed recent evidence that CO_2 might modulate photophosphorylation in guard cells. Indol-3-ylacetic acid (IAA) is known to activate H^+ extrusion, and the similarity of the actions of fusicoccin and of IAA in overcoming the differences between adaxial and abaxial stomata would suggest that both compounds could modulate proton pumping in guard cells. Further support for this hypothesis is the fact that fusicoccin can overcome ABA-induced stomatal closure (Squire and Mansfield, 1972), as IAA can also (Fig. 13.3).

It is thus likely that the observed interplay between CO_2, ABA, and IAA may be the result of direct or indirect modulation of ionic fluxes in guard cells. Although the ability of IAA to eliminate the response to CO_2 and of ABA to restore it (Fig. 13.4) might suggest that all three exert their effects on the same process, this is not necessarily the case. For example, IAA may exert control over K^+ influx, and ABA over K^+ efflux, guard cell turgor being the result of the net balance. Carbon dioxide could modulate energy supply, as Zeiger (1983) suggested, and the energy relationships for influx and efflux may differ to the extent that they are differentially affected by CO_2. The importance of metabolic energy for responses to ABA involving loss of solutes has been shown by Weyers et al. (1982).

The mode of action of ABA on guard cells. Hartung (1983) found that there was no uptake of ABA into the guard cells of *Valerianella locusta* at pH 8.0, yet ABA induced stomatal closure as effectively at that pH as at lower pH values (e.g. 5.0), at which it entered the cytosol readily. He concluded that ABA need not enter the interior of the guard cells in order to induce their loss of turgor, and he proposed that the site of action must be either at the surface of the plasma membrane or at a point readily accessible from the apoplast.

Recently, De Silva, Hetherington, and Mansfield (1985) have shown that the action of ABA is dependent on the presence of calcium ions. There was no detectable effect of ABA when a chelating agent was used to remove the apoplastic calcium, and there was a pronounced synergism between external Ca^{2+} and ABA. It was proposed that ABA might act on the guard cells by increasing the passage of Ca^{2+} through the plasma membrane. Calcium ions could then bind to the small protein calmodulin, which is known to activate several key enzymes in plant cells (Dieter, 1984).

Interactions between growth regulators and temperature. We have concentrated above on interactions between plant growth regulators and CO_2 in their effects on stomata, but it has become increasingly apparent that other factors also interact to determine the final influence of a single variable. Variation in these additional influences may often explain the poor correlation between the variation in the endogenous concentration of an individual regulator and stomatal behavior. Recently, Rodriguez and Davies (1982) showed that although ABA was effective in closing maize stomata when applied to leaves incubated at temperatures above 12° C, it was not effective at lower temperatures, at which the application of 5.6×10^{-1} mol m^{-3} (560 μM) ABA did not significantly reduce stomatal aperture and may even have enhanced it. Observations of this type have been substantially confirmed by Eamus and Wilson (1983) working with intact leaves of *Phaseolus*. We may speculate that the unusually large stomatal apertures shown by plants experiencing water deficit at cool temperatures may be explained by such an effect of temperature on the stomatal response to ABA. For example, Loftfield (1921) has commented that when "thin-leaved mesophytes" experience water deficit, stomata close partially during the middle of the day, and when such midday closure occurs, night opening appears and increases in degree and extent with it. Presumably, cool night temperatures might promote an increasing degree of stomatal opening against an increasing concentration of ABA. In addition, chilling temperatures will often lock open the stomata of chilling-sensitive species (Wilson, 1976).

When introduced through the petiole, IAA will promote the opening of

Phaseolus stomata even at chilling temperatures (Eamus and Wilson, 1984). These authors proposed that at low temperatures the guard cell membrane may be damaged so that binding sites that were previously specific for IAA may become accessible to ABA. Eamus and Wilson (1984) suggest that if the site of attachment on the membrane rather than the nature of the binding molecule determines the stomatal response, then binding of ABA to the IAA site at low temperature might result in K^+ accumulation and stomatal opening such as Rodriguez and Davies (1982) observed. Unless the proposed membrane damage is readily reversible, our own observations would not necessarily support this model, since we have observed that maize stomata incubated in ABA and held at low temperature will close rapidly as the incubation solution is warmed (Rodriguez and Davies, 1982). Astle and Rubery (1983) have reported on the nature of carriers for ABA and IAA in primary roots. Carrier-mediated uptake of the two compounds in this system seems to be independent.

Are Different Growth Regulators Operating in Different Situations?

There is little firm evidence of a specific relationship between plant water status and endogenous auxin concentrations, but most work points to a reduction in auxin content with increasing soil water deficit. This may be because of an increase in IAA oxidase activity stimulated by the water deficit (Darbyshire, 1971) or because of an inhibition of auxin transport (Davenport, Morgan, and Jordan, 1980). Interestingly, the wilty mutant of tomato, which shows high stomatal conductance despite water deficit, contains higher than normal amounts of IAA-like substances, apparently because of a high rate of IAA synthesis (Tal et al., 1979). This high rate is recorded despite the development of low water potentials and turgors. One study reports that IAA concentrations increase considerably as a result of soil drying (Hall et al., 1977).

Considerable evidence suggests that in many plants cytokinins may be another important endogenous promoter of stomatal opening (Blackman and Davies, 1984b; Incoll and Jewer, this volume), and Blackman and Davies (1983) have reported that increasing concentrations of either zeatin or kinetin can reverse ABA-stimulated stomatal closure in maize leaves (Fig. 13.8). This effect is remarkably similar to that of IAA on *Commelina* stomata (Fig. 13.4). There is evidence that xylem sap from roots subjected to drought or osmotic stress contains only low concentrations of cytokinin-like substances (Itai and Vaadia, 1965; Itai, Richmond and Vaadia, 1968), suggesting that drought-induced variation in cytokinin concentrations may influence stomatal behavior. Nevertheless, few data of this kind have been obtained with modern analytical techniques to determine endogenous concentrations of cytokinin or auxin, and

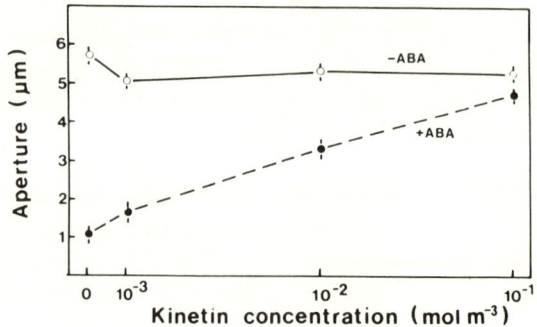

Fig. 13.8. The effects of kinetin on the apertures of maize stomata. Leaf pieces were incubated on a range of kinetin concentrations, with (●---●) or without (○——○) ABA (10^{-1} mol m^{-3}). Data represent means of 60 observations ± standard error. (Redrawn from Blackman and Davies, 1983: 1621.)

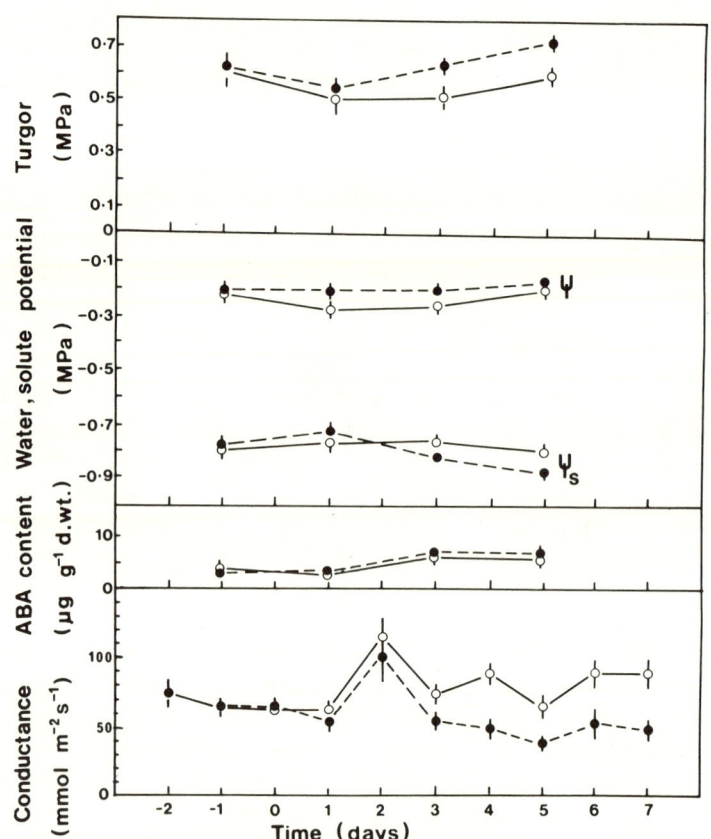

Fig. 13.9. Water potential (Ψ), solute potential (Ψ_s), ABA content, and stomatal behavior of maize leaves in plants when water was withheld from part of the root system. Control plants had both parts of the root system well watered (○——○). Plants in which part of the root system was dried (●---●) were last watered on day 0. Points are means of 8 observations ± standard error. (Redrawn from Blackman and Davies, 1985: 42.)

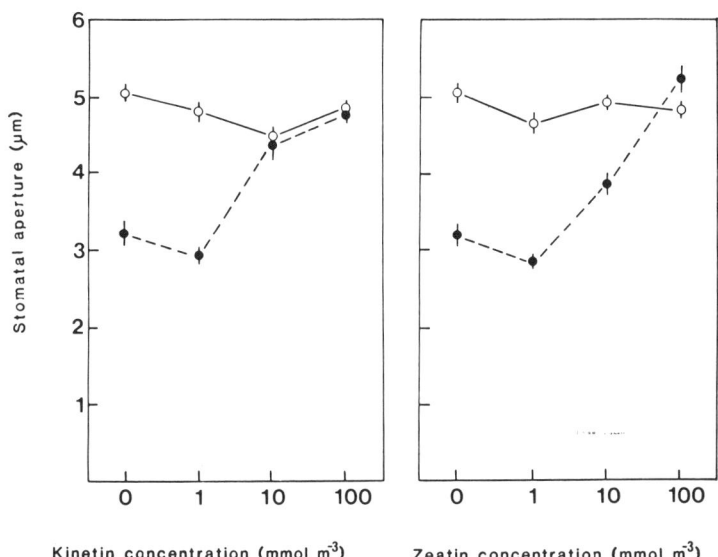

Fig. 13.10. Apertures of stomata of maize leaves incubated in the light in CO_2-free air on distilled water, kinetin, or zeatin solutions. Leaves that were taken from well-watered (○——○) or partially dried (●---●) plants were sampled after 8 h of light on day 5 after water was withheld from part of the root system of half the plants. Points are means of 60 observations ± standard error. (Redrawn from Blackman and Davies, 1985: 45.)

there is a great need to use such methods to identify the specific compounds involved in the regulation of stomatal behavior in different situations. Information is needed on the kinetics of transport and on concentration/deficit relationships.

One piece of circumstantial evidence that the deficit-induced lack of a specific compound in shoots can restrict stomatal opening is provided by a simple split-root experiment (Blackman and Davies, 1985). Drying a portion of the root system of a maize plant can restrict stomatal opening, even though both the water potential and the turgor increase (Fig. 13.9) and despite the fact that there is no compelling evidence for an increase in concentration or a redistribution of ABA. Leaves removed from the plants with roots in partly dried soil and incubated under ideal conditions for stomatal opening continued to show restricted stomatal apertures until cytokinins were added to the incubation medium (Fig. 13.10). This suggests that plants with some roots in dry soil may show restricted stomatal apertures because a promoter of stomatal opening, in this case cytokinin, is present in low concentration in the leaves.

Recently, several authors have considered the possibility that under certain circumstances a signal moving from roots experiencing anoxia because of flooding, chilling, or water deficit may override the water relations of the leaf and control stomatal behavior (Davies and Sharp, 1981; Bradford and Hsiao,

1982; Cowan, 1982). There is considerable attraction in the view that variation in the supply of cytokinins may constitute such a signal, since much cytokinin is thought to be synthesized in the roots (Van Staden and Davey, 1979).

We have appreciated for some time how wide a range of environmental factors can elicit a response from a single pair of cells, but to date much research on plant water deficits has concentrated upon characterization of the effects of a deficit-induced increase in a single inhibitor of stomatal opening, namely, ABA. It now seems likely that water deficit may also reduce the concentration in the leaves of promoters of stomatal opening, and thus it is more realistic to expect that the balance of regulators in the leaves will help to determine the degree of stomatal opening. Stomatal behavior in different species under varying environmental conditions must be interpreted in terms of variation in one or more components of an interactive system linking external and endogenous factors. Changes in the levels of single components may or may not elicit a response, and the interactive system operates so that a change in a single component may be promotive or inhibitive, depending on the prevailing conditions.

REFERENCES

Astle, M. C., and P. H. Rubery. 1983. Carriers for abscisic acid and indole-3-acetic acid in primary roots: Their regional localisation and thermodynamic driving forces. *Planta*, 157: 53–63.

Blackman, P. G., and W. J. Davies. 1983. The effects of cytokinins and ABA in stomatal behaviour of maize and *Commelina*. *Journal of Experimental Botany*, 34: 1619–26.

———. 1984a. Modification of the CO_2 responses of maize stomata by abscisic acid and by naturally occurring and synthetic cytokinins. *Journal of Experimental Botany*, 35: 174–79.

———. 1984b. Age-related changes in stomatal response to cytokinins and abscisic acid. *Annals of Botany*, 54: 121–25.

———. 1985. Root to shoot communication in maize plants of the effects of soil drying. *Journal of Experimental Botany*, 36: 39–48.

Boysen-Jensen, P. 1936. *Growth hormones in plants*. New York: McGraw-Hill.

Bradbury, D., and W. B. Ennis. 1952. Stomatal closure in kidney bean plants treated with ammonium 2,4-dichlorophenoxyacetate. *American Journal of Botany*, 39: 324–28.

Bradford, K. J., and T. C. Hsiao. 1982. Stomatal behavior and water relations of waterlogged tomato plants. *Plant Physiology*, 70: 1508–13.

Brown, J. W. 1946. Effect of 2,4-dichlorophenoxyacetic acid on the water relations, the accumulation and distribution of solid matter, and the respiration of bean plants. *Botanical Gazette*, 107: 332–43.

Cowan, I. R. 1982. Regulation of water use in relation to carbon gain in higher plants.

In O. L. Lange, P. S. Nobel, C. B. Osmond, and H. Ziegler, eds., *Physiological plant ecology* 2, Encyclopedia of Plant Physiology, n.s., 12B: 589–614. Berlin: Springer-Verlag.
Darbyshire, B. 1971. Changes in indolacetic acid oxidase activity associated with plant water potential. *Physiologia plantarum,* 25: 80–84.
Davenport, T. L., P. W. Morgan, and W. R. Jordan. 1980. Reduction of auxin transport capacity with age and internal water deficits in cotton petioles. *Plant Physiology,* 65: 1023–25.
Davies, W. J., and R. E. Sharp. 1981. Water stress, abscisic acid and assimilate distribution. In J. Kralovic, ed., *Mechanisms of assimilate distribution and plant growth regulators*: 53–67. Bratislava: Slovak Academy of Sciences.
De Silva, D. L. R., A. M. Hetherington, and T. A. Mansfield. 1985. Synergism between calcium ions and abscisic acid in preventing stomatal opening. *New Phytologist,* 100: 473–82.
Dieter, P. 1984. Calmodulin and calmodulin-mediated processes in plants. *Plant, Cell and Environment,* 7: 371–80.
Eamus, D., and J. M. Wilson. 1983. ABA levels and effects in chilled and hardened *Phaseolus vulgaris. Journal of Experimental Botany,* 34: 1000–1006.
———. 1984. A model for the interaction of low temperature, ABA, IAA and CO_2 in the control of stomatal behaviour. *Journal of Experimental Botany,* 35: 91–98.
Farquhar, G. D., and T. D. Sharkey. 1982. Stomatal conductance and photosynthesis. *Annual Review of Plant Physiology,* 33: 317–45.
Ferri, M. G., and A. Lex. 1948. Stomatal behavior as influenced by treatment with β-naphthoxyacetic acid. *Contributions from Boyce Thompson Institute,* 15: 283–90.
Ferri, M. G., and M. Rachid. 1949. Further information on the stomatal behavior as influenced by treatment with hormone-like substances. *Anais da Academia brasiliera de ciências,* 21: 155–66.
Grace, J. 1977. *Plant response to wind.* London: Academic.
Hall, M. A., J. A. Kapuya, S. Sivakumaran, and A. John. 1977. The role of ethylene in the response of plants to stress. *Pesticide Science,* 8: 217–23.
Hartung, W. 1983. The site of action of abscisic acid at the guard cell plasmalemma of *Valerianella locusta. Plant, Cell and Environment,* 6: 427–28.
Itai, C., A. Richmond, and Y. Vaadia. 1968. The role of root cytokinins during water and salinity stress. *Israel Journal of Botany,* 17: 187–95.
Itai, C., and Y. Vaadia. 1965. Kinetin-like activity in root exudate of water-stressed sunflower plants. *Physiologia plantarum,* 18: 941–45.
Jarvis, R. G., and T. A. Mansfield. 1980. Reduced stomatal responses to light, carbon dioxide and abscisic acid in the presence of sodium ions. *Plant, Cell and Environment,* 3: 279–83.
Letham, D. S., P. B. Goodwin, and T. J. V. Higgins, eds. 1978. *Phytohormones and related compounds: A comprehensive treatise* 1. Amsterdam: Elsevier.
Loftfield, J. V. G. 1921. *The behavior of stomata.* Publication of the Carnegie Institution of Washington, no. 314.
MacRobbie, E. A. C. 1981. Ionic relations in stomatal guard cells. In P. G. Jarvis and T. A. Mansfield, eds., *Stomatal physiology*: 51–70. Cambridge: Cambridge University Press.
Mansfield, T. A. 1967. Stomatal behaviour following treatment with auxin-like substances and phenylmercuric acetate. *New Phytologist,* 66: 325–30.
———. 1976. Delay in the response of stomata to abscisic acid in CO_2-free air. *Journal of Experimental Botany,* 27: 559–64.
Mansfield, T. A., and J. A. Wilson. 1981. Regulation of gas exchange in water-stressed

plants. In C. B. Johnson, ed., *Physiological processes limiting plant productivity*: 237–51. London: Butterworths.

Morison, J. I. L., and P. G. Jarvis. 1983. Direct and indirect effects of light on stomata, II. In *Commelina communis* L. *Plant, Cell and Environment*, 6: 103–9.

Ogunkanmi, A. B., D. J. Tucker, and T. A. Mansfield. 1973. An improved bioassay for abscisic acid and other antitranspirants. *New Phytologist*, 72: 277–82.

Pemadasa, M. A. 1979. Stomatal responses to two herbicidal auxins. *Journal of Experimental Botany*, 30: 267–74.

———. 1981. Abaxial and adaxial stomatal behaviour and responses to fusicoccin on isolated epidermis of *Commelina communis* L. *New Phytologist*, 89: 373–84.

———. 1982a. Differential abaxial and adaxial stomatal responses to indole-3-acetic acid in *Commelina communis* L. *New Phytologist*, 90: 209–19.

———. 1982b. Abaxial and adaxial stomatal responses to light of different wavelengths and to phenylacetic acid on isolated epidermes of *Commelina communis* L. *Journal of Experimental Botany*, 33: 92–99.

———. 1982c. Effects of phenylacetic acid on abaxial and adaxial stomatal movements and its interaction with abscisic acid. *New Phytologist*, 92: 21–30.

———. 1983. Sodium ions can eliminate the normal disparity in abaxial and adaxial stomatal opening on isolated epidermes. *New Phytologist*, 94: 201–9.

Raschke, K. 1975a. Simultaneous requirement of carbon dioxide and abscisic acid for stomatal closing in *Xanthium strumarium* L. *Planta*, 125: 243–59.

———. 1975b. Stomatal action. *Annual Review of Plant Physiology*, 26: 309–40.

Rodriguez, J. L., and W. J. Davies. 1982. The effects of temperature and ABA on stomata of *Zea mays* L. *Journal of Experimental Botany*, 33: 977–87.

Schneider, E. A., and F. Wightman. 1978. Auxins. In D. S. Letham, P. B. Goodwin, and T. J. V. Higgins, eds., *Phytohormones and related compounds: A comprehensive treatise* 1: 29–106. Amsterdam: Elsevier.

Snaith, P. J., and T. A. Mansfield. 1982a. Stomatal sensitivity to abscisic acid: Can it be defined? *Plant, Cell and Environment*, 5: 309–11.

———. 1982b. Control of the CO_2 responses of stomata by indol-3-ylacetic acid and abscisic acid. *Journal of Experimental Botany*, 33: 360–65.

Squire, G. R., and T. A. Mansfield. 1972. Studies of the mechanism of action of fusicoccin, the fungal toxin that induces wilting, and its interaction with abscisic acid. *Planta*, 105: 71–78.

Tal, M., D. Imber, A. Erez, and E. Epstein. 1979. Abnormal stomatal behaviour and hormonal imbalance in *flacca*, a wilty mutant of tomato, V. Effect of abscisic acid on indolacetic acid metabolism and ethylene evolution. *Plant Physiology*, 63: 1044–48.

Tucker, D. J., and T. A. Mansfield. 1971. A simple bioassay for detecting "antitranspirant" activity of naturally occurring compounds such as abscisic acid. *Planta*, 98: 157–63.

Turner, N. C. 1970. Responses of abaxial and adaxial stomata to light. *New Phytologist*, 69: 647–53.

Van Staden, J., and J. E. Davey. 1979. The synthesis, transport and metabolism of endogenous cytokinins. *Plant, Cell and Environment*, 2: 93–106.

Wardle, K., and K. C. Short. 1981. Induced stomatal responses in epidermal strips of *Vicia faba* L. *Journal of Biological Education*, 15: 117–22.

Weyers, J. D. B., and J. R. Hillman. 1980. Effects of abscisic acid on $^{86}Rb^+$ fluxes in *Commelina communis* L. leaf epidermis. *Journal of Experimental Botany*, 31: 711–20.

Weyers, J. D. B., N. W. Paterson, P. J. Fitzsimons, and J. M. Dudley. 1982. Metabolic inhibitors block ABA-induced stomatal closure. *Journal of Experimental Botany*, 33: 1270–78.

Wilson, J. A. 1981. Stomatal responses to applied ABA and CO_2 in epidermis detached from well-watered and water-stressed plants of *Commelina communis* L. *Journal of Experimental Botany*, 32: 261–69.

Wilson, J. M. 1976. The mechanism of chill- and drought-hardening of *Phaseolus vulgaris* leaves. *New Phytologist*, 76: 257–70.

Zeiger, E. 1983. The biology of stomatal guard cells. *Annual Review of Plant Physiology*, 34: 441–75.

Zelitch, I. 1963. The control and mechanisms of stomatal movement. In I. Zelitch, ed., *Stomata and water relations in plants*: 18–42. Connecticut Agricultural Experiment Station, Bulletin 664. New Haven.

14

Stomatal Responses to Air Humidity and to Soil Drought

E.-D. Schulze, N. C. Turner, T. Gollan, and K. A. Shackel

Drought is a major environmental factor influencing plant growth and performance in both natural vegetation and managed crops. Thus, information about plant responses to drought and their underlying mechanisms will improve our understanding of plant adaptations to climatic extremes and will have relevance for plant management and breeding. There are two major levels at which plants regulate water loss and carbon gain. At the whole-plant level, plants regulate carbon partitioning to roots and shoots in response to water relations parameters (Schulze, Schilling, and Nagarajah, 1983). Carbon partitioning into leaves has positive feedback on plant carbon relations but it has negative effects on plant water balance. A second level of regulation occurs at the single-leaf level. Plants regulate the opening of the stomatal pore in relation to plant-internal factors, such as plant water stress and photosynthetic capacity, and in relation to climatic factors. But the effect of stomatal response on canopy water loss may largely depend on the roughness of the vegetation surface. Both levels of operation are reviewed in detail by Schulze (1986a, b). In the following we are summarizing the results of Turner, Schulze, and Gollan (1984, 1985) and Gollan, Turner, and Schulze (1985) on responses of stomata to drought, which is either a dry atmosphere or a dry soil, or both.

Effects of Interactions between Air Humidity and Soil Water Status on Stomatal Conductance

Exposure of a single leaf or a whole plant to dry air will initially cause a change in transpiration (Fig. 14.1), which in turn will lead to a change in the water status of the leaf. It is important to note that in this case leaf water potential is a dependent variable. But leaf water status may also affect stomatal conductance and thus in turn alter transpiration rate. In that case, leaf water potential is an independent variable. Apart from the transpiration/leaf water potential feedback loop (Stålfelt, 1956), stomata may respond directly to the humidity of the air (Lange et al., 1971). This will cause transpiration not to rise proportionally with the vapor concentration difference.

312 Schulze et al.

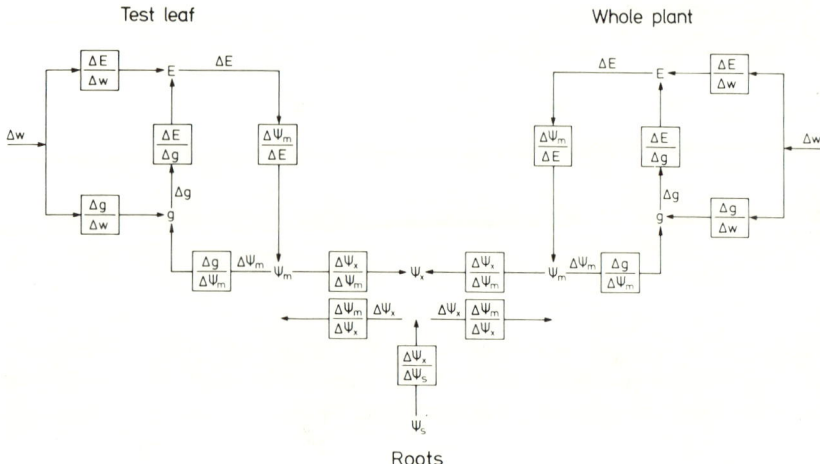

Fig. 14.1. Schematic representation of feedback and feedforward loops relating stomatal functioning to air humidity and plant water status in both single test leaves and whole plants. E, transpiration; g, stomatal conductance; Δw, air humidity; Ψ_m, mesophyll water potential; Ψ_x, xylem water potential; Ψ_s, soil water potential; Δ, change.

The principal ways of stomatal regulation appear quite clearly from Figure 14.1, but for most situations it is quite difficult to identify various response types. Figure 14.2 gives an example of concurrent changes in leaf conductance, transpiration, and leaf water potential resulting from a change in water vapor concentration. Species of very different leaf morphology show stomatal closure with dry air, but in most such cases the response of the stomata is not strong enough to cause a decrease in transpiration with increasing leaf to air vapor concentration difference. Therefore leaf water potentials decline, as a consequence of high transpiration rates. If leaf conductance is related to the resulting level of leaf water potential, it is indeed very difficult to decide whether plants respond to air humidity or to leaf water potential. Some species (see Fig. 14.2, *Vigna*) maintain an almost constant leaf water potential at varying levels of transpiration and air humidity, which could be interpreted as a strong feedback response of stomata to internal leaf water status.

The system of interrelated loops depicted in Figure 14.1 may be analyzed by perturbations. One possible perturbation operates through the xylem water potential. Since there is hydraulic continuity throughout the plant, changes in the leaf water potential of the whole plant will cause corresponding changes in the xylem water potential of the plant, and vice versa. Thus, changes in the transpiration of the plant can induce changes in the water potential of a single leaf enclosed in a different environment. Another possible perturbation operates via the root. Drying the soil causes a decrease in the water potential of the entire plant and in a single test leaf.

In the remainder of this paper, experiments on responses of stomata to the aforementioned perturbations are described. From these experiments it is possible to evaluate responses to atmospheric drought at different levels of leaf water status, responses to changes in leaf water potential in a single test leaf produced by changes in transpiration of the entire plant, and responses to changes in leaf water potential produced by changes in the water status of the soil.

Stomatal responses to dry air and to leaf water potential. Lange et al. (1971) postulated that the turgor relations of stomatal cells depend on the pattern of the water flow to and within the epidermis. Thus, gradients in water potential induced by stomatal and peristomatal transpiration could regulate the degree of stomatal opening. Figure 14.3 shows for *Tradescantia virginiana* that a step change in air humidity caused rapid changes in epidermal turgor but little or no change in xylem water potential. Under steady-state conditions, the turgor of all epidermal cells was similar; that is, no consistent differences in turgor were

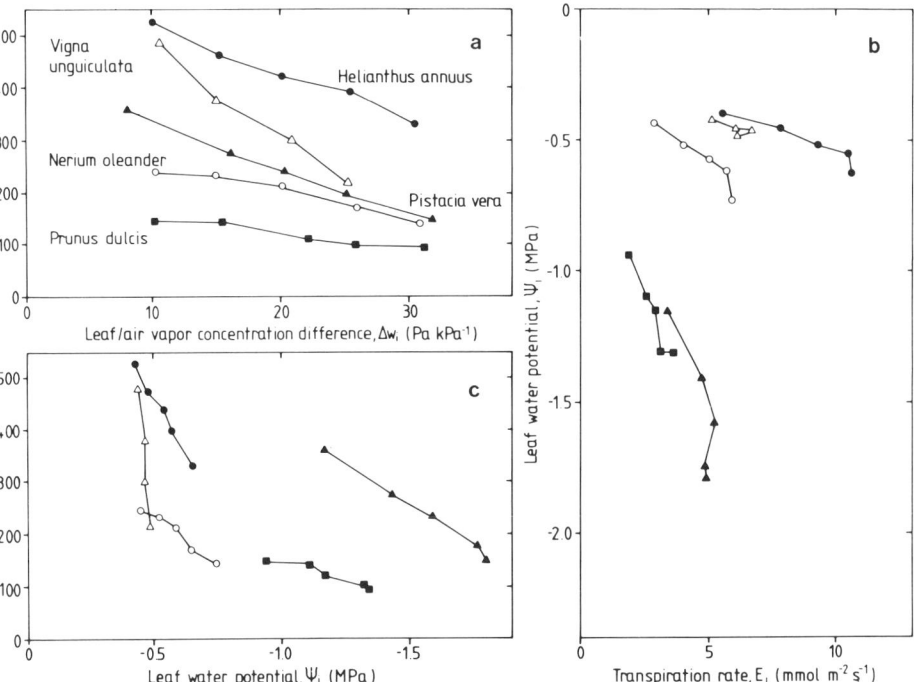

Fig. 14.2. Relationship between (a) vapor concentration difference and leaf conductance, (b) transpiration and leaf water potential, and (c) leaf water potential and leaf conductance for various species. Measurements were made on a leaf or leaves in a cuvette while the vapor concentration difference over the remainder of the plant in a surrounding cabinet was maintained at 10 Pa kPa^{-1}. (Adapted from Turner, Schulze, and Gollan, 1984.)

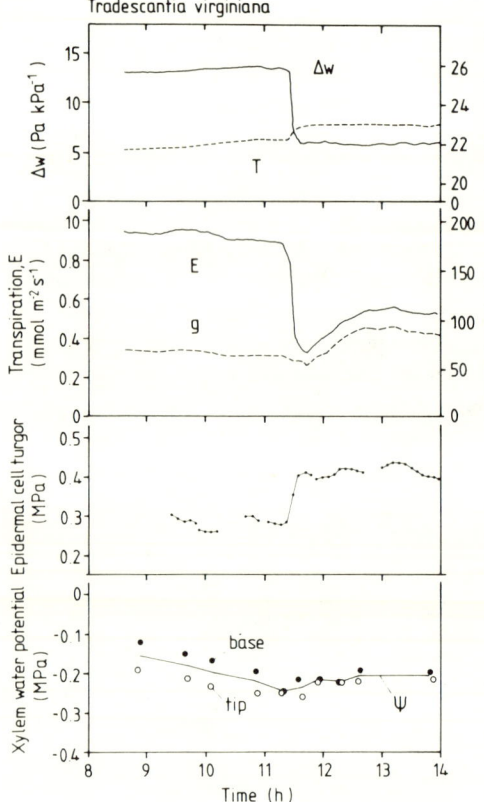

Fig. 14.3. Changes in transpiration (E), leaf conductance (g), turgor pressure of epidermal cells (P), xylem water potential (Ψ), and leaf temperature (T) during changes of air humidity (VPD) in *Tradescantia virginiana*. (Adapted with permission from Shackel and Brinckmann, 1984.)

found between epidermal cells closer to and those farther from the stomata (Shackel and Brinckmann, 1984). It was shown that water potential gradients within the leaf were associated with transpirational water flux, and that the water potential of epidermal cells was more sensitive to changes in evaporative demand than that of the whole leaf. It is interesting to note that the change in stomatal conductance caused by a step change in humidity began after the change in epidermal cell turgor was completed. Thus, the mechanism of the stomatal response to humidity may be not a simple hydraulic but rather a metabolic mechanism, which is triggered by changes in leaf turgor (Schulze, 1986a, b).

On the whole-plant level interactions between air humidity and leaf water status make interpretations complicated. When the entire plant and a single test leaf of *Helianthus nuttallii* were exposed to constant high air humidity, and the soil was drying slowly, leaf water potential increased almost linearly with decreasing transpiration rates (Fig. 14.4a). This means that there is also a linear relationship between stomatal conductance and leaf water potential (Fig.

14.4c, curve *C*). One could interpret this relationship as a classic feedback response to leaf water status. That this was not the case was indicated by experiments in which the water relations of either the entire plant or the single leaf were perturbed by changes in air humidity and the resultant change in transpiration.

Changing the transpiration of a test leaf by decreasing air humidity while keeping the entire plant in a constant humid environment caused an increase in transpiration and a decrease in leaf water potential of the test leaf (Fig. 14.4a, curve *A*). The slope of curve *A* in Figure 14.4a represented the resistance for the liquid flow of water to the leaf. Stomata closed in response to dry air (Fig.

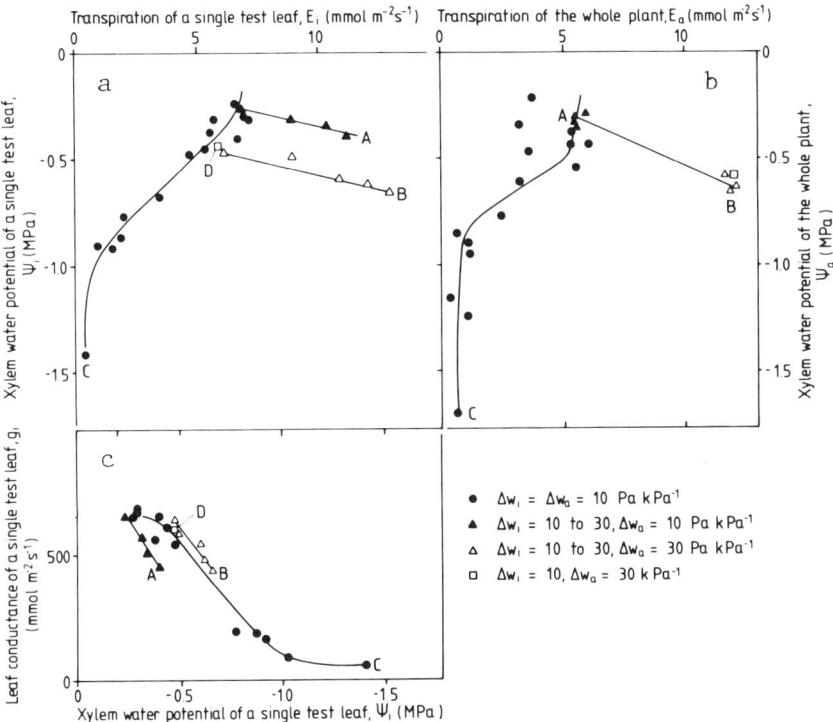

Fig. 14.4. Relationship between (*a*) transpiration (E_i) and xylem water potential (Ψ_i) of a test leaf, (*b*) transpiration (E_a) and leaf water potential (Ψ_a) of the whole plant, and (*c*) leaf conductance (g_i) and leaf water potential (Ψ_i) of a test leaf in *Helianthus nuttallii*: *A*, at high soil water content, vapor concentration difference for the test leaf was changed stepwise from 10 to 30 Pa kPa^{-1} while that for the whole plant was maintained at 10 Pa kPa^{-1}; *B*, same as curve *A*, but the whole plant was maintained at a vapor concentration difference of 30 Pa kPa^{-1}; *C* (drawn line), in drying soil, vapor concentration deficit for both the whole plant and the test leaf were maintained at 10 Pa kPa^{-1}; *D* (Fig. 14.4a and c only), transpiration and leaf conductance of a test leaf at a vapor concentration difference of 10 Pa kPa^{-1} at high soil water content following the increase in transpiration of the whole plant (shift from points *A* to *B* in Fig. 14.4b).

14.4c, curve *A*), but the relation between leaf conductance and leaf water potential had changed; that is, when the air surrounding the test leaf was dried, the same degree of stomatal closure occurred at a higher leaf water potential than when the soil was dried (compare curves *C* and *A* in Fig. 14.4c). In this experiment leaf water potential and transpiration in the rest of the plant remained unaffected (Fig. 14.4b, data points at *A*).

When the vapor concentration difference of the air surrounding the entire plant was increased, the water potential of the entire plant decreased about 0.2 MPa (compare points *A* and *B* in Fig. 14.4b). This decrease in whole-plant water potential caused the water potential in the single test leaf, which was kept at a high air humidity, to decrease as well without any change in leaf conductance (Figs. 14.4a and 14.4c, point *D*). Following this shift in leaf water potential the vapor pressure difference of the air surrounding the test leaf was also increased. This caused an increase in transpiration and a decrease in leaf water potential as a consequence of the liquid flow resistance for water to the test leaf (Fig. 14.4a, parallel slopes of curves *A* and *B*). Stomata closed in dry air, and the sensitivity to dry air was unchanged compared with the previous humidity experiment (compare slope and range of curves *A* and *B* in Fig. 14.4c). But in the second case the response occurred at an even lower water potential than was observed with drying soil (Fig. 14.4c, curve *C*). When the air humidity surrounding the test leaf was again changed from dry to humid conditions, the stomatal response was reversible; that is, leaf conductance increased, and transpiration decreased, to the level of point *D* in Figure 14.4a and c. Increasing the humidity of the air surrounding the entire plant caused an increase of the leaf water potential of both the whole plant and the test leaf to the starting conditions. Although stomatal closure in response to dry air and concurrent changes in leaf water potential and stomatal closure in response to dry soil showed similar slopes (Fig. 14.4c, curves *A*, *B*, and *C*), there were important differences between the effects of dry air and the effects of dry soil. When stomata closed in response to high vapor concentration difference, the flux of water through the plant was high, and the stomatal response was reversible when the vapor concentration difference was decreased. When stomata closed at a low vapor concentration difference in response to dry soil, however, the flux of water through the plant was low. It was shown in separate experiments that in the latter case the stomatal response was not easily reversible (Schulze and Hall, 1982).

Figure 14.4 shows for *Helianthus nuttallii* that the stomatal response to changes in air humidity at an initial stage of drying soil is not altered by changes in leaf water potential due to an increase in the transpiration of the whole plant. This phenomenon appears to be more general; it is observed not only in mesophytic herbaceous plants but also in sclerophyllous woody species. Figure 14.5 shows the relation between leaf conductance and leaf water potential of a test leaf in *Nerium oleander*. While the soil was drying, leaf water potential and

Stomatal Responses to Air Humidity and to Soil Drought   317

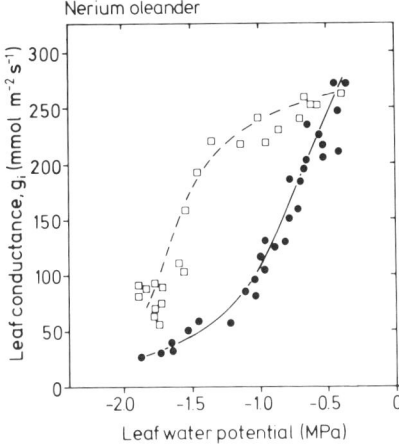

Fig. 14.5. Relationship between leaf conductance (g_i) and leaf water potential (Ψ_i) in *Nerium oleander* as soil water content decreased and vapor concentration difference (Δw_i) for the test leaf was maintained at 10 Pa kPa^{-1} while the whole plant was either maintained at an atmosphere with a constant vapor concentration difference of 10 Pa kPa^{-1} (●) or with a vapor concentration deficit varied daily from 10 to 30 Pa kPa^{-1} (□). (Adapted from Gollan, Turner, and Schulze, 1985.)

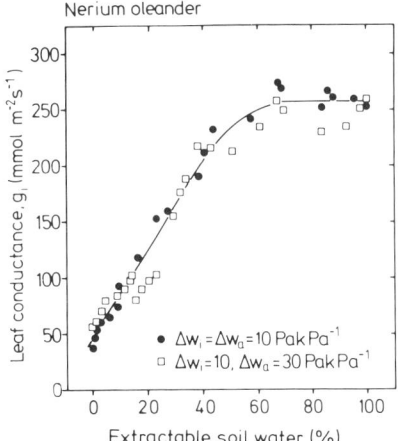

Fig. 14.6. Same data as Figure 14.5 but showing the relationship between the percentage of extractable soil water and leaf conductance. (Adapted from Gollan, Turner, and Schulze, 1985.)

leaf conductance decreased. But increasing the whole-plant transpiration changed the relationship between leaf conductance and leaf water potential in a test leaf at constant high humidity over the entire range of water potentials, compared with the relationship when the transpiration of the whole plant was low (Gollan, Turner, and Schulze, 1985). Thus, changing the humidity of the air surrounding the plant can change the relationship between leaf conductance and leaf water potential in a test leaf under constant humidity conditions without altering the leaf conductance in this leaf. We conclude that in this type of experiment leaf water potential *per se* did not affect stomata.

Stomatal responses to soil drought. When the soil was drying, leaf conductance in a test leaf of *Nerium oleander* showed a very divergent relationship to leaf water potential, depending on the transpiration of the remainder of the

318 *Schulze et al.*

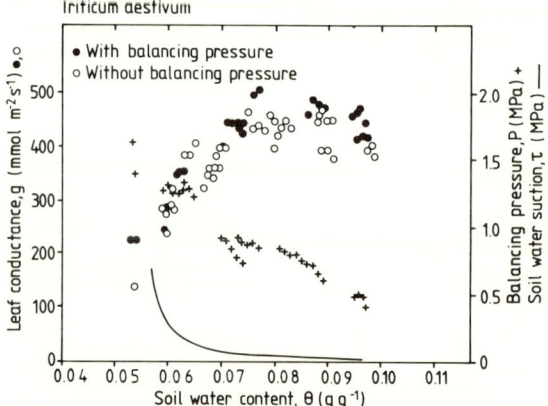

Fig. 14.7. Relationship between soil water content (θ) and leaf conductance (g), soil water suction (τ), and the balancing pressure (P) in *Triticum aestivum*. (○) leaf conductance of control plants, to which no balancing pressure was applied; (●) leaf conductance of plants that were maintained at high leaf water potential by applying the balancing pressure (+). The solid line shows the increase in soil water suction at decreasing soil water content. (Adapted from Gollan, Passioura, and Munns, 1986.)

plant. But when the same data shown in Figure 14.5 were plotted as a function of soil water status (Fig. 14.6), stomata closed at identical levels of soil water depletion (Gollan, Turner, and Schulze, 1985). Similar results were found in *Helianthus annuus* (Turner, Schulze, and Gollan, 1985). Thus, it appears that stomata respond directly to soil drought. This hypothesis was tested by Gollan, Passioura, and Munns (1986), using an experimental approach that excludes any effects of leaf water potential on leaf conductance when soil water potential is decreasing. Plants of *Triticum aestivum* were grown in pots that could be enclosed in a pressure chamber. By using an electronic device the pressure in the chamber could be controlled so as to maintain the pressure in the xylem of the shoot at zero. Thus the decrease in soil water potential was balanced by the pressure applied to the soil, and leaf water potential in the shoot was unaffected by the drying soil. Details of the system are described by Passioura and Munns (1985) and Passioura and Tanner (1985). At a constant vapor concentration difference, leaf conductance of plants that were maintained at zero water potential by applying the balancing pressure showed a similar response to drying soil, compared with leaf conductance of control plants whose leaf water potential was allowed to decrease (Fig. 14.7). Stomata of leaves that never experienced a decrease in leaf water potential closed in response to drying soil.

Since no relationship was found between leaf water potential and leaf conductance (Fig. 14.6), and since leaf conductance decreased in fully turgid plants (Fig. 14.7) as a response to drying soil, it seems likely that the stomatal response to dry soil is triggered by a metabolic signal from the roots. The possible nature of that signal is discussed by Schulze (1986b).

Effects of Changes in Air Humidity on CO_2 Assimilation

In addition to their effect on stomatal conductance, decreases in air humidity result in reduced rates of CO_2 uptake (Ball and Farquhar, 1984; Sharkey, 1984; Tenhunen et al., 1984). When air humidity was decreased stepwise and was then returned to its initial level, stomatal conductance of *Prunus armeniaca* returned to its initial level, but CO_2 assimilation did not (Fig. 14.8). Since the water content of the leaves actually increased during the exposure to dry air, this could not have been an effect of increased water stress. Recently, this observation was further investigated by Sharkey (1984). He suggested that high rates of transpiration and low soil water potentials had analogous effects on the photosynthetic capacity of leaves of *Xanthium strumarium*. The important observation is, assimilation and stomatal conductance appear to change in concert, such that the mesophyll-internal CO_2 concentration remains constant (Ball and Farquhar, 1984; Tenhunen et al., 1984). But mesophyll and stomata also have the ability to respond independently to environmental signals. Stomata of isolated epidermis, without remnants of mesophyll cells, show the

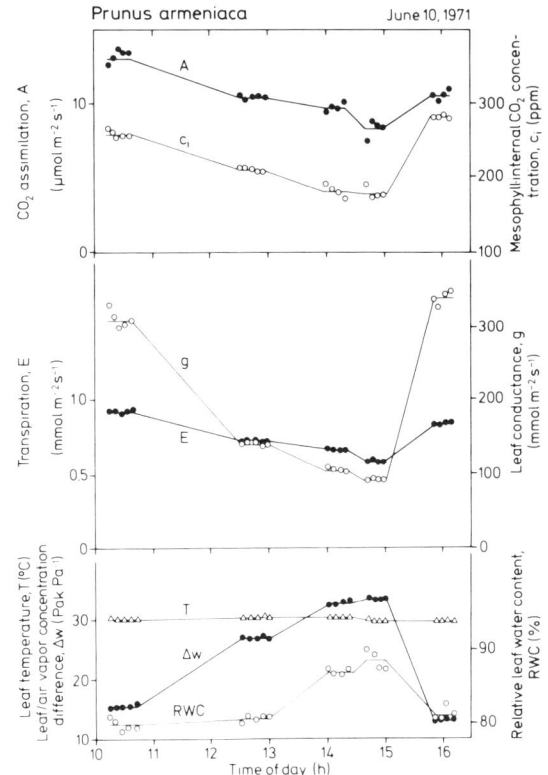

Fig. 14.8. Changes in CO_2 assimilation (A), mesophyll-internal CO_2-concentration (c_i), transpiration (E), leaf conductance (g), leaf temperature (T), and relative water content (RWC) during changes of air humidity (Δw) in *Prunus armeniaca*. (Adapted from Schulze et al., 1972.)

same response to humidity as stomata attached to the mesophyll (Lösch, 1977). A direct and independent response in the epidermis and in the mesophyll is conceptually simpler, and it is possible that both are controlled by a common trigger (Raschke and Hedrich, 1985).

Conclusion

Experimental evidence is given that stomata respond to changes in air humidity and soil water status, but there is no response in relation to leaf water status in the range of water potentials that can be obtained by perturbations via changes of the transpiration rate. Stomata respond to soil water status in such a way that the decrease in soil water status is not mediated by a decrease in xylem water potential. The changes in CO_2 assimilation that were observed during stomatal responses to air humidity seem to be a consequence of rather than a control factor for stomatal closure. There is an aftereffect of low air humidity on CO_2 assimilation. Stomata also close in response to dry air in isolated epidermes. At the cellular level, water potential gradients within the leaf cause epidermal cell turgor to be very sensitive to changes in air humidity. Stomata are triggered by epidermal turgor but have their own metabolic time constant and response of opening and closing. Besides a feedforward response of stomata to air humidity, there seems to be an additional feedforward response of stomata to soil drying. Evidence has been presented for a direct response of stomata to dry soil.

REFERENCES

Ball, M. C., and G. D. Farquhar. 1984. Photosynthetic and stomatal responses of two mangrove species, *Aegiceras corniculatum* and *Avicennia marina*, to long term salinity and humidity conditions. *Plant Physiology*, 74: 1–6.

Gollan, T., J. B. Passioura, and R. Munns. 1986. Soil water status affects the stomatal conductance of fully turgid wheat and sunflower leaves. *Australian Journal of Plant Physiology*, 13: 459–64.

Gollan, T., N. C. Turner, and E.-D. Schulze. 1985. The responses of stomata and leaf gas exchange to vapour pressure deficits and soil water content, III. In the sclerophyllous woody species *Nerium oleander*. *Oecologia*, 65: 356–62.

Lange, O. L, R. Lösch, E.-D. Schulze, and L. Kappen. 1971. Responses of stomata to changes in humidity. *Planta*, 100: 76–86.

Lösch, R. 1977. Responses of stomata to environmental factors—Experiments with isolated epidermal strips of *Polypodium vulgare*, I. Temperature and humidity. *Oecologia*, 29: 85–97.

Passioura, J. B., and R. Munns. 1985. Hydraulic resistance of plants, II. Effects of rooting medium, and time of day, in barley and lupin. *Australian Journal of Plant Physiology*, 11: 341–50.

Passioura, J. B., and C. B. Tanner. 1985. Oscillations in apparent hydraulic conductance of cotton plants. *Australian Journal of Plant Physiology*, 12: 455–62.
Raschke, K., and R. Hedrich. 1985. Simultaneous and independent effects of abscisic acid on stomata and the photosynthetic apparatus in whole leaves. *Planta*, 163: 105–18.
Schulze, E.-D. 1986a. Whole-plant responses to drought. *Australian Journal of Plant Physiology*, 13: 127–41.
———. 1986b. Carbon dioxide and water vapor exchange in response to drought in the atmosphere and in the soil. *Annual Review of Plant Physiology*, 37: 247–74.
Schulze, E.-D., and A. E. Hall. 1982. Stomatal responses, water loss and CO_2 assimilation rates of plants in contrasting environments. In O. L. Lange, P. S. Nobel, C. B. Osmond, and H. Ziegler, eds., *Physiological plant ecology* 2, Encyclopedia of Plant Physiology, n.s., 12B: 181–230. Berlin: Springer-Verlag.
Schulze, E.-D., O. L. Lange, U. Buschbom, L. Kappen, and M. Evenari. 1972. Stomatal responses to changes in humidity in plants growing in the desert. *Planta*, 108: 259–70.
Schulze, E.-D., K. Schilling, and S. Nagarajah. 1983. Carbohydrate partitioning in relation to whole plant production and water use of *Vigna unguiculata* (L.) Walp. *Oecologia*, 58: 169–77.
Shackel, K., and E. Brinckmann. 1984. In-situ measurement of epidermal cell turgor, leaf water potential and gas exchange in *Tradescantia virginiana* (L.). *Plant Physiology*, 78: 66–70.
Sharkey, T. D. 1984. Transpiration-induced changes in the photosynthetic capacity of leaves. *Planta*, 160: 143–50.
Stålfelt, M. G. 1956. Die stomatäre Transpiration und die Physiologie der Spaltöffnungen. In W. Ruhland, ed., *Pflanze und Wasser*, Handbuch der Pflanzenphysiologie 3: 351–426. Berlin: Springer-Verlag.
Tenhunen, J. D., O. L. Lange, J. Gebel, W. Beyschlag, and J. A. Weber. 1984. Changes in photosynthetic capacity, carboxylation efficiency, and CO_2 compensation point associated with midday stomatal closure and midday depression of net CO_2 exchange of leaves of *Quercus suber*. *Planta*, 162: 193–203.
Turner, N. C., E.-D. Schulze, and T. Gollan. 1984. The responses of stomata to vapour pressure deficits and soil water content, I. Species comparison at high soil water contents. *Oecologia*, 63: 338–42.
———. 1985. The responses of stomata to vapour pressure deficits and soil water content, II. In the mesophytic species *Helianthus annuus*. *Oecologia*, 65: 348–55.

15

Diurnal Variations in Leaf Conductance and Gas Exchange in Natural Environments

J. D. Tenhunen, R. W. Pearcy, and O. L. Lange

In terrestrial vascular plants, stomata control both the transfer of carbon dioxide from the ambient atmosphere to the sites of carboxylation in the leaf mesophyll and the loss of water from the wet surfaces within the leaf to the ambient air. These highly sensitive valves respond to internal as well as external factors. Their function is a compromise. By opening, they facilitate photosynthetic carbon fixation in the leaf, and because of the cooling associated with transpiration, they help avoid thermal damage to the leaf if ambient temperature is high. By closing, they conserve water and reduce the risk that the plant may become dehydrated. Understanding the significance of stomatal control in different species and in different environments requires a knowledge of stomatal responses to environmental factors, such as light, temperature, humidity, and water availability, varied both separately and in combination. This paper will demonstrate the influence of such factors by describing characteristic diurnal time courses of stomatal conductance for water vapor and rates of leaf gas exchange in plants growing in arid and humid habitats (see also Schulze and Hall, 1982). We shall try to interpret these time courses in terms of their significance for the growth of particular species in particular habitats.

Reliable measurements of gas exchange and stomatal conductance under field conditions can be obtained only with special apparatus that ensures minimal disturbance of leaf microclimate. Earlier in this century, liquid infiltration, viscous flow porometry, and leaf-surface impressions were some of the methods used to obtain information about stomatal aperture and its variation during the course of the day (see review of Meidner, 1981, and this volume). Reliable techniques for measuring rates of leaf transpiration and CO_2 assimilation and the use of leaf transpiration rates to estimate stomatal conductance were first applied under field conditions about 60 years ago. In his investigations of water relations and primary production in North African desert plants, one of the pioneers in this field, Otto Stocker, utilized two different measuring systems. First, he enclosed a leaf in a small transparent chamber, drew an air stream through the chamber with a hand-operated pump, and determined the amount of CO_2 that had been removed from a known volume of air (CO_2 assimilation) by recording the conductivity change occurring in an absorbing solution (Fig. 15.1A). Then he measured the rate of leaf water loss by weighing excised leaves before and after a short period of exposure to natural conditions (*Momentanmethode*: see Huber, 1927; Stocker, 1929). Stocker is seen in Figure 15.1B with the torsion balance he used in Mauritania in the southern Sahara. His experiments gave us our first comprehensive insight into regulation of gas exchange and stomatal conductance by plants that remain metabolically active under desert conditions. Figure 15.1C illustrates with an original drawing the type of information on CO_2 assimilation and transpiration rate that was obtained during this early period. Stomata of *Capparis spinosa* were shown to close during midday to such an extent that transpiration decreased, despite the increase in potential rate of evaporation (Stocker, 1954). Rate of CO_2 uptake was strongly depressed; sometimes there was a net loss of CO_2 during the hottest part of the day. These first observations were exciting to early investigators, but a number of technical problems caused concern about the validity of the measurements. For example, enclosure of leaves in transparent chambers without temperature control can result in heating of the leaf tissue (Bosian, 1960), or transpiration can increase humidity inside the cuvette and thereby can lead to an increase in stomatal aperture (Lange, Koch, and Schulze, 1969); or again, sudden changes in leaf water relations (*Ivanoff-effect*: see Stocker, 1956) due to leaf excision can affect stomatal conductance and therefore the rate of transpiration.

Modern techniques have overcome such problems. To study plant material within the cuvette, systems have been developed for continuous control of the environment and for measuring water vapor and CO_2 exchange. The material may be a single leaf, a branch, or even all the above-ground parts of a plant. With leaves, diffusion conductance is calculated from the rate of transpiration and from differences in the leaf-to-air water vapor concentration. Reference signals for controlling temperature and humidity in the cuvette are provided by sensors in the ambient atmosphere, so that environmental variations inside mimic the outside (Mooney et al., 1971; Koch, Lange, and Schulze, 1971). A cuvette being used in this way to investigate gas exchange in leaves of Mediterranean evergreen sclerophyll shrubs is seen in Figure 15.1D. The

---

Fig. 15.1 (facing page). Historical and recent apparatus and methods for measuring leaf gas exchange and leaf conductance for water vapor. *A*, The Darmstadt apparatus of Stocker and Vieweg (1960) for determining leaf CO_2 uptake by measuring changes in the conductivity of NaOH solutions: *K*, cuvette; *G*, conductivity meter; *P*, pump; *WB*, registration device. (Reproduced with permission from Stocker and Vieweg, 1960.) *B*, Transpiration being measured in *Acacia tortilis* with the torsion-balance weighing method by Stocker in Mauritania. (Photo: O. L. Lange.) *C*, Diurnal time courses of leaf CO_2 assimilation rate (*Ass*), dark respiration rate (*At*), transpiration rate (*Tr*), and potential evaporation rate from wetted blotting paper (*Ev*), obtained for leaves of *Capparis spinosa* in southern Algeria on May 19, 1953, with the methods illustrated in Figs. 15.1A and B. (Reproduced with permission from Stocker, 1954.) *D*, Humidity- and temperature-controlled gas-exchange chamber (Walz Mess- und

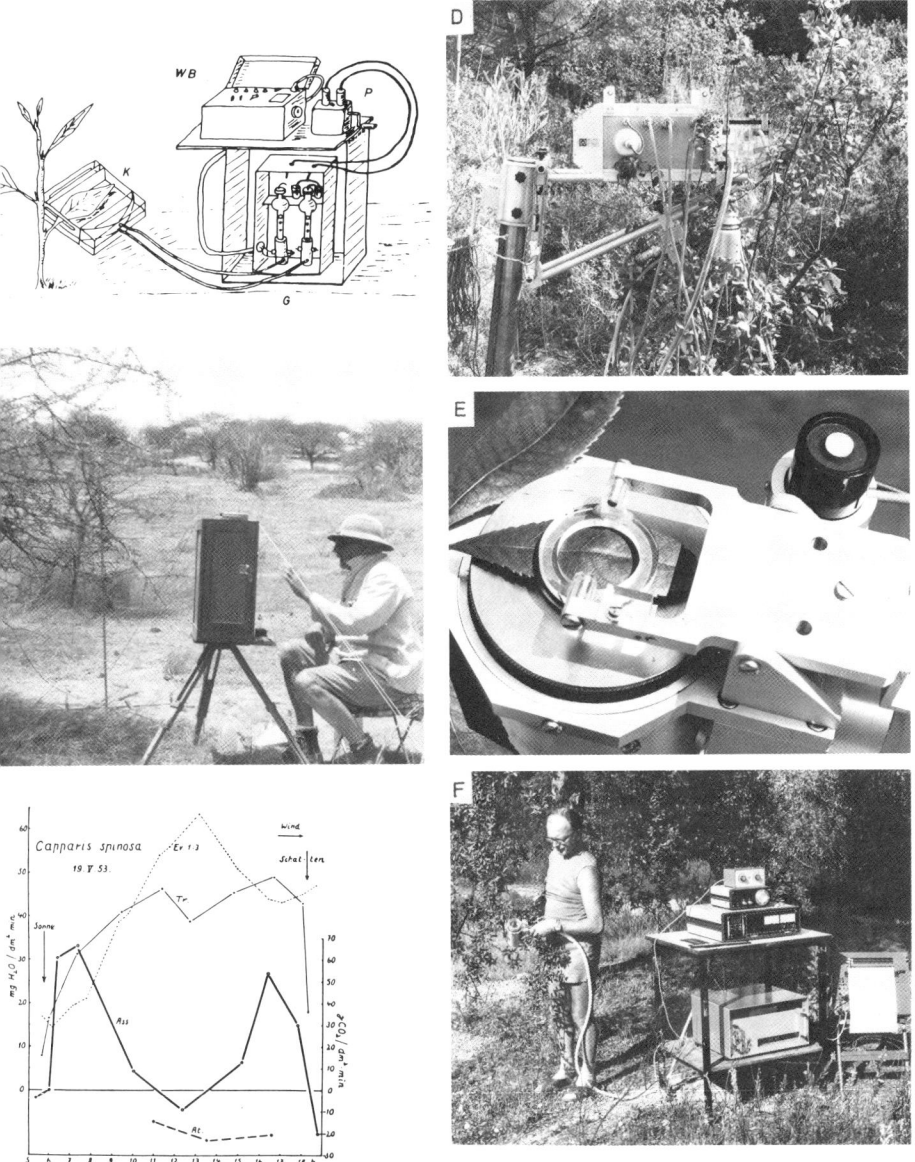

Regeltechnik, Effeltrich, West Germany) being used for continuous automatic recordings of gas exchange, conductance, and environment of leaves of *Arbutus unedo* growing in a maquis in Sobreda, Portugal (see also Tenhunen et al., 1984b). (Photo: O. L. Lange.) *E*, A leaf of *Arbutus unedo* clamped into position for measurement in the cuvette of an H_2O/CO_2 porometer. (Photo: Walz Mess- und Regeltechnik.) *F*, The porometer shown in Fig. 15.1E being used to measure gas exchange and conductance in leaves of *Quercus suber* in Portugal: shown are the measurement cuvette (held by the investigator), the air pumping unit (on the table, above), incoming humidity control unit (below), and line recorder (at right). (Photo: O. L. Lange.)

associated control systems, gas analyzers, and data acquisition system are housed in a mobile laboratory.

More compact, portable systems are now available to measure rates of leaf gas exchange and stomatal conductance. They use cuvettes so constructed that the internal environment, though uncontrolled, changes little during the minute or two when the leaf is enclosed. The first examples of such equipment were the transpiration porometers, so called because they were used to determine rate of transpiration and stomatal conductance (see Beardsell, Jarvis, and Davidson, 1972; Slavik, 1974; Tenhunen et al., 1980). Progress in the technology of CO_2 measurement has led to a second generation of such instruments, which measure the rate of CO_2 fixation also (Griffiths and Jarvis, 1981; Field, Berry, and Mooney, 1982; Williams, Gurner, and Austin, 1982; Schulze et al., 1982; McPherson, Green, and Rollinson, 1983; Carbonneau, 1983). An H_2O/CO_2 porometer is shown in Figure 15.1E. Only 400 mm^2 of the lower surface of the leaf is exposed in the cuvette during the determination of gas exchange. Figure 15.1F depicts the measuring system used for the experiments with the sclerophyllous cork oak, *Quercus suber*, described in this paper. The use of such instrumentation has broadened our knowledge about natural stomatal response under fluctuating environmental conditions.

Patterns in Arid Habitats

Characteristic diurnal patterns. The variation in leaf conductance on a projected leaf-area basis (G) and the rate of net photosynthesis (NP) in leaves of the cork oak, *Quercus suber*, shown in Figure 15.2, are typical for plant species adapted to habitats with a prolonged dry season. Throughout this paper, "leaf conductance" (G) refers to the conductance for vapor transfer across the epidermis and external boundary layer of the leaf. Because the rate of ventilation in the cuvettes used for these measurements is constant and large, and because cuticular conductance in these leaves is small, variations in G are chiefly determined by variations in stomatal aperture. The daily patterns shown for *Quercus suber* were recorded on clear days in an area of natural Mediterranean scrub vegetation dominated by sclerophyll species in Portugal (Sobreda da Caparica, near Lisbon; see Tenhunen et al., 1981), while tracking ambient temperature and humidity. Also shown are time courses for NP observed under similar conditions, but with ambient partial pressure of CO_2 (P_a) increased to 2,500 μbar. There is a marked similarity between the patterns of NP and G measured with normal P_a. Undoubtedly G limits NP somewhat, but since the patterns in NP with high P_a show the same diurnal and seasonal trends, variations in light, temperature, and plant water relations must significantly influence photosynthetic activity. This leads to the question of whether G is in some way directly influenced by photosynthesis in the leaf mesophyll (see Heath, 1948; Farquhar, Dubbe, and Raschke, 1978; Cowan and

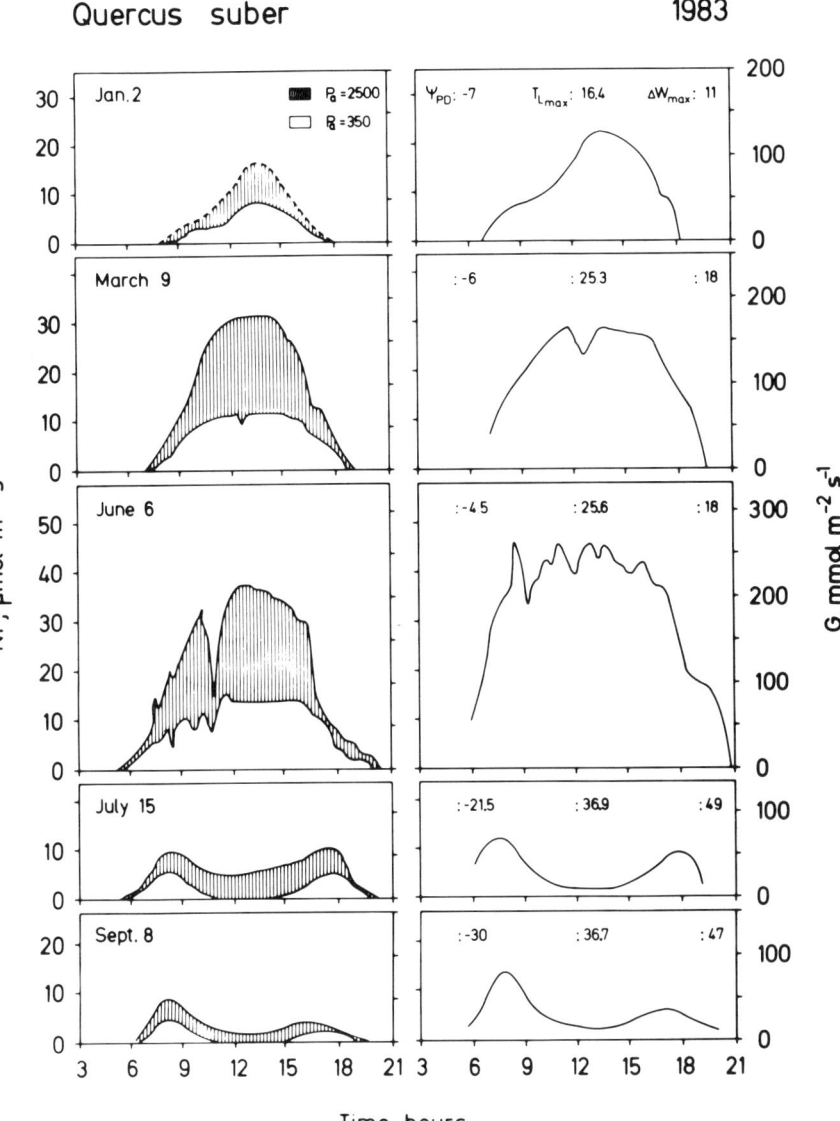

Fig. 15.2. Diurnal time courses of leaf conductance (*G*) and net photosynthesis rate (*NP*) in leaves of *Quercus suber* at Sobreda da Caparica, near Lisbon, Portugal, observed with a cuvette system of the type shown in Figure 15.1D. The predawn leaf water potential (Ψ_{PD}) in bar, maximum leaf temperature ($T_{L_{max}}$) in °C, and maximum humidity difference between leaf and external air (ΔW_{max}) in mbar bar^{-1} on each day is shown. Also shown are time courses of *NP* observed on subsequent days with similar light intensity, leaf temperature, and cuvette humidity, but with 2,500 μbar rather than normal ambient CO_2 partial pressure (P_a) maintained in the cuvette. The high-CO_2 time course on January 2 (dashed line) was constructed for ambient light and temperature conditions from light and temperature response curves of *NP* measured at 2,500 μbar CO_2.

Farquhar, 1977; Wong, Cowan, and Farquhar, 1979; for an alternative view, see Jarvis and Morison, 1981).

During cool weather with predawn plant water potentials (Ψ_{PD}) near zero (Fig. 15.2, January to June), and even during cool periods with moderate plant water stress (Tenhunen, Lange, and Jahner, 1982), leaf responses in *Quercus suber* resemble those typical of more mesophytic species under mesic conditions. Stomata open rapidly with increase in light and temperature during the early morning and remain open for the remainder of the day. Stomatal conductance, net photosynthesis rate, and CO_2-saturated photosynthesis rate are maximum at midday, when light intensity and leaf temperature are greatest; both these rates and stomatal conductance are depressed by cold in the winter, sometimes severely, depending on the sensitivity of individual species (see Larcher and Bauer, 1981). Leaf age or stress history, or both, also affect the maxima, as may be seen from the data obtained with year-old leaves on March 9 and with present-year leaves (which developed in April) on June 6, both exposed to similar environmental conditions (Figure 15.2; see also Schulze and Hall, 1982).

During periods of moderate water stress, such as those at the beginning of the summer drought, when Ψ_{PD} begins to drop and there is a large atmospheric humidity deficit at midday, plants of arid habitats conserve water by means of midday stomatal closure. The daily patterns in G, as well as in NP and transpiration rates (Tr, not shown), are characterized by morning and afternoon maxima with a decrease at midday (Fig. 15.2, July 15). When soil water stress is more severe, stomata open widest during the morning, close at midday, and tend less to reopen during the afternoon (Fig. 15.2, September 8). Morning opening of stomata and the substantial photosynthetic capacity of the mesophyll at leaf temperatures below 30° C ensure a moderate rate of CO_2 uptake, at least during the early hours of the day. The carbon thus fixed is important if the plant is to maintain a favorable carbon balance during periods of stress. With high air temperature and a large humidity deficit, and under extreme water stress, NP at midday may be near compensation or even negative. An efflux of CO_2 may continue throughout the afternoon that diminishes with decreasing temperature, then increases again with decreasing light intensity late in the day (Schulze and Hall, 1982).

The characteristics of leaf gas exchange described above have been documented for a large variety of plant types and species. The two-peak pattern associated with moderate leaf water stress and the less symmetrical patterns, with a predominant morning peak, that are associated with more severe stress have been observed in leaves of sclerophyll shrubs (von Guttenberg and Buhr, 1935; Rouschal, 1938; Poljakoff, 1946; Oppenheimer, 1947; Mooney and Dunn, 1970; Mooney and Kummerow, 1971; Morrow and Mooney, 1974; Dunn, 1975; Eckardt et al., 1975; Mooney, Harrison, and Morrow, 1975; Tenhunen et al. 1980, 1981, 1982, 1984a, b; Tenhunen, Lange, and Braun,

1981; Lösch et al., 1982; Lange, Tenhunen, and Braun, 1982), of woody species of the Australian arid zones (Hellmuth, 1971; Schulze et al., 1982; Pereira et al., 1985), of C_3 and C_4 desert species (Stocker, 1954; Lange, Koch, and Schulze, 1969; Schulze et al., 1980; Schulze and Hall, 1982), of cultivated and crop species growing under desert conditions (Schulze et al., 1974, 1975), and of the gymnosperm *Welwitschia mirabilis* (von Willert et al., 1982).

Midday stomatal closure. Midday stomatal closure should be a well-developed characteristic in species that are regularly confronted with drought stress. We have found midday stomatal closure in practically all woody species of Mediterranean maquis vegetation. But species-specific differences in sensitivity to high leaf temperatures and to large humidity deficits, and species-specific differences in the tendency to exhibit midday stomatal closure, have been found both in sclerophylls growing in the natural habitat (Tenhunen and Lange, unpublished) and in potted plants studied under simulated habitat conditions in a growth chamber (Lange, Tenhunen, and Braun, 1982).

The typical environmental conditions in Sobreda under which plants exhibited midday stomatal closure and depression of gas exchange are illustrated in Figure 15.3. Midday stomatal closure commonly occurs during periods of high light intensity, which contributes to increases in leaf temperature (T_L) and consequently to increases in the humidity difference between leaf and air (ΔW). In some plant species in arid ecosystems, midday stomatal closure has been correlated with change in leaf water potential (Roy and Mooney, 1983). We have not observed this; rather, stomata appear to close with little change in water potential (Ψ; Fig. 15.3). Often the closure results in an increase in leaf water potential that persists until the stomata reopen in response to reduced T_L and ΔW (Tenhunen et al., 1984b).

We have investigated midday stomatal closure as illustrated in Figure 15.3 by subjecting well-watered potted plants of *Arbutus unedo* and other sclerophyll shrubs to a time course of temperature and humidity typical of hot, dry Mediterranean summer conditions. Under these simulated habitat conditions in the growth chamber, the plants responded in the same way as under conditions in the field (Tenhunen et al., 1980; Tenhunen, Lange, and Braun, 1981; Lange, Tenhunen, and Braun, 1982). Soil water stress is therefore not required to elicit midday stomatal closure and midday depression of gas exchange; atmospheric conditions play the primary role (see Schulze et al., this volume). For four successive days we exposed leaves of *Quercus ilex* and other sclerophyllous species to a stepwise increase in light intensity that reached saturating values for photosynthesis around 9 A.M., reached 1,000 μmol m^{-2} s^{-1} photosynthetically active radiation (PAR) at midday, then decreased in the afternoon (i.e. with time-course characteristics similar to those in Fig. 15.3); we also increased the maximum temperature and humidity deficit (the dew point of the air was 15° C) on each day. Under these conditions we

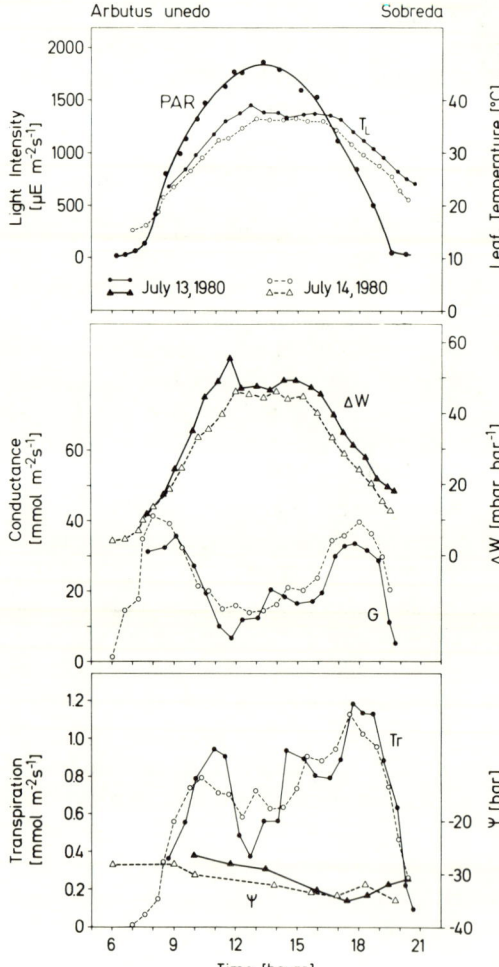

Fig. 15.3. Diurnal time course of light intensity incident on the leaves (*PAR*), leaf temperature (T_L), leaf water potential (Ψ), leaf conductance for water vapor (G), humidity gradient between leaf and air (ΔW) and transpiration rate (*Tr*) in leaves of *Arbutus unedo* in a maquis near Lisbon, Portugal (see Tenhunen et al., 1981). Symbols indicate average values for eight leaves. Measurements conducted with an H_2O porometer (LiCor, Lincoln, Nebraska) with 200 mm² of lower leaf surface clamped to the cuvette. (Redrawn from Tenhunen et al., 1980.)

recorded a greater midday decrease in G, *NP*, and *Tr* on each day (Fig. 15.4). The transpiration ratio, *Tr/NP*, remained below 200 mol mol^{-1} until very high temperature was reached at midday. Then, despite strong stomatal closure, it increased (i.e. water-use efficiency decreased) as *NP* decreased to near zero.

In another series of experiments, the influence of light intensity on midday closure was examined. As in the first series, air temperature (T_A) was gradually increased (by approximately 4° C per hour) after the lights were turned on in the morning. When T_L reached 38° C, in the late morning, T_A was held constant for several hours. Light intensity was held constant at different levels on different days (above and below the saturation level for photosynthesis, approximately 500 μmol m^{-2} s^{-1}). As seen in Figure 15.5, light intensity has little effect on

Quercus ilex

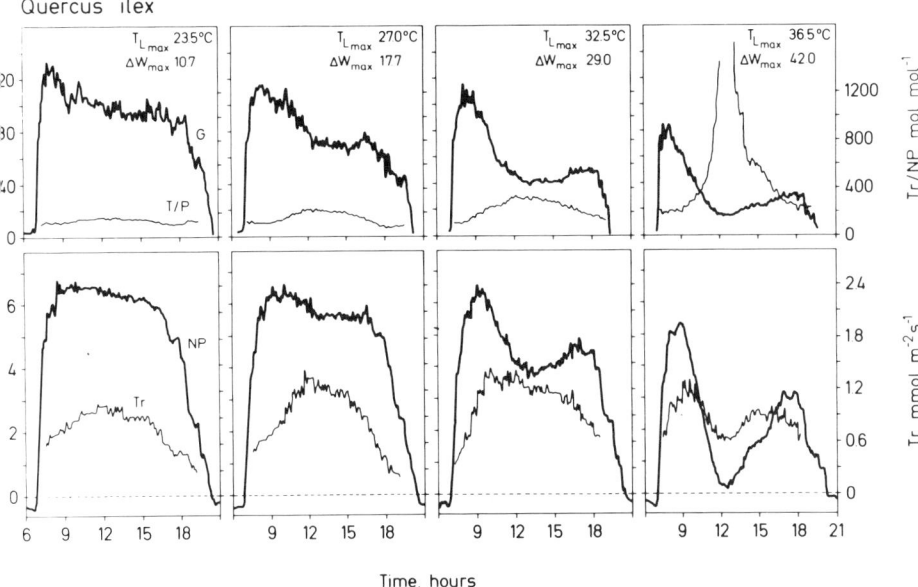

Fig. 15.4. Diurnal time courses of leaf conductance (*G*), rate of net photosynthesis (*NP*), rate of transpiration (*Tr*), and the ratio *Tr/NP* in *Quercus ilex* on four consecutive days in a growth chamber. On each day, the air temperature was increased from 19° C to the maximal values shown in the body of the figure at midday ($T_{L_{max}}$) and was decreased again to 19° C by the end of the photoperiod. Dew point of the air was 15° C, resulting in the maximal values of leaf-to-air humidity difference (ΔW_{max}) indicated in mbar bar^{-1}. Light was changed stepwise and was maximal at midday (see Tenhunen, Lange, and Braun, 1981). Methods as described for Figure 15.2. (Redrawn from Tenhunen, Lange, and Braun, 1981.)

the time courses of conductance. When light intensity is low, however, T_L (and therefore ΔW) is considerably lower, because of the smaller radiation load at the time closure begins (Fig. 15.6). The closure cannot be a response to leaf-internal water stress, because *Tr* at any time during the morning is much smaller on days with low light intensity than on days with high light intensity. The temperature optimum for photosynthesis in C_3 leaves is known also to decrease with decreasing light intensity. Again we are led to ask whether the similarity between responses of *G* and of photosynthesis indicates that the controls of stomatal aperture and photosynthetic activity are interrelated.

It is also suggested in Figure 15.6 that increased light intensity stimulates stomatal opening at temperatures between 25° and 35° C, whereas increased T_L or increased ΔW, or both, will lead to stomatal closure. Direct control on stomatal aperture by ambient humidity is well documented (Lösch and Tenhunen, 1981; see Schulze et al., this volume). During alternate stepwise increases of light and temperature in growth-chamber experiments (Tenhunen et al., 1980; Tenhunen, Lange, and Braun, 1981; Lange, Tenhunen, and Braun,

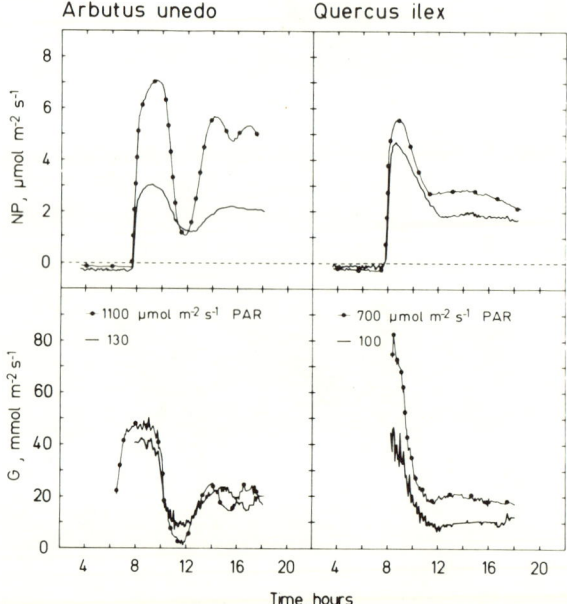

Fig. 15.5. Time course of conductance (G) and net photosynthesis rate (NP) in leaves of *Arbutus unedo* and *Quercus ilex*. Light (PAR) was turned on at 7:45 A.M. and was constant at the intensities indicated. Air temperature was initially 19° C, was increased 4° per hour until a leaf temperature of approximately 38° C was attained, then was held constant. Dew point was 15° C. Methods as described for Figure 15.2.

1982), stomata tend to close with each increase in temperature, then reopen with each increase in light until sensitivity to light is lost; at that point they continue to close with further increases in temperature and ΔW.

Photosynthetic parameters during midday stomatal closure. Tenhunen et al. (1984b) determined the CO_2-response curve of photosynthesis of leaves of *Quercus suber* as a function of time of day during the summer dry season in the natural habitat (see Fig. 15.7). Under conditions that elicited strong midday stomatal closure, CO_2- and light-saturated photosynthetic capacity (P_M) decreased by 65 percent, carboxylation efficiency (CE) decreased similarly, and partial pressure of CO_2 at compensation (Γ) increased from 40 μbar during the early morning to 160 μbar at noon. Leaf conductance (G) decreased from 80 to 20 mmol m^{-2} s^{-1} and in approximately the same proportion as NP at midday, so that leaf-internal partial pressure of CO_2 (P_i) remained almost constant. Schulze et al. (1974) showed that leaves of *Prunus armeniaca* growing at Avdat in the Negev Desert, which commonly exhibit midday depression of CO_2 uptake, showed no depression of NP when the air was saturated with water vapor and even though soil water was depleted at the end of the dry season. Presumably stomata do not close under these conditions. Likewise, laboratory experiments with *Arbutus unedo* (Beyschlag, unpublished) show that decreases in CO_2-saturated photosynthetic capacity, which normally occur at midday if temperature is high, are also eliminated if ambient humidity is kept high.

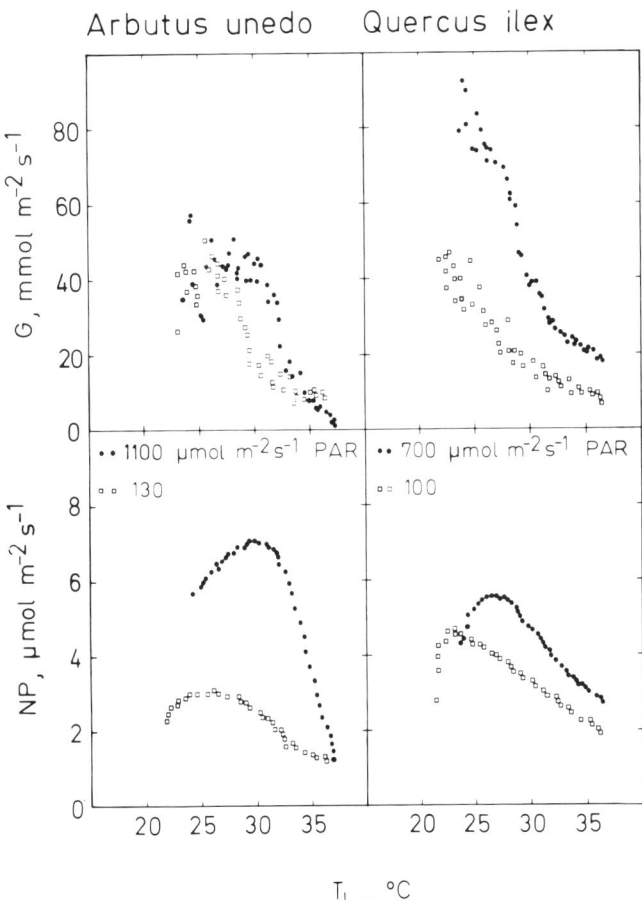

Fig. 15.6. Conductance (G) and net photosynthesis rate (NP) in leaves of *Arbutus unedo* and *Quercus ilex* (same data as in Fig. 15.5) plotted as functions of leaf temperature.

Raschke (1982) has found that applying the hormone abscisic acid (ABA) to leaves of monocots and dicots can lead to depression of P_M, decrease in CE, stomatal closure, and constant or slightly elevated P_i. Possibly the regulation of midday depression involves hormonal effects on photosynthesis and on stomata, whether simultaneous or sequential. In any case, the sensing of high temperature or high ΔW, or both, is rapid and leads to complex reversible metabolic changes.

Water stress and midday stomatal closure. Soil water stress modifies the characteristics of midday closure (Schulze et al., 1980). Tenhunen, Lange, and

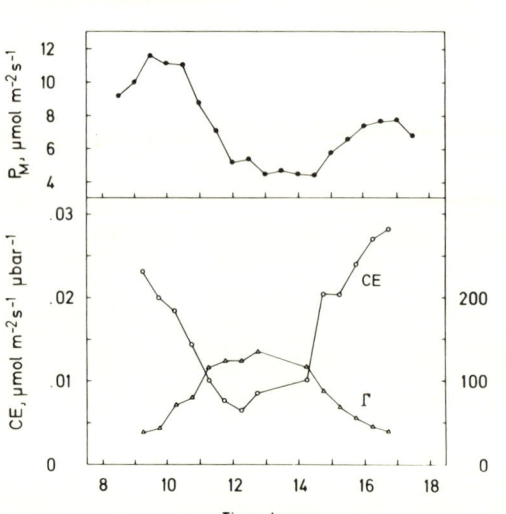

Fig. 15.7. Diurnal time course of CO_2- and light-saturated rate of net photosynthesis (P_M), the initial slope of the CO_2 response of net photosynthesis or carboxylation efficiency (CE), and partial pressure of CO_2 at compensation or when net CO_2 uptake is zero (Γ) in leaves of *Quercus suber* during a day with strong midday stomatal closure and depression of gas exchange. Measurements with shrubs growing in the natural habitat at Sobreda, near Lisbon, Portugal. (Redrawn from Tenhunen et al., 1984b.)

Jahner (1982) measured G in *Arbutus unedo* plants subjected to different watering regimes during the summer dry season at Sobreda. Midday closure in well-watered plants, that is, plants with a predawn leaf water potential (Ψ_{PD}) between -10 and -25 bar, occurred with T_L greater than 30° C and ΔW greater than 30 mbar bar^{-1}. Moderate water stress (Ψ_{PD} between -25 and -35 bar) decreased maximum values of G and resulted in greater sensitivity to high values of T_L and ΔW. Stomata on leaves of moderately stressed plants closed earlier and reopened later in the day (i.e. both reactions occurred at much lower temperature and ΔW values), and stomata remained closed longer at midday. Severe water stress (Ψ_{PD} of -50 bar) resulted in a continuous decline in G following a single early-morning peak. Similar results were obtained with potted plants of sclerophylls that were allowed to become dry while experiencing simulated summer-day conditions (Fig. 15.8). The rate of net photosynthesis varied in the same manner as leaf conductance. Further drying experiments conducted with high P_a (Fig. 15.9) suggest that in most species the variations in NP reflect real variations in the photosynthetic metabolism of the leaf mesophyll and are not determined by variation in G alone. Changes in photosynthetic capacity (NP at CO_2 saturation) observed during drying probably indicate a similar pattern of change in CE (slope of the CO_2 response), since these two parameters seem most often to change together and are proportionally related (cf. Mooney, Björkman, and Collatz, 1977; Björkman, Downton, and Mooney, 1980; Matthews and Boyer, 1984; Tenhunen et al.,

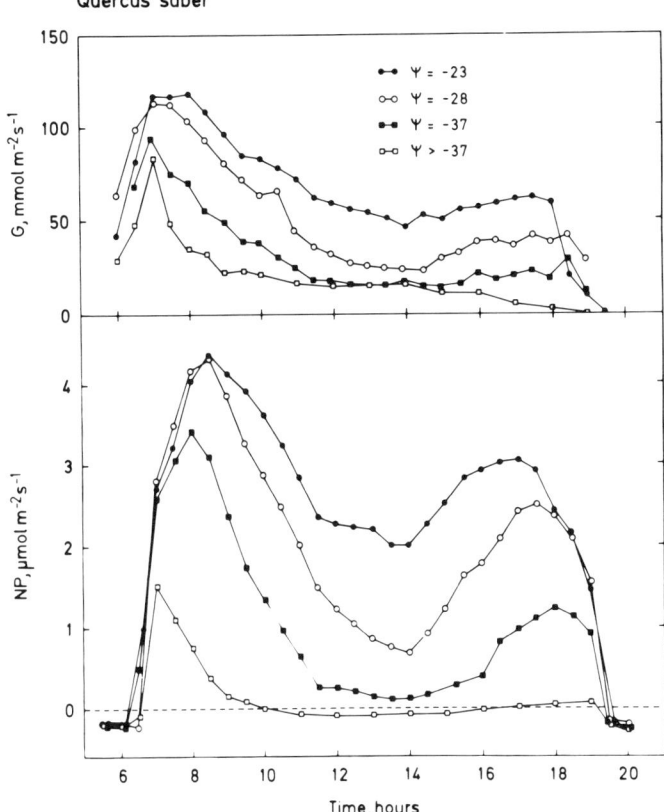

Fig. 15.8. Diurnal time courses of leaf conductance (G) and net photosynthesis rate (NP) in leaves of potted *Quercus suber* plants on the third (solid circles), fourth (open circles), fifth (solid squares), and sixth (open squares) days of a drying cycle with standard day temperature and humidity conditions as discussed in the text and with normal air supplied to the measurement cuvette. Light intensity was increased from 0 to 1,000 μmol m^{-2} s^{-1} between 6 A.M. and 9 A.M., was constant between 9 A.M. and 5 P.M., and then was decreased. Leaf water potentials (Ψ) were obtained at 11 A.M., at the beginning of the constant temperature phase at midday. Methods as described for Figure 15.2. (Redrawn from Gebel, 1983.)

1984b). Schulze et al. (1980) described decreases in leaf photosynthetic metabolism under water stress in the field in response to decreases in Ψ and to length of exposure to decreased Ψ ("bar days").

Importance of midday stomatal closure. Theoretical studies (Cowan and Farquhar, 1977; Cowan, 1982) have suggested that a leaf fixing the maximum possible amount of CO_2 while expending a set amount of available water must regulate stomatal conductance in accordance with the time course of the

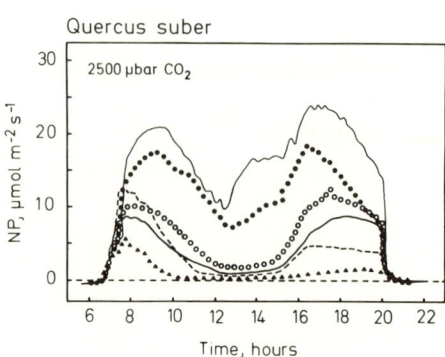

Fig. 15.9. Daily time courses of light-saturated (1,000 μmol m^{-2} s^{-1} PAR) and CO$_2$-saturated (2,500 μbar CO$_2$ external to the leaf) net photosynthesis rate (*NP*) on the first (upper solid line), second (solid circles), fourth (open circles), fifth (lower solid line), sixth (dashed line), and seventh (triangles) days of a drying cycle with standard day temperature and humidity conditions as discussed in the text. Predawn water potentials were -17, -22, -19, -20, -26, and -33 bar, respectively; midday water potentials were -17, -19, -24, no measurement, -27, and -40, respectively. Methods as described for Figure 15.2. (Redrawn from Gebel, 1983.)

weather. As water becomes less available and midday evaporative demand increases, regulation that results in optimal CO$_2$ fixation should increasingly lead to strong midday stomatal closure. Whether optimal regulation actually occurs under field conditions remains to be established. It seems clear, however, that the physiological mechanisms of midday stomatal closure and of reopening in response to atmospheric conditions allow effective water use in a variety of situations in which T_L decreases from temperatures higher than the optimum for *G* and *NP* to temperatures close to this optimum. For example, in some coastal Mediterranean areas, cloud formation is common during the dry season, even though precipitation seldom occurs. The observations depicted in Figure 15.10 illustrate how interruptions of PAR at midday, when stomata have closed, cause decreases in T_L and in ΔW, which then quickly lead to increases in *G* and *NP*. Changes in air circulation under similar circumstances can also markedly affect leaf function. Leaves of *Pistacia lentiscus* were observed during extremely hot weather with still air at midday (Fig. 15.11). Leaf temperatures in the ventilated cuvettes increased to approximately 43° C. In the early afternoon on June 15, a cool sea breeze was sensed by the measurement devices; it resulted in a decrease of several degrees in T_A and T_L within the cuvettes. Although T_L remained well above 35° C, stomata reopened and *NP* increased. Since wind speed within the cuvette remained constant, the effect of this breeze on leaf function in the natural situation was undoubtedly even more dramatic.

The phenomena known as midday stomatal closure and midday depression of leaf gas exchange have been characterized mainly through study of plants adapted to extended drought. Data obtained with temperate-zone species suggest that a similar diurnal variation in *G* and *NP* occurs when these plants

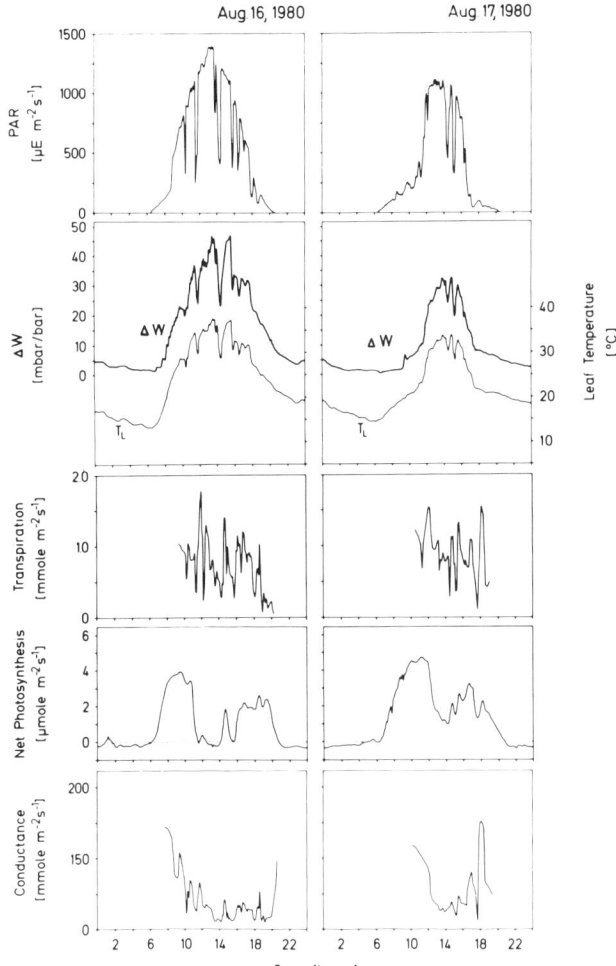

Fig. 15.10. Diurnal time course of light intensity incident on leaves (*PAR*), leaf temperature (T_L), humidity gradient between leaf and air (ΔW), transpiration rate, net photosynthesis rate, and leaf conductance for water vapor in leaves of *Arbutus unedo* on August 16 and August 17, 1980, under hot summer conditions in an experimental garden in Würzburg. Methods as described for Figure 15.2.

experience hot and dry atmospheric conditions and water stress (cf. results obtained with *Glycine max* by Turner et al., 1978; and with *Vitis vinifera* by Lange and Meyer, 1979). Further investigations are needed to define the importance of midday stomatal closure for photosynthetic production and regulation of water use of naturally occurring temperate-zone species.

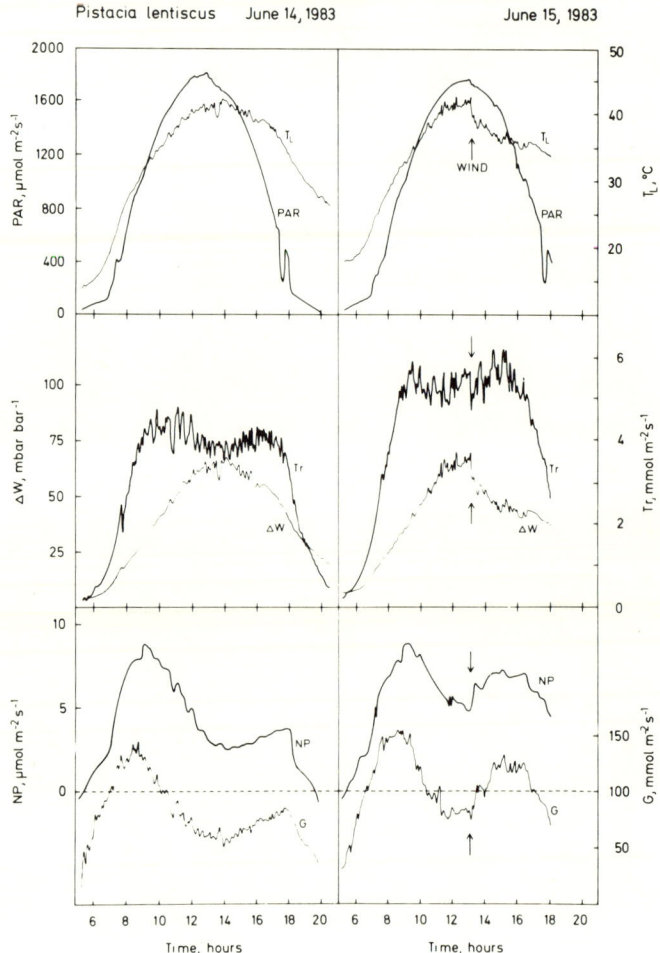

Fig. 15.11. Diurnal time course of light intensity incident on leaves (*PAR*), leaf temperature (T_L), humidity gradient between leaf and air (ΔW), transpiration rate (*Tr*), net photosynthesis rate (*NP*), and leaf conductance for water vapor (*G*) in leaves of *Pistacia lentiscus* on June 14 and 15, 1983, in the natural habitat in Sobreda, near Lisbon, Portugal. Methods as described for Figure 15.2.

Patterns in Humid Environments

Characteristic diurnal patterns. Humid environments, defined as those in which annual precipitation exceeds annual potential evapotranspiration, lack the extended drought characteristic of deserts and of the Mediterranean climate. In some temperate regions and tropical forests, precipitation exceeds 4 m per year. Nevertheless, for periods of a few weeks to a few months during the normal climate cycle, there may be little or no precipitation, and in unusual years prolonged drought may occur. Because of the high leaf-area indexes of plant communities in these regions, soil water deficits can develop rapidly and can markedly affect plant performance. For example, Ψ in an evergreen tropical forest in Panama was as low as -40 bar during a dry season (Robichaux et al., 1984). Under these conditions, the priorities of stomatal behavior—maximizing carbon gain, minimizing transpirational water loss, and preventing excessive water deficits—are likely to be similar in humid-zone and arid-zone vegetation. Differences in stomatal behavior between humid-zone and arid-zone plants, therefore, relate to differences in the natural environments and to quantitative rather than qualitative differences in responses to specific environmental changes.

Observations of the diurnal time course of gas exchange rates and leaf conductances have been made in a variety of humid and semihumid ecosystems, including broad-leafed deciduous forests (Schulze, 1970; Kaufmann, 1982; Weber et al., 1985), deciduous hedgerows (Küppers, 1984a), coniferous forests (Jarvis, 1976; Running, 1976), and tropical forests (Pearcy and Calkin, 1983; Pearcy, 1986). Figure 15.12 shows the responses for the European beech, *Fagus sylvatica*. Both G and NP were strongly light-dependent in the early morning and early evening. At midday, the variation in G and the smaller variation in NP corresponded to variations in T_L and ΔW. However, since photosynthesis of *F. sylvatica* is independent of temperature over the range from 10° to 24° C (Schulze, 1970), the variations of G and NP are due to variations in ΔW. On warmer, drier days humid-zone plants can show a more substantial stomatal closure: for example, observations of Küppers (1984a) on *Prunus spinosa*, a deciduous hedgerow shrub in Central Europe, show that G reached a maximum at 9 A.M. and then declined by 50 percent by 3 P.M. in response to T_L reaching 28° C and ΔW reaching 22 mbar bar^{-1}. The sensitivity of stomata to humidity does not appear in principle to differ between humid- and arid-zone plants (Schulze and Hall, 1982).

Response to humidity. Humidity is much less important in limiting photosynthesis in temperate-zone plants, primarily because large values for ΔW are rare. Schulze (1970) estimated that a large ΔW limits photosynthetic carbon gain of *F. sylvatica* due to stomatal closure by only 2 percent over the growing season and by only 13 percent on the driest days. However, large ΔWs sometimes occur in the tropics, particularly in the seasonally dry forests, because of high T_L. Whitehead, Okali, and Fasehun (1981) used a diffusion porometer to observe diurnal variations in G in two tropical forest trees, *Gmelina arborea* and *Tectonia grandis*, in a southern Nigerian plantation during the dry season. Leaf temperatures were not reported, but air vapor-pressure deficits were as high as 40 mbar bar^{-1}. Leaf conductances peaked in the morning, then declined in the afternoon in response to increasing air vapor-pressure deficits.

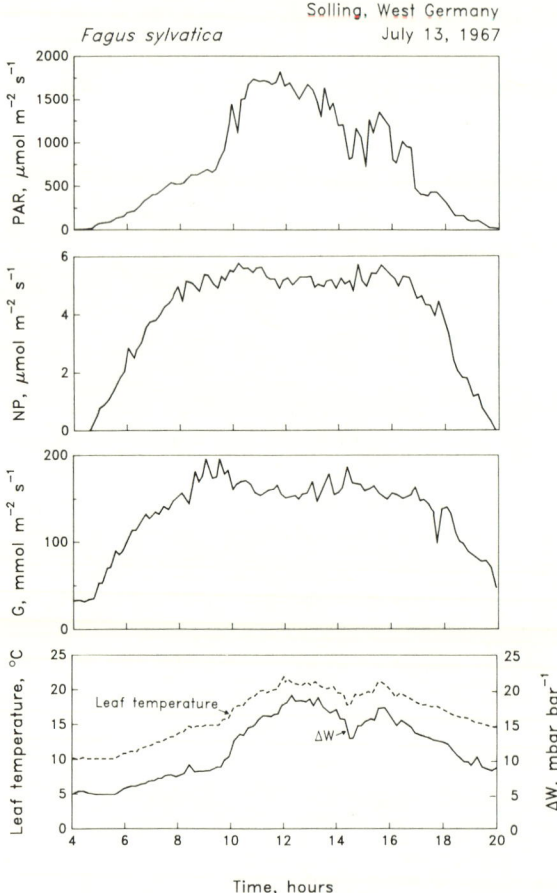

Fig. 15.12. Diurnal time course of light intensity incident on leaves (*PAR*), leaf temperature, humidity difference between leaf and air (ΔW), net photosynthesis rate (*NP*), and leaf conductance for water vapor (*G*) measured for leaves of *Fagus sylvatica*. Methods as described for Figure 15.2. (Redrawn from Schulze, 1970.)

On different days, the greater ΔW was, the earlier G began to decline. Measurements made in the same plantation by Grace, Okali, and Fasehun (1983) during the wet season revealed maximum vapor pressure deficits of only 12 mbar bar^{-1} and higher maximum G. At this time, G was highest at midday.

Response to water potential. The role of Ψ in regulating stomatal behavior has been studied less in humid-zone plants than in arid-zone plants. Leaf water potential did not appear to be an important variable in explaining either the diurnal course or the seasonal changes in G observed in *Tectonia grandis* growing in the tropical Nigerian plantations discussed above. There was little difference in Ψ between the wet and dry seasons, yet maximum G was considerably less in the latter. Küppers (1984b) investigated the effects of Ψ on G in plants growing in hedgerows in central Germany. He kept the light and atmospheric conditions of individual leaves constant over the day, but Ψ

varied, because *Tr* varied in the rest of the plant. Measurements showed that *NP* and *G* often declined slightly over the day but did not vary in a direct relation to Ψ.

Response to fluctuations in light. Light is frequently variable in the humid zone because of cloudiness and shading of leaves within the canopy. Figure 15.13 shows how *NP* and *G* in leaves of *Prunus spinosa* closely follow the changes in light under passing clouds. Because the variations in *NP* and *G* are proportional, P_i remains nearly constant. A comparison of the leaf responses in the upper canopy and in the understory for *Argyrodendron peralatum*, an Australian tropical-forest tree, is shown in Figures 15.14 and 15.15. Light in the upper canopy was variable during the morning because of intermittent clouds and shading by other branches. Consequently, both *G* and *NP* were quite

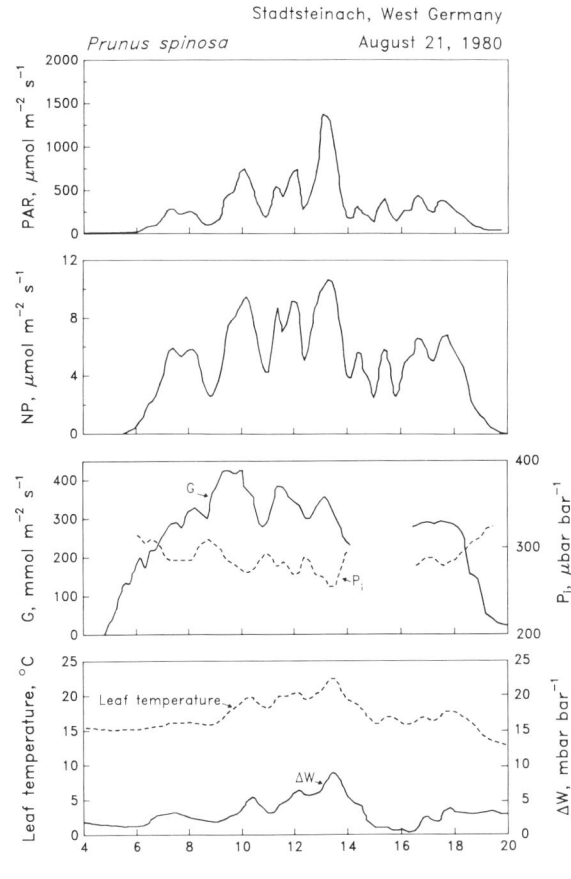

Fig. 15.13. Diurnal time course of light intensity incident on leaves (*PAR*), leaf temperature, humidity difference between leaf and air (ΔW), net photosynthesis rate (*NP*), leaf conductance for water vapor (*G*), and leaf-internal CO_2 partial pressure (P_i) measured for leaves of *Prunus spinosa* on a partly cloudy day. Methods as described for Figure 15.2. (Redrawn from Küppers, 1984a.)

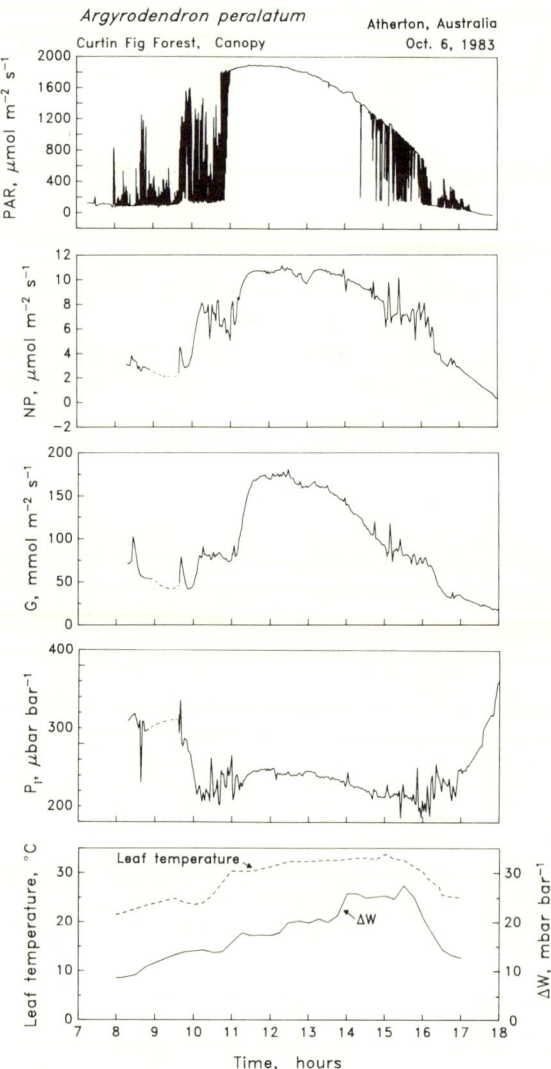

Fig. 15.14. Diurnal time course of light intensity incident on leaves (*PAR*), leaf temperature, humidity difference between leaf and air (ΔW), net photosynthesis rate (*NP*), leaf conductance for water vapor (*G*), and leaf-internal CO_2 partial pressure (P_i) for leaves of *Argyrodendron peralatum* in the canopy, 32 m above ground in an Australian tropical rainforest. (Redrawn from Pearcy, 1985.)

variable. After 11 A.M., *NP* and *G* increased to a maximum when irradiance remained high; these values then declined through the rest of the afternoon, presumably in response to the declining light intensity. In the understory, *NP* and *G* of leaves of the same species measured at saturating light intensities were only half those found in the canopy. These differences reflect parallel long-term adjustments of photosynthetic capacity and of *G* to prevailing light conditions, which have been documented for many species (Björkman, 1982; Schulze and Hall, 1982). Under natural diurnal conditions, *NP* was on the average only

Diurnal Variations in Leaf Conductance and Gas Exchange   343

about one-tenth and G about one-fourth of the values measured at the top of the canopy. Photon flux density was extremely low, ranging from 10 to 20 µmol m^{-2} s^{-1}, except during a few brief sunflecks, when it was 200 to 500 µmol m^{-2} s^{-1}. The dynamic response of photosynthetic CO_2 assimilation to these sunflecks is obvious. Although G was much lower in the understory than in the canopy, it seldom limited the photosynthetic rate. This is evident since P_i was only a few microbars less than P_a when the light level was 10 to 20 µmol m^{-2} s^{-1}. When a sunfleck occurred, stomata may have transiently limited

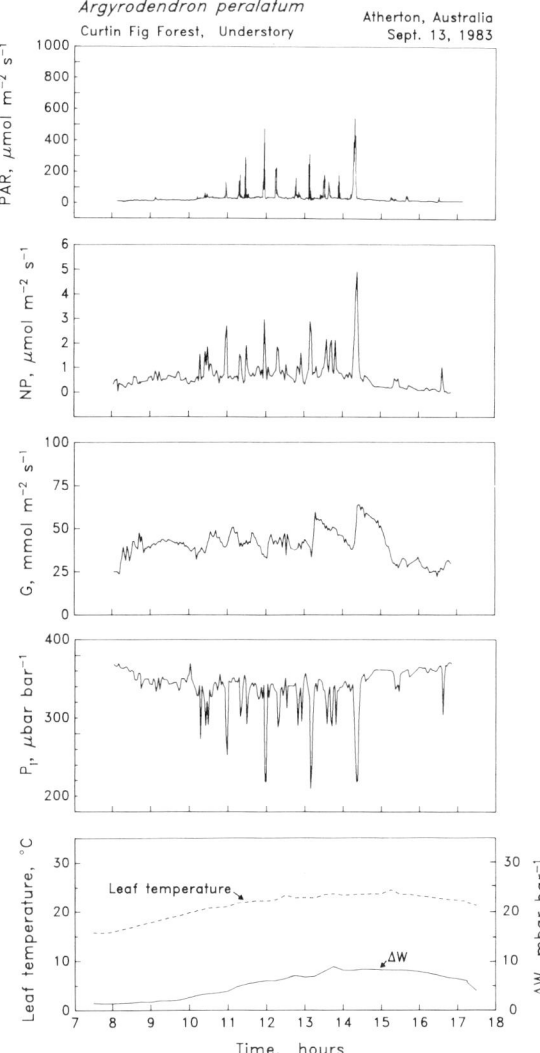

Fig. 15.15. Diurnal time course of light intensity incident on leaves (PAR), leaf temperature, humidity difference between leaf and air (ΔW), net photosynthesis rate (NP), leaf conductance for water vapor (G), and leaf-internal CO_2 partial pressure (P_i) for leaves of *Argyrodendron peralatum* on an understory sapling in an Australian tropical rainforest. (Redrawn from Pearcy, 1985.)

the rate of CO_2 fixation. Even so, the limitation was not large; P_i did not drop below 220 μbar.

The data of Björkman, Ludlow, and Morrow (1972) for *Alocasia macrorrhiza*, a herbaceous understory species common in Australian tropical rainforests, also showed very high P_i, implying that stomata play only a minor role in limiting CO_2 uptake. Küppers (1984a) found for shade leaves of the temperate-zone species *Acer campestre* and *Prunus spinosa* that stomata impose little limitation on photosynthesis. Intercellular CO_2 partial pressure remained above 300 μbar except in the presence of sunflecks, when it dropped to 250 μbar. In contrast, Weber et al. (1985) found that P_i in shade leaves of the forest tree *Acer saccharum* remained at approximately 75 percent of P_a during shady periods and decreased further during sunflecks. This suggests a significant stomatal limitation on photosynthetic CO_2 fixation in this species.

The amount of CO_2 taken up per mole of absorbed photons, or the quantum yield, of C_3 plants depends on P_i because oxygen competitively inhibits carboxylation. At 25° C, the quantum yield for CO_2 uptake by C_3 plants is 14 percent higher at a P_i of 320 μbar than at 240 μbar (Berry and Raison, 1981). Where leaves are in extreme shade and the rate of CO_2 fixation is limited by light nearly all day, the advantage of maintaining open stomata and high P_i could be substantial.

Since G of *Argyrodendron* clearly depends on light intensity and decreases during the low-light periods between sunflecks, it may seem surprising that the stomatal restriction of photosynthesis during sunflecks is not greater. Measurements during step changes in light have shown that one reason it is not is that light-activated photosynthetic enzymes during sunflecks play a greater role than does stomatal opening in limiting CO_2 fixation (Pearcy, 1986). As is evident in Figure 15.15, stomata opened rapidly in response to sunflecks but closed much more slowly. The data of Björkman, Ludlow, and Morrow (1972) for *Alocasia macrorrhiza* show this also. This rapid stomatal opening is not sufficient to prevent a decrease in P_i during the one- or two-minute duration of a sunfleck. The slow closure, however, may result in more efficient use of light energy in subsequent sunflecks, if they occur within a short time. Pearcy (1983) has shown that several sunflecks often occur in close sequence, followed by longer low-light intervals.

The stomatal behavior of extreme shade leaves, as illustrated here with *Argyrodendron*, is not consistent with optimal CO_2 fixation in relation to water use. Low photosynthetic rates and high G result in high P_i. In shade, NP of *A. peralatum* is nearly independent of P_i between 200 and 600 μbar bar^{-1} (Pearcy, 1986). Thus, water-use efficiency could be substantially improved with no penalty in photosynthetic rate if the stomata were narrow in low light. Of course a penalty would be incurred during sunflecks unless the stomata responded rapidly. However, Tr is low, because low vapor pressure deficits are characteristic of these habitats and because water requirements can usually be met. Moreover, it is probably unrealistic to expect that stomata could respond to rapid changes in light intensity in a way consistent with the hypothesis of Cowan and Farquhar (1977), since stomatal opening and closing are clearly much slower processes. In environments where light intensity is usually very low, maximal carbon gain rather than optimal water use should set priorities for stomatal behavior. That stomata remain wide in such environments suggests the former is indeed more important.

Conclusion

Although there are few detailed studies of stomatal behavior in plants in their natural environments, basic similarities have emerged in diurnal time-course characteristics and in the nature of responses to individual environmental changes. Under natural conditions, light and humidity deficit are of great importance in determining stomatal aperture. Much evidence indicates that a direct relationship between stomatal aperture and photosynthetic metabolism may be important in regulating leaf function in the long term (i.e. in relation to seasonal changes in the environment and to changes in the plant's phenological state) and in the short term (i.e. in relation to diurnal variation). This coupling of stomatal and photosynthetic behavior needs to be studied in more detail in the natural environment. The importance of direct effects of light, temperature, and humidity deficit on stomatal conductance has been recognized for some time. New interest in the possible role of processes linked to photosynthesis in the leaf mesophyll was stimulated by Wong, Cowan, and Farquhar (1979) and Körner, Scheel, and Bauer (1979), who emphasized how closely NP and G are correlated, irrespective of whether change in their magnitude might be due to changes in growth conditions or to differences among plant types. Further studies from contrasting environments—for example, of closely related *Euphorbia* species (Pearcy, Osteryoung, and Randall, 1982), of desert annuals (Werk et al., 1983), and of cultivated crop and forest species and naturally occurring perennial species in arid zones (see review of Schulze and Hall, 1982)—also indicate a close relationship between NP and G. Examples reported in this paper of the maintenance of constant P_i despite large changes in G and NP illustrate this interesting phenomenon. It remains to be seen whether there is a mechanistic link between stomatal and mesophyll functions, perhaps involving hormonal regulation, or the two functions only appear to be linked because they respond similarly but independently to external environmental signals.

Under natural conditions, diurnal changes in Ψ seem not to exert a major influence on stomatal aperture: there is no clear correlation between Ψ and G. However, changes in Ψ during the day depend on changes in several factors that control Tr, such as humidity, light intensity, and T_L. Therefore, identifying independent effects of Ψ would be exceedingly difficult. Schulze and Küppers

(1979) attempted to manipulate Ψ and air humidity independently in the laboratory and found no relationship between G and short-term changes in Ψ.

Nevertheless, water status does substantially affect the diurnal course of G in the long term. Maximum G decreases with decreasing Ψ_{PD}, and the response of G to other environmental variables is modified. Again, the close parallel between changes in photosynthetic properties and the effects of long-term water stress on maximum G suggests that an overall metabolic control is acting to regulate carbon fixation and water use. Bates and Hall (1981) interpreted the change in G of *Vigna unguiculata* as a response to soil water content rather than to bulk Ψ and took this to indicate hormonal control and transmission of information from the roots to the shoot. But the nature of controls that alter leaf metabolism in response to drought is unknown.

Cowan and Farquhar (1977) suggested that stomata might tend to function in such a way that the diurnal course of conductance allows maximal carbon gain for a given amount of water lost. They have shown, in accordance with certain assumptions, that this would be achieved if stomatal movement were such that it would hold the gain ratio ($\partial E/\partial A$, where E is the rate of transpiration and A is the rate of assimilation) constant over the course of the day. In some situations, stomata appear to function in a manner consistent with this hypothesis (see e.g. Farquhar, Schulze, and Küppers, 1980; Schulze and Hall, 1982; Williams, 1983). However, these studies have also demonstrated the extreme difficulty of determining reliable values of $\partial E/\partial A$ and, therefore, of determining whether behavior is really optimal. Perhaps most profitably the hypothesis of optimal water use should be regarded as only one of a number of possibilities to be considered when stomatal response and leaf function are being examined in the natural environment.

Considerable progress in understanding stomatal behavior has been achieved since the pioneering research of Stocker, but there remain many questions concerning the nature of stomatal response to daily environmental variations. There is a need to clarify how differences in stomatal responses determined under laboratory conditions relate to leaf performance in the natural environment. Experimentation should be undertaken with cloned plants transplanted to differing environments in order to determine how long-term acclimation to new environmental conditions affects the regulation of stomatal response. Also, detailed studies of selected species transplanted to a common environment would enlarge our understanding of how heredity and environment affect stomatal behavior under fluctuating conditions. Thus, the monitoring of daily time courses of leaf response in gas exchange has played and will continue to play an extremely important role in elucidating how stomatal response influences photosynthetic production, plant growth, and, ultimately, plant success.

REFERENCES

Bates, L. M., and A. E. Hall. 1981. Stomatal closure with soil water depletion not associated with changes in bulk leaf water status. *Oecologia*, 50: 62–65.
Beardsell, M. F., P. G. Jarvis, and B. Davidson. 1972. A null-balance diffusion porometer suitable for use with leaves of many shapes. *Journal of Applied Ecology*, 9: 677–90.
Berry, J. A., and J. K. Raison. 1981. Responses of macrophytes to temperature. In O. L. Lange, P. S. Nobel, C. B. Osmond, and H. Ziegler, eds., *Physiological plant ecology* 1, Encyclopedia of plant physiology, n.s., 12A: 277–338. Berlin: Springer-Verlag.
Björkman, O. 1982. Responses to different quantum flux densities. In O. L. Lange, P. S. Nobel, C. B. Osmond, and H. Ziegler, eds., *Physiological plant ecology* 1, Encyclopedia of plant physiology, n.s., 12A: 57–107. Berlin: Springer-Verlag.
Björkman, O., W. J. S. Downton, and H. A. Mooney. 1980. Response and adaptation to water stress in *Nerium oleander*. *Carnegie Institution of Washington Yearbook*, 79: 150–57.
Björkman, O., M. M. Ludlow, and P. A. Morrow. 1972. Photosynthetic performance of two rainforest species in their native habitat and analysis of their gas exchange. *Carnegie Institution of Washington Yearbook*, 71: 94–102.
Bosian, G. 1960. Zum Küvettenklimaproblem: Beweisführung für die Nichtexistenz 2-gipfeliger Assimilationskurven bei Verwendung von klimatisierten Küvetten. *Flora*, 149: 167–88.
Carbonneau, A. 1983. Measurement of gross photosynthesis under natural conditions by using a $^{14}CO_2$ diffusion porometer, I. Description of some convenient techniques. *Photosynthetica*, 17: 235–39.
Cowan, I. R. 1982. Regulation of water use in relation to carbon gain in higher plants. In O. L. Lange, P. S. Nobel, C. B. Osmond, and H. Ziegler, eds., *Physiological plant ecology* 2, Encyclopedia of Plant Physiology, n.s., 12B: 589–613. Berlin: Springer-Verlag.
Cowan, I. R., and G. D. Farquhar. 1977. Stomatal function in relation to leaf metabolism and environment. In D. H. Jennings, ed., *Integration of activity in the higher plant*: 471–505. Symposia of the Society for Experimental Biology, 31. Cambridge: Cambridge University Press.
Dunn, E. 1975. Environmental stresses and inherent limitations affecting CO_2 exchange in evergreen sclerophylls in Mediterranean climates. In D. M. Gates and R. Schmerl, eds., *Perspective of biophysical ecology*: 159–81. Ecological Studies, 12. Berlin: Springer-Verlag.
Eckardt, F. E., G. Heim, M. Methy, and R. Sauvezon. 1975. Interception de l'énergie rayonnante, échanges gazeux et croissance dans une forêt méditerranéenne à feuillage persistant (*Quercetum ilicis*). *Photosynthetica*, 9: 145–56.
Farquhar, G. D., D. R. Dubbe, and K. Raschke. 1978. Gain of the feedback loop involving carbon dioxide and stomata: Theory and measurement. *Plant Physiology*, 62: 406–12.
Farquhar, G. D., E.-D. Schulze, and M. Küppers. 1980. Responses to humidity by stomata of *Nicotiana glauca* L. and *Corylus avellana* L. are consistent with the optimisation of carbon dioxide uptake with respect to water loss. *Australian Journal of Plant Physiology*, 7: 315–27.
Field, C., J. A. Berry, and H. A. Mooney. 1982. A portable system for measuring

carbon dioxide and water vapour exchange of leaves. *Plant, Cell and Environment,* 5: 179–86.

Gebel, J. 1983. Der Einfluss von Bodenaustrocknung auf die Photosynthesekapazität, die mittägliche Gaswechseldepression und den Wasserhaushalt bei *Quercus suber*—Labor- und Freiland-Untersuchungen. Diplom Thesis, University of Würzburg.

Grace, J., D. U. U. Okali, and F. E. Fasehun. 1983. Stomatal conductance of two tropical trees during the wet season in Nigeria. *Journal of Applied Ecology,* 19: 659–70.

Griffiths, J. H., and P. G. Jarvis. 1981. A null balance carbon dioxide and water vapour porometer. *Journal of Experimental Botany,* 32: 1151–68.

von Guttenberg, H., and H. Buhr. 1935. Studien über die Assimilation und Atmung mediterraner Macchiapflanzen während der Regen- und Trockenzeit. *Planta,* 24: 163–265.

Heath, O. V. S. 1948. Studies in stomatal action. Control of stomatal movement by a reduction in the normal carbon dioxide content of the air. *Nature,* 161: 179–81.

Hellmuth, E. 1971. Eco-physiological studies on plants in arid and semi-arid regions in Western Australia, III. Comparative studies on photosynthesis, respiration and water relations of ten arid zone and two semi-arid zone plants under winter and late summer climatic conditions. *Journal of Ecology,* 59: 225–59.

Huber, B. 1927. Zur Methodik der Transpirationsbestimmung am Standort. *Berichte der Deutschen botanischen Gesellschaft,* 45: 611–18.

Jarvis, P. G. 1976. The interpretation of the variations in leaf water potential and stomatal conductance found in canopies in the field. *Philosophical Transactions of the Royal Society,* ser. B, 273: 593–610.

Jarvis, P. G., and J. I. L. Morison. 1981. The control of transpiration and photosynthesis by the stomata. In P. G. Jarvis and T. A. Mansfield, eds., *Stomatal Physiology:* 247–79. Cambridge: Cambridge University Press.

Kaufmann, M. R. 1982. Leaf conductance as a function of photosynthetic photon flux density and absolute humidity difference from leaf to air. *Plant Physiology,* 69: 1018–22.

Koch, W., O. L. Lange, and E.-D. Schulze. 1971. Ecophysiological investigations on wild and cultivated plants in the Negev Desert, I. Methods: A mobile laboratory for measuring carbon dioxide and water vapour exchange. *Oecologia,* 8: 296–309.

Körner, C., J. A. Scheel, and H. Bauer. 1979. Maximum leaf diffusive conductance in vascular plants. *Photosynthetica,* 13: 45–82.

Küppers, M. 1984a. Carbon relations and competition between woody species in a central European hedgerow, I. Photosynthetic characteristics. *Oecologia,* 64: 332–43.

⸻. 1984b. Carbon relations and competition between woody species in a central European hedgerow, II. Stomatal responses, water use, and hydraulic conductivity in the root/leaf pathway. *Oecologia,* 64: 344–54.

Lange, O. L., W. Koch, and E.-D. Schulze. 1969. CO_2-gas exchange and water relationships of plants in the Negev Desert at the end of the dry period. *Berichte der Deutschen botanischen Gesellschaft,* 82: 39–61.

Lange, O. L., and A. Meyer. 1979. Mittäglicher Stomataschluss bei Aprikose (*Prunus armeniaca*) und Wein (*Vitis vinifera*) im Freiland trotz guter Bodenwasserversorgung. *Flora,* 168: 511–28.

Lange, O. L., J. D. Tenhunen, and M. Braun. 1982. Midday stomatal closure in Mediterranean type sclerophylls under simulated habitat conditions in an environmental chamber, I. Comparison of the behavior of various European Mediterranean species. *Flora,* 172: 563–79.

Larcher, W., and H. Bauer. 1981. Ecological significance of resistance to low temperature. In O. L. Lange, P. S. Nobel, C. B. Osmond, and H. Ziegler, eds., *Physiological plant ecology* 1, Encyclopedia of Plant Physiology, n.s., 12A: 401–37. Berlin: Springer-Verlag.

Lösch, R., and J. D. Tenhunen. 1981. Stomatal responses to humidity—Phenomenon and mechanism. In P. G. Jarvis and T. A. Mansfield, eds., *Stomatal physiology*: 137–61. Cambridge: Cambridge University Press.

Lösch, R., J. D. Tenhunen, J. S. Pereira, and O. L. Lange. 1982. Diurnal courses of stomatal resistance and transpiration of wild and cultivated Mediterranean perennials at the end of the summer dry season in Portugal. *Flora*, 172: 138–60.

McPherson, H. G., A. E. Green, and P. L. Rollinson. 1983. The measurement, within seconds, of apparent photosynthetic rates using a portable instrument. *Photosynthetica*, 17: 395–406.

Matthews, M. A., and J. S. Boyer. 1984. Acclimation of photosynthesis to low leaf water potentials. *Plant Physiology*, 74:161–66.

Meidner, H. 1981. Measurements of stomatal aperture and responses to stimuli. In P. G. Jarvis and T. A. Mansfield, eds., *Stomatal physiology*: 25–49. Cambridge: Cambridge University Press.

Mooney, H. A., O. Björkman, and G. J. Collatz. 1977. Photosynthetic acclimation to temperature and water stress in the desert shrub *Larrea divaricata*. *Carnegie Institution of Washington Yearbook*, 76: 328–35.

Mooney, H. A., and E. L. Dunn. 1970. Convergent evolution of Mediterranean sclerophyll shrubs. *Evolution*, 24: 292–303.

Mooney, H. A., E. L. Dunn, A. T. Harrison, P. A. Morrow, B. Bartholomew, and R. L. Hays. 1971. A mobile laboratory for gas exchange measurements. *Photosynthetica*, 5: 128–32.

Mooney, H. A., A. T. Harrison, and P. A. Morrow. 1975. Environmental limitations of photosynthesis on a California evergreen shrub. *Oecologia*, 19: 293–301.

Mooney, H. A., and J. Kummerow. 1971. The comparative water economy of representative evergreen sclerophyll and drought deciduous shrubs of Chile. *Botanical Gazette*, 132: 245–52.

Morrow, P., and H. A. Mooney. 1974. Drought adaptations in two Californian sclerophylls. *Oecologia*, 15: 205–22.

Oppenheimer, H. 1947. Studies on the water balance of unirrigated woody plants. *Palestine Journal of Botany*, 6: 63–77.

Pearcy, R. W. 1983. The light environment and growth of C_3 and C_4 tree species in the understory of a Hawaiian forest. *Oecologia*, 58: 19–25.

———. 1986. Carbon dioxide exchange of tropical tree species in different microenvironments of an Australian tropical rainforest. [In preparation.]

Pearcy, R. W., and H. Calkin. 1983. Carbon dioxide exchange of C_3 and C_4 tree species in the understory of a Hawaiian forest. *Oecologia*, 58: 26–32.

Pearcy, R. W., K. Osteryoung, and D. Randall. 1982. Carbon dioxide exchange characteristics of C_4 Hawaiian *Euphorbia* species native to diverse habitats. *Oecologia*, 55: 333–41.

Pereira, J. S., J. D. Tenhunen, O. L. Lange, W. Beyschlag, and A. Meyer. 1985. Seasonal and diurnal patterns in leaf gas exchange of *Eucalyptus globulus* trees growing in Portugal. *Canadian Journal of Forest Research*, 16: 177–84.

Poljakoff, A. 1946. Ecological investigations in Palestine, I. The water balance of some Mediterranean trees. *Palestine Journal of Botany*, 3: 138–50.

Raschke, K. 1982. Involvement of abscisic acid in the regulation of gas exchange: Evidence and inconsistencies. In P. F. Wareing, ed., *Plant growth substances, 1982*: 581–90. London: Academic.

Robichaux, R. H., P. W. Rundel, L. Stemmerman, J. E. Canfield, S. R. Morse, and W. E. Freidman. 1984. Tissue water deficits and plant growth in wet tropical environments. In E. Medina, H. A. Mooney, and C. Vasquez-Yanes, eds., *Physiological ecology of the wet tropics*: 99–112. The Hague: Junk.

Rouschal, E. 1938. Zur Ökologie der Macchien, I. Der sommerliche Wasserhaushalt der Macchienpflanzen. *Jahrbücher für wissenschaftlicher Botanik*, 87: 436–523.

Roy, J., and H. A. Mooney. 1983. Physiological adaptation and plasticity to water stress of coastal and desert populations of *Heliotropium curassavicum* L. *Oecologia*, 52: 370–75.

Running, S. W. 1976. Environmental control of leaf water conductance in conifers. *Canadian Journal of Forest Research*, 6: 104–12.

Schulze, E.-D. 1970. Der CO_2-Gaswechsel der Buche (*Fagus silvatica* L.) in Abhängigkeit von den Klimafaktoren im Freiland. *Flora*, 159: 177–232.

Schulze, E.-D., and A. E. Hall. 1982. Stomatal responses, water loss and CO_2 assimilation rates of plants in contrasting environments. In O. L. Lange, P. S. Nobel, C. B. Osmond, and H. Ziegler, eds., *Physiological plant ecology* 2, Encyclopedia of Plant Physiology, n.s., 12B: 181–230. Berlin: Springer-Verlag.

Schulze, E.-D., A. E. Hall, O. L. Lange, and H. Walz. 1982. A portable steady-state porometer for measuring the carbon dioxide and water vapour exchanges of leaves under natural conditions. *Oecologia*, 53: 141–45.

Schulze, E.-D., and M. Küppers. 1979. Short-term and long-term effects of plant water deficits on stomatal response to humidity in *Corylus avellana* L. *Planta*, 146: 319–26.

Schulze, E.-D., O. L. Lange, M. Evenari, L. Kappen, and U. Buschbom. 1974. The role of air humidity and leaf temperature in controlling stomatal resistance of *Prunus armeniaca* L. under desert conditions, I. A simulation of the daily time course of stomatal resistance. *Oecologia*, 17: 159–70.

―――. 1975. The role of air humidity and leaf temperature in controlling stomatal resistance of *Prunus armeniaca* L. under desert conditions, II. The significance of leaf water status and internal carbon dioxide concentration. *Oecologia*, 18: 219–33.

―――. 1980. Long-term effects of drought on wild and cultivated plants in the Negev Desert, II. Diurnal patterns of net photosynthesis and daily carbon gain. *Oecologia*, 45: 19–25.

Slavik, B. 1974. *Methods of studying plant water relations*. Berlin: Springer-Verlag.

Stocker, O. 1929. Eine Feldmethode zur Bestimmung der momentanen Transpirations- und Evaporationsgrösse. *Berichte der Deutschen botanischen Gesellschaft*, 47: 126–36.

―――. 1954. Der Wasser- und Assimilationshaushalt südalgerischer Wüstenpflanzen. *Berichte der Deutschen botanischen Gesellschaft*, 67: 288–99.

―――. 1956. Messmethoden der Transpiration. In W. Ruhland, ed., *Pflanze und Wasser*, Handbuch der Pflanzenphysiologie 3: 291–311. Berlin: Springer-Verlag.

Stocker, O., and G. H. Vieweg. 1960. Die darmstädter Apparatur zur Momentanmessung der Photosynthese unter ökologischen Bedingungen. *Berichte der Deutschen botanischen Gesellschaft*, 73: 198–208.

Tenhunen, J. D., O. L. Lange, and M. Braun. 1981. Midday stomatal closure in Mediterranean type sclerophylls under simulated habitat conditions in an environmental chamber, II. Effect of the complex of leaf temperature and air humidity on gas exchange of *Arbutus unedo* and *Quercus ilex*. *Oecologia*, 50: 5–11.

Tenhunen, J. D., O. L. Lange, M. Braun, A. Meyer, R. Lösch, and J. S. Pereira. 1980. Midday stomatal closure in *Arbutus unedo* leaves in a natural macchia and under simulated habitat conditions in an environmental chamber. *Oecologia*, 47: 365–67.

Tenhunen, J. D., O. L. Lange, J. Gebel, W. Beyschlag, and J. A. Weber. 1984b.

Changes in photosynthetic capacity, carboxylation efficiency, and CO_2 compensation point associated with midday stomatal closure and midday depression of net CO_2 exchange of leaves of *Quercus suber*. *Planta*, 162: 193–203.

Tenhunen, J. D., O. L. Lange, and D. Jahner. 1982. The control by atmospheric factors and water stress of midday stomatal closure in *Arbutus unedo* growing in a natural macchia. *Oecologia*, 55: 165–69.

Tenhunen, J. D., O. L. Lange, J. S. Pereira, R. Lösch, and F. Catarino. 1981. Midday stomatal closure in *Arbutus unedo* leaves: Measurements with a steady-state porometer in the Portuguese evergreen scrub. In N. S. Margaris and H. A. Mooney, eds., *Components of productivity of Mediterranean regions—Basic and applied aspects*: 61–69. The Hague: Junk.

Tenhunen, J. D., H. P. Meister, M. M. Caldwell, and O. L. Lange. 1984a. Environmental constraints on productivity of the Mediterranean sclerophyll shrub *Quercus coccifera*. Proceedings of INTECOL workshop—Rates of natural primary production and agricultural production. Zaragoza, Spain. *Options méditerranéennes*, 84: 33–53.

Turner, N. C., J. E. Begg, H. M. Rawson, S. D. English, and A. B. Hearn. 1978. Agronomic and physiological responses of soybean and sorghum crops to water deficits, III. Components of leaf water potential, leaf conductance, $^{14}CO_2$ photosynthesis, and adaptation to water deficits. *Australian Journal of Plant Physiology*, 5: 179–94.

Weber, J. A., T. W. Jurik, J. D. Tenhunen, and D. M. Gates. 1985. Analysis of gas exchange in seedlings of *Acer saccharum*: Integration of field and laboratory studies. *Oecologia*, 65: 338–47.

Werk, K. S., J. Ehleringer, I. N. Forseth, and C. S. Cook. 1983. Photosynthetic characteristics of Sonoran Desert winter annuals. *Oecologia*, 59: 101–5.

Whitehead, D., U. U. Okali, and F. E. Fasehun. 1981. Stomatal responses to environmental variables in two tropical forest species during the dry season in Nigeria. *Journal of Applied Ecology*, 18: 571–87.

von Willert, D. J., B. M. Eller, E. Brinckmann, and R. Baasch. 1982. CO_2-gas exchange and transpiration of *Welwitschia mirabilis* Hook. fil. in the Central Namib Desert. *Oecologia*, 55: 21–29.

Williams, B. A., P. J. Gurner, and R. B. Austin. 1982. A new infra-red gas analyser and portable photosynthesis meter. *Photosynthesis Research*, 3: 141–51.

Williams, W. E. 1983. Optimal water use efficiency in a California shrub. *Plant, Cell and Environment*, 6: 145–51.

Wong, S. C., I. R. Cowan, and G. D. Farquhar. 1979. Stomatal conductance correlates with photosynthetic capacity. *Nature*, 282: 424–26.

16

Stomata in Plants with Crassulacean Acid Metabolism

Irwin P. Ting

Stomata represent the major portals for gas exchange in plants. They are of interest to botanists because, when open, they allow the diffusion of CO_2 into the plant and, as a consequence, the loss of great quantities of water. Unlike stomata of most plants, those in succulent plants showing crassulacean acid metabolism (CAM) are mostly open at night and closed during the day (Kluge and Ting, 1978). As a consequence, the ratio of water loss to carbon gain is greatly reduced, since the stomata are closed during the day, when the evaporative demand is high. The ratio of water loss to carbon gain in succulent plants that have CAM can be as little as one-tenth that of typical C_3 species (Joshi, Boyer, and Kramer, 1965; Erler, 1969; Ting, Johnson, and Szarek, 1972; Meinzer and Rundel, 1973; Neales, 1973; Szarek and Ting, 1975a; Nobel, 1976). In this brief review, I will discuss the known features of stomata in CAM plants and attempt to integrate them in the context of nocturnal stomatal opening.

Reverse-phase stomatal opening in succulent plants has been known for many years through the early work of Coville, MacDougal, Livingston, and Shreve at the Carnegie Desert Laboratory, in Tucson, Arizona. In 1907, Livingston studied transpiration of cacti relative to evaporation and noted that, unlike nonsucculent plants, cacti exhibit relative transpiration that is high at night and low during the day. Livingston was able to conclude that there was a mechanism associated with gas exchange in succulents that was different from that in ordinary plants. Later, Shreve (1916) found that the stomata of cacti were in fact open wider at night than during the day.

Even though it was known as early as 1804 through experiments of DeSaussure that cacti could take up CO_2 in the dark (see Kluge and Ting, 1978), it remained for detailed studies of Nishida (1963), Joshi, Boyer, and Kramer (1965), Ekern (1965), and Aubert (1971) to confirm nocturnal stomatal opening in succulent plants. In 1975, Neales divided CAM succulents into three groups on the basis of CO_2 uptake. In the first of these, he noted that in some plants with weak CAM, for example, *Portulacaria afra*, CO_2 was taken up toward the end of the night and throughout the day. (As will be shown below, this pattern is highly modified in *P. afra* by water stress.) In the second group are those CAM plants that take up CO_2 during most of the night and not during the day, which frequently have a brief early morning "burst" of CO_2 uptake, with little or no CO_2 uptake toward the end of the light period. This pattern appears to be the most typical for succulent plants showing CAM. The third pattern, characterized by cacti, shows CO_2 uptake throughout the dark period and not during the light period, but it is not substantially different from the pattern of the crassulacean succulents. Data discussed below show that patterns of stomatal opening are generally similar to those of CO_2 uptake in these plants, and that CO_2 uptake is correlated with stomatal conductance. The graphs of Neales (1975) showing the three CAM patterns mentioned above also show a correlation of CO_2 uptake with water loss and hence with stomatal opening.

Morphology

The general appearance of stomata in CAM succulents is largely the same, regardless of whether they are stem succulents or leaf succulents. It was calculated for 15 randomly selected CAM succulents that their mean stomatal density was 27 mm^{-2} (Ting et al., 1972). This number is lower than for the average mesophytic plants (and considerably lower than that for a deciduous tree, which may have a thousand or more stomata per square millimeter). Although not sunken, stomata in CAM succulents are frequently associated with deep stomatal chambers, primarily because of a hypodermis; or they may occur in crypts, as in *Nolina* (LaPre, 1979). Thomson and De Journett (1970) observed that the stomatal guard cells of *Opuntia ficus-indica* were slightly sunken below the surface of the leaf and were suspended over large substomatal cavities. They did not observe plasmodesmata between subsidiary cells and guard cells.

An electron micrograph of the guard cells of the CAM succulent *Bryophyllum daigremontiana* Hamet et Perr shows no special features attributable to CAM metabolism (Fig. 16.1). The cells have a large central nucleus, typical but small plastids with starch grains, and numerous mitochondria, endoplasmic reticula, microbodies, and vacuoles. A comparison of these features and their counterparts in *O. ficus-indica* (Kluge and Ting, 1978) reveals little difference between the two genera. Furthermore, a comparative study of guard cells from C_3, C_4, and CAM plants by Faraday, Thomson, and Platt-Aloia (1982) did not reveal consistent differences among the three types. Thus, there seem to be no discernible differences between the guard cells and stomatal apparatus of CAM succulents and those of more typical non-CAM mesophytic plants. A remarkable feature, however, is the presence of thick depositions of a waxy material on the epidermis (Fig. 16.2). It seems that this waxy material, coupled with tight stomatal closure, accounts for the extremely low water loss in these

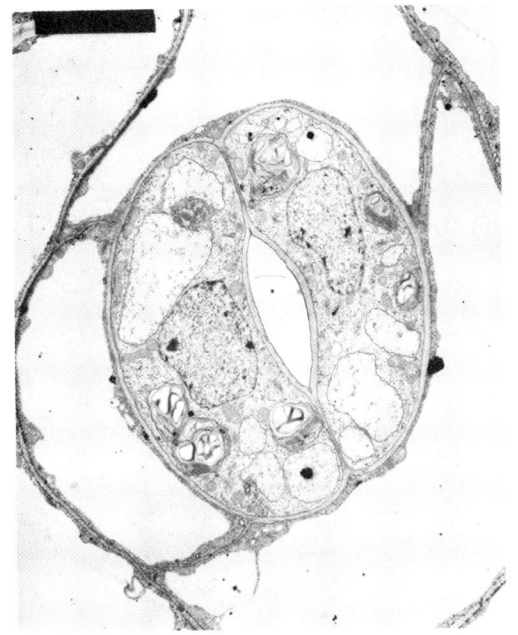

Fig. 16.1. Electron micrograph of guard cells from *Bryophyllum daigremontianum* Hamet et Perr ($\times$ 1,400). The micrograph shows typical plant guard cells containing nuclei, endoplasmic reticulum, plastids, mitochondria, and other organelles. (Micrograph courtesy of David Gibeaut and W. W. Thomson, Department of Botany and Plant Sciences, University of California, Riverside.)

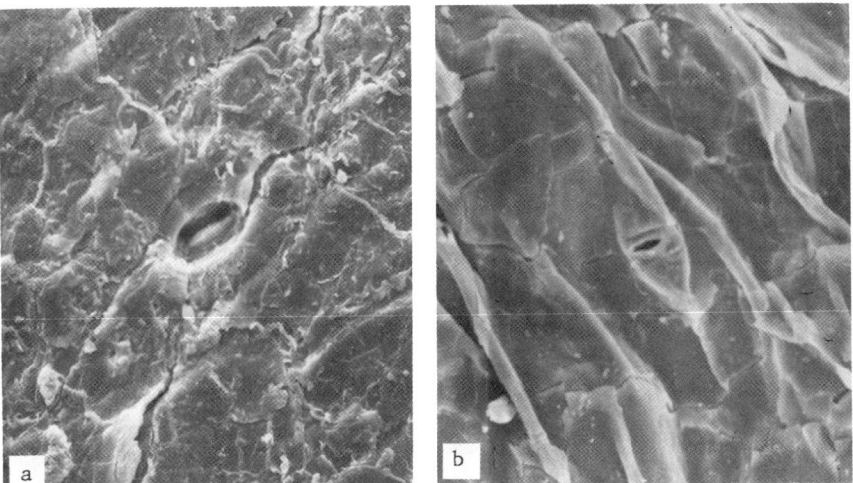

Fig. 16.2. Scanning electron micrographs of the surface of two cactus species: *a*, *Ferocactus acanthodes* (Lemaire) Britton et Rose ($\times$ 320); *b*, *Mammalaria dioica* K. Brandegee ($\times$ 320). Both micrographs show heavy deposits of wax on the surface.

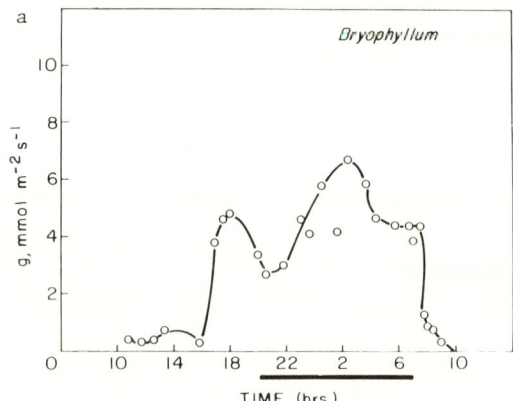
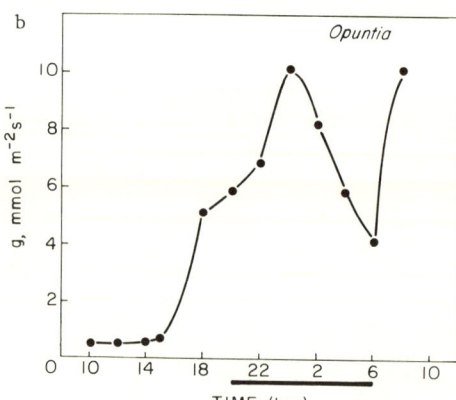
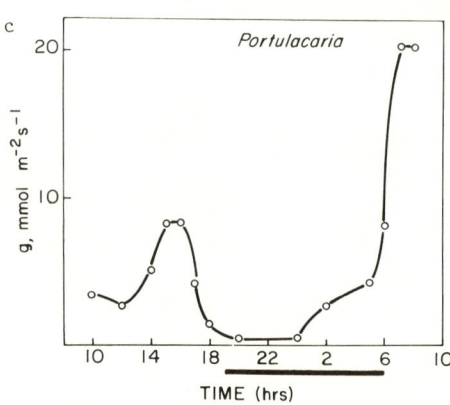

Fig. 16.3. Diurnal patterns of stomatal conductance for *Bryophyllum daigremontianum*, *Opuntia basilaris* Engelmann et Bigelow, and *Portulacaria afra* (L.) Jacq. The patterns for *Bryophyllum* (*a*) and *Opuntia* (*b*) are quite similar, with measurable conductance (*g*) toward the end of the light period, conductance throughout the dark period (indicated by solid bar on the time axis), and a burst at the very beginning of the light period. The pattern for unstressed *Portulacaria* (*c*) is distinctly different; there is stomatal conductance throughout the day, with midday closure, conductance during the afternoon, then stomatal closure again at the beginning of the night, followed by measurable conductance toward the end. When *Portulacaria* is water-stressed (not shown), there is no stomatal conductance during the day. The data for *Bryophyllum* are from a growth-chamber experiment by C. B. Osmond (unpublished), and recalculated from Kluge and Ting (1978). The time scale is experimental. The data for *Opuntia basilaris* are from Szarek (1974), and the data for *Portulacaria* are from Hanscom (1977).

species during periods of drought (Szarek, Johnson, and Ting, 1973; Hanscom and Ting, 1977).

Diurnal Course of Stomatal Opening

Typical diurnal patterns of stomatal conductance for CAM succulents are shown in Figure 16.3 (see also Osmond, 1978). Stomata are open throughout the dark period and show a peak opening at dawn, followed by closure throughout most of the day and then slight opening in the afternoon, toward the end of the light period (Fig. 16.3a). The early-morning stomatal opening is quite typical for many CAM plants; however, it is not always observed. Figure 16.3b shows a typical pattern for a cactus. Here it can be seen that the pattern is not markedly different from that in *Bryophyllum*. The pattern for *P. afra*, a species that shifts from C_3 to CAM, is depicted in Figure 16.3c. Under well-watered conditions, the stomata of *P. afra* are open throughout the daylight period, with some closure at midday and occasionally toward the end of the day. Stomata tend to remain closed in the early dark period and open toward the end of the dark period. In contrast, when these plants are moderately water-stressed, stomata close during the day but remain open during the dark period. It is not clear, however, if water stress increases the levels of nocturnal opening (Ting, 1981). From these diurnal curves of stomatal conductance, it can be concluded that nocturnal stomatal opening is characteristic of CAM succulent plants, a feature consistent with their CO_2 exchange patterns (Kluge and Ting, 1978; Osmond, 1978).

A characteristic feature of stomatal opening in CAM plants is the burst that occurs in early morning (Osmond, 1978; LaPre, 1979; see also Winter and Tenhunen, 1982). LaPre (1979) has studied in some detail the initial-light stomatal opening in *Yucca* (Fig. 16.4). He observed stomatal opening that began at sunrise and reached a maximum about an hour after sunrise. Subsequently, stomatal closure was very rapid. This transient stomatal opening at dawn has been interpreted as a response to transient changes in internal CO_2 concentration, resulting from light effects in photosynthesis (Cockburn, Ting, and Sternberg, 1979).

Another important characteristic of stomatal conductance in CAM plants is their considerably lower values as compared with most mesophytic plants, with the observed maxima being of the order of 40 to 200 mmol $m^{-2} s^{-1}$. These low conductances are in part the result of lower stomatal densities. In addition, a lower overall conductance is to be expected because of a frequently greater boundary-layer resistance ensuing from the spherical shapes of leaves and stems from these succulents (Nobel, 1975; Ting and Szarek, 1975).

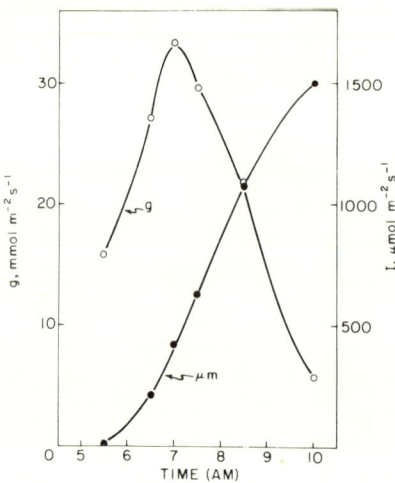

Fig. 16.4. Stomatal conductance (g) in *Yucca schidigera* Roezl ex Ortgies measured during sunrise (I, light flux). The data are recalculated from LaPre (1979). The diurnal pattern of stomatal conductance for *Yucca schidigera* is similar to those of *Opuntia* and *Bryophyllum* (see Fig. 16.3).

Factors Affecting Stomatal Opening in CAM Plants

Age. Similarly as for other plants, there are extensive reports indicating that leaf age is a factor in stomatal opening of CAM plants. By and large, young leaves tend to show greater stomatal conductances than older, mature leaves (Ting, Dean, and Dugger, 1967; Jones, 1975). Extensive studies by Queiroz (see Queiroz and Brulfert, 1982) of photoperiod and CAM induction in *Kalanchoe blossfeldiana* support this observation. Further, it is clear that the entire CAM phenomenon is greatly influenced by age in these plants (Queiroz and Brulfert, 1982).

Carbon dioxide. Nishida (1963) suggested that nocturnal stomatal opening in succulent plants was the result of organic acid accumulation. He showed that dark CO_2 fixation occurred in stomatal guard cells, and not in adjacent epidermal cells, and suggested that accumulated organic acids induced stomatal opening by either increasing the osmotic pressure or lowering the pH. Meidner and Mansfield (1968) disagreed with his interpretation, arguing that nocturnal stomatal opening in CAM plants was most likely a normal response to low tissue CO_2 concentration. Data of Bruinsma (1958), however, obtained at different levels of CO_2, did not support the suggestion of Meidner and Mansfield. Nor did Nobel and Hartsock (1979) find a stomatal response in CO_2 concentrations up to 800 μl l^{-1}. However, at constant temperature and in the dark, Neales (1970) found that variation in ambient CO_2 concentration caused a corresponding inverted variation of transpiration rate in *Agave americana*. The CO_2 regulation of CAM stomata is nevertheless an attractive hypothesis, because during the decarboxylation phase of CAM during the light period, internal carbon dioxide reaches concentrations of 1 percent or more (Cockburn, Ting, and Sternberg, 1979).

Photoperiod and thermoperiod. The extensive studies by Queiroz and co-workers (Queiroz and Brulfert, 1982) have revealed that CAM is induced by photoperiod in some plants. In *K. blossfeldiana*, for example, short days that induce flowering also induce CAM in young leaves. This induction of CAM is accompanied by the appearance of reverse-phase stomatal opening, consistent with CAM metabolism.

Additional work with *K. blossfeldiana* revealed that the stomatal opening pattern was markedly affected by thermoperiod, as well as by photoperiod (Ting, Thompson, and Dugger, 1967). When plants were grown on cool days and cool nights with little diurnal variation in temperature, stomatal opening was typical of C_3 species. However, when the plants were grown on warm days (30° C) and cool nights (15° C), CAM was induced, and nocturnal stomatal opening occurred. It is not known if the thermoperiod effect is unique to *K. blossfeldiana* and related to the photoperiod response, or if it is a more general phenomenon associated with CAM induction.

Nobel and Hartsock (1979) found that leaf temperature was a significant factor in stomatal opening of *Agave deserti*. When leaf temperature was high, stomatal conductance tended to be low. Complicating the interpretation of temperature effects on stomatal conductance is the observation that higher nocturnal temperatures, as compared with day temperature, caused a shift to greater daytime water loss in *A. americana* (Neales, 1973).

Abscisic acid. The stomata of isolated epidermal strips of *K. daigremontiana* close when floated on buffers with low concentrations of abscisic acid (ABA; Jewer, Incoll, and Howarth, 1981). Because ABA concentrations associated with stomatal closure were lower than those required for non-CAM plants tested, the authors concluded that in *Kalanchoe* the stomata are unusually sensitive to ABA, making them very sensitive to closure during water stress that results in ABA production. When well-watered *P. afra* plants that do not show CAM are treated with ABA, the stomata tend to close during the day, but remain open at night (Ting, 1981). This daytime stomatal closure is accompanied by other CAM characteristics such as diurnal fluctuations of organic acids and nocturnal CO_2 uptake. These experiments could implicate an ABA involvement in CAM induction, although, because measurements were taken several weeks after ABA treatment, it was not possible to ascertain if the ABA effect was primarily on stomata or if there was a more complicated growth response. Subsequent unpublished studies of my own have shown that the level of ABA increases very markedly in these plants when they are subjected to water stress.

Similarly as in other plants, the stomata of *K. daigremontiana* open in the presence of the cytokinins kinetin and zeatin when the epidermis is floated on buffer in the dark (Jewer and Incoll, 1981).

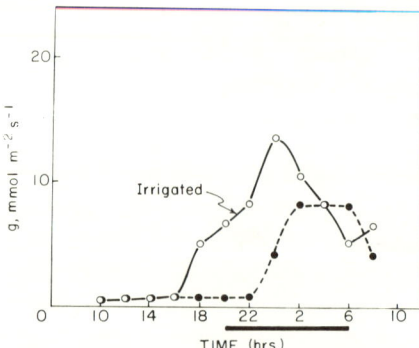

Fig. 16.5. Effect of irrigation on stomatal conductance in *Opuntia basilaris* growing at the Deep Canyon Desert Research Center, Palm Desert, California. The control plants (nonirrigated: solid circles) were not completely stressed and showed stomatal conductance (g) throughout most of the night (indicated by solid bar on the time axis). When the test plants were irrigated (open circles), stomatal conductance increased in the midafternoon and reached its peak shortly after midnight. This response is similar to that to natural precipitation. (Data calculated from Hanscom, 1977.)

Potassium. It is assumed that potassium plays a role in the opening of the stomata of CAM plants as it does in other plants. Dayanandan and Kaufman (1975), studying potassium fluxes and stomatal opening, reported potassium uptake by guard cells was correlated with stomatal opening in *Kalanchoe* and *Crassula* (see also Jewer, Incoll, and Howarth, 1981).

Water. Studies have shown that under natural conditions, stomata of CAM plants remain closed during drought (Szarck and Ting, 1974, 1975b). Such observations have been mimicked by artificial drought and irrigation under both laboratory and field conditions (Hanscom and Ting, 1977). Water deficits in CAM plants tend to bring about complete stomatal closure and totally override other regulating factors (Szarek, Johnson, and Ting, 1973). Since most of the CAM succulents have large quantities of tissue water, stomatal closing in response to drought may take several weeks (Nobel, 1977; Rayder and Ting, 1983).

There appears to be a good correlation between tissue water potential and stomatal opening in CAM plants. Figure 16.5 shows the response of stomata to irrigation under field conditions. Under water-stress conditions, stomatal conductance is virtually nil. Upon irrigation, stomata open during the dark period. In Figure 16.5, control plants had recently received precipitation. There was no stomatal conductance during the light period, but stomatal conductance increased toward the end of the dark period. Irrigation changed the pattern so that stomata were open throughout the dark period and also toward the end of the light period. In agreement with this observation, Osmond, Nott, and Zinth (1979) reported afternoon stomatal opening after precipitation in *Opuntia inermis*.

Some data indicate that stomata of CAM plants also respond to vapor pressure deficits (VPD; Conde and Kramer, 1975). At a constant cladode temperature in *O. inermis*, stomatal conductance decreased when VPD increased at night but not when it was increased in the day (Osmond et al., 1979). In contrast, Nobel and Hartsock (1979) did not find a significant stomatal effect of VPD on *A. deserti*. Comins and Farquhar (1982) developed a model for stomatal regulation and water economy for CAM plants that can predict the extent of C_3 or CAM gas exchange relative to water stress.

Stomatal Opening under Natural Conditions

There are many measurements of the activity of CAM plants growing in natural habitats. Although most of these studies have been conducted in the deserts of the southwestern United States (Dinger and Patten, 1974; Szarek and Ting, 1974; Nobel, 1976), beginning with the early investigations at the Carnegie Desert Laboratory, Tucson, Arizona, data are also available on succulents growing in the deserts of Australia (Osmond, 1978), the Middle Eastern deserts (Lange et al., 1975), in South Africa (Schutte, Steyn, and van der Westhuizen, 1967), and South America (Coutinho, 1969; Medina et al., 1977). Underlying all of these observations is the theme that water availability largely regulates stomatal opening in CAM plants. In one of the first modern and thorough investigations, Bartholomew (1973) showed with *Dudleya farinosa* growing in coastal California that dark CO_2 fixation continued during the dry season and was largely unaffected by temperature. Daytime CO_2 uptake was drought-dependent in that during the dry season it was diminished. We presume that stomatal opening followed the same pattern. In some cases, however, stomatal opening continued into the dry season, presumably because of extensive water storage by the tissues (Nobel, 1977). Similar observations were reported for *A. americana* and *O. basilaris* and have been substantiated by laboratory experiments with other CAM succulents (Rayder and Ting, 1983). Figure 16.6 illustrates an annual pattern of stomatal opening at the Deep Canyon Desert Research Center, Palm Desert, California, correlated with water availability (Szarek, 1974). Stomatal conductance was high when water was present and low or zero when water was less sufficient. Stomatal closure under drought can be reversed by irrigation at any time during the year (Hanscom and Ting, 1978). Studies by Osmond (1978) in eastern Australia with *O. inermis* also indicate the importance of water. In March, during the fall, stomata were open at the end of the light period and throughout the dark period, whereas in the dry midsummer stomata were only open toward the end of the dark period.

Studies in the Negev Desert with *Caralluma negevensis* by Lange et al. (1975) showed that dark fixation was minimal during the high temperatures and drought of the dry season. We can presume that stomatal opening followed CO_2 fixation.

There are few studies directed at the effect of photoperiod on stomatal opening and CAM under natural environmental conditions. However, seasonal studies such as those of Szarek (1974) shed some light on the question.

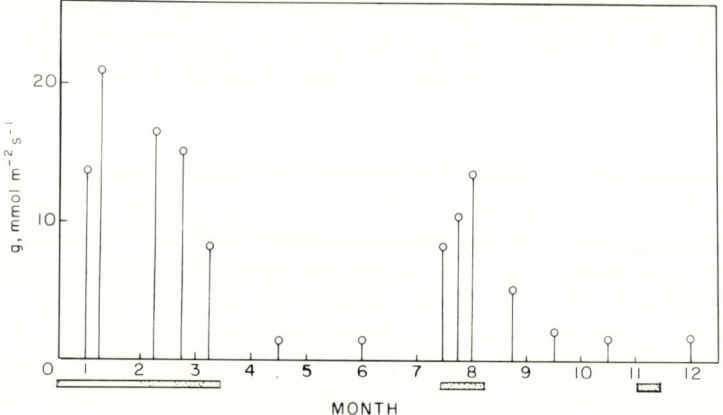

Fig. 16.6. Mean stomatal conductance for *Opuntia basilaris* for selected days throughout an annual cycle at the Deep Canyon Desert Research Center, Palm Desert, California. The bars along the abscissa indicate the periods of precipitation. Stomatal conductance (g) is highly correlated with precipitation but not with temperature or season. Irrigation during the dry months will promote stomatal conductance in this species. (Data calculated from Szarek, 1974.)

Mechanism of Stomatal Opening in CAM Plants

Morphological and physiological data on the stomata of succulent plants do not indicate significant differences between those exhibiting and those lacking CAM metabolism. Hence, despite the fact that the CAM plants are characterized by a reverse-phase stomatal opening, there is no basis to conclude that their stomata are basically different. Rather, it is more reasonable to conclude that the reverse-phase stomatal opening of CAM plants reflects environmental signals and control mechanisms that differ because of CAM. The stomata of CAM plants appear to close more tightly than stomata of most plants, yet it seems that differences are quantitative rather than qualitative.

The most attractive hypothesis for stomatal opening in CAM plants is that it is directly linked to CAM through changes in internal CO_2 concentrations. During the dark period, CAM plants accumulate malic acid through a very active carboxylating system mediated by phosphoenolpyruvate carboxylase. The initial product of this reaction is oxaloacetate, which is reduced to malate and stored in vacuoles as malic acid.

During the malic acid accumulation period, internal CO_2 concentration remains low because of active carboxylation. Stomatal opening is correlated with this low internal CO_2 concentration. Toward the end of the dark period, when malic acid accumulation tends to diminish, stomata may open somewhat. With the onset of the subsequent light period, malic acid is rapidly decarboxylated, resulting in high internal CO_2 concentrations that may exceed 1 percent on a volume basis. Stomata rapidly close. It is presumed that decarboxylation of malic acid during the initial light period generates CO_2 more rapidly than it is assimilated through photosynthesis. Hence, stomata remain closed in response to the high CO_2. When malic acid is depleted, CO_2 is no longer generated, and photosynthesis reduces CO_2 internal concentration below ambient, resulting in stomatal opening. This hypothesis is consistent with results obtained in experiments in which plants were placed in artificially high CO_2 (Cockburn, Ting, and Sternberg, 1979).

The hypothesis that internal CO_2 concentrations regulate stomatal opening in CAM plants does not apply under all conditions. When water is deficient, the stomata close regardless of the internal CO_2 concentrations as shown by the experiments of Cockburn, Ting, and Sternberg (1979). Experiments by Jewer et al. (1981) with isolated epidermal strips of *K. daigremontiana* showed that stomata did not open in response to light when in CO_2-free air. Thus, at least under these conditions, the stomata are not responsive to light, an observation seemingly inconsistent with the hypothesis that stomata open in the afternoon in response to low CO_2 concentrations. No explanation is apparent, but the phenomenon observed could be related to the high sensitivity of these stomata to ABA (Jewer et al., 1981).

Other factors that influence stomatal opening in CAM plants, such as temperature, photoperiod, hormones, and age, could be integrated into the scheme of CO_2 regulation because of their overall effect on CAM metabolism. Their effects on stomatal opening may thus be indirect, through an effect on CAM. This hypothesis is attractive because it does not require intrinsic qualitative assumptions about the stomata of CAM plants; it only assumes that specific control features are a basic aspect of CAM metabolism.

Although future studies of the physiology of CAM stomata should provide new information on their functional properties, it seems likely that a fuller understanding of reverse-phase stomatal opening will require a more thorough understanding of the mechanism of stomatal operation.

REFERENCES

Aubert, B. 1971. Effets de la radiation globale sur la synthèse d'acides organiques et la regulation stomatique des plantes succulentes, exemple d'*Ananas comosus* (L.) Merr. *Oecologia plantarum*, 6: 25–34.

Bartholomew, B. 1973. Drought response in the gas exchange of *Dudleya farinosa* (Crassulaceae) grown under natural conditions. *Photosynthetica*, 7: 114–20.

Bruinsma, J. 1958. Studies on the crassulacean acid metabolism. *Acta botanica Neerlandica*, 7: 531–88.

Cockburn, W., I. P. Ting, and L. O. Sternberg. 1979. Relationships between stomatal

behavior and internal carbon dioxide concentration in crassulacean acid metabolism plants. *Plant Physiology*, 63: 1029–32.
Comins, H. N., and G. D. Farquhar. 1982. Stomatal regulation and water economy in crassulacean acid metabolism plants: An optimization model. *Journal of Theoretical Biology*, 99: 263–84.
Conde, L. F., and P. J. Kramer. 1975. The effect of vapor pressure deficit and diffusion resistance in *Opuntia compressa*. *Canadian Journal of Botany*, 53: 2923–26.
Coutinho, L. M. 1969. Novas observacões sôbre a ocorrência do "efeito de Saussure" e suas relacões com a suculência, a temperatura folhear e os movimentos estomáticos. *Botanica*, 24: 44–102.
Dayanandan, P., and P. B. Kaufman. 1975. Stomatal movements associated with potassium fluxes. *American Journal of Botany*, 62: 221–31.
Dinger, B. E., and D. T. Patten. 1974. Carbon dioxide exchange and transpiration in species of *Echinocereus* (Cactaceae), as related to their distribution within the Pinaleno Mountains, Arizona. *Oecologia*, 14: 389–411.
Ekern, P. C. 1965. Evapotranspiration of pineapple in Hawaii. *Plant Physiology*, 40: 736–39.
Erler, W. L. 1969. Daytime stomatal closure in *Agave americana* as related to enhanced water-use efficiency. In C. C. Hoff and M. L. Rudesch, eds., *Physiological systems in semi-arid environments*: 239–47. Albuquerque: University of New Mexico Press.
Faraday, C. D., W. W. Thomson, and K. A. Platt-Aloia. 1982. Comparative ultrastructure of guard cells of C_3, C_4 and CAM plants. In I. P. Ting and M. Gibbs, eds., *Crassulacean acid metabolism*: 18–26. Rockville, Md.: American Society of Plant Physiologists.
Hanscom, Z. 1977. Plant water status affects acid metabolism and gas exchange in succulents. Ph.D. dissertation, University of California, Riverside.
Hanscom, Z., and I. P. Ting. 1977. Physiological responses to irrigation in *Opuntia basilaris* Engelm. & Bigel. *Botanical Gazette*, 138: 159–67.
_____. 1978. Irrigation magnifies CAM-photosynthesis in *Opuntia basilaris* (Cactaceae). *Oecologia*, 33: 1–15.
Jewer, P. C., and L. D. Incoll. 1981. Promotion of stomatal opening in detached epidermis of *Kalanchoe daigremontiana* Hamet et Perr. by natural and synthetic cytokinins. *Planta*, 153: 317–18.
Jewer, P. C., L. D. Incoll, and G. L. Howarth. 1981. Stomatal responses in isolated epidermis of the crassulacean acid metabolism plant *Kalanchoe daigremontiana* Hamet et Perr. *Planta*, 153: 238–45.
Jones, M. B. 1975. The effect of leaf age on leaf resistance and CO_2 exchange of the CAM plant *Bryophllum fedtschenkoi*. *Planta*, 123: 91–96.
Joshi, M. C., J. S. Boyer, and P. J. Kramer. 1965. Growth, carbon dioxide exchange, transpiration and transpiration ratio of pineapple. *Botanical Gazette*, 126: 174–79.
Kluge, M., and I. P. Ting. 1978. *Crassulacean acid metabolism: Analysis of an ecological adaptation*. Berlin: Springer-Verlag.
Lange, O. L., E.-D. Schulze, L. Kappen, M. Evenari, and U. Buschbom. 1975. CO_2 exchange pattern under natural conditions of *Caralluma negevensis*, a CAM plant of the Negev Desert. *Photosynthetica*, 9: 318–26.
LaPre, L. 1979. Physiological ecology of *Yucca schidigera*. Ph.D. dissertation, University of California, Riverside.
Livingston, B. E. 1907. Relative transpiration in cacti. *Plant World*, 10: 110–14.
Medina, E., M. Delgado, J. H. Troughton, and J. D. Medina. 1977. Physiological ecology of CO_2 fixation in Bromeliaceae. *Flora*, 166: 137–52.

Meidner, H., and T. A. Mansfield. 1968. *Physiology of stomata*. London: McGraw-Hill.

Meinzer, F. C., and P. W. Rundel. 1973. Crassulacean acid metabolism and water use efficiency in *Echeveria pumila*. *Photosynthetica*, 7: 358–64.

Neales, T. F. 1970. Effect of ambient carbon dioxide concentration on the rate of transpiration of *Agave americana* in the dark. *Nature*, 228: 880–82.

_____. 1973. The effect of night temperature on CO_2 assimilation, transpiration, and water use efficiency in *Agave americana* L. *Australian Journal of Biological Sciences*, 26: 705–14.

_____. 1975. The gas exchange pattern of CAM plants. In R. Marcelle, ed., *Environmental and biological control of photosynthesis*: 299–310. The Hague: Junk.

Nishida, K. 1963. Studies on stomatal movement of crassulacean plants in relation to the acid metabolism. *Physiologia plantarum*, 16: 281–98.

Nobel, P. S. 1975. Effective thickness and resistance of the air boundary layer adjacent to spherical plant parts. *Journal of Experimental Botany*, 26: 120–30.

_____. 1976. Water relations and photosynthesis of a desert CAM plant, *Agave deserti*. *Plant Physiology*, 58: 576–82.

_____. 1977. Water relations and photosynthesis of a barrel cactus, *Ferocactus acanthodes*, in the Colorado desert. *Oecologia*, 27: 117–33.

Nobel, P. S., and T. L. Hartsock. 1979. Environmental influences on open stomates of a crassulacean acid metabolism plant, *Agave deserti*. *Plant Physiology*, 63: 63–66.

Osmond, C. B. 1978. Crassulacean acid metabolism. A curiosity in context. *Annual Review of Plant Physiology*, 29: 379–414.

Osmond, C. B., M. M. Ludlow, R. Davis, I. R. Cowan, S. B. Powles, and K. Winter. 1979. Stomatal responses to humidity in *Opuntia inermis* in relation to CO_2 and H_2O exchange patterns. *Oecologia*, 41: 65–79.

Osmond, C. B., D. L. Nott, and P. M. Zinth. 1979. Carbon assimilation patterns and growth of the introduced CAM plant *Opuntia inermis* in eastern Australia. *Oecologia*, 40: 331–50.

Queiroz, O., and J. Brulfert. 1982. Photoperiod-controlled induction and enhancement of seasonal adaptation to drought. In I. P. Ting and M. Gibbs, eds., *Crassulacean acid metabolism*: 208–30. Rockville, Md.: American Society of Plant Physiologists.

Rayder, L., and I. P. Ting. 1983. CAM-idling in *Hoya carnosa*. *Photosynthesis Research*, 4: 203–11.

Schutte, K. H., R. Steyn, and M. van der Westhuizen. 1967. Crassulacean acid metabolism in South African succulents: A preliminary investigation into its occurrence in various families. *Journal of South African Botany*, 33: 107–10.

Shreve, E. B. 1916. An analysis of the causes of variations in the transpiring power of cacti. *Physiological Research*, 2: 73–127.

Szarek, S. R. 1974. Physiological mechanisms of drought adaptation in *Opuntia basilaris* Engelm. Ph.D. dissertation, University of California, Riverside.

Szarek, S. R., H. B. Johnson, and I. P. Ting. 1973. Drought adaptation of *Opuntia basilaris*. Significance of recycling carbon through crassulacean acid metabolism. *Plant Physiology*, 52: 539–41.

Szarek, S. R., and I. P. Ting. 1974. Seasonal patterns of acid metabolism and gas exchange in *Opuntia basilaris*. *Plant Physiology*, 54: 76–81.

_____. 1975a. Photosynthetic efficiency of CAM plants in relation to C_3 and C_4 plants. In R. Marcelle, ed., *Environmental and biological control of photosynthesis*: 289–98. The Hague: Junk.

_____. 1975b. Physiological responses to rainfall in *Opuntia basilaris*. *American Journal of Botany*, 62: 602–9.

Thomson, W. W., and R. De Journett. 1970. Studies on the ultrastructure of the guard cells of *Opuntia*. *American Journal of Botany*, 57: 309–16.

Ting, I. P. 1981. Effects of abscisic acid on CAM in *Portulacaria afra*. *Photosynthesis Research*, 2: 39–48.

Ting, I. P., M. L. Dean, and W. M. Dugger, Jr. 1967. Leaf resistance in succulent plants. *Nature*, 213: 526–27.

Ting, I. P., H. B. Johnson, and S. R. Szarek. 1972. Net CO_2 fixation in crassulacean acid metabolism plants. In C. C. Black, ed., *Net carbon dioxide assimilation in higher plants*: 26–53. Proceedings of the Joint Symposium of the Southern Section of the American Society of Plant Physiologists and Cotton, Inc. Raleigh, N.C.

Ting, I. P., and S. R. Szarek. 1975. Drought adaptation in crassulacean acid metabolism plants. In N. F. Hadley, ed., *Environmental physiology of desert organisms*: 152–67. Stroudsburg, Pa.: Dowden, Hutchinson, and Rass.

Ting, I. P., M. L. Thompson, and W. M. Dugger, Jr. 1967. Leaf resistance to water vapor transfer in succulent plants: Effect of thermoperiod. *American Journal of Botany*, 54: 245–51.

Winter, K., and J. D. Tenhunen. 1982. Light-stimulated burst of carbon dioxide uptake following nocturnal acidification in the crassulacean acid metabolism plant *Kalanchoe daigremontiana*. *Plant Physiology*, 70: 1718–22.

17

Leaf-Age Effects on Stomatal Conductance

Christopher B. Field

Many features of leaf structure and physiology vary with leaf age. As leaves age, they function first as sinks for carbon and mineral nutrients, later as sources of carbon, and finally as sources of mineral nutrients, mined for redistribution throughout the plant.

An individual plant is an aggregate; its leaves are a population of multiple ages distributed through an array of microsites. In order to develop an integrated picture of whole-plant growth, photosynthesis, or water consumption from the responses of single leaves, we must understand the consequences both of leaf aging and of environmental heterogeneity within the canopies of individual plants. This chapter summarizes existing information concerning the effects of leaf age on stomatal conductance, emphasizing the interaction between stomatal conductance and other leaf functions, especially photosynthesis.

Two recent reviews have summarized the effects of leaf age on stomatal conductance (Solarova and Pospisilova, 1983) and on the density and size of stomata (Ticha, 1982). I will not attempt another comprehensive review but will focus on three questions. What happens to stomatal conductance during leaf aging? What factors are responsible for the changes? And do the age-dependent changes in stomatal conductance represent functional responses to changing circumstances, or do they represent deterioration, "wearing out"?

Techniques for Studying Leaf Aging

The effects of leaf age on stomatal conductance have been studied with a number of techniques, including controlled-environment systems for the simultaneous measurement of CO_2 and water-vapor exchange, a wide variety of porometers (steady-state, transient, and viscous flow), and direct observation

---

The author wishes to thank the organizers of the United States–Australia workshop on stomatal function both for assistance in improving this paper and for providing a rare opportunity to examine a topic from the biochemical to the ecological level; N. Chiarello and E.-D. Schulze for helpful ideas and discussions; and the University of Utah for supporting the experiments that led to many ideas presented here.

of stomata or stomatal impressions. Here, I emphasize the results of studies of the first kind but include some studies using other techniques. Controlled-environment gas-exchange systems offer high accuracy and an opportunity to consider conductance and photosynthesis together. Experiments based on these systems, however, rarely explore the behavior of plants in natural habitats, which presents one of the biggest challenges for future studies.

Experiments with controlled-environment gas-exchange systems typically yield values for leaf conductance or resistance calculated from the rate of transpiration, the water content of the air surrounding the leaf, and the water content of the air spaces within the leaf. This last parameter is not measured but is calculated, utilizing the assumption that the substomatal cavities are saturated with water vapor (see Ball, this volume). Conductance (the inverse of resistance) is used here because it is positively and almost linearly related to transpiration (Burrows and Milthorpe, 1976). Leaf conductance is usually expressed over a projected (one-sided) leaf area, even though the measurement includes gas exchange across both leaf surfaces. Leaf conductance includes stomatal and cuticular conductances in parallel and usually also includes a large boundary-layer conductance in series with the other terms. Specific values reported here are leaf conductances to water vapor, in the units proposed by Cowan (1977). Water-vapor conductance differs from CO_2 conductance by a multiplicative constant. Intercellular partial pressures of CO_2 (p_i) are calculated from independent measurements of leaf conductance and photosynthesis (Sharkey et al., 1982).

Many studies of leaf aging report photosynthesis (A) and conductance (g) under a loosely defined set of standard conditions. For measurements under saturating light intensity, the CO_2 concentration typical of normal air, the optimum temperature for photosynthesis at normal CO_2, and some constant difference between the water-vapor concentration in the leaf and in the air (Δw), I will use the terms photosynthetic capacity (A_{max}) and conductance at photosynthetic capacity (g_{Amax}).

The Phenomenon of Leaf Aging

I will consider aging in normal attached leaves. Much of the literature on leaf aging, or senescence, especially studies of the effects of plant-growth substances, concerns aging in excised leaves or leaf segments, a phenomenon I will not discuss here. Studies of aging in excised leaves have made important contributions to our understanding of aging in attached leaves, but these studies necessarily disregard the integration of leaf aging with plant development and plant-environment interactions.

Leaf longevity varies from 30 years or longer for needles of bristlecone pine to a few weeks for leaves of many annuals, but in all species thus far studied the leaves pass through a series of comparable life-history stages. Initially, leaves are engaged in intensive biosynthesis but are sustained by imported

Leaf-Age Effects on Stomatal Conductance

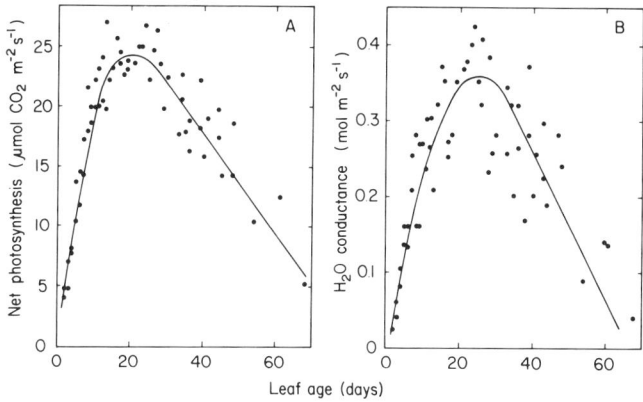

Fig. 17.1. Photosynthesis (A) and stomatal conductance (B) measured at saturating light intensity, a CO_2 partial pressure of 33 Pa, a leaf temperature of 27.8° C, and a vapor concentration gradient of 21 mmol mol^{-1} in leaves of cotton varying in age. Photosynthesis and conductance in these experiments were not influenced by leaf insertion level. (Redrawn from Constable and Rawson, 1980.)

carbon and accomplish little net photosynthesis (the expansion stage). Photosynthetic capacity typically reaches a maximum (A_{max}) about the time that the leaves reach full expansion (Woolhouse, 1978), though it occurs well before full expansion in tobacco (Rawson and Hackett, 1974) and somewhat before full expansion in cotton (Constable and Rawson, 1980) and papaya (Lin and Ehleringer, 1982). As leaves age further, A_{max}, protein content, and RNA content gradually decline (the aging phase). Leaves of some plants subsequently enter a phase in which an increase in dark respiration signals accelerated disassembly of cellular components (the senescence phase: Woolhouse, 1978; Thomas and Stoddart, 1980).

All leaves pass through similar stages; but the details vary, largely in response to two phenomena. First, many leaves age in a continually changing environment—seasonal change, for example, or the successive shading of older leaves by younger leaves. Second, the life histories of leaves are not independent of plant phenology: leaves produced at different stages in a plant's life frequently differ dramatically in size and longevity, or reproductive and vegetative phases may provide very different physiological environments.

Stomatal conductance during leaf aging. The effects of leaf age on g_{Amax} and on A_{max} are generally parallel (Fig. 17.1). Leaf conductance at photosynthetic capacity (g_{Amax}) may begin to decline before A_{max} (Catsky, Ticha, and Solarova, 1976); but the opposite pattern, with A_{max} declining first, has also been reported (Turner, 1969; Vaclavic, 1975). The overall pattern of a steep initial rise in A_{max} and g_{Amax} followed almost immediately by a gradual decline is modified in some plants to include a single broad plateau or multiple peaks in A_{max} and g_{Amax}. For example, the maintenance of approximately constant

A_{max} over much of the life of the leaf is common in woody deciduous species that flush an annual leaf population over a short period. Leaf conductance at photosynthetic capacity (g_{Amax}) closely parallels A_{max} in some but not all species with this phenology (Heichel and Turner, 1983; Küppers, 1984b). Asynchrony of changes in A_{max} and g_{Amax} may reflect the different sensitivities of these parameters to seasonal change. Many variations in a general paradigm of leaf aging are temporally related to either seasonal or phenological change. They may also result from artificial manipulations. Overall, decreases in A_{max} and g_{Amax} are delayed or temporarily reversed by life-history events or artificial treatments that increase whole-plant demands for photosynthate, that decrease whole-plant transpiration, or that increase nutrient availability. Treatments with the opposite effects tend to accelerate decreases in A_{max} and g_{Amax}. For example, g_{Amax} and A_{max} transiently increase in aging leaves of *Xanthium strumarium* (Krizek and Milthorpe, 1973), bean (Fraser and Bidwell, 1974), and soybean (Woodward and Rawson, 1976) in response to flowering or pod filling. In bell pepper, A_{max} and g_{Amax} normally increase at the onset of fruiting, but defloration prevents these responses (Hall and Brady, 1977). If leaves are removed from other parts of the plant, leaves of many species can be rejuvenated, displaying increased g_{Amax} and A_{max} (Hodgkinson, 1974; Nowak and Caldwell, 1984; von Caemmerer and Farquhar, 1984), increased protein and RNA synthesis, and regreening (Thomas and Stoddart, 1980). Age-related decreases in A_{max} and g_{Amax} may be accelerated by seasonal change or by water stress (Ludlow, 1975). Shading after leaf expansion tends to accelerate the loss of photosynthetic capacity (Pearce and Lee, 1969; Miyaji, 1984), but shading often increases the longevity of the leaf (Bazzaz and Harper, 1977; Jurik, Chabot, and Chabot, 1979; Miyaji, 1984).

The responses of stomatal conductance to environmental factors are modified by leaf aging (see Schulze et al., Sharkey and Ogawa, and Morison, this volume). The nature, mechanisms, and implications of the effects of leaf age on stomatal responses to humidity, light intensity, CO_2 concentration, and water stress are areas requiring much more research, but some patterns have emerged. Leaf age mainly affects stomatal responses by altering the magnitudes of the responses or the range of environmental factors over which responses occur. Stomatal responses to individual environmental variables do not appear to be restricted to leaves in particular age classes or life-history phases.

Leaf aging does not in general alter the response of g to light intensity, but the details of the response may vary. In the chaparral shrub *Lepechinia calycina*, the initial slope of the light response is nearly constant during the aging phase, and the light intensity required for saturation decreases (Fig. 17.2). In bean, the initial slope decreases with aging, but the intensity for saturation is approximately constant (Catsky and Ticha, 1980; Solarova, 1980). Compared with that of leaves near the age of full expansion, minimum

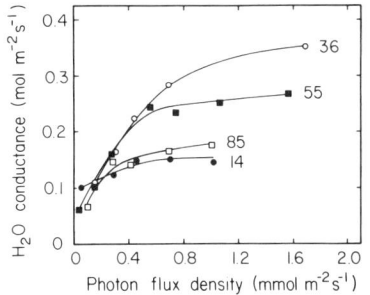

Fig. 17.2. The response of stomatal conductance to light intensity in leaves differing in age. The numbers at the right of each response curve are leaf ages in days. Data are from a naturally growing drought-deciduous shrub of the California chaparral, *Lepechinia calycina*. Maximum leaf lifespans are approximately 120 days during the normal growing season. (Data from Field, 1981b.)

conductance in the dark may be higher both in very young leaves (Fig. 17.2; perhaps as a result of high cuticular conductance) and in very old leaves (Burrows and Milthorpe, 1976).

The response of g to humidity, or, more specifically, to the difference between the water-vapor concentration in the leaf and that in the air (Δw), appears to be qualitatively altered by leaf aging in some species but largely unaffected in others. In a study of five European hedgerow species, Küppers (1984b) found that, though leaf aging did not alter the Δw response of g in three species, it decreased the sensitivity of g to Δw in *Ribes uva-crispa* but increased the sensitivity in *Crataegus* × *macrocarpa*.

The most widely reported response of conductance to decreasing water potential is sequential closure of stomata, progressing from older to younger leaves (Lawlor and Milford, 1975; Jordan, Brown, and Thomas, 1975) or from lower to upper leaves (Turner, 1974); but the opposite trend, with initial closure in the younger leaves, has been observed in orange trees (Syvertsen, Smith, and Allen, 1981). Aging alters the critical leaf-water potential required to initiate closure but does not change the overall form of the response. Though abscisic acid is implicated in drought-induced stomatal closure (see Raschke, this volume), several species, including cotton (Jordan, Brown, and Thomas, 1975), pearl millet, rice, and wheat (Quarrie and Henson, 1981), respond to drought with maximum accumulation of abscisic acid in the youngest leaves, which are in general least susceptible to drought-induced stomatal closure. The varying sensitivity of conductance to abscisic acid during aging requires further research.

Stomatal conductance and leaf-insertion level. Leaves produced at different phases of a plant's life history, especially on short-lived plants, operate in the context of changing demands, destinations, and priorities for mineral nutrients, water, and fixed carbon, as well as in a potentially changing physical environment. Because every leaf on a plant may have a slightly different life history from every other, one cannot safely assume that simultaneous observations on successive leaves yield the same picture of leaf aging as sequential observations on a single leaf (Richards, 1934).

Age dependence of photosynthesis and transpiration does not vary with leaf-insertion level in cotton (Constable and Rawson, 1980). In bean, A_{max} and g_{Amax} change at different rates for leaves varying in insertion level or in degree of shading, and the differences may (Davis, van Bavel, and McCree, 1977) or may not (Fraser and Bidwell, 1974) disappear when leaf age is expressed as a proportion of the time from maturity to abscision. Soybean and tobacco leaves from a sequence of insertion levels differ not only in longevity but also in peak photosynthetic capacity (Rawson and Hackett, 1974; Vaclavic, 1975; Woodward, 1976) though in tobacco, all leaves produced during the plant's mature phase have similar photosynthetic histories (Rawson and Hackett, 1974).

The causes of this variation with insertion level are not clear, but they are probably tied to two major factors. First, the schedule and pattern of leaf aging is altered if leaves are shaded (Davis, van Bavel, and McCree, 1977; Field, 1981a), and shading varies with insertion level. Leaves produced early in the life of an annual plant or early in the active season of a herbaceous perennial may be heavily shaded by higher leaves, but leaves produced later may never be shaded. Second, reproduction imposes demands for both carbon and mineral nutrients. Reproduction may prolong active photosynthesis in the leaves that serve as carbon sources (Hall and Brady, 1977), but it might also accelerate aging in the leaves that serve as sources of minerals.

Age Dependence in Leaf Structure

One possible explanation for the age dependence of stomatal conductance is that the patterns result from ontogenetic changes in guard cell viability, in stomatal density (number of stomata per unit of leaf area), or in physical limitations on the dimensions of the pores. Though the available evidence is not conclusive, these physical factors are probably not the primary determinants of the effects of leaf age on stomatal conductance.

Results of many studies (reviewed by Ticha, 1982) indicate that stomatal density increases after unfolding because the rate of stomatal differentiation initially exceeds the rate of leaf expansion. Stomatal density usually peaks before a leaf reaches 25 percent of its final size, though the maximum does not occur until 60 percent of final size in bell pepper (Schoch, 1972). Stoma mother cells are present in tobacco until the leaves reach 80 percent of their final size (Rawson and Craven, 1975), a value that probably marks the latest production of stomata.

After the initial increase, stomatal density declines steadily until the leaf reaches its final size. However, increases in stomatal size may be sufficient to offset decreases in density, resulting in a constant proportion of the leaf area occupied by stomatal complexes (Fig. 17.3; Brown and Rosenberg, 1970).

The constancy of stomatal area during expansion argues against control of conductance by changes in leaf structure, but it does not address the possibility

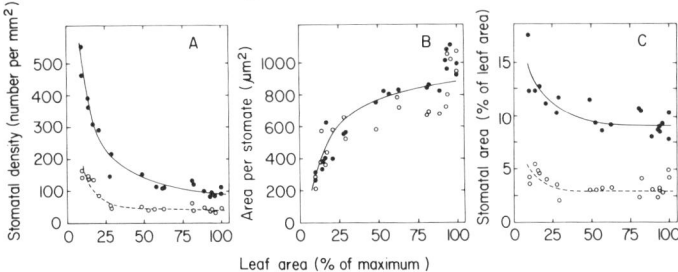

Fig. 17.3. Stomatal density (A), stomatal size (B), and the total area occupied by stomatal structures (guard cells and pores, C) in tobacco leaves during expansion. Data from adaxial (open circles) and abaxial (solid circles) surfaces of leaves 11 and 18, counting from the base of the plant. The plants were grown in a glasshouse under natural illumination. (Redrawn from Rawson and Craven, 1975.)

that decreases in conductance beyond the age of full expansion are related to the loss of functional stomata. Available evidence indicates that this possibility too is unlikely to be generally important. In many species, if not in general, guard cells appear to remain functional longer than mesophyll cells in the same leaf. Two kinds of evidence document the characteristics necessary (but not sufficient) for longer functionality in guard cells. First, guard cells retain chlorophyll longer than mesophyll cells, as judged by the ability to reduce silver nitrate (Molisch, 1918) and by chlorophyll fluorescence (Kenda, Thaler, and Weber, 1953; Zeiger and Schwartz, 1982). Second, fluorescence transients from aging leaves of *Ginkgo biloba* indicate that the guard cells are capable of performing electron transport and photophosphorylation after mesophyll cells have lost that ability (Zeiger and Schwartz, 1982).

Another possible explanation for the effects of age on leaf conductance is that the changes are caused by nonstomatal components of the diffusion path. Leaf cuticle often continues to develop during leaf expansion, but changes in cuticular conductance are unlikely to be quantitatively important during the aging phase. Increasing deposition of the leaf cuticle probably accounts for the observation that, although minimum leaf conductances in darkness decrease during leaf expansion (Ludlow and Wilson, 1971), minimum stomatal conductances appear to remain essentially constant (Burrows and Milthorpe, 1976). Cuticular effects could significantly affect leaf conductance during the aging phase if stomata were increasingly plugged with cuticular wax. Wax-filled stomatal antechambers are common in conifers (Rhine, 1924), though detailed studies of sitka spruce reveal that the stomata are not sealed but are covered with a complex mat of interwoven wax tubes (Jeffree, Johnson, and Jarvis, 1971). Cuticular wax substantially decreases leaf conductance in sitka spruce, but with similar effects in old and young leaves alike (Jeffree, Johnson, and Jarvis, 1971).

The Integration of Leaf Function during Aging

Unraveling the consequences of the age dependence of stomatal conductance requires an integrated view of leaf function. Stomata are regulatory structures, and we cannot assess their function except in the context of the processes they influence and are influenced by. If the primary role of stomata is to allow sufficient inward diffusion of CO_2 while preventing excessive water loss, we should focus on the relationship between stomatal conductance and photosynthesis during leaf aging.

Stomatal conductance and photosynthesis. The effects of leaf age on stomatal conductance at light saturation generally parallel their effects on photosynthetic capacity (Fig. 17.1). The coordination between g_{Amax} and A_{max} during leaf aging is demonstrated by the strong correlation between these parameters (Fig. 17.4; Ludlow and Wilson, 1971; Woodward and Rawson, 1976). In leaves of *Lepechinia calycina* varying in age, this correlation is also preserved across changes in light intensity (Fig. 17.5). Independent of leaf age, leaves operating at one photosynthetic rate and grown under one regime of water availability have the same or nearly the same conductance, even though they may be operating at substantially different light intensities and at different fractions of their light-saturated photosynthetic capacities. The maintenance of a linear A-to-g relationship over many light levels characterizes leaves of plants grown at differing nitrogen availabilities (Wong, Cowan, and Farquhar, 1985) as well as leaves of different ages, broadening the evidence that conductance and photosynthesis vary in parallel.

Not only are A and g linearly related, but the regression of one parameter on the other often yields a y-intercept near zero. A consequence of this relationship is that the partial pressure of the CO_2 in the intercellular spaces (p_i) is nearly constant during leaf aging (Davis and McCree, 1978; Constable and Rawson,

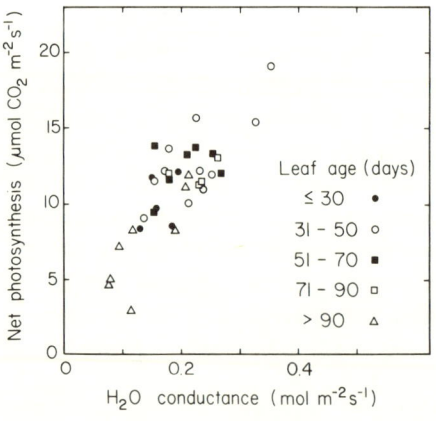

Fig. 17.4. The relation between photosynthesis and stomatal conductance in leaves of naturally growing *Lepechinia calycina* varying in age. The data are equilibrium values measured at saturating light intensity, a CO_2 partial pressure of 34 Pa, and a vapor concentration gradient of 15 mmol mol^{-1}. The arrangement of the axes is not meant to imply that conductance is the independent variable. (Unpublished data from the study of Field, 1981b.)

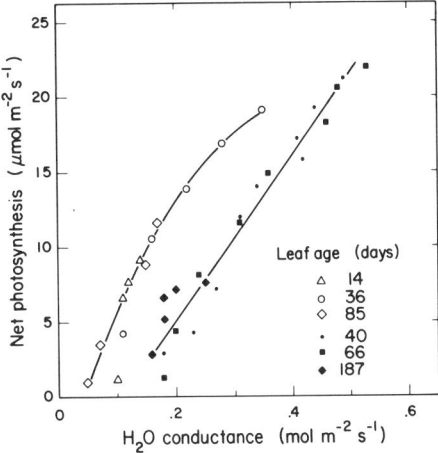

Fig. 17.5. The relationship between photosynthesis and stomatal conductance in three leaves from naturally growing (open symbols) and three from irrigated plants (solid symbols) of *Lepechinia calycina* subjected to a range of light intensities. Maximum leaf lifespans on irrigated plants are approximately 200 days. For each leaf, points at low conductance and photosynthesis represent measurements at low light intensity. Points at high photosynthesis and conductance represent measurements at high light intensity. The arrangement of the axes is not meant to imply that conductance is the independent variable. (Redrawn from Field, 1981b.)

1980; Field and Mooney, 1983) and at all but the lowest light intensities. Some studies do report an age dependence of p_i (see Morison, this volume), though it is small in most cases. A common pattern is that p_i increases gradually with aging (Rawson and Constable, 1980; Evans, 1983), but it may decrease and later increase (Lin and Ehleringer, 1982) or increase and later decrease (cf. *Prunus spinosa* in Küppers, 1984).

At a given Δw and with a large boundary-layer conductance, a linear A-to-g relationship with a y-intercept near zero (or a constant p_i) indicates that the ratio of photosynthesis to transpiration is also constant. This ratio and the ratio of photosynthesis to stomatal conductance have been widely used as measures of instantaneous water-use efficiency. These simple measures provide only a snapshot and not the dynamics of water-use efficiency, but they are useful measures of physiological potential. Some studies that do not report p_i do provide evidence of constant A-to-g ratios during aging, but other results document some variation. Woodward and Rawson (1976) and Rawson and Woodward (1976) found that the A-to-g ratio varied substantially during leaf aging in soybean and tobacco.

An absence of age dependence in these instantaneous measurements of water-use efficiency indicates that the stomata establish a consistent compromise between water loss and CO_2 gain; this constitutes some of the strongest evidence that stomatal function is not degraded during leaf aging but is adjusted to preserve the integration of leaf function. Those results reporting slight increases in p_i or in the A-to-g ratio during leaf aging suggest the possibility of even finer tuning in the A-to-g relationship. To the extent that older leaves are typically more shaded than younger leaves (Field, 1981a, 1983), older leaves may be cooler and may operate at lower Δw's than younger leaves. As a result, older leaves may often have slightly lower transpiration rates than younger leaves with the same conductance. Thus, under natural conditions an increasing p_i during aging may yield a constant ratio of photosynthesis to transpiration.

Cowan and Farquhar (1977) provide a more sophisticated approach to analyzing water-use efficiency. Their theory specifically addresses the allocation of water to a single leaf over the course of a day, but the analysis for all the leaves on a plant is formally identical. Optimality is defined as the condition under which any redistribution of a fixed total amount of transpiration (through changes in conductance) results in a decrease in total photosynthesis from all the leaves. To date, no study of leaf aging has included an analysis of water-use efficiency according to this criterion. The evidence for a constant p_i or A-to-g ratio during leaf aging, though insufficient to demonstrate maximization of water-use efficiency, is not inconsistent with that paradigm. Rigorous assessment of water-use efficiency during leaf aging will require detailed information on responses of photosynthesis and of stomatal conductance to p_i, Δw, light, and temperature, as well as a thorough analysis of the microsites of leaves of different ages.

Leaf Aging as Acclimation

Probably the question most frequently addressed in studies of the effects of leaf age on photosynthesis is whether the changes, especially the declines during the aging phase, are caused by changes in stomatal conductance or by changes in the biochemical capacity of the photosynthetic machinery. The question can be asked at two levels—what are the physiological causes, and what are the environmental factors and evolutionary responses that have shaped the physiological patterns? Most studies have addressed only the first question, but our understanding of leaf aging will not be complete until we can provide answers to both.

One powerful approach to differentiating stomatal limitations from biochemical limitations on photosynthesis is the method described by Farquhar and Sharkey (1982), based on analyzing the p_i response of photosynthesis. Few studies of leaf aging have included p_i responses; most of the existing interpretations are based on the linear analysis originated by Gaastra (1959). Errors in the linear analysis tend to overestimate stomatal limitations (Farquhar and Sharkey, 1982). Even with this bias, a majority of the studies attempting to separate the effects of leaf age on photosynthesis into stomatal and nonstomatal components have concluded that the latter are dominant. Because the magnitude of the bias may change during aging, this conclusion cannot be interpreted as confirming the dominance of nonstomatal limitations; it does, however, demonstrate their great importance. The nonstomatal causes of age-dependent decline in photosynthetic capacity probably involve coordinated decreases in the levels of several photosynthetic components, as is indicated by the declines in the concentrations of leaf nitrogen (Field and Mooney, 1983), protein (Molisch, 1930), RNA, and chlorophyll (Hardwick, Wood, and Woolhouse, 1968), as well as in the total activity of the primary carboxylating enzyme, ribulose-1,5-bisphosphate carboxylase-oxygenase (Friedrich and Huffaker, 1980; Wittenbach, 1983).

As a first approximation, the dependence of p_i on leaf age specifies the relative importance of stomatal and nonstomatal effects on photosynthesis during leaf aging. This approximation is strictly correct only if leaf aging alters the response of A to p_i by scaling A with a simple multiplicative constant, an effect largely though not totally supported by experiments on wheat (Evans, 1983) and on several woody species (Küppers, 1984a). Assuming that the approximation is correct, an increasing p_i during aging indicates that nonstomatal factors are dominant. If p_i decreases, stomatal factors dominate. A constant p_i indicates that stomatal and nonstomatal limitations are changing in concert. Concerted change is a functionally useful response because the stomata prevent old leaves from becoming "holes in the dike" where the meager returns in CO_2 fixation cannot balance the large outflows of water. Though p_i is not always constant during leaf aging, variation is usually small. Photosynthetic capacity declines during leaf aging in response to stomatal and nonstomatal factors.

What controls the age-dependent changes in photosynthesis and conductance? The answer is far from clear, but several interesting hypotheses have been presented. Thimann and Satler (1979) observed that detached leaves lose chlorophyll more slowly when stomata are open than when they are closed or artificially occluded. From this, they suggested that senescence (chlorophyll loss) in detached leaves may be caused by a continuously produced gaseous compound that accumulates to active levels inside leaves only when conductance is very low. Such a mechanism could also regulate aging in attached leaves, though the control sufficient to stimulate increasing A_{max} in developing leaves and decreasing A_{max} in aging leaves, both of which have low g_{Amax}, would be complex.

Ýcas (1984) modified the hypothesis of Thimann and Satler (1979) to suggest that the factor controlling senescence is an inhibitor rather than a promoter. He argues that a senescence inhibitor produced in the roots and passively distributed in the transpiration stream will accumulate in the leaves with the highest transpiration rates. If either of these hypotheses were correct, age-dependent decreases in conductance should always precede decreases in photosynthesis. But in fact, age-dependent declines in A_{max} and g_{Amax} are temporally variable. The changes are often closely synchronous, though either factor may change first.

An opposite interpretation, that g_{Amax} is controlled by A_{max}, was suggested by Wong, Cowan, and Farquhar (1979), who observed that a linear relationship between A_{max} and g_{Amax} is preserved over a large range of treatments, including chemical inhibition of A_{max} with 3-(3,4-dichlorophenyl)-1,1-dimethylurea

(DCMU), an inhibitor of photosynthetic electron transport. Though much evidence supports it, this hypothesis cannot explain the observation that age-dependent decreases in g_{Amax} may precede declines in A_{max}.

Available data do not allow positive identification of the parameter or parameters controlling leaf aging. To some extent, the failure to identify any one control reflects the extremely robust integration of photosynthesis and conductance. Perhaps the integration has been shaped by natural selection to function with multiple controls, including direct and indirect leaf-age dependence on both photosynthesis and conductance.

The suggestion that leaf aging has been molded by natural selection implies that aging confers some advantage. It is difficult to envision how an age-dependent decrease in photosynthesis or conductance would confer any direct advantage, but it is important to place aging leaves in the context of whole plants, where resources removed or withheld from an aging leaf are potentially available for investment in other leaves or other plant parts.

Leaf microhabitats. The growth form of many plants places older leaves in microsites where they are increasingly shaded by younger leaves. Even in plants with open canopies, younger leaves are consistently exposed to higher light intensities than older leaves (Field, 1981a, 1983). From the top of a crop canopy downward, the decrease in light availability may be dramatic; leaves at the bottom may receive only a small percentage of the light falling on the highest leaves.

Resource allocation within the plant. Photosynthetic capacity (A_{max}) is not a good index of photosynthesis under reduced illumination. High A_{max} is usually associated with a high rate of respiration in the dark and a high light-compensation point, resulting in decreased low-light photosynthesis relative to leaves with lower maximum capacities (Björkman, 1981). This trade-off is clear during leaf aging. Under the shaded conditions encountered by older leaves, age-dependent decreases in A_{max} are largely irrelevant (Fig. 17.6; Aslam, Lowe, and Hunt, 1977; Field, 1981a). Under very shady conditions, aging may even increase daily photosynthesis (Field, 1981a).

If the factors limiting photosynthesis are in short supply or can be obtained only at the expense of photosynthate potentially useful for other functions, natural selection should favor features that enable efficient use of the limiting factors. Given that a high A_{max} yields marginal benefits under shaded conditions, efficient use may entail shifting photosynthetic capacity from shaded leaves, where it is largely unusable, to leaves exposed to higher light. In order for a plant to maintain effective integration of photosynthesis and transpiration, any reallocation should produce balanced effects on stomatal and nonstomatal limits to photosynthesis.

The nonstomatal or biochemical determinants of photosynthetic capacity are probably complex (Farquhar, von Caemmerer, and Berry, 1980), but the strong

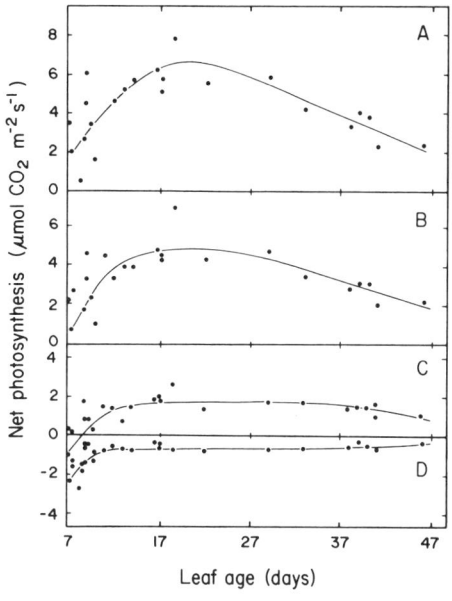

Fig. 17.6. The relationship between photosynthesis and leaf age at photon flux densities corresponding to light saturation (1,490 μmol m^{-2} s^{-1}: A), at 245 μmol m^{-2} s^{-1} (B), at 68 μmol m^{-2} s^{-1} (C), and in darkness (0 μmol m^{-2} s^{-1}: D). The data are from *Fragaria virginiana* grown in growth chambers under low light. (Redrawn from Jurik, Chabot, and Chabot, 1979.)

correlation between photosynthetic capacity and leaf nitrogen content (Field and Mooney, 1986) argues that leaf nitrogen can be used to summarize total investment in the biochemistry of photosynthesis. Nitrogen frequently limits plant growth, and assimilating nitrogen from the environment entails significant costs (Bloom et al., 1985). Natural selection should, therefore, favor features that enable efficient use of nitrogen. Reallocation of nitrogen out of older leaves, where its investment yields little photosynthesis, to younger leaves in higher light or to other plant structures may represent an evolutionary advantage, but with the consequence that aging leaves rarely if ever maintain high photosynthetic capacities. Numerical simulations based on this hypothesis indicate that whole-plant photosynthesis is increased when nitrogen is reallocated among leaves during aging. Actual nitrogen contents in aging leaves of *Lepechinia calycina* approximate the optimal pattern (Field, 1983). Further support for this hypothesis comes from the observation that leaf aging has smaller effects on photosynthesis in the infrequently shaded leaves of low-growing annuals in Death Valley than in the typically shaded leaves of annuals growing in old fields (Mooney et al., 1981).

Why should age be the cue for nitrogen reallocation, when direct environmental cues are available? First, environmental cues do influence conductance during aging, acting in addition to intrinsic cues (Davis, van Bavel, and McCree, 1977). Second, redistribution of nitrogen is probably expensive, requiring ATP for active transport (Radin and Elmore, 1980). Tying nitrogen distribution solely to environmental cues could lead to excessive reallocation, for which the transport costs would outweigh the benefits of increased photosynthesis. Thus leaf aging as molded by natural selection may present a conservative but functional solution to the problem of changing microsites. According to this line of reasoning, the effects of leaf age on photosynthesis are due to both stomatal and nonstomatal factors. Intrinsic constraints on the schedule of reallocating biochemical components act in combination with environmental factors and plant phenology to set the tempo for aging.

Acclimation or deterioration. During the expansion phase and during the gradual decline of photosynthetic capacity through the aging phase, stomata function within fairly narrow limits to maintain a consistent balance between photosynthesis and transpiration. This is precisely their function when the leaf is at its physiological peak. From the whole-plant perspective, the maintenance of this integration is a better index of functional competence than is maximum conductance. Stomata do not deteriorate as the leaf ages, but they do adjust.

REFERENCES

Aslam, M., S. B. Lowe, and L. A. Hunt. 1977. Effect of leaf age on photosynthesis and transpiration of cassava (*Manihot esculenta*). *Canadian Journal of Botany*, 55: 2288–95.

Bazzaz, F. A., and J. L. Harper. 1977. Demographic analysis of the growth of *Linum usitatissimum*. *New Phytologist*, 78: 193–208.

Björkman, O. 1981. Responses to different quantum flux densities. In O. L. Lange, P. S. Nobel, C. B. Osmond, and H. Ziegler, eds., *Physiological plant ecology 1, Responses to the physical environment*, Encyclopedia of Plant Physiology, n.s., 12A: 57–107. Berlin: Springer-Verlag.

Bloom, A. J., F. S. Chapin III, and H. A. Mooney. 1985. Resource limitation in plants—An economic analogy. *Annual Review of Ecology and Systematics*, 16: 363–92.

Brown, K. W., and N. J. Rosenberg. 1970. Influence of leaf age, illumination, and upper and lower surface differences on stomatal resistance of sugar beet (*Beta vulgaris*) leaves. *Agronomy Journal*, 62: 20–24.

Burrows, F. J., and F. L. Milthorpe. 1976. Stomatal conductance in the control of gas exchange. In T. T. Kozlowski, ed., *Water deficits and plant growth 4, Soil water measurement, plant responses, and breeding for drought resistance*: 103–52. New York: Academic.

von Caemmerer, S., and G. D. Farquhar. 1984. Effects of partial defoliation, changes of irradiance during growth, short-term water stress and growth at enhanced $p(CO_2)$ on the photosynthetic capacity of leaves of *Phaseolus vulgaris* L. *Planta*, 160: 320–29.

Catsky, J., and I. Ticha. 1980. Ontogenetic changes in the internal limitations to bean-leaf photosynthesis, V. Photosynthetic and photorespiration rates and conductances for CO_2 transfer as affected by irradiance. *Photosynthetica*, 14: 392–400.

Catsky, J., I. Ticha, and J. Solarova. 1976. Ontogenetic changes in the internal limitations to bean-leaf photosynthesis, I. Carbon dioxide exchange and conductances for carbon dioxide transfer. *Photosynthetica*, 10: 394–402.

Constable, G. A., and H. M. Rawson. 1980. Effect of leaf position, expansion, and age on photosynthesis, transpiration and water use efficiency of cotton. *Australian Journal of Plant Physiology*, 7: 89–100.

Cowan, I. R. 1977. Stomatal behaviour and environment. *Advances in Botanical Research*, 4: 117–228.

Cowan, I. R., and G. D. Farquhar. 1977. Stomatal function in relation to leaf metabolism and environment. In D. H. Jennings, ed., *Integration of activity in the higher plant*: 471–505. Symposia of the Society for Experimental Biology, 31. Cambridge: Cambridge University Press.

Davis, S. D., C. H. M. van Bavel, and K. J. McCree. 1977. Effect of leaf aging upon stomatal resistance in bean plants. *Crop Science*, 17: 640–45.

Davis, S. D., and K. J. McCree. 1978. Photosynthetic rate and diffusion conductance as a function of age in leaves of bean plants. *Crop Science*, 18: 280–82.

Evans, J. R. 1983. Nitrogen and photosynthesis in the flag leaf of wheat (*Triticum aestivum* L.). *Plant Physiology*, 72: 297–302.

Farquhar, G. D., S. von Caemmerer, and J. A. Berry. 1980. A biochemical model of photosynthetic CO_2 assimilation in the leaves of C_3 species. *Planta*, 149: 78–90.

Farquhar, G. D., and T. D. Sharkey. 1982. Stomatal conductance and photosynthesis. *Annual Review of Plant Physiology*, 33: 317–45.

Field, C. 1981a. Leaf age effects on the carbon gain of individual leaves in relation to microsite. In N. S. Margaris and H. A. Mooney, eds., *Components of productivity of Mediterranean-climate regions: Basic and applied aspects*: 41–50. The Hague: Junk.

———. 1981b. Carbon gain consequences of leaf aging in a California shrub. Ph.D. dissertation, Stanford University, Stanford, Calif.

———. 1983. Allocating leaf nitrogen for the maximization of carbon gain: Leaf age as a control on the allocation program. *Oecologia*, 56: 341–47.

Field, C., and H. A. Mooney. 1983. Leaf age and seasonal effects on light, water, and nitrogen use efficiency in a California shrub. *Oecologia*, 56: 348–55.

———. 1986. The photosynthesis-nitrogen relationship in wild plants. In T. J. Givnish, ed., *On the economy of plant form and function*: 25–55. Cambridge: Cambridge University Press.

Fraser, D. E., and R. G. S. Bidwell. 1974. Photosynthesis and photorespiration during the ontogeny of the bean plant. *Canadian Journal of Botany*, 52: 2562–70.

Friedrich, J. W., and R. C. Huffaker. 1980. Photosynthesis, leaf resistances, and ribulose-1,5-bisphosphate carboxylase degradation in senescing barley leaves. *Plant Physiology*, 65: 1103–7.

Gaastra, P. 1959. Photosynthesis of crop plants as influenced by light, carbon dioxide, temperature and stomatal diffusion resistance. *Mededelingen van de Landbouwhogeschool*, Wageningen, 59: 1–68.

Hall, A. J., and C. J. Brady. 1977. Assimilate source-sink relationships in *Capsicum annuum* L., II. Effects of fruiting and defloration on the photosynthetic capacity and senescence of leaves. *Australian Journal of Plant Physiology*, 4: 771–83.

Hardwick, K., M. Wood, and H. W. Woolhouse. 1968. Photosynthesis and respiration in relation to leaf age in *Perilla frutescens* (L.) Britt. *New Phytologist*, 67: 79–86.

Heichel, G. H., and N. C. Turner. 1983. CO_2 assimilation of primary and regrowth foliage of red maple (*Acer rubrum* L.) and red oak (*Quercus rubra* L.): Response to defoliation. *Oecologia*, 57: 14–19.

Hodgkinson, K. C. 1974. Influence of partial defoliation on photosynthesis,

photorespiration and transpiration by lucerne leaves of different ages. *Australian Journal of Plant Physiology*, 1: 561–78.
Jeffree, C. E., R. P. C. Johnson, and P. G. Jarvis. 1971. Epicuticular wax in the stomatal antechamber of Sitka spruce and its effects on the diffusion of water vapour and carbon dioxide. *Planta*, 98: 1–10.
Jordan, W. R., K. W. Brown, and J. C. Thomas. 1975. Leaf age as a determinant in stomatal control of water loss from cotton during water stress. *Plant Physiology*, 56: 595–99.
Jurik, T. W., J. F. Chabot, and B. F. Chabot. 1979. Ontogeny of photosynthetic performance in *Fragaria virginiana* under changing light regimes. *Plant Physiology*, 63: 542–47.
Kenda, G., I. Thaler, and F. Weber. 1953. Schliesszellen-Chloroplasten vergilben nicht. *Protoplasma*, 42: 246–49.
Krizek, D. T., and F. L. Milthorpe. 1973. Effect of photoperiodic induction on the transpiration rate and stomatal behaviour of debudded *Xanthium* plants. *Journal of Experimental Botany*, 24: 76–86.
Küppers, M. 1984a. Carbon relations and competition between woody species in a central European hedgerow, I. Photosynthetic characteristics. *Oecologia*, 64: 322–43.
———. 1984b. Carbon relations and competition between woody species in a central European hedgerow, II. Stomatal responses, water use, and hydraulic conductivity in the root/leaf pathway. *Oecologia*, 64: 344–54.
Lawlor, D. W., and G. F. J. Milford. 1975. The control of water and carbon dioxide flux in water-stressed sugar beet. *Journal of Experimental Botany*, 26: 657–65.
Lin, Z. F., and J. Ehleringer. 1982. Effects of leaf age on photosynthesis and water use efficiency of papaya. *Photosynthetica*, 16: 514–19.
Ludlow, M. M. 1975. Effect of water stress on the decline of leaf net photosynthesis with age. In R. Marcelle, ed., *Environmental and biological control of photosynthesis*: 123–34. The Hague: Junk.
Ludlow, M. M., and G. L. Wilson. 1971. Photosynthesis of tropical pasture plants, III. Leaf age. *Australian Journal of Biological Sciences*, 24: 1077–87.
Miyaji, K.-I. 1984. Longevity and productivity of leaves of a cultivated annual, *Glycine max* Merrill, II. Productivity of leaves in relation to their longevity, plant density and sowing time. *New Phytologist*, 97: 479–88.
Molisch, H. 1918. Über die Vergilbung der Blatter. *Sitzungsberichte der Österreichischen Akademie der Wissenschaften in Wien, Mathematisch-naturwissenschaftliche Klasse*, 127: 3–34.
———. 1930. *The longevity of plants*. Transl. E. H. Fulling. New York: published by the translator.
Mooney, H. A., C. Field, S. L. Gulmon, and F. A. Bazzaz. 1981. Photosynthetic capacity in relation to leaf position in desert versus old-field annuals. *Oecologia*, 50: 109–12.
Nowak, R. S., and M. M. Caldwell. 1984. A test of compensatory photosynthesis in the field: Implications for herbivory tolerance. *Oecologia*, 61: 311–18.
Pearce, R. B., and D. R. Lee. 1969. Photosynthetic and morphological adaptation of alfalfa leaves to light intensity at different stages of maturity. *Crop Science*, 9: 791–94.
Quarrie, S. A., and I. E. Henson. 1981. Abscisic acid accumulation in detached cereal leaves in response to water stress, II. Effects of leaf age and position. *Zeitschrift für Pflanzenphysiologie*, 101: 439–46.
Radin, J. W., and C. D. Elmore. 1980. Concepts of translocation with special reference

to the assimilation of nitrogen and its movement into fruits. In J. D. Hesketh and J. W. Jones, eds., *Predicting photosynthesis for ecosystem models* 2: 143–54. Boca Raton, Fla. CRC Press.

Rawson, H. M., and G. A. Constable. 1980. Carbon production of sunflower cultivars in field and controlled environments, I. Photosynthesis and transpiration of leaves, stems, and heads. *Australian Journal of Plant Physiology*, 7: 555–73.

Rawson, H. M., and C. L. Craven. 1975. Stomatal development during leaf expansion in tobacco and sunflower. *Australian Journal of Botany*, 23: 253–61.

Rawson, H. M., and C. Hackett. 1974. An exploration of the carbon economy of the tobacco plant, III. Gas exchange of leaves in relation to position on the stem, ontogeny and nitrogen content. *Australian Journal of Plant Physiology*, 1: 551–60.

Rawson, H. M., and R. G. Woodward. 1976. Photosynthesis and transpiration in dicotyledonous plants, I. Expanding leaves of tobacco and sunflower. *Australian Journal of Plant Physiology*, 3: 245–56.

Rhine, J. B. 1924. Clogging of stomata of conifers in relation to smoke injury and disturbance. *Botanical Gazette*, 78: 226–32.

Richards, F. J. 1934. On the use of simultaneous observations on successive leaves for the study of physiological change in relation to leaf age. *Annals of Botany*, 48: 497–504.

Schoch, P. G. 1972. Effect of shading on structural characteristics of the leaf and yield of fruit in *Capsicum annuum* L. *Journal of the American Society for Horticultural Science*, 97: 461–64.

Sharkey, T. D., K. Imai, G. D. Farquhar, and I. R. Cowan. 1982. A direct confirmation of the standard method of estimating intercellular partial pressure of CO_2. *Plant Physiology*, 69: 657–59.

Solarova, J. 1980. Diffusive conductances of adaxial (upper) and abaxial (lower) epidermes: Response to quantum irradiance during development of primary *Phaseolus vulgaris* L. leaves. *Photosynthetica*, 14: 523–31.

Solarova, J., and J. Pospisilova. 1983. Photosynthetic characteristics during ontogenesis of leaves, VIII. Stomatal diffusive conductance and stomatal reactivity. *Photosynthetica*, 17: 101–51.

Syvertsen, J. P., M. L. Smith, Jr., and J. C. Allen. 1981. Growth rate and water relations of citrus leaf flushes. *Annals of Botany*, 47: 97–105.

Thimann, K. V., and S. O. Satler. 1979. Relation between leaf senescence and stomatal closure: Senescence in light. *Proceedings of the National Academy of Sciences*, 76: 2295–98.

Thomas, H., and J. L. Stoddart. 1980. Leaf senescence. *Annual Review of Plant Physiology*, 31: 83–111.

Ticha, I. 1982. Photosynthetic characteristics during ontogenesis of leaves, VII. Stomata density and sizes. *Photosynthetica*, 16: 375–471.

Turner, N. C. 1969. Stomatal resistance to transpiration in three contrasting canopies. *Crop Science*, 9: 303–7.

———. 1974. Stomatal behaviour and water status of maize, sorghum and tobacco under field conditions, II. At low soil water potential. *Plant Physiology*, 53: 360–65.

Vaclavic, J. 1975. Comparison of the changes in net photosynthetic CO_2 uptake and water vapour efflux during leaf ontogenesis with the differences between the leaves according to their descending insertion level. *Biologia plantarum*, 17: 411–15.

Wittenbach, V. A. 1983. Effect of pod removal on leaf photosynthesis and soluble protein composition of field-grown soybeans. *Plant Physiology*, 73: 121–24.

Wong, S. C., I. R. Cowan, and G. D. Farquhar. 1979. Stomatal conductance correlates with photosynthetic capacity. *Nature*, 282: 424–26.

———. 1985. Leaf conductance in relation to rate of CO_2 assimilation, II. Effects of short-term exposures to different photon flux densities. *Plant Physiology*, 78: 826–29.
Woodward, R. G. 1976. Photosynthesis and expansion of leaves of soybean grown in two environments. *Photosynthetica*, 10: 274–79.
Woodward, R. G., and H. M. Rawson. 1976. Photosynthesis and transpiration in dicotyledonous plants, II. Expanding and senescing leaves of soybean. *Australian Journal of Plant Physiology*, 3: 257–67.
Woolhouse, H. W. 1978. Senescence processes in the life cycle of flowering plants. *Bioscience*, 28: 25–31.
Ýcas, J. W. 1984. The effect of nutrient distribution and senescence on whole-canopy productivity. Ph.D. dissertation, Cornell University, Ithaca, N.Y.
Zeiger, E., and A. Schwartz. 1982. Longevity of guard cell chloroplasts in falling leaves: Implication for stomatal function and cellular aging. *Science*, 218: 680–82.

18

Transfer Processes in Plant Canopies in Relation to Stomatal Characteristics

J. J. Finnigan and M. R. Raupach

Earlier papers in this book have concentrated on single cells, single leaves, or, at most, single plants. However, the physiological state of a plant community substantially influences the microclimate within it; in turn, the microclimate influences the physiological state, so that neither is independent of the other. Plant communities differ in this respect from isolated plants, for which the microclimate and the physiology are independent. In this paper we concentrate on the interaction between the aerial canopy environment and the plants' behavior.

We begin by deriving the basic equations for conservation of mass that apply within a canopy. These equations provide the essential analytical framework both for predictive modeling and for interpretation of measurements. Next, we describe the character of the turbulent airstream within the canopy, drawing heavily upon results obtained in the last few years that demonstrate the intermittent nature of canopy turbulence. Third, we consider transport processes at individual leaf surfaces in the context of unsteady external conditions. Finally, we treat three methods of modeling the behavior of a whole canopy. These three models—one of them probably familiar to many readers of this book, the others less so—illustrate the way the various parts of the transfer mechanism interact in a complete description of the plant-air continuum.

The Meteorological Environment of a Plant Canopy

Turbulent transport, eddy fluxes, and the scalar conservation equation. The lowest layers of the atmosphere are usually turbulent, with rapid fluctuations in wind speed, temperature, humidity, and CO_2 concentration superimposed on more slowly varying background conditions. Within a plant canopy, the characteristic unsteadiness of all turbulent flow is further complicated by extreme spatial variability. We cope with both problems in essentially the same way: by averaging in time we reduce the unsteadiness in the flow to manageable proportions and by averaging in space we subdue heterogeneity.

An important consequence of the form of the equations describing instantaneous values of concentrations in a moving fluid is that equations for the evolution of average values contain mean products of fluctuations. We begin this section, therefore, by deriving the time-averaged, point-valued equation of conservation of a scalar constituent of the atmosphere (henceforth, such constituents will be referred to generically, as "scalars") and then introduce the less familiar operation of volume averaging to obtain an equation set that adequately describes the situation within a plant canopy.

Let $c(x_j, t)$ be the instantaneous mole fraction, or concentration, of some scalar such as water vapor or CO_2. The conservation equation for c is

$$\frac{\partial c}{\partial t} + u_j \frac{\partial c}{\partial x_j} = \frac{\partial}{\partial x_j}\left[k_c \frac{\partial c}{\partial x_j}\right] \tag{1}$$

where k_c is the molecular diffusivity of the scalar, u_j the instantaneous velocity vector, x_j a position coordinate, and t time. We employ the tensor summation convention

$$u_j \frac{\partial c}{\partial x_j} = u_1 \frac{\partial c}{\partial x_1} + u_2 \frac{\partial c}{\partial x_2} + u_3 \frac{\partial c}{\partial x_3}.$$

Equation 1 is exact, provided that the molar volume of the air (or the air density, if c were to represent mass fraction rather than mole fraction) may be taken as constant. This usually is an excellent assumption in canopy flow, except very close to the surface of very hot leaves, where it may lead to errors of the order of one or two percent. An exhaustive treatment of the equations of motion, energy, and mass conservation in turbulent flow is provided by Businger (1982), where approximations such as these are discussed at length. The assumption of constant density enables the equation of conservation of total air mass, or the "continuity equation," to be written in the simple form

$$\frac{\partial u_j}{\partial x_j} = 0. \tag{2}$$

We define a time average operator, denoted by an overbar, thus:

$$\bar{c}(t) = \frac{1}{T}\int_{t-T/2}^{t+T/2} c(t')\,dt' \tag{3}$$

where T is the averaging time, so that u_j and c can be separated into a time-averaged part and a part due to turbulent fluctuations:

$$u_j = \bar{u}_j + u'_j, \quad c = \bar{c} + c'. \tag{4}$$

Plant-Canopy Transfer and Stomatal Characteristics

This separation is called the "Reynolds decomposition." Finally, by substituting Equation 4 into Equation 1 and time-averaging the resulting equation, we obtain

$$\frac{\partial \bar{c}}{\partial t} + \bar{u}_j \frac{\partial \bar{c}}{\partial x_j} + \frac{\partial}{\partial x_j} \overline{u'_j c'} - \frac{\partial}{\partial x_j} \left[k_c \frac{\partial \bar{c}}{\partial x_j} \right] = 0 \qquad (5)$$

where we have used the continuity equation

$$\frac{\partial u_j}{\partial x_j} = \frac{\partial \bar{u}_j}{\partial x_j} = \frac{\partial u'_j}{\partial x_j} = 0 \qquad (2a)$$

to rearrange the third term in Equation 5.

Equation 5 can be thought of as a balance between processes acting to change $\bar{c}$ in an infinitesimal control volume. The first two terms represent, respectively, the change of $\bar{c}$ on time scales longer than T and the "advection" of $\bar{c}$ out of the control volume by the mean wind $\bar{u}_j$. The fourth term, the divergence of the molecular flux $-k_c \partial \bar{c}/\partial x_j$, changes $\bar{c}$ by molecular diffusion out of the control volume. The third term is the divergence of the so-called eddy flux, the process that distinguishes turbulent transport from transport in laminar flows. The "eddy flux" term appeared when the Reynolds decomposition was applied to the product term $u_j \partial c/\partial x_j$ of Equation 1 and the result time-averaged. It represents a real physical process: when fluctuations in c and u_j are correlated, then the covariance $\overline{u'_j c'}$ is the average flux of scalar in the direction of x_j caused by the turbulent motion. As long as we are not very close to a solid surface (in practice, within a few millimeters), then the eddy flux $\overline{u'_j c'}$ is at least two or three orders of magnitude greater than the molecular flux, and the latter can be safely neglected. For a further discussion of these points, see Tennekes and Lumley (1972).

Equation 5 describes the conservation of mass of scalar in any infinitesimal control volume in the air spaces between the leaves. In the free atmosphere, Equation 5 and other "single point" equations like it provide a satisfactorily complete description of the dynamics of scalar conservation. However, this is not so inside a plant canopy, for two reasons. First, time-averaged statistics of the turbulent wind and concentration fields (means, standard deviations, and covariances) vary enormously with position in the canopy. Although aspects of this variability have been measured systematically in a few simple cases (Seginer et al., 1976; Mulhearn and Finnigan, 1978; Raupach, Thom, and Edwards, 1980; Schuepp, 1982; Raupach, Coppin, and Legg, 1985), it is not usually practical to describe its every detail, either in theory or in experiment. Second, the turbulent transport process for a scalar within the canopy is closely connected with the presence of sources or sinks of the scalar at leaf surfaces. Like the time-averaged statistics of the turbulence, these sources or sinks have a heterogeneous distribution in the canopy.

To introduce both these characteristics of the plant-air layer into the conservation equations, we use the concept of volume averaging, which is the spatial counterpart of averaging in time. By choosing an averaging volume of appropriate size, we can examine systematic variability on larger scales while averaging out heterogeneity on smaller scales. For example, in the common case of a homogeneous canopy that varies systematically with height but randomly or quasi-periodically in the horizontal directions, the correct averaging volume is a thin horizontal slab through the canopy, encompassing sections of numerous plants. Further, activity at leaf surfaces, such as scalar sources or sinks, is decribed automatically in the volume-averaged equations by the appearance of extra terms.

The formal application of volume averaging proceeds in exactly the same way as the time averaging and Reynolds decomposition we have already discussed. We begin by defining the volume averaging operator, $\langle \ \rangle$:

$$\langle c \rangle (x_j) = \frac{1}{R} \iiint_R c(x_j + x'_j) \, dx'_1 \, dx'_2 \, dx'_3 \qquad (6)$$

where R is the averaging volume and is taken to encompass the air space only, and not plant parts. Each time-averaged quantity can be decomposed into a volume-averaged part plus a departure from the volume average, so that

$$\bar{u}_j = \langle \bar{u}_j \rangle + \bar{u}''_j \qquad (7)$$

and similarly for $\bar{c}$. Now, considering Equation 5 as an example, we take the volume average, as we previously took the time average, to obtain (with some mathematical manipulation given in detail by Finnigan, 1985):

$$\underbrace{\frac{\partial \langle \bar{c} \rangle}{\partial t}}_{\text{I}} + \underbrace{\langle \bar{u}_j \rangle \frac{\partial \langle \bar{c} \rangle}{\partial x_j}}_{\text{II}} + \underbrace{\frac{\partial}{\partial x_j} \overline{\langle u'_j c' \rangle}}_{\text{III}} + \underbrace{\frac{\partial}{\partial x_j} \langle \bar{u}''_j \bar{c}'' \rangle}_{\text{IV}} -$$

$$\underbrace{\frac{\partial}{\partial x_j} \left\langle k_c \frac{\partial \bar{c}}{\partial x_j} \right\rangle}_{\text{V}} + \underbrace{\frac{1}{R} \sum_{i=1}^{n} \iint_{s_i} k_c \frac{\partial \bar{c}}{\partial n} \, ds}_{\text{VI}} = 0. \qquad (8)$$

Here the surface s_i over which the surface integral is taken is the surface of the ith member of the n leaves or plant parts in the averaging volume R. On s_i, n_j is the unit normal vector pointing into the air, and $\partial/\partial n$ denotes differentiation in the n_j direction.

Plant-Canopy Transfer and Stomatal Characteristics

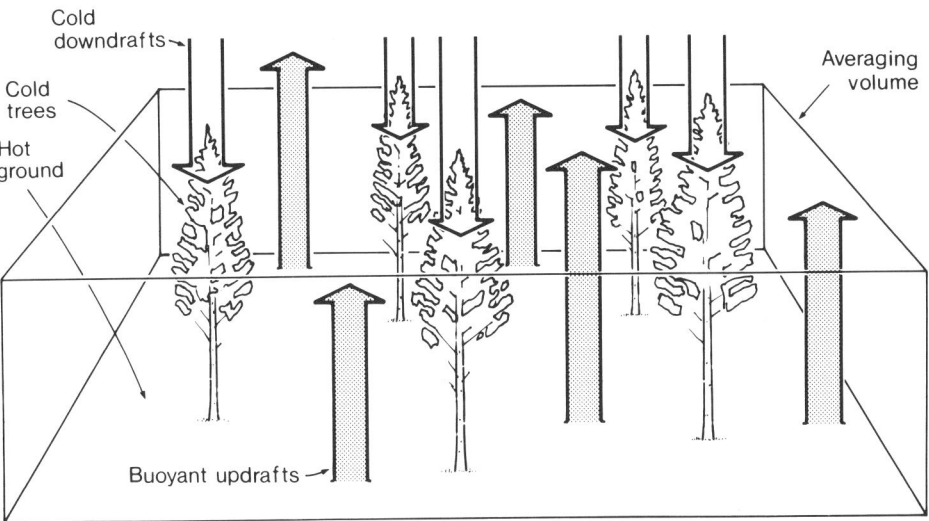

Fig. 18.1. A schematic diagram of a situation that could give rise to a significant dispersive flux effect. Well-watered trees are assumed to be freely transpiring and therefore colder than the dry ground. In these circumstances a pattern of updrafts and downdrafts might be locked in to the distribution of foliage.

In Equation 8, the terms I (time rate-of-change), II (advection), III (eddy flux divergence), and V (molecular flux divergence) have immediate counterparts in Equation 5, the only difference between the two equations being that Equation 8 is averaged in space as well as in time. However, terms IV and VI in Equation 8 are new; they arise from the volume-averaging operation and hence have no counterparts in Equation 5. Of fundamental importance is term VI, the scalar source term, which accounts for the introduction or removal of $\bar{c}$ from the averaging volume by molecular diffusion to or from plant surfaces. It is the sum of the fluxes of a scalar across the surfaces of all plant parts in the averaging volume and arises naturally from volume-averaging the last term on the left-hand side of Equation 5. The other new term, IV, is the divergence of a "dispersive flux" $\langle \bar{u}_j'' \bar{c}'' \rangle$, which bears the same relationship to volume averaging that the familiar eddy flux has to time averaging; it describes transport due to any persistent combination of deviations from spatially averaged behavior. (If mean flow advection is not negligible, the dispersive flux includes other, lower-order contributions as well; see Finnigan, 1985.) A stylized example of a process leading to a nonzero dispersive flux is shown in Figure 18.1. The dispersive flux has generally been ignored hitherto; its existence has been appreciated only recently, after rigorous derivation of the conservation equations, starting with Wilson and Shaw (1977).

In Equation 8, term V (molecular flux divergence) is almost always negligible in practice. If the canopy is horizontally uniform and the flow is steady, then terms I and II also vanish, and Equation 8 becomes simply a balance between the source term (VI) and the divergences of the eddy and dispersive fluxes (III and IV). The dispersive flux is most likely to be important in canopies that are significantly heterogeneous (e.g. in natural forests) or distinctly periodic (e.g. in row crops); it will be unimportant where plants have grown together to form a continuous, fairly uniform foliage layer. Direct measurements of dispersive flux in a completely regular model canopy in a wind tunnel showed that there it was only a few percent of the eddy flux (Raupach, Coppin, and Legg, 1985).

Equation 8 is the basic statement of conservation of a scalar within plant-canopy air layers and contains terms describing all the important processes; as such it forms a rational basis for mathematical models of canopy behavior. However, it displays a characteristic common to all equations for mean moments of turbulent quantities: it is not "closed." The mean concentration $\langle \bar{c} \rangle$ is a first moment, and in its equation appear extra unknowns, namely, the eddy flux and the dispersive flux, which are second moments. Similarly, equations for second moments such as $\langle \overline{u_j' c'} \rangle$ contain third moments such as $\langle \overline{u_i' u_j' c'} \rangle$, and so on, to any order. A central problem of turbulence modeling is to find closure assumptions, or "parameterizations," so that the $(n+1)$th-moment terms appearing in an equation for an nth moment can be written in terms of nth moments only. If this is done, the equations are, at least in principle, soluble.

The most intractable problem in the description of canopy transport is that of understanding and parameterizing the eddy flux, term III. In the simplest and oldest parameterization, known as "K theory" and dating back to Ludwig Prandtl, the eddy flux is written as the product of the mean scalar gradient and an eddy diffusivity K_c:

$$\langle \overline{u_j' c'} \rangle = -K_c \frac{\partial}{\partial x_j} \langle \bar{c} \rangle. \qquad (9)$$

The eddy diffusivity is defined in direct analogy to its molecular counterpart and is supposed to absorb all the details of the turbulent transport mechanism in a single-valued function; it remains the most widely used model of transport in a canopy. Descriptions of the several procedures for assigning a value to K_c can be found in Monteith (1975–76). However, in recent years both the conceptual foundations of K theory and the practical consequences of using it have been widely questioned. To understand why, we must look more closely at the structure of the random motions of the atmosphere that go to make up u_j' and c'.

The structure of the transporting eddies. The air layers within and just above vegetation are among the most turbulent regions of the atmosphere. This is

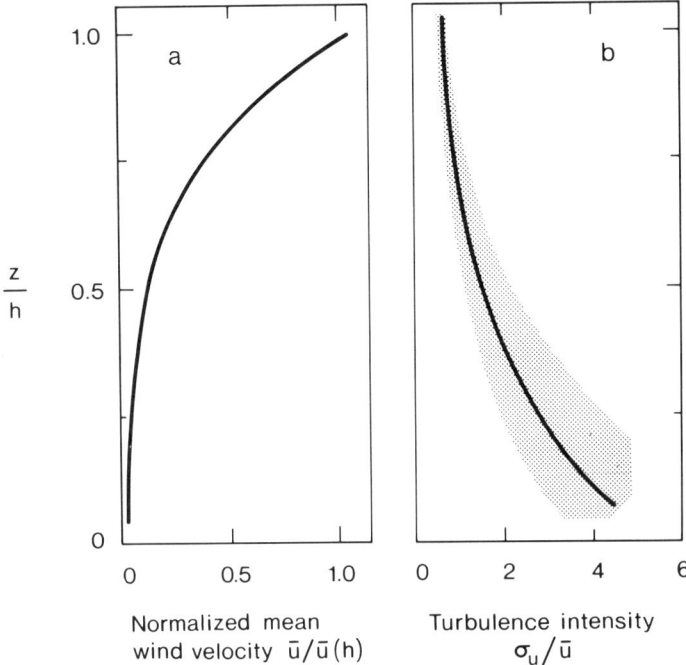

Fig. 18.2. a, The mean wind speed within an extensive uniform corn canopy normalized with the mean velocity $\bar{u}_h$ at the top of the canopy. (Adapted from Wilson et al., 1982.) b, The streamwise turbulence intensity $\sigma_u/\bar{u}$ measured in the same corn canopy. The shaded area denotes the range of scatter of the experimental points.

because within the canopy the mechanisms that universally generate atmospheric turbulence—the dynamic instabilities that accompany gradients in mean velocity or in mean density—are augmented because the foliage directly intercepts the wind. In exerting a drag force on the plants, the wind loses momentum; hence wind velocity decreases rapidly as one descends into the canopy. At the same time, the kinetic energy lost from the mean wind is converted into kinetic energy of turbulent fluctuations. These effects are shown clearly in Figure 18.2. Figure 18.2a displays a profile of normalized mean horizontal wind, $\bar{u}(z)/\bar{u}(h)$, measured in an extensive corn canopy of height h by Wilson et al. (1982). The profile shows the characteristic decay of $\bar{u}$ typical of uniform canopies. Figure 18.2b is the turbulence intensity (normalized standard deviation) of streamwise wind fluctuations, $\sigma_u(z)/\bar{u}(z)$. A $\sigma_u/\bar{u}$ value of 0.25 in the atmospheric surface layer would be regarded as very turbulent indeed; this helps to place the values of between 1 and 4 from within the canopy into their true perspective. Note also that $\sigma_u/\bar{u}$ increases with depth, suggesting that the turbulent wind fluctuations penetrate the lower canopy more effectively than does the mean wind.

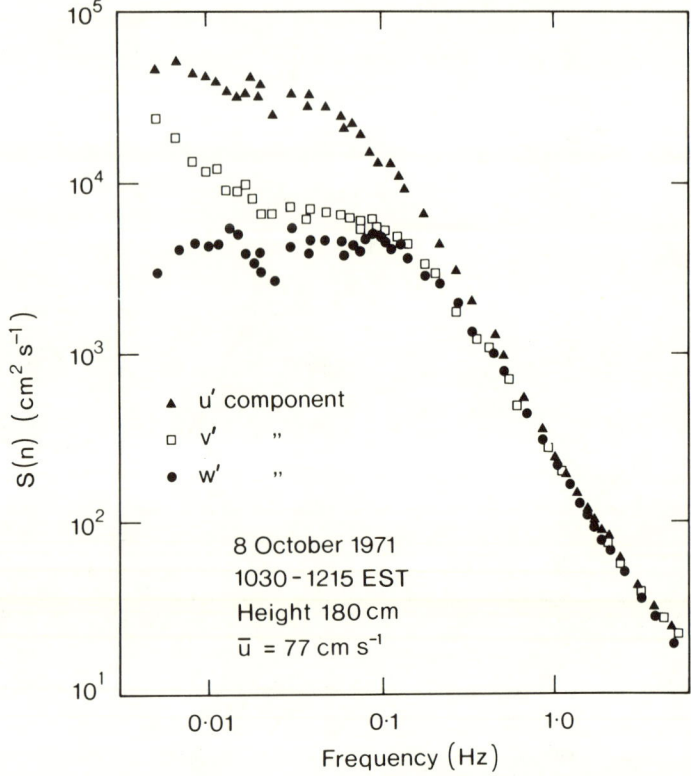

Fig. 18.3. Power spectra of streamwise (u'), lateral (v'), and vertical (w') velocity fluctuations measured at height 1.80 m within a corn canopy 2.6 m high. (Redrawn from Shaw, Silversides, and Thurtell, 1974.)

Turbulence can be thought of as a superposition of eddies—unsteady motions—with a wide range of length and time scales. Information about the range of time scales is conveniently displayed in power spectra. In a power spectrum the amount of kinetic energy in wind fluctuations of a given frequency is plotted against the frequency; this gives a picture of the way the total energy in the turbulence is distributed among all scales of motion, from brief, transient gusts to gradual changes in wind speed. Figure 18.3 shows typical spectra of horizontal and vertical velocity fluctuations measured in and above a corn canopy (2.6 m high) by Shaw, Silversides, and Thurtell (1974). Both within and above the canopy, there is significant energy in the fluctuations between 5 and 0.01 Hz, a very wide frequency range. By a rough rule of thumb known as "Taylor's hypothesis" these frequencies (ω) can be converted into length scales (l_x) through the formula

$$l_x = \bar{u}/\omega \tag{10}$$

where $\bar{u}$ is the local mean wind speed. Taylor's hypothesis assumes that the eddies are convected past the sensor at the mean wind speed in a frozen state. This is a good assumption for small high-frequency eddies but not for the larger-scale motions. For the spectra of Figure 18.3, $\bar{u} \approx 1$ m s^{-1} and the frequency range becomes a length scale range of 0.20 to 100 m.

Many of the qualitative features of turbulence spectra are identical with those of random stochastic processes, and turbulence itself was for many years regarded as an essentially chaotic, structureless phenomemon. Beginning in the 1960's, however, a complete reappraisal of this view has taken place. Numerous laboratory experiments on turbulent flows, supported more recently by field observations, have shown that turbulence exhibits both deterministic and random features and that turbulent transport is an organized and structured process. In particular, the events making up the eddy flux $\overline{u'_j c'}$ are highly intermittent rather than continuous. Willmarth (1975) reviewed the material leading to this reappraisal; here we give a few examples from work on real plant canopies.

The first illustration of the intermittent nature of canopy transport is provided by Figure 18.4, which shows daytime measurements made in the trunk space of a ponderosa pine forest of height 16 m (Denmead and Bradley, 1985). Since CO_2 was being released by the soil, air moving upward tended to be CO_2-enriched, whereas air that had moved down through the canopy was depleted of CO_2 by assimilation. Figure 18.4 shows filtered time traces of vertical velocity (w), temperature (θ), humidity (q), and CO_2 concentration (c) recorded at height 3 m during a 15-minute period together with the corresponding vertical eddy fluxes of heat, water vapor, and CO_2. (We use the meteorological convention whereby $u_j = (u,v,w)$ and $x_j = (x,y,z)$, with w and z upward, and u and x in the mean wind directions.) Most of the transport occurred during five events. One of these was associated with an updraft in which cool, moist, CO_2-enriched air was displaced from below, but four were associated with gusts that penetrated the canopy from above and carried down warmer, drier, CO_2-depleted air. Very little transport occurred in the interim periods, which, of course, formed most of the averaging time.

Similar examples of the intermittency of momentum transport in a waving wheat crop are given by Finnigan (1979), but in order to proceed beyond qualitative illustrations we must introduce the ideas of conditional sampling and conditional statistics.

One of the best ways of quantifying the intermittency is through "quadrant analysis," which we describe using the vertical heat flux $\overline{w'\theta'}$ as an example. The heat flux is measured by first recording the fluctuations $w'(t)$ and $\theta'(t)$ over some time period with instruments fixed at a single point in space, and then

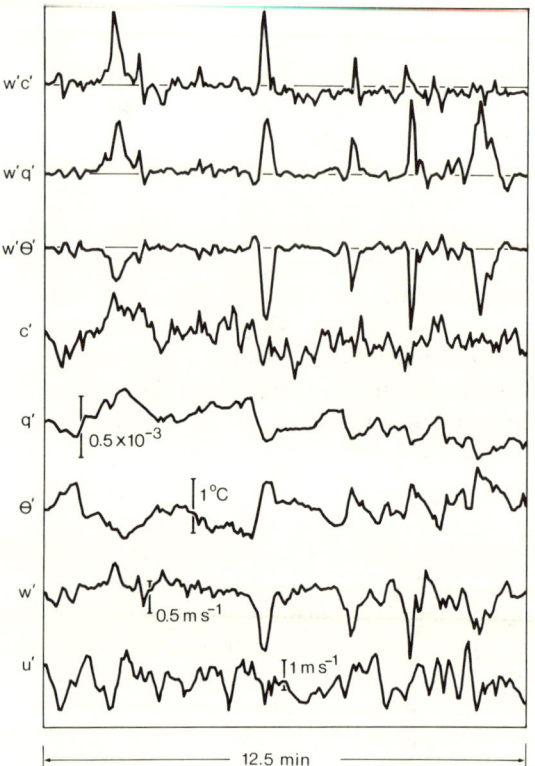

Fig. 18.4. Filtered time traces of horizontal wind speed (u'), vertical wind (w'), temperature (θ'), specific humidity (q'), CO_2 concentration (c'), and instantaneous vertical fluxes of heat ($w'\theta'$), water vapor ($w'q'$), and CO_2 ($w'c'$) at 3 m above ground in the trunk space of Uriarra Forest, an extensive plantation of ponderosa pine about 16 m high. (Adapted from Denmead and Bradley, 1985.)

calculating the covariance $\overline{w'\theta'}$. Quadrant analysis calculates the contributions to $\overline{w'\theta'}$ from various types of events, defined as quadrants in the $\theta'w'$ plane shown in Figure 18.5 (note that the w' is conventionally the ordinate and θ' the abscissa). Each instantaneous value of $[\theta'(t), w'(t)]$ represents a point on the plane that can be categorized according to the quadrant in which it falls. For example, when $\overline{w'\theta'}$ is positive (upward heat flux), then contributions to $\overline{w'\theta'}$ from quadrants 1 and 3 of the $\theta'w'$ plane must exceed those of the other two quadrants, so we expect the dominant events to be ejections ($\theta' > 0$, $w' > 0$; upward movement of warm air) and sweeps ($\theta' < 0$, $w' < 0$; downward movement of cold air). To refine the analysis, we add a hyperbolic exclusion zone bounded by the curves $|w'\theta'|$ = constant, the "hole," to permit classification of events according to the magnitude of their contribution to $\overline{w'\theta'}$ as well as according to their quadrant.

The conditional heat flux $f_{i,H}$ is defined as

$$f_{i,H} = \frac{1}{\sigma_w \sigma_\theta T} \int_0^T w'(t)\, \theta'(t)\, I_{i,H}(w', \theta')\, dt \qquad (11)$$

where $I_{i,H}(w',\theta')$ is an indicator function that is 1 when the point $[\theta'(t), w'(t)]$ lies in quadrant i of the $\theta'w'$ plane and at the same time lies outside the hole, so that $|w'(t)\,\theta'(t)| > H\sigma_w\sigma_\theta$. The parameter H is the hole size, and the standard deviations σ_w and σ_θ are used to normalize w' and θ' to permit data from a wide range of conditions to be compared on the same diagram.

When $H = 0$, so that exclusion zone vanishes, we have

$$\sum_{i=1}^{4} f_{i,0} = r_{w\theta} \tag{12}$$

where $r_{w\theta}$ is the correlation coefficient $\overline{w'\theta'}/\sigma_w\sigma_\theta$.

The first applications of quadrant analysis to real plant canopies examined the momentum flux $-\rho_a\overline{u'w'}$ (where ρ_a is the air density), using a formalism similar to that for the heat flux. Finnigan (1979) did this for a canopy of wheat; Shaw, Tavangar, and Ward (1983) repeated the exercise in a corn crop. With minor differences that could be ascribed to the greater flexibility and coherent waving of the wheat, they obtained the same results; momentum transfer to the canopy was accomplished primarily by gusts of high-speed air that originated in the boundary layer well above the surface and penetrated into the canopy. These gusts were very intermittent occurrences. For instance, in the upper half of Shaw's corn crop more than 50 percent of the momentum flux came from gusts during which $|u'w'| > 5|\overline{u'w'}|$, yet these events occupied only 6 percent of the total time. For our purposes, however, we are more interested in scalar tranport, and for this we must turn to two different experiments.

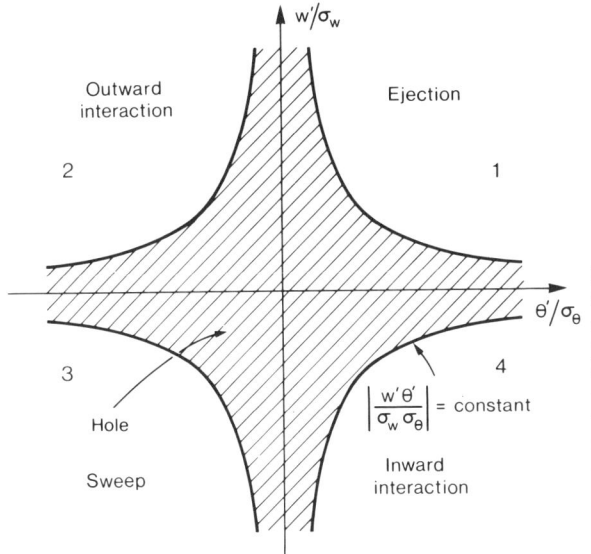

Fig. 18.5. Definition of the four regions or quadrants of the $\theta'w'$ plane. The parabolic exclusion zone, the hole, is shown also, providing five criteria by which to classify instantaneous contributions to the total flux $\overline{w'\theta'}$.

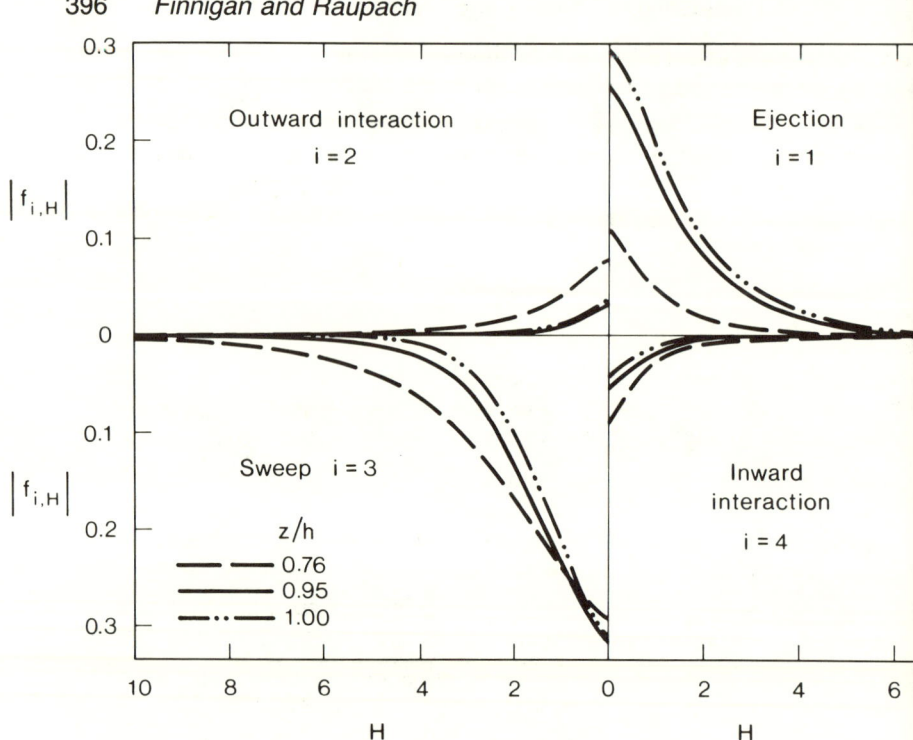

Fig. 18.6. The conditional heat flux $|f_{i,H}|$ in each quadrant plotted against hole size for three different levels within the upper part of a wind-tunnel model crop of height h. The plane heat source was at a height $z/h = 0.80$. (Adapted from Coppin, Raupach, and Legg, 1986.)

The first of these is a wind-tunnel experiment described by Raupach, Coppin, and Legg (1986) and by Coppin, Raupach, and Legg (1986). Their model canopy consisted of thin, bluff aluminum strips, each 10 mm wide, 1 mm thick, and 60 mm high, arranged in a regular diamond array with cross-stream spacing of 60 mm and downwind spacing of 44 mm, so that the frontal area index was 0.23. The model was placed in a boundary-layer wind tunnel with a good simulation of the turbulent wind in the atmospheric surface layer. Heat was supplied via a grid of fine nichrome wires strung through the canopy at a height of 50 mm, the notional mean level of absorption of radiation in this idealized canopy. The flux of heat supplied was small enough that it behaved as a passive scalar, and measurements were made far enough from the leading edge of the model canopy that streamwise derivatives were negligible. The situation was therefore essentially one-dimensional, the measured mean and turbulent wind fields being typical of those measured in natural canopies.

The conditional flux $f_{i,H}$ at three heights within and above the canopy is plotted against hole size in Figure 18.6. It is immediately apparent that the dominant contributions to mean heat flux, $\overline{w'\theta'}$, come from ejections ($w' > 0$,

Plant-Canopy Transfer and Stomatal Characteristics

$\theta' > 0$) and sweeps ($w' < 0$, $\theta' < 0$), the interaction quadrants being negligible. Further, sweeps exceed ejections, particularly within the canopy, and decay somewhat less rapidly with hole size than ejections; sweep contributions tend to come in larger installments. The ratio of sweeps to ejections $f_{3,0}/f_{1,0}$ is plotted in Figure 18.7 and shows that, close to the source, sweeps contribute about two and a half times more than ejections to $\overline{w'\theta'}$. The picture can be sharpened somewhat by defining a hole size H' above which half of the flux is contributed. The ratio $f_{3,H'}/f_{1,H'}$ is also plotted in Figure 18.7; when we restrict our attention to these more energetic transport events, sweeps are even more dominant. Lastly, we can examine the time fraction occupied by the energetic events for which $|w'\theta'| > H'\sigma_w\sigma_\theta$. Figure 18.8 shows that, although these events transport half the total flux $\overline{w'\theta'}$, they occupy 12 percent of the time at most, and only 5 percent of the time at heights near the source.

The physical picture that these data convey is of intermittent, intense displacement of warm canopy air by cooler boundary-layer air, and hence of the intermittent renewal of the air within the canopy. The main displacement events are strong sweeps, which make large but short-lived contributions to the total flux. Supplementing these events, and linked dynamically to them, are upward movements of warmer air which contribute less flux overall and occupy a larger fraction of the total time. Remember that half the flux is transported in 10 percent of the time, but that during this 10 percent sweeps dominate.

This wind-tunnel experiment also showed that, when the flow is neutrally stratified (i.e. when buoyancy makes no contribution to the turbulence), heat and momentum transfer are directly related: downward-moving air masses are

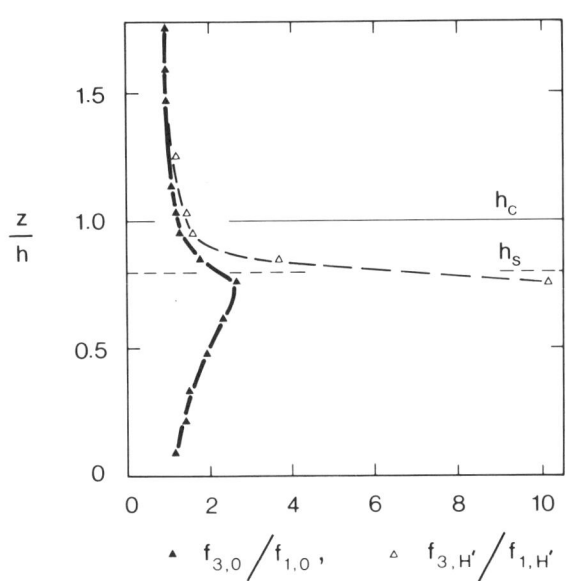

Fig. 18.7. The ratio of the contributions of sweeps and ejections, $f_{3,0}/f_{1,0}$, to the total heat flux. Also plotted is $f_{3,H'}/f_{1,H'}$, the sweep/ejection ratio at a hole size H' above which half the total flux is transmitted. Solid line, $f_{3,0}/f_{1,0}$; broken line, $f_{3,H'}/f_{1,H'}$; h_s, source height; h_c, crop height. (Adapted from Coppin, 1985.)

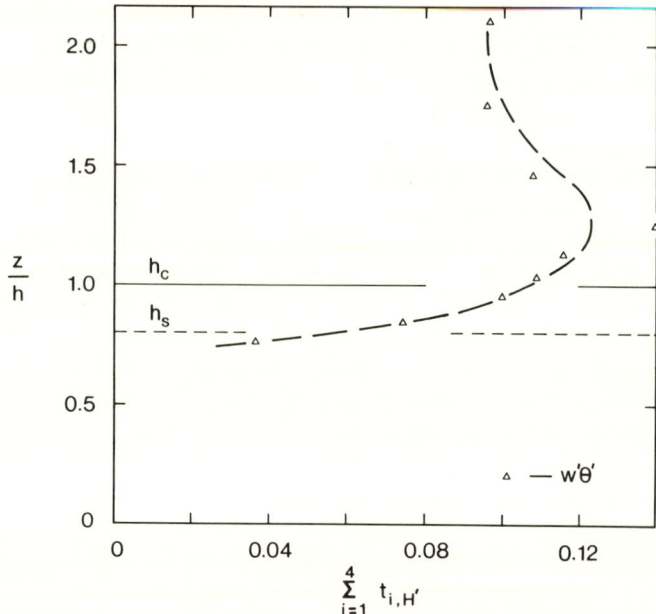

Fig. 18.8. The intermittency of the flux contributions, $\sum_{i=1}^{4} t_{i,H'}$, that is, the time fraction taken to transmit half of the total heat flux. (Adapted from Coppin, 1985.)

also faster-moving air masses. In other words, the sweeps that displace warm canopy air are simultaneously the gusts of fast-moving air that dominate momentum transport. Finnigan (1979) showed through the technique of space-time correlation that such gusts were also "large structures," turbulent eddies much larger than the height of the canopy.

The second experiment we discuss was carried out recently over a field of rye grass 50 cm high (Coppin, 1985). Data were gathered above the canopy only, at heights of 0.75, 1.5, and 4.0 m. This work extends the conclusions from the wind tunnel (which are necessarily limited to neutral conditions) to unstable conditions. The field results in near-neutral conditions agree with the wind-tunnel results in every respect; in particular, heat and momentum transfer are closely coupled. However, for unstable conditions, quadrant analysis reveals substantial differences between heat and momentum transfer. Heat transfer becomes strongly ejection-dominated, the more so with increasing height and increasing instability, conditions in which ejections can be identified with rising buoyant plumes. Momentum transfer, however, remains strongly sweep- or gust-dominated close to the canopy and does not become ejection-dominated as height increases. In unstable conditions, there is no longer an identity between sweeps for heat and gusts for momentum, and the simple correspondence of mechanisms for heat and momentum transfer is no longer observed.

Plant-Canopy Transfer and Stomatal Characteristics

What are the consequences of all this for the interpretation and prediction of canopy microclimate? First, the observation that a major part of the transport is effected by displacement events or "large structures" much bigger than the canopy height contradicts one of the basic tenets of K theory. The definition of an eddy diffusivity K_c, in analogy with the molecular diffusivity, carries the assumption that the eddies doing the mixing are much smaller than the length scale of changes in concentration gradient (Corrsin, 1974). This length scale is of the order of the canopy height, which is of the same order as the vertical dimension of the transporting eddies. The consequences can be seen in Figure 18.9, where simultaneous flux and gradient measurements are presented from the same pine forest that provided the data of Figure 18.4. Within the trunk space, upward heat and moisture fluxes coexisted with a positive gradient of temperature and a zero humidity gradient, whereas at the base of the crown space a negative CO_2 gradient is accompanied by a negative flux. The heat and CO_2 fluxes would each require a negative K_c to fit the model of Equation 9, while the vapor flux implies an infinite K_c. The answer, of course, lies in the fact that the transport is driven by large-scale eddies, which have no simple connection to the local gradient. We can see this in Figure 18.10 where short time profiles of temperature and humidity depict a gust of fast, dry, cool air from above displacing warm, moist, slow air from within the canopy. Unfortunately, the alternatives to closing Equation 8 by use of K theory neither are simple nor yield analytical solutions. We consider them when we discuss canopy modeling in more detail, toward the end of this paper.

The profiles in Figure 18.10 lead us into a second consequence of our conditional sampling investigations; individual leaves exist in an unsteady

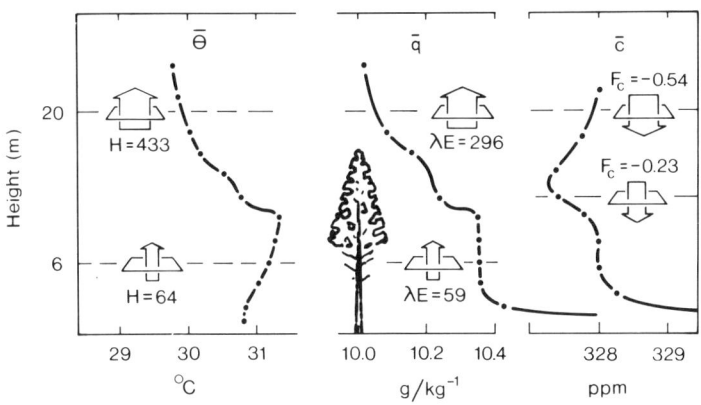

Fig. 18.9. Profiles of average temperature ($\bar{\theta}$), mass fraction of water vapor ($\bar{q}$), and volume fraction of CO_2 ($\bar{c}$) in Uriarra Forest over 1 hour, with simultaneous measured fluxes of sensible heat (H), latent heat (λE), and CO_2 (F_c) at two heights. (Adapted from Denmead and Bradley, 1985.)

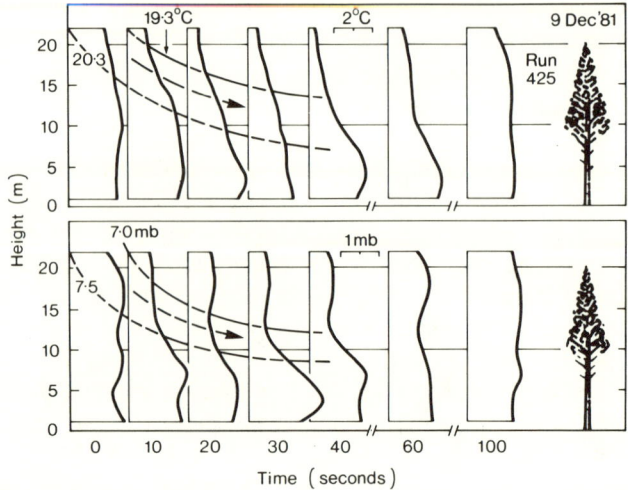

Fig. 18.10. Profiles of temperature (top) and vapor pressure (bottom) in Uriarra Forest during and after the passage of a gust. Base line for temperature is 18.5° C, and for vapor pressure, 6.5 mb. Broken lines are contours of constant temperature and vapor pressure. Arrows depict the penetration of the gust. (Reproduced from Denmead and Bradley, 1985.)

environment where intense, intermittent renewal of canopy air punctuates relatively quiescent periods. During these interim periods, turbulent mixing within the canopy occurs on a much smaller scale than during the renewals. If the previous renewal event was accomplished by a gust, for example, small-scale energetic turbulence will be generated in the wakes of fluttering leaves, stems, and branches. In the period until the next sweep, this small-scale mixing will ensure that concentration profiles build up that roughly reflect the spatial source distributions in the canopy. This process can also be clearly seen in the last few profiles of Figure 18.10.

Exchange processes at the leaf surface. We now direct attention to the source term, VI, in Equation 8. In particular, we concentrate on the response of leaves to the fluctuating environment we have described. However, we must first review the physics of heat and mass transfer at plant surfaces.

Vapor transfer from a leaf with a dry cuticle proceeds almost entirely through the stomatal openings. From the leaf surface, vapor and sensible heat are diffused through the boundary layer to the relatively well-mixed air between the leaves. In comparison with the molecular diffusion, the small normal velocity accompanying mass transfer across the epidermis is negligible, although within the passages of partially opened stomata it can have an appreciable effect; in particular, the inward movement of CO_2 molecules may be hindered by the outward flow of H_2O molecules (Leuning, 1983).

Since the transfer to the air adjacent to the surface is entirely molecular, we can apply Fick's law of molecular diffusion at the surface, writing

$$F_c(0) = -\frac{k_c}{V}\frac{\partial c}{\partial n}\bigg|_0 \tag{13}$$

where n is a coordinate normal to the surface and positive outwards, V the molar volume, and $F_c(0)$ the surface flux of a scalar with mole fraction c. Equation 13 applies at any point on the plant-air interface, but for our purposes it is not particularly useful as written. It is usually rewritten in terms of averages over the entire leaf surface as

$$\langle\langle F_c \rangle\rangle(0) = [\langle\langle c \rangle\rangle(b) - \langle\langle c \rangle\rangle(0)] g_{bc} \tag{14}$$

where g_{bc} is a dimensional transfer coefficient called the leaf boundary-layer conductance, $c(b)$ is the concentration at a point $n = b$ just outside the leaf boundary layer, and the double pointed brackets denote averages over the entire leaf surface. To understand g_{bc}, we note that

$$g_{bc} = \frac{k_c}{V\ell_c}, \quad \ell_c = \frac{\langle\langle c \rangle\rangle(0) - \langle\langle c \rangle\rangle(b)}{\left\langle\left\langle \frac{\partial c}{\partial n}\big|_0 \right\rangle\right\rangle}$$

so that g_{bc} is the quotient of the molecular diffusivity and a length ℓ_c that describes the average normal gradient of c at the leaf surface. It is important to note that g_{bc} is a derived quantity that depends on the distribution of c about the leaf; most of this section is devoted to determining its value. Henceforth we refer simply to g_b, identifying the scalar c only where necessary.

The flux from the leaf is often expressed in dimensionless form as the Sherwood number, Sh:

$$Sh = \frac{Vd\, F_c(0)}{k_c[c(b) - c(0)]} \tag{15}$$

where d is a length characteristic of the leaf. We have dropped the double pointed brackets for simplicity, but the values of F_c and c should be assumed to be averages over the leaf surface. When the quantity being transferred is sensible heat, the Sherwood number takes a special name and becomes the Nusselt number, Nu:

$$Nu = \frac{Vd\, F_\theta(0)}{C_p k_\theta[\theta(b) - \theta(0)]} \tag{16}$$

with θ the temperature, C_p the molar heat capacity of air, and k_θ the thermal diffusivity. The relationship between the dimensionless flux (Sh or Nu) and g_b is therefore

$$g_{bc} = \frac{k_c Sh}{Vd}, \quad g_{b\theta} = \frac{k_\theta Nu}{Vd}. \tag{17}$$

Hence the problem of determining g_b is equivalent to that of finding Sh or Nu.

In addressing this problem we must distinguish at the outset between several fundamentally different cases: the leaf boundary layer may be laminar or turbulent, and may be driven by buoyancy forces arising from the temperature and humidity gradients close to the leaf (the regime of free convection) or by inertial forces arising from the incident airflow (the regime of forced convection). In free convection the heat and moisture fluxes and the wind field around the leaf are coupled. In forced convection they are independent. In some very simple situations, such as forced convection maintained by steady laminar flow parallel to a leaf considered as a thin, flat plate, g_b and Sh can be calculated by analytic solution of the conservation equation with molecular transport, Equation 1 (Cowan, 1972). Such solutions are useful in determining the form of the functional dependence of Sh and g_b on wind speed and diffusivity, but they cannot provide quantitative results for real leaves, for reasons to be considered below.

When forced convection dominates, a combination of analysis and experience shows that the dimensionless fluxes Sh and Nu depend on wind speed, body size, and scalar diffusivities in the following way:

$$Sh = A\, Re^n\, Sc^{1/3}$$
$$Nu = A\, Re^n\, Pr^{1/3} \tag{18}$$

where A and n are constants, $Sc = \nu/k_c$ is the Schmidt number, and $Pr = \nu/k_\theta$ is the Prandtl number, ν being the kinematic viscosity of air (1.50×10^{-5} m^2 s^{-1} at 20° C); $Re = Ud/\nu$ is the Reynolds number, U being a velocity characteristic of the wind outside the leaf boundary layer. The exponent n is ½ when the leaf boundary layer is laminar (an analytical result), and is about 0.8 when the boundary layer is turbulent. The transition from a laminar to a turbulent boundary layer occurs as the Reynolds number is increased beyond a critical value, about 5×10^4 for a leaf in a plant canopy. (This critical value of Re is lower than is usually quoted for objects in wind tunnels because canopy air flows are very turbulent and transition to turbulence in the leaf boundary layer occurs at a lower Re value.)

Are leaf boundary layers laminar or turbulent? For a leaf with $d = 5$ cm, the velocity U would have to exceed 15 m s^{-1} to achieve completely turbulent flow in the leaf boundary layer; smaller leaves would require proportionately greater

Plant-Canopy Transfer and Stomatal Characteristics 403

speeds. In real canopies, leaf boundary layers are almost invariably laminar, so that molecular diffusion dominates the transport within them. However, they are equally invariably very unsteady. In practice the distinction between laminar and turbulent boundary layers in such cases is rather hazy; but if, after applying the Reynolds decomposition and time averaging, the molecular flux remains comparable to the eddy flux, we will call such layers unsteady laminar rather than turbulent. This is usually the case on leaves. Fully developed turbulent boundary layers, in contrast, display a set of characteristics not observed in leaf boundary layers.

The "constant" A depends on leaf geometry, leaf fluttering, and both the angle of attack and turbulence content of the approaching airstream. In the unlikely event of steady laminar flow parallel to a leaf, A is 0.7 for transfer from one side of the leaf only.

When the wind speed in the canopy is low enough and the temperature difference between leaf and the air large enough, free convection rather than forced convection dominates. The precise requirement is that the Grashof number, Gr, be much larger than the square of the Reynolds number:

$$Gr = g \left| \frac{\Delta \rho}{\rho_a} \right| \frac{d^3}{v^2} >> Re^2 \qquad (19)$$

where ρ_a is the ambient air density (outside the leaf boundary layer), and $\Delta \rho$ the density change across the boundary layer. Can free convection ever be important? Consider a leaf of size $d = 5$ cm, with a leaf-air temperature difference of 10° C in a sea-level atmosphere of ambient temperature 25° C. Neglecting possible contributions to $\Delta \rho$ from the water-vapor flux from the leaf, Equation 19 gives the requirement $U << 0.13$ m s^{-1} for free convection to dominate the leaf boundary layer. This rough calculation, based on a guessed value for leaf-air temperature difference, ignores the fact that the temperature difference is actually a derived quantity, controlled by the canopy microclimate. Nevertheless, more sophisticated calculations give much the same result; free convection can dominate only during periods of very low wind speed within the canopy. However, in the unsteady canopy environment we can expect free convection to be important between gusts even when calculations based on mean velocities suggest that it should be negligible.

For free convection, an analysis of the heat transfer from a heated vertical plate where the density-driven boundary layers remain laminar (Bird, Stewart, and Lightfoot, 1960) predicts a functional form for the Nusselt number, and hence also for the Sherwood number:

$$Nu = B \, Pr^{1/4} \, Gr^{1/4}$$
$$Sh = B \, Sc^{1/4} \, Gr^{1/4}. \qquad (20)$$

For transfer from one side of a heated vertical flat plate the value of B is 0.52. Other analytical results can be found for horizontal plates of finite extent, but as in the case of forced convection, their usefulness is in establishing functional forms rather than numerical values. Grashof numbers for leaves in natural plant canopies are not high enough to give turbulent free convection in leaf boundary layers, so this case need not concern us.

To obtain more practically useful numbers for the coefficients A and B, three avenues have been explored. First of all, there exists a considerable body of data in the engineering literature on heat and mass transfer from bodies of simple shape such as rectangular plates, cylinders, and spheres. Second, transfer from real and model leaves has been measured. Here, leaf geometry at least is correctly reproduced, and occasionally leaf flapping or fluttering is also allowed. Finally, realistic model leaves have been placed in real canopies (or real canopy leaves have been modified to prevent transpiration), and heat or mass transfer measured directly. The first two methods give consistent and systematic results but show puzzling differences when compared to the third method.

Schuepp (1972) reports a series of measurements of forced convection on simple flat plates, leaf-shaped plate models, and actual leaves. Mass transfer with a constant-concentration boundary condition was modeled by an electrochemical analog in a water tunnel. He also reviews a variety of data from other authors, collected mainly in wind tunnels. He found that leaf-shaped flat plates held parallel to the flow closely obeyed Equation 18 with $n = 0.5$ and $A = 0.7$, but that if the leaves were placed crosswind and allowed to flutter, then increases occurred in Sh and g_b, ranging from 20 percent for round discs or broad leaves to 80 percent for elongated narrow leaves. The average increase for a range of fluttering leaves and leaf models was 40 percent ± 10 percent over the value for a rigid flat plate. Schuepp also tested clusters of rigid, leaf-shaped plates of realistic geometry and found they had g_b values about half that of a fluttering real-leaf cluster of equal size and shape. His water tunnel had a free-stream turbulence intensity of only 4 percent (turbulence intensity being the ratio of standard deviation of wind fluctuations to mean wind speed), and the Reynolds numbers based on leaf dimensions were between 2×10^3 and 4×10^4, so that all of the leaf boundary layers were laminar.

Schuepp (1973) also provides valuable information on free convective transfer from flat plates and model leaves. In these cases for nonvertical, rough leaves he found increases of 50 percent in g_b over the value predicted for a vertical flat plate by Equation 19 with $B = 0.52$, but he also found that a 25 percent increase is a more reliable estimate for an average plant.

In summary, measurements on realistic fluttering-leaf models in the laboratory, in both free and forced convection, show g_b values enhanced by the order of 50 percent over engineering formulas for flat plates. On the other hand, leaves in real canopies are not isolated; their mutual proximity can induce a

Plant-Canopy Transfer and Stomatal Characteristics

shelter effect, in which (for forced convection) conductances of properly clustered plant parts are lower than those measured on isolated plant parts in wind tunnels. For example, Landsberg and Thom (1971) reported that the conductance for water vapor from a single spruce needle in its natural position on a spruce shoot was only half that for the same needle in an isolated position in a wind tunnel.

There are important differences between even the best laboratory simulations and the flow in real canopies. This is immediately apparent from measurements made in real canopies either on model or on real leaves. Some field studies (summarized by Grace, 1980) found g_b values up to two and a half times higher than the engineering formulas, or nearly twice as large as wind tunnel simulations of the type we have mentioned. To explain these results, we must consider the turbulent environment of the canopy: the turbulence intensity, and especially the turbulence scale. The eddies within a canopy vary in size from a fraction of the characteristic leaf dimension to many thousands of times that size. This cannot be reproduced in a wind-tunnel model, where the upper limit of scale usually is only a few multiples of the leaf size. The smallest eddies, those about the same size as the leaf boundary-layer thickness, increase conductance through the laminar leaf boundary layer by a process known as "vortex amplification" (Sutera, 1965). Eddies of the size of the leaf itself are likely to set the leaf flapping, so that the leaf is not even approximately parallel to the oncoming airstream, and transfer from the downwind side of the leaf is into a separated wake rather than an attached boundary layer. Heat and mass transfer into separated wakes is much more efficient than through attached boundary layers at Reynolds numbers typical of leaves in canopies (Finnigan and Longstaff, 1982). This effect is probably responsible for the doubling of g_b observed by Schuepp (1972) when he compared fluttering leaf clusters with their rigid counterparts. In short, eddies smaller than a few leaf sizes augment transfer. It is important to note that a substantial fraction of the turbulence in a plant canopy is tuned to the leaf scale, where it has maximum effect in increasing g_b.

Large eddies appear to the leaf as variations in mean velocity, because the time scale over which the wind speed changes in these eddies is much larger than that required for the leaf boundary layer to adjust to the changing speed, or for a vortex to be shed from the leaf, or for the leaf to flutter through a cycle. We expect the large eddies therefore to have a relatively weak effect on A and n. However, because the dependence of conductance on velocity is nonlinear, the time-averaged magnitude of g_b does not correspond to the time-averaged velocity. Rather, by first-order expansion of Equation 18 with $n = \frac{1}{2}$ (that is, $Re < 5 \times 10^4$), we obtain

$$\overline{g_b(U)} = g_b(\bar{U})(1 + \tfrac{1}{2}\sigma_U/\bar{U})$$

where σ_u is the large-eddy standard deviation of U. The g_b on the left applies in a real canopy with large eddies, while the one on the right is the smaller value measured in a wind tunnel from which large eddies are absent.

There seem to be no reliable data for free convective heat and mass transfer under field conditions. Measurements on live leaves in a wind tunnel by Kumar and Barthakur (1971) produced conductances up to four times as large as those measured in Schuepp's electrochemical tank, but their experimental techniques seem less reliable than his. Schuepp did note that groups of closely spaced leaves, freely convecting, can exhibit much higher conductances than their individual members because of a cooperative effect. What the actual consequences would be in the field is, however, not known.

Let us distill all this data to suggest numerical values of g_b for forced and free convection from leaves in natural canopies, considering transfer from both sides of a leaf (hence the factor 2 in the following equations). For forced convection, Equations 17 and 18 with $n = \frac{1}{2}$ (that is, $Re < 5 \times 10^4$) give, for heat transfer

$$g_{b\theta} = \frac{2Ak_\theta^{2/3}}{V\nu^{1/6}}\left[\frac{U}{d}\right]^{1/2}.$$

We take $A = 1.3 \pm 50$ percent (in contrast to the flat-plate engineering value of 0.7) as a likely typical value in a real canopy, bearing in mind all that has been said. Then

$$g_{b\theta} = 0.54 \left[\frac{U}{d}\right]^{1/2}, \pm 50\% \tag{21}$$

where $g_{b\theta}$ is in mol m^{-2} s^{-1}, U/d in s^{-1} and $k_\theta = 2.2 \times 10^{-5}$ m^2 s^{-1}, $Pr = 0.71$, and $V = 24.1 \times 10^{-3}$ m^3 mol^{-1}, appropriate to a temperature of 20° C. The uncertainty of ± 50 percent is meant to emphasize that this value can only be taken as a rough guide and that all the effects we have discussed may conspire to increase or decrease it. Conductances for scalars other than heat are related to $g_{b\theta}$ through

$$g_{bc} = g_{b\theta} \, Le^{-2/3} \tag{22}$$

where $Le = k_\theta/k_c$ is the Lewis number. For water vapor Le is 0.84, and for CO_2 Le is 1.36.

For free convection, we use Equation 20 to write

$$g_{b\theta} = \frac{2Bk_\theta^{3/4}g^{1/4}}{\nu^{1/4}V}\left[\frac{\Delta\rho}{\rho_a d}\right]^{1/4}$$

and take $B = 0.9 \pm 50$ percent in a real canopy, obtaining

$$g_{b\theta} = 0.68 \left[\frac{\Delta\rho}{\rho_a d}\right]^{1/4}, \pm 50\% \tag{23}$$

with d in m, $g = 9.8$ m s^{-2}, and other units and quantities as for Equation 21. When the density gradient is due to simultaneous temperature and moisture gradients, an approximate expression is

$$\frac{\Delta\rho}{\rho_a} \cong -\frac{\Delta T}{T} - 0.38\,\Delta q \tag{24}$$

where Δq is the difference in mole fraction of water vapor between the leaf surface and the well-mixed air outside its boundary layer. Conductances for other scalars are related to $g_{b\theta}$ through

$$g_{bc} = g_{b\theta}\,Le^{-3/4}. \tag{25}$$

What value should be assigned to g_b when both free and forced convection are important is less easily stated. If some value must be obtained, an empirical approach, generally engendering errors of no more than 25 percent, is to calculate g_b by both Equations 21 and 23, and choose the larger value (Monteith, 1973). In actual fact, if a calculation based on time-average values of $\Delta\rho/\rho_a$ suggests a regime of mixed convection, then what the leaf will encounter is periods of extremely low wind speed in which free convection dominates, punctuated by short periods of active exchange in which forced convection is the mode of transfer. The balance between the two depends primarily upon the intermittency of the gusts, a quantity that increases with atmospheric instability.

The combination equation. The formulas we have derived for g_b, while completing the prescription (Equation 14) of heat or vapor transfer, still fall short of the expressions needed to close Equation 8 (that is, to express the scalar source VI in terms of $\langle \bar{c} \rangle$ and other known quantities) so that the equation provides a complete description of the concentration field $\langle \bar{c} \rangle$. The reason is that Equation 14 introduces a further unknown, the surface concentration $\langle\langle c \rangle\rangle(0)$. We can overcome this problem by combining the expressions for heat and moisture flux from the leaf with the condition of conservation of energy and by making a simplifying assumption. The fluxes of sensible heat H and of latent heat LE (where L is the latent heat of vaporization per mole, and E the molar flux density of water vapor) are not independent. They are linked by the condition of energy balance at the leaf surface:

$$\phi = H + LE + CP + S \tag{26}$$

where ϕ is the incoming net irradiance or net radiation flux, CP the flux of radiation absorbed in photosynthesis, and S the energy flux that goes into heating the plant tissue. All these quantities have units W m^{-2} (throughout this paper, we adopt the micrometeorological habit of using the term "flux" loosely, in place of the formally correct "flux density," to mean a transfer rate per unit area of surface). Equation 26 represents an average over both sides of the leaf surface, so its terms should strictly be written as $\langle\langle\phi\rangle\rangle$, and so on; however, as before we treat the double pointed brackets as understood.

In practice, CP rarely exceeds 1 or 2 percent of ϕ. For thin leaves S is usually less than 0.05ϕ, although S may be important in the energy balance of massive plants like trees (Stewart and Thom, 1973). For simplicity we neglect both these effects.

For ϕ we write

$$\phi = [\phi_0 - \sigma(\theta_0^4 - \theta_a^4)] \tag{27}$$

where σ is the Stefan-Boltzmann constant (5.75×10^{-8} W m^{-2} K^{-4}) and the leaf emissivity is assumed to be unity; ϕ_0 is the net radiation flux that would be received by the leaf if the leaf surface temperature θ_0 were the same as ambient air temperature θ_a. Equation 27 illustrates the contribution radiative exchange may make to the total energy balance. In all that follows we assume that $\theta_a = \langle\theta\rangle$.

We now combine Equation 26 with the transfer equations for H and E. The necessary particular forms of Equation 14 are

$$H = C_p\, g_{b\theta}\, (\theta_0 - \theta_b) \tag{28}$$

$$E = g_{bq}\, (q_0 - q_b). \tag{29}$$

Here θ_0 and q_0 denote $\langle\langle\theta\rangle\rangle(0)$ and $\langle\langle q\rangle\rangle(0)$, where q is the mole fraction of water vapor. From Equations 22 and 25 we can calculate that g_{bq} exceeds $g_{b\theta}$ by 12 percent in forced convection and by 14 percent in free convection; it is usual to neglect this small difference, writing $g_b \approx g_{b\theta} \approx g_{bq}$.

The equation for transfer through the epidermis is written, similarly to Equation 29, as

$$E = g_s\, [q^*(\theta_0) - q_0] \tag{30}$$

where q^* is the mole fraction of water vapor at saturation, so that we are assuming the substomatal cavities to be saturated at leaf surface temperature θ_0, and g_s, of course, is the stomatal conductance.

With the approximation that $q^*(\theta_0)$ may be represented by the first term of its Taylor series expansion about $q^*(\theta_b)$ (with the assumptions, in other words, that $\theta_b - \theta_0$ is not too large), Equations 26, 28, 29, and 30 can be combined to eliminate the explicit dependence of E on surface values of temperature and humidity. The resulting expression is:

$$E = \frac{\phi \, \epsilon/L + \delta_b g_b}{(\epsilon + 1) + g_b/g_s} \tag{31}$$

where $\delta_b = q^*(\theta_b) - q_b$ is the saturation deficit of the ambient air just outside the leaf boundary layer, while $\epsilon = (L/C_p)(dq^*/d\theta)$ is the rate of change of latent heat content of saturated air with change of sensible heat content. Equation 31 is the well-known combination equation of evaporation. For air at 20° C and 1,000 mb pressure, $\epsilon = 2.2$ and $\epsilon/L \simeq 5.10^{-5}$ mol J^{-1}.

The actual rate of evaporation from a leaf, then, is controlled by the environmental variables ϕ and δ_b in linear combination and by the physiological and boundary layer conductances g_s and g_b. Analytical solutions of Equation 31 are not usually possible in practical situations (see Cowan, 1972); instead we can explore three physically meaningful limiting cases. Loosely, these are the limits of large g_s, small g_b, and large g_b, but to be more precise it is useful to rewrite Equation 31 in the dimensionless form

$$\frac{LE}{\phi} = \frac{\epsilon + g_b/g_i}{\epsilon + 1 + g_b/g_s} \tag{31a}$$

where the left-hand side is the ratio of the latent heat flux to the net radiation. The quantity g_i, given by

$$g_i = \phi/(L\delta_b) \tag{31b}$$

is a quasi conductance—a group of terms with the dimensions of a conductance—that includes all the effects of the environmental variables δ_b and ϕ on the ratio LE/ϕ. The three special cases can now be defined precisely as limits of the ratios g_b/g_s and g_b/g_i.

The first case occurs when g_s becomes very large compared with g_b, as would happen if the leaf surface were wet. Under this condition Equations 31a and 31 simplify to

Case a, $(g_b/g_s \to 0)$

$$\frac{LE}{\phi} \to \frac{\epsilon + g_b/g_i}{\epsilon + 1}, \quad E \to \frac{\phi \, \epsilon/L + \delta_b g_b}{\epsilon + 1}. \tag{32}$$

Equation 32 describes what is called "potential evaporation." The physiological resistance (resistance being the reciprocal of conductance) is negligible and therefore has no influence, so that the evaporation rate is controlled by ϕ and $\delta_b g_b$ in linear combination.

The second case occurs when g_b is very small compared with both g_s and g_i, as happens when winds are very light. Then, Equations 31a and 31 become

Case b, $(g_b/g_i \to 0, g_b/g_s \to 0)$

$$\frac{LE}{\phi} \to \frac{\epsilon}{\epsilon + 1}, \quad E \to \frac{\phi\epsilon}{L(\epsilon + 1)}. \tag{33}$$

This situation is known as "equilibrium evaporation." The physiological conductance has no control here either, not because $1/g_s$ is zero but because $1/g_b$ is the dominant resistance to transpiration. Water vapor accumulates next to the leaf until saturation is reached there, irrespective of the value of g_s, as long as g_s is not so small that it is comparable to g_b. Equilibrium evaporation has been proposed as a natural state for well-watered crops in the absence of severe horizontal advection (Denmead and McIlroy, 1970) and also to describe the quiescent periods between gusts, when g_b is small and is probably dominated by free convection, such as on warm, still days in broad-leafed canopies (McNaughton and Jarvis, 1983). We will examine the latter proposal shortly.

The third case occurs when g_b becomes large and g_s very small, as would occur in a dry canopy on windy days or, in times of more moderate average wind, during a gust or sweep event. Now, Equations 31a and 31 reduce to:

Case c, $(g_b/g_i \to \infty, g_b/g_s \to \infty)$

$$\frac{LE}{\phi} = \frac{g_s}{g_i}, \quad E = \delta_b g_s. \tag{34}$$

Here the influence of stomatal conductance is paramount, and the evaporation rate is closely coupled to the ambient saturation deficit δ_b.

The actual situation of a leaf within a canopy, of course, will correspond exactly to none of these cases. The leaves are continually responding to changes in ambient conditions on a diurnal time scale, on the time scale of changes in meteorological conditions, and on a turbulent time scale. This last scale is set by the durations of the intermittent transfer events and quiescent periods we have discussed, and we now proceed to investigate their effect. *Inter alia* we want to know whether the intermittency of the turbulence can ever lead to a condition of equilibrium evaporation during quiescent periods.

Consider the following idealized gust sequence in an idealized canopy:

1. The cycle consists of a gust, in which g_b is high, followed suddenly by a quiescent period, in which g_b takes a low, constant value.
2. ϕ and g_s do not change throughout the cycle.
3. The gust renews the air in the canopy so that at $t = 0$, when the gust ends and the quiescent period begins, the canopy contains air with a uniform vapor deficit, $\delta_b(0)$, characteristic of the air above the canopy.
4. During quiescence, small-scale mixing occurs so that the buildup of water vapor and heat within the canopy reflects the fluxes from groups of adjacent

leaves with similar properties. Thus, the vapor lost in time t during quiescence from a typical leaf of area a can be regarded as increasing the vapor content of an air volume a/l, where l is the leaf-area density or leaf area per unit volume. Apart from this small-scale mixing, groups of leaves are isolated from the flow above, and from each other, until the next gust arrives.

Under these assumptions, we find the time evolutions of δ_b and E in the quiescent period by differentiating Equation 31 with respect to t, using assumption 4 with Equations 26, 28, and 29 to reexpress $d\delta_b/dt$, and obtaining a differential equation for $E(t)$:

$$\frac{dE}{dt} = \frac{g_b lV [\phi\epsilon/L - (\epsilon + 1)E]}{\epsilon + 1 + g_b/g_s}$$

which has the solution

$$\frac{LE(t)}{\phi} = \left[\frac{LE(0)}{\phi} - \frac{\epsilon}{\epsilon + 1}\right] e^{-t/\tau} + \frac{\epsilon}{\epsilon + 1} \qquad (35)$$

where

$$\tau = \frac{1}{lV}\left[\frac{1}{g_b} + \frac{1}{g_s(\epsilon + 1)}\right]. \qquad (36)$$

Here $E(0)$ is the evaporation rate at the end of the gust and the start of quiescence, and g_b takes its constant, quiescent value.

Equation 35 shows that the evaporation during quiescence approaches exponentially the equilibrium rate, Equation 33. This happens because of the aerodynamic isolation imposed during the quiescent period, which causes the air within the canopy to approach saturation. Such isolation can only occur when the wind speed is low and g_b is small. However, small values of g_b imply large values of the equilibration time constant τ, from Equation 36. Hence, the smaller we make g_b in an effort to bring about aerodynamic isolation, the longer we must wait for the equilibrium rate to be achieved. These contradictory requirements suggest that the equilibrium rate is not even approximately attained during typical quiescent periods.

Let us quantify this assertion. In a canopy with leaf dimension $d = 5$ cm, with zero wind speed during quiescence, and with $\Delta T = 5°$ C at ambient temperature 20° C, Equation 23 (for free convection) gives $g_b = 0.52$ mol m^{-2} s^{-1}. If, also, $l = 0.2$ m^{-1} (for a eucalypt sclerophyll forest, say) and $g_s = \infty$, then τ is 400 s (more as $g_s \to 0$). This is much longer than the typical time between gusts, which might be of the order of 50 s for such a canopy. With

$t = 50$ s and $\tau = 400$ s, $e^{-t/\tau} = 0.88$, so the leaves in the canopy can never travel far toward equilibrium evaporation. Except possibly deep within a dense canopy, where only the most energetic sweeps might penetrate, equilibrium evaporation seems unattainable and leaf evaporation will be biased toward times of active exchange.

Models of Canopy Behavior

So far in this paper we have focused upon two of the three main determinants of canopy behavior: turbulent transport and the energy balance at leaf surfaces. The third factor, radiation absorption within the canopy, we have taken as given. In the next section we will concentrate on the synthesis of these processes, and we will do this by describing three models of whole-canopy performance. Two of them are based essentially on Equation 8 and may serve either as predictive models, with values assumed for the controlling parameters, or as frameworks within which measurements on real plants may be analyzed.

The single-leaf model. Let us begin with Equation 8 for water vapor in the case of an extensive horizontally uniform canopy, assume steady nonadvective (horizontally uniform) conditions, and allow the averaging volume R to contain the full canopy height h over an extensive horizontal area a. If the ground surface is evaporating, we include it in the averaging volume and treat it just like an extra set of plant leaves. Then, multiplying Equation 8 by $VR/a = Vh$, we obtain

$$E_h = \sum_{k=1}^{n} \frac{a_k}{a} \left[\frac{\phi_k \epsilon/L + \delta_{bk} g_{bk}}{\epsilon + 1 + g_{bk}/g_{sk}} \right] \tag{37}$$

where E_h is the average water vapor flux in mol m^{-2} s^{-1} through the plane $z = h$ at the top of the canopy. In the source term VI of Equation 8, the surface integral over an individual leaf of area a_k has been replaced by its individual combination equation, the values ϕ_k, δ_{bk}, g_{bk}, and g_{sk} being those for that leaf. Hence, the right-hand side of the equation is total water vapor flux from all n leaves in the averaging volume. On the left-hand side, the divergence theorem has been used to replace the volume-averaged divergence of total flux (eddy plus dispersive) by the integrated flux across the walls of the averaging volume. In steady, nonadvective conditions, this reduces to simply the average flux out of the top of the canopy. Equation 37 says that, in these simple conditions, what goes in (from leaves) must come out (at the top of the canopy).

The single-leaf model (Monteith, 1965) is obtained by replacing the right-hand side of Equation 37 by the expression

$$E_h = \frac{\phi' \epsilon / L + \delta_h g_a'}{\epsilon + 1 + g_a'/g_s'}. \tag{38}$$

Equation 38 treats the entire canopy as a single giant leaf, the primed quantities ϕ', g_a', and g_s' being equivalent to the corresponding leaf quantities ϕ_k, g_{bk}, and g_{sk} in Equation 37. The vapor deficit δ_h refers to a convenient reference level, often $z = h$, the top of the canopy, while the bulk available energy ϕ' is equal to the net radiant energy flux entering the canopy less the energy flux into photosynthesis and storage. Hence, both ϕ' and δ_h are easily measurable. The quantities g_a', the bulk aerodynamic conductance, and g_s', the bulk stomatal conductance, are the critical parameters that must be prescribed before the single-leaf model can be applied to a canopy.

How can we estimate g_a' and g_s'? They are generally thought of intuitively (even if they are not measured this way) as area-weighted combinations of the leaf conductances g_{bk} and g_{sk}. However, as Cowan (1968) pointed out 17 years ago, the nonlinearity of the combination equation suggests that the correspondence between g_s', g_a' and g_{sk}, g_{bk} might not be so simple. We can test this proposition by deriving expressions for g_s', g_a' in terms of the leaf variables g_{sk}, g_{bk}, through rigorous comparison of Equations 37 and 38.

The first step is to make a change of variables in Equation 37 so that the reference level for the measurement of vapor deficit for each leaf is at $z = h$, rather than just outside the leaf's boundary layer. This involves replacing the leaf boundary layer conductance g_{bk} with a total aerodynamic conductance g_{ak}, the series sum of g_{bk} and a turbulent conductance g_{tk} that accounts for transfer from the outer edge of the boundary layer of leaf k up to the reference level $z = h$. If F_c is the flux of c from a single leaf, then Equation 14, which defines the boundary-layer conductance for c, can be extended to define the turbulent and aerodynamic conductances as well:

$$F_c = g_b(c_0 - c_b) = g_t(c_b - c_h) = g_a(c_0 - c_h) \tag{39}$$

where the subscript h denotes a concentration at $z = h$, and the leaf subscript k is omitted. These equations imply that g_a is the parallel sum of g_b and g_t:

$$\frac{1}{g_a} = \frac{1}{g_b} + \frac{1}{g_t}. \tag{40}$$

Unfortunately, g_t is much harder to describe than g_b, because there exists no simple expression for the turbulent flux equivalent to Equation 13, Fick's law for the molecular flux. Therefore, g_t is known for each leaf only when we have solved the problem of determining the concentration distribution throughout the canopy, or, in other words, when we have modeled the turbulent transport. The first sections of this paper have already shown that this problem cannot be

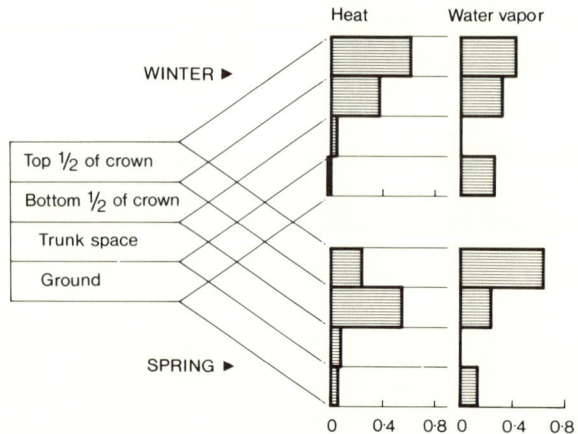

Fig. 18.11. The vertical distribution of the source strengths (term VI of Eq. 8) for heat and water vapor measured in Uriarra Forest during winter and spring of 1975. The source strengths were inferred from the divergences of the directly measured eddy fluxes. (Adapted from Denmead, 1984.)

solved by defining an eddy diffusivity, because of the domination of transport within the canopy by large eddies. Equation 39 hints also that g_t depends on the source distribution F_c, so that g_t is not necessarily the same for two scalars with different source distributions; another of our canopy models will confirm this. In practice, source distributions for different scalars may differ widely, as Figure 18.11 shows. In short, knowledge of g_t and hence also of g_a can be obtained only by solving a far more complicated model than the single-leaf model we are considering now.

One way to avoid these difficulties is to limit ourselves to situations in which $g_t \gg g_b$ for all leaves so that $g_a \cong g_b$. The canopy is then well enough stirred by turbulence that $(c_0 - c_b) \gg (c_b - c_h)$, so $c_b \cong c_h$ everywhere, and a detailed description of turbulent transport is irrelevant. Since g_b is proportional to $(U/d)^{1/2}$ in forced convection and g_t is (very roughly speaking) proportional to U, the condition $g_t/g_b \gg 1$ is better met in broad-leafed than in narrow-leafed canopies and where wind speeds are higher rather than lower—in other words, near the top of the canopy rather than the bottom. It is therefore a good approximation on windy days in the crown of a eucalypt forest and a poor one in still conditions deep within a pine forest.

If we write Equation 39 for both heat and water vapor and assume that g_a is the same for both species (which, if g_t is not negligible, requires at the least that their source distributions coincide), then a new leaf combination equation can be derived, with δ_h as the vapor deficit rather than δ_b, and with g_a replacing g_b. Equation 37 then becomes

$$E_h = \sum_{k=1}^{n} \frac{a_k}{a} \left[\frac{\phi_k \epsilon/L + \delta_h g_{ak}}{\epsilon + 1 + g_{ak}/g_{sk}} \right] = \frac{\phi' \epsilon/L + \delta_h g'_a}{\epsilon + 1 + g'_a/g'_s} \quad (41)$$

where the right-hand side of Equation 38 has been included to facilitate the comparison that follows. We will use Equation 41 to define the bulk properties ϕ', g'_a, and g'_s in terms of the corresponding leaf properties, ϕ_k, g_{ak}, and g_{sk}. Equation 41 can be regarded as one equation in the three unknowns ϕ', g'_a, and g'_s, so that two other equations are needed to complete the solution. These extra relationships can be chosen in a variety of ways; our choice is to write Equation 41 in the equilibrium limit (Equation 33), the potential limit (Equation 32), and, last, in its most general form, as above. Successive exact definitions are then obtained for ϕ', g'_a, and g'_s that may be compared with the intuitive definition of each bulk property as an area-weighted average of its leaf counterparts.

In the equilibrium limit, where the evaporation from each leaf depends only on ϕ_k (Equation 33), it is reasonable to demand that the evaporation from the whole canopy will likewise depend only on ϕ'. Then Equation 41 becomes

$$E_h = \frac{\phi' \epsilon/L}{\epsilon + 1} = \sum_{k=1}^{n} \frac{a_k}{a} \left[\frac{\phi_k \epsilon/L}{\epsilon + 1} \right]$$

which gives the definition of ϕ':

$$\phi' = \sum_{k=1}^{n} \frac{a_k}{a} \phi_k. \quad (42)$$

Hence, ϕ' is simply the sum of the area-weighted net radiation fluxes for each leaf, or the total net radiation flux incident upon the canopy. This is an entirely plausible definition that accords with intuition; we need only note that if heat storage and photosynthesis terms (S and CP in Equation 26) are important for individual leaves within the canopy, then they need to be subtracted from ϕ_k and likewise from ϕ'.

Next we write Equation 41 in the limit of potential evaporation, when $g_{sk} = \infty$. It is sensible to set $g'_s = \infty$ in this limit also; otherwise the bulk stomatal resistance of the canopy would not be zero when the stomatal resistances for each individual leaf were zero. With this condition, and also with ϕ' as already defined from Equation 42, we obtain

$$g'_a = \sum_{k=1}^{n} \frac{a_k}{a} g_{ak} \quad (43)$$

so g'_a is defined as an area-weighted average or parallel sum of the leaf aerodynamic conductances g_{ak}. Once again, this is the intuitively natural definition.

Last, we define g'_s from Equation 41, using Equations 42 and 43 for ϕ' and g'_a. The full expression for g'_s is very messy, but its structure can be seen by rearranging the single-leaf model (the second line of Equation 41) to give

$$\frac{1}{g'_s} = \frac{1}{g'_a}\left[\frac{\epsilon\phi'}{LE_h} - (\epsilon + 1)\right] + \frac{\delta_h}{E_h} \quad (44)$$

and then substituting for ϕ', g'_a, and E_h in terms of leaf variables, from Equations 42 and 43 and the first line of Equation 41, respectively. It follows that this rigorous definition of g'_s is not equal to the intuitive definition

$$g'_{sI} = \sum_{i=1}^{n} \frac{a_k}{a} g_{sk} \quad (45)$$

which is the parallel, area-weighted sum of all the leaf stomatal conductances in the averaging volume. Instead, g'_s contains information about ϕ_k and g_{ak} as well as g_{sk}, so it is no longer a purely physiological parameter. Conversely, we cannot expect to derive the correct value of g'_s in Equation 41 by purely physiological measurements.

To explore quantitatively the relationship between g'_s and g'_{sI}, we will use some plausible, empirical forms for ϕ_k, g_{ak}, and g_{sk}. It is convenient to replace the summations by integrals over z, and ϕ_k, g_{ak}, and g_{sk} by the smooth functions $\phi(z)$, $g_a(z)$, and $g_s(z)$ obtained by volume averaging over extensive, thin, horizontal slabs of area a and vertical extent z to $z + dz$. The area weighting a_k/a becomes $l(z)\,dz$, where $l(z)$ is the leaf area per unit volume (leaf-area density) averaged over the thin slab. Our empirical forms are

$$\phi(z) = \phi(h)\ e^{-c_1\zeta}$$
$$g_a(z) = g_a(h)\ e^{-c_2\zeta}$$
$$g_s(z) = g_s(h)\ e^{-c_3\zeta} \quad (46)$$

where ζ is the leaf-area index above z:

$$\zeta(z) = \int_z^h l(z')\,dz'. \quad (47)$$

We will take the total leaf-area index, $\zeta(0)$, to be 3, and c_1 and c_2 to be 0.5 and 1.0, respectively, which would be typical of an extensive, mature cereal crop. Denmead (1976) has shown by measurements in a wheat canopy that g_s

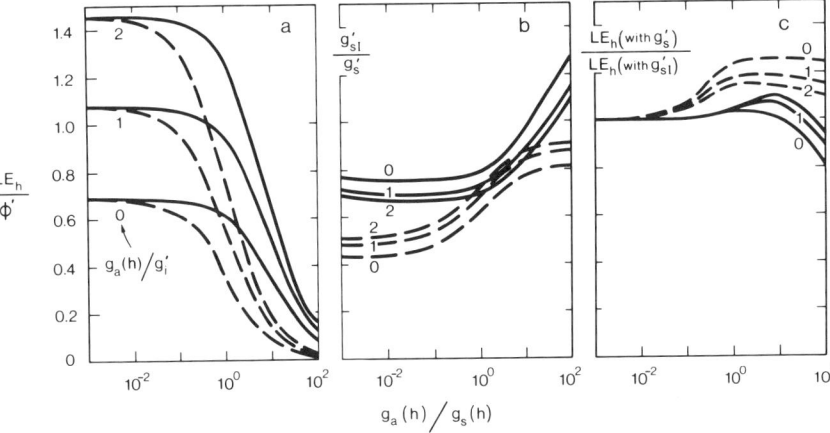

Fig. 18.12. *a*, The dependence of the ratio of emitted latent heat to net radiation absorbed in the canopy upon the two controlling parameters of the single leaf model: $g_a(h)/g_i'$ and $g_a(h)/g_s(h)$. The curves correspond to the exponential forms of $g_a(z)$, $g_s(z)$, and $\phi(z)$ given by Equation 46. *b*, The dependence of the ratio of intuitive bulk stomatal conductance g_{sI}' to the exact value g_s'. *c*, The ratio of the latent heat flux from the canopy calculated with the exact value of g_s' to that calculated using g_{sI}'.

is roughly proportional to ϕ when the canopy is not water-stressed, and that stress causes lower leaves to close their stomata first; these observations suggest that $c_3 \gtrsim c_1$.

Figure 18.12a shows the evaporation from the canopy, normalized as LE_h/ϕ' and plotted against $g_a(h)/g_s(h)$, a parameter that sets the overall relative levels of aerodynamic and stomatal conductances. Curves are plotted for $c_3 = 0$, representing constant stomatal conductance with height, and $c_3 = 2$, which concentrates transpiration at the top of the canopy. The effect of vapor deficit is shown by plotting the curves at three values of $g_a(h)/g_i'$, where g_i' is $\phi'/(L\delta_h)$, a bulk quasi conductance analogous to the leaf g_i in Equation 31b.

For small $g_a(h)/g_s(h)$ the stomata have little effect, but once $g_s(h)$ becomes comparable to $g_a(h)$, transpiration falls rapidly; this occurs sooner of course for $c_3 = 2$ than for $c_3 = 0$. The combination of $g_a(h)/g_i' = 0$ and $g_a(h)/g_s(h) \to 0$ corresponds to the limiting case of equilibrium evaporation. These curves are included only to place the next two diagrams in context.

Figure 18.12b shows the difference between g_{sI}', the intuitive value of g_s' given by Equation 45, and g_s', the correct value from Equation 44. The ratio g_s'/g_{sI}' is a complicated function of $g_a(h)/g_i'$ and $g_a(h)/g_s(h)$; for realistic, working values of $g_a(h)/g_s(h)$, the intuitive and the correct values can easily differ by a factor of two. Figure 18.12c reveals, however, that the consequences of this error are not as bad as they might be, at least in a predictive sense. This shows the transpiration calculated from Equation 38 with the intuitive choice g_{sI}', as a fraction of evaporation calculated using the correct value. The errors are not particularly large, most of the discrepancies in the working range of $g_a(h)/g_s(h)$ being less than 20 percent. This is both good and bad news for stomatal physiologists applying their measurements to real canopies. It suggests (given the overall limitations of the specification of g_a and the fact that we have chosen a simple canopy structure) that a g'_s concocted from direct stomatal measurements will predict an evaporation rate that is not grossly in error. Conversely, if one attempts to reverse this process and deduce the true stomatal conductance from a measurement of total transpiration, then a factor of two is the likely error. This curious state of affairs is a direct consequence of the nonlinearity of the process, which causes errors to diverge in one direction while converging in the other.

We indicated (following Equation 41) that it is possible to define ϕ', g'_a, and g'_s in a number of ways, of which we have analyzed only one. There are other, equally plausible alternative schemes, for example, to use the intuitive definitions for ϕ' and g'_s (Equations 42 and 45) and then to solve for g'_a from Equation 41. However, all these schemes have in common the important central feature that it is not possible simultaneously to make ϕ' depend only on ϕ_k, g'_a only on g_{ak}, and g'_s only on g_{sk}. In other words, we cannot distill the radiative, aerodynamic, and physiological properties of the canopy into three separate parameters ϕ', g'_a, and g'_s. This is caused by the inescapable fact that the leaf combination equation is highly nonlinear in both g_a and g_s.

Attempts to evaluate g'_a and g'_s in practice are largely empirical and rely on the fact that some lack of uniqueness in the definitions can be tolerated if an acceptable answer for E_h is all that is required—see Figure 18.12c, for example. One method of finding g'_a does need mentioning, however, because it is often used with insufficient care: this is the technique of deducing g'_a from the bulk aerodynamic conductance for momentum,

$$g_{aM} = \frac{u_*^2}{V\bar{u}(h)}$$

where u_* is the friction velocity $(-\overline{u'w'})^{1/2}$. The downward flux of momentum is $\rho_a \overline{u'w'}$, and the molar volume V is included to give g_{aM} the same dimensions as other conductances defined in this book. At least three difficulties must be recognized and met before the method can be used successfully. First, momentum absorption at individual leaves occurs through both pressure and viscous drag forces, the pressure drag having no analogue in scalar transfer. Therefore, there is no correspondence between the leaf conductances g_{bM} (for momentum) and g_{bc} (for a scalar). This is called the "bluff-body effect" (Thom, 1972), and although it is now well known, existing formulas for describing it (Thom, 1975) are essentially empirical and probably of limited validity. The second problem is the deduction of u_* from the slope of the mean velocity profile above the surface (the usual means of obtaining u_* for this purpose). In the region just above a rough vegetated surface, where these measurements must usually be made, the logarithmic law, which relates u_* to the slope of $\bar{u}(z)$ in neutral conditions, is known to fail and to lead to meaningless u_* values (Raupach, 1979; Garratt, 1980). The third difficulty is that values of g_t (and therefore of g_a) will certainly be different for momentum and a scalar if the sink distribution for momentum differs much from the source distribution for the scalar.

Second-order-closure models. Although it has not been presented this way in the past, we have seen that the single-leaf model effectively consists of an application of Equation 8 over the entire canopy volume. As such it can afford no information on the variation of concentrations, source strengths, and fluxes within the canopy. When these variations are important, as they are, for example, in relating physiology to the canopy microclimate, then a much more detailed model is required. Essentially, it is necessary to solve Equation 8 with a sufficiently fine averaging volume to resolve the necessary detail.

This requires a direct attack on the problem of describing the eddy and dispersive flux divergences, terms III and IV in Equation 8, to close the equation. We will concentrate here on the eddy flux, because little is known about the dispersive flux, except that it is negligible in a few very simple situations (Raupach, Coppin, and Legg, 1986). We have already shown that K-theory or first-order closure, involving the eddy diffusivity defined by Equation 9, is an unacceptable way of describing the eddy flux. In aeronautical and mechanical engineering a widely used but more complicated alternative is second-order closure. In this method, instead of parameterizing the eddy flux in terms of mean concentrations, the equation for eddy flux itself is solved simultaneously with Equation 11. This means that the third-order quantities appearing in the eddy flux equation must be written in terms of the eddy flux itself. It must be said immediately that while some of these parameterizations work well, others present problems as severe in their own way as the use of K-theory. Nevertheless, if we wish to employ an Eulerian description of the canopy, that is, to use velocities and concentrations referred to axes fixed in space, then second-order closure is the minimum level of complexity that must be embraced. We will discuss the method briefly below and concentrate upon the differences that appear when this type of model is applied to plant canopies rather than the free air above.

The governing equations for the eddy flux at a point are obtained from primitive equations in much the same way as Equation 5 was derived from Equation 1 (see e.g. Tennekes and Lumley, 1972), while the extra steps involving volume averaging in a canopy are covered by Finnigan (1985). To discuss the method we will need some indication of the structure of the equation, but we need not write it out formally; a description of the various terms will suffice. The equation for $\langle \overline{w'c'} \rangle$ is

$$\frac{\partial}{\partial t} \langle \overline{w'c'} \rangle = GP + PD + BP + TT + (DSP + WP). \qquad (48)$$

A balance exists between a collection of terms competing to change the vertical eddy flux of c in a control volume. The first four terms on the right-hand side of Equation 48 represent processes that are present both in the canopy air spaces and in the free air above; and the last two terms, in parentheses, represent effects confined to the canopy. There are three kinds of terms in Equation 48: production, destruction, and transport. Production terms describe processes that act on average to associate fluctuations in velocity and concentration, while destruction terms act to destroy this correlation. They both operate *in situ*, creating or destroying eddy flux at a point. Transport terms, in contrast, export any excess of production over destruction at one point to other places where destruction exceeds production.

Taken in turn the labels in Equation 48 refer to:

1. Gradient Production (GP). If there is a mean concentration gradient, then vertical wind-speed fluctuations tend to be associated with parcels of air arriving from levels where the concentration is different from the local value. This effects a correlation between w' and c'.

2. Pressure Destruction (PD). The effect of fluctuating pressure in the turbulent airstream is, on average, to destroy the $\overline{w'c'}$ correlation.

3. Buoyant Procution (BP). Hot air rises; cold air sinks. If the hot rising air is associated with c' fluctuations of one sign, if it is more moist, say, while the cool sinking air is drier, then a contribution is made to $\langle \overline{w'c'} \rangle$.

Well above the canopy, only these three processes are significant. The eddy flux then results from a balance between the three of them, and we have a situation known as local equilibrium. If buoyant production is also negligible, and only the balance between the GP and PD terms is important, then we find that we can derive and use an eddy diffusivity (Wyngaard, 1982). This is the situation in neutral conditions well above the canopy. Within and just above canopies, however, the third type of term, transport, becomes important.

4. Turbulent Transport (TT). Just as turbulent fluctuations can transport concentration through the eddy flux process, so also can they transport the eddy flux itself. Measurements made by Coppin, Raupach, and Legg (1985) and Raupach, Coppin, and Legg (1986) show that within and just above canopies TT is always important, so that even when buoyancy effects are absent, there is at least a three-way balance between GP, PD, and TT. Under these conditions an eddy diffusivity cannot be defined (Finnigan, 1985). The reason is that a large TT term inevitably signals the presence of large-scale turbulent fluctuations, much larger than the scale of the concentration gradient, and we have already seen that this invalidates an eddy diffusivity.

Terms 1 to 4 are present both within the canopy and above it. The last two terms, however, represent mechanisms for associating c' and w' peculiar to the foliage air spaces.

5. Drag-Source Production (*DSP*). This is the average product of the fluctuating drag force and the surface concentration on a leaf or branch. The aerodynamic drag on a leaf removes momentum from the airstream and reduces the velocity in the wake. At the same time this wake contains fluid that has just flowed over the leaf and so is marked by the surface concentration there; if the leaf is hot, for example, the wake contains warmer, slower air than average.

6. Waving Production (*WP*). This term is the average product of the flux of c from a leaf and the velocity of leaf motion. A moving leaf drags along with it a mass of air, and this is marked by the flux of c through the leaf surface.

The contribution to the eddy flux made by terms 5 and 6 differs from that of the free-air terms in two important respects. While the free-air terms produce flux with the same length scales as the turbulent velocity and gradients of mean concentration, the canopy-interaction terms produce flux with length and time scales matched to individual plant wakes and vibrational frequencies. Similarly, while the spatial distribution of flux produced by the free-air terms follows the distribution of mean concentration gradient and turbulent velocity fields, the canopy terms produce flux coupled directly to the scalar source distribution.

Of the six terms we have described, two, pressure destruction and turbulent transport, are third-order terms that must be rewritten in terms of other, second-order quantities to close Equation 48. A description of the procedures usually adopted for this, together with references to other work, may be found in Wyngaard (1982). We will not discuss them here. So far, only one application of second-order closure to canopy flows has appeared in the literature. This is the model of Wilson and Shaw (1977), which describes momentum rather than scalar transport within the canopy. Nevertheless it successfully incorporates turbulent transport and terms analogous to the *DSP* term, and correctly predicts the countergradient flux of momentum observed deep within dense canopies. A second-order-closure model for scalars, developed recently by one of the present authors, shows countergradient heat and moisture transfer similar to that shown in Figure 18.8, and its predictions prove to be sensitive to the inclusion or exclusion of *DSP* and *WP* terms.

Both these models comprise complex groups of equations that must be solved on a computer. With some simplifying assumptions, however, the parameterizations used to close Equation 48 can be used to derive an instructive analytic expression for $\langle \overline{w'c'} \rangle$. When the canopy is horizontally uniform, we can write

$$GP = -\langle \overline{w'^2} \rangle \frac{\partial \langle \bar{c} \rangle}{\partial z}, \quad PD = -\frac{\langle \overline{w'c'} \rangle}{\tau}. \tag{49}$$

The formula for *GP* is exact; that for *PD* is a common, simple parameterization in which τ is a time scale for the turbulence. If only *PD* were to appear on the right-hand side of Equation 48, this parameterization would cause $\overline{\langle w'c' \rangle}$ to decay exponentially to zero, with the time scale τ. Putting both formulas into the full Equation 48, assuming steady conditons ($\partial \overline{\langle w'c' \rangle}/\partial t = 0$), and rearranging, we obtain (Wyngaard, 1982)

$$\overline{\langle w'c' \rangle} = - \left\{ \overline{\langle w'^2 \rangle} \tau \right\} \frac{\partial \langle \bar{c} \rangle}{\partial z} + (BP + TT + DSP + WP)\tau. \quad (50)$$

The quantity in braces is, in effect, an eddy diffusivity K_c. By comparing Equations 50 and 9, we see how the extra terms in the eddy flux equation, in addition to *GP* and *PD*, prevent a dependence of eddy flux solely upon the mean gradient, as implied by Equation 9. Of the extra terms, *TT* is of critical importance in a canopy, as it is fundamentally linked with the large-scale nature of the transporting eddies and with the gusting process. The terms *DSP* and *WP*, associated with turbulence on wake and waving scales, are more important in quiescent periods than in gusts. They also show how the eddy flux is directly linked with the source distribution, and this is one of the reasons that in the last section we emphasized that the minimum requirement for equality of turbulent conductances g_t for different species was similarity of source distributions.

Lagrangian models. The second-order-closure approach raises several difficulties in parameterizing the terms *PD*, *TT*, *DSP*, and *WP* in Equation 48. The biggest problem is *TT*: most existing ways of parameterizing this term link *TT* to the gradient of eddy flux through an eddy diffusivity and are unacceptable in plant canopies for the same reasons that cause the downfall of the eddy diffusivity concept at first order, Equation 9. Lagrangian models provide a different viewpoint on the turbulent transport of scalars and are nowhere subject to these restrictions.

One way of describing the turbulent dispersion of a scalar is to track the wandering motion of scalar-carrying fluid elements or marked fluid particles, so called because they have been marked with scalar by previous contact with a scalar source. By a "fluid element" we mean a tiny connected lump of fluid containing very many molecules but smaller than the smallest eddies. In a continuous fluid, such a lump remains materially connected throughout its motion. Any property that is referred to an individual wandering fluid element (such as its position, velocity, or scalar concentration) is called a Lagrangian property, and models of scalar dispersion that depend upon descriptions of the behavior of wandering fluid elements are called Lagrangian models; in contrast, Eulerian fluid properties—velocities, scalar concentrations, and so on—are measured at fixed points in space, and Eulerian models of scalar dispersion make use of fixed-point scalar conservation equations, such as Equations 1, 5, or 8. All the models discussed so far in this paper are Eulerian.

Plant-Canopy Transfer and Stomatal Characteristics

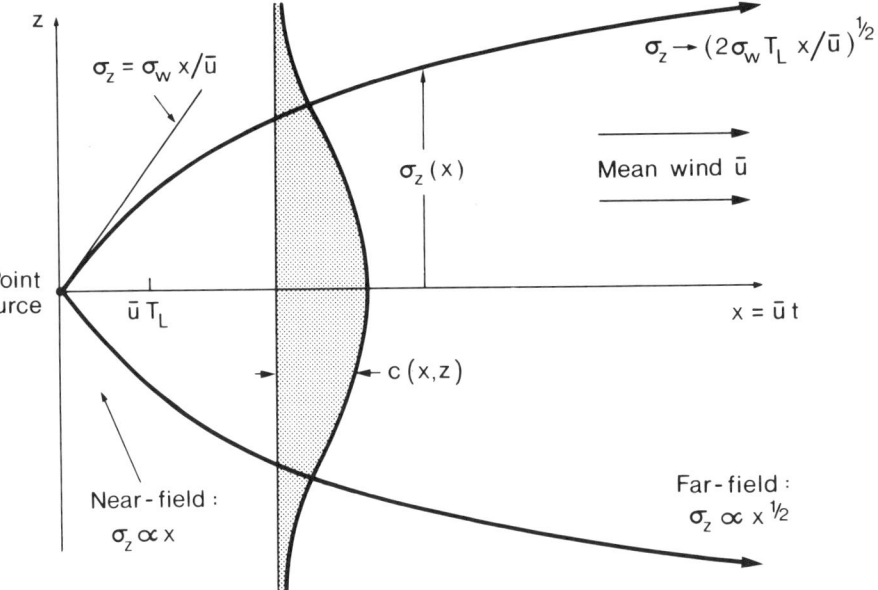

Fig. 18.13. A schematic diagram of the spread of a concentration plume from a source in homogeneous turbulence with a mean wind speed $\bar{u}$.

To gain a first insight into Lagrangian models, consider a small, isolated scalar source (say, a small leaf emitting water vapor) in a homogeneous turbulent flow with a constant mean velocity $\bar{u}$ in the x-direction. The problem of describing the evolution of the emitted cloud of scalar is a very old one that was solved in essence by Taylor (1921) and with some elaboration by Batchelor (1949). The result of the Taylor-Batchelor analysis is that the scalar concentration distribution is Gaussian with a half-width in one direction (along the z-axis, say) $\sigma_Z(t)$, which obeys

$$\sigma_Z(t) \approx \sigma_W t, \quad t \ll T_L$$
$$\sigma_Z(t) \approx (2\sigma_W^2 T_L t)^{1/2}, \quad t \gg T_L \quad (51a)$$

where t is the time after release of a marked particle, or since the constant mean velocity allows us to convert time after release into distance from the source:

$$\sigma_Z(x) \approx \sigma_W x/\bar{u}, \quad x \ll \bar{u} T_L$$
$$\sigma_Z(x) \approx (2\sigma_W^2 T_L x/\bar{u})^{1/2}, \quad x \gg \bar{u} T_L \quad (51b)$$

so that the plume of marked particles has the shape sketched in Figure 18.13. In Equations 51a and 51b σ_W is the standard deviation of the vertical

Lagrangian velocity $W(t)$ (that is, the vertical velocity of a marked fluid particle), and T_L is the Lagrangian integral time-scale, defined by

$$T_L = \frac{1}{\sigma_w^2} \int_0^\infty \overline{W(0)W(t)}\, dt,$$

which is a measure of the time over which $W(t)$ remains correlated, or the "memory time" of $W(t)$. Equation 51 shows that σ_Z grows in two quite different ways, depending on whether the travel time t of the marked particles forming the plume of scalar is much less than or much greater than T_L. When $t \ll T_L$, a particle moves away from the source without significant change in its initial velocity $W(0)$, so σ_Z grows linearly with time. When $t \gg T_L$, the particle has "forgotten" its initial velocity and can be regarded as undergoing a random walk with a sequence of independent steps, so σ_Z grows with $t^{1/2}$; this growth law is characteristic of any homogeneous random walk after many steps (Monin and Yaglom, 1971). The two travel-time regimes $t \ll T_L$ and $t \gg T_L$ are called the "near-field" and the "far-field" regimes, respectively.

The existence of near-field and far-field regimes is a general property of turbulent dispersion, not confined to homogeneous turbulence—though, of course, the exact form of Equation 51 is limited to homogeneous turbulence. Of critical importance is the fact that a K-theory, such as Equation 9, can only ever describe the far-field regime because K-theory is formally equivalent to a random-walk model (Monin and Yaglom, 1971). Therefore, whenever near-field effects are important, K-theory will fail. This statement is the equivalent, from a Lagrangian viewpoint, of our earlier Eulerian length-scale arguments (around Figures 18.8 and 18.9) and dynamical arguments (following Equation 50) for the failure of K-theory.

In a canopy, near-field effects are very important. Legg, Raupach, and Coppin (1986) showed from a wind-tunnel experiment that T_L is about $0.3h/u_*$ within a neutrally stratified canopy of height h with a friction velocity u_*. Since $\bar{u}(h)/u_*$ is typically about 3, the length $\bar{u}(h)T_L$ is about equal to h; in other words, scalar dispersing away from a leaf near the top of the canopy must travel for a distance of about h before its travel time is sufficient to reach the transition ($t = T_L$) between the near-field and far-field regimes. Much of the heat or water vapor measured at any point within a canopy emanates from nearby upwind leaves and is therefore subject to near-field dispersion.

In practice, the construction of a Lagrangian model for turbulent transport and dispersion within a canopy proceeds by first modeling the dispersion from any one leaf and then superposing the calculated concentration and flux distributions for numerous leaves. The source strengths—the rates at which the leaves emit marked particles—would be modeled in the same way as in an Eulerian scheme using Equation 31, the individual-leaf combination equation. The ambient concentrations that this equation requires are given by the

Plant-Canopy Transfer and Stomatal Characteristics 425

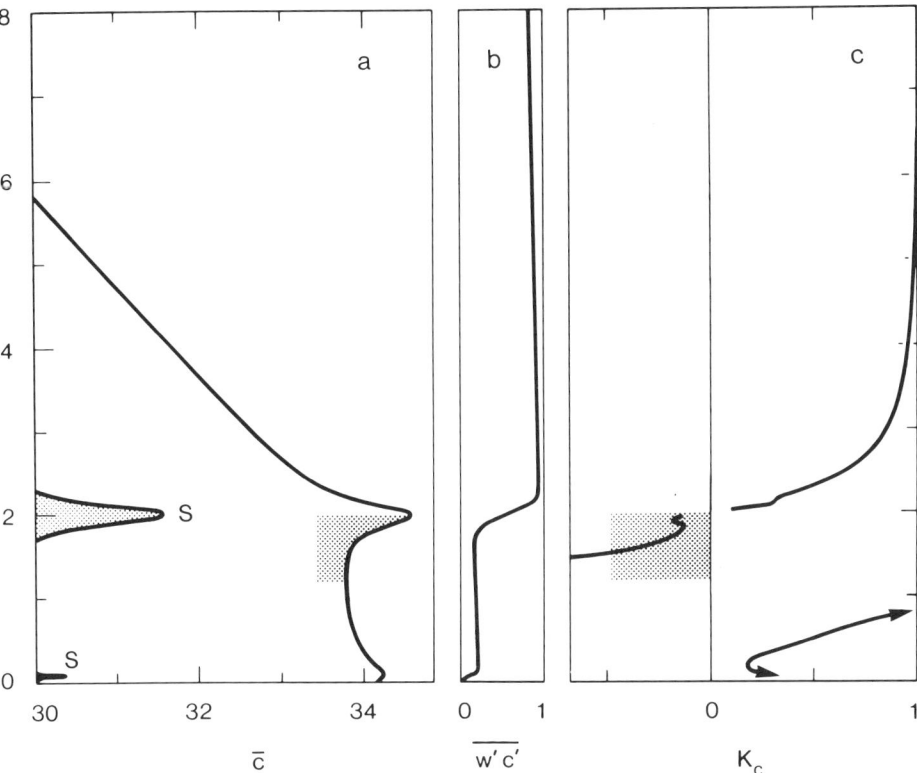

Fig. 18.14. a, The source strength S and resulting mean concentration profile $\bar{c}(z)$ calculated from a Lagrangian dispersion model. The shaded areas in this figure and in Figure 18.14c mark the region of countergradient flux. The height axis is in units of $\sigma_w T_L$; the upper and lower sources are of strengths 0.8 and 0.2, respectively, giving a total source strength of 1. b, The vertical flux $\overline{w'c'}$ calculated from the same model. c, The effective eddy diffusivity K_c obtained by dividing the flux by the gradient. The negative K_c in the countergradient region is clear, and the fact that K_c is undefined when the gradient is zero is indicated by a break in the line. (Adapted from Raupach, 1986.)

emissions from all upwind leaves whose plumes cross that particular source. Figure 18.14 shows the results of a simple analytic model of this kind for a single scalar of mole fraction c (Raupach, 1986). The turbulence in the canopy is assumed to be homogeneous, with a uniform, height-independent wind profile, while the ground ($z = 0$) reflects the dispersing scalar. (These assumptions enable the analytic Taylor-Batchelor theory to be used directly.) The scalar sources are at two heights, an upper, "crown," source of relative strength 0.8, and a lower, "ground," source of relative strength 0.2. The concentration ($\bar{c}$) and vertical flux ($\overline{w'c'}$) profiles shown are those from a long, uniform fetch of canopy such that the travel time of scalar from the leaves farthest upwind is $1,000 T_L$. What is seen is a substantial height range just below the upper source where the flux is countergradient, just as observed in Uriarra Forest (Fig. 18.8). This is essentially due to near-field dispersion from nearby leaves in the upper source whose plumes have not yet spread very much and therefore dominate the local mean concentration field.

Although based on the severely oversimplified assumption of homogeneous turbulence with a reflecting boundary, Figure 18.14 demonstrates three important points. First, Lagrangian models can successfully predict countergradient fluxes, as observed in the field. Second, the concentration and flux fields are extremely dependent on source distribution (for example, there is no countergradient flux if the small ground source in Fig. 18.14 is removed); this confirms our earlier assertion that the turbulent conductance g_t in Equation 39 is highly dependent on source distribution. Third, the apparently naive assumption that the turbulent wind field is homogeneous gives surprisingly realistic concentration and flux fields, suggesting that source inhomogeneity is more important than wind inhomogeneity.

When the wind field is inhomogeneous, Lagrangian models must be solved by numerical, random-flight techniques (recent work is referenced by Durbin, 1983; and Thomson, 1984). It is important to note that all Lagrangian models developed so far predict scalar dispersion only, and require Lagrangian velocity statistics, like σ_W and T_L, as input.

We can sum up the relative strengths of the Eulerian and Lagrangian approaches in this way. Eulerian theories, essentially second-order-closure models or elaborations of them, can predict both the wind and scalar fields within the one formalism and can incorporate subtle interactions between the sources and the turbulence (such as the *DSP* and *WP* terms in Equation 48), but they have fundamental weaknesses in describing turbulent transport. Lagrangian theories easily and adequately describe the turbulent transport, but they cannot readily handle effects such as those represented by *DSP* and *WP*, and they require information about Lagrangian turbulent velocity statistics, which are notoriously hard to measure. However, early indications are that a rather simple assumption about the turbulent velocity field gives surprisingly good predictions.

Future models, which will aim to be both robust and physically sound, will probably involve both approaches.

REFERENCES

Batchelor, G. K. 1949. Diffusion in a field of homogeneous turbulence. *Australian Journal of Science Research*, 2: 437–50.

Plant-Canopy Transfer and Stomatal Characteristics 427

Bird, R. B., W. E. Stewart, and E. N. Lightfoot. 1960. *Transport phenomena*. New York: Wiley.

Businger, J. A. 1982. Equations and concepts. In F. T. M. Nieuwstadt and H. van Dop, eds., *Atmospheric turbulence and air pollution modelling*: 1–36. Dordrecht: Reidel.

Coppin, P. A. 1985. Heat and mass transfer mechanisms above and within plant canopies. In *Proceedings of the 3rd Australasian conference on heat and mass transfer, Melbourne, May 1985*: 465–72. Melbourne: University of Melbourne Press.

Coppin, P. A., M. R. Raupach, and B. J. Legg. 1986. Experiments on scalar dispersion within a plant canopy, II. An elevated plane source. *Boundary-Layer Meteorology*, 35: 167–91.

Corrsin, S. 1974. Limitations of gradient transport models in random walks and in turbulence. *Advances in Geophysics*, 18A: 25–60.

Cowan, I. R. 1968. Mass, heat and momentum exchange between stands of plants and their atmospheric environment. *Quarterly Journal of the Royal Meteorological Society*, 94: 318–32.

———. 1972. Mass and heat transfer in laminar boundary layers with particular reference to assimilation and transpiration in leaves. *Agricultural Meteorology*, 10: 311–29.

Denmead, O. T. 1976. Temperate cereals. In J. L. Monteith, ed., *Vegetation and the atmosphere* 2: 1–31. London: Academic.

———. 1984. Plant physiological methods for studying evapotranspiration: Problems in telling the forest from the trees. *Agricultural Water Management*, 8: 167–89.

Denmead, O. T., and E. F. Bradley. 1985. Flux-gradient relationships in a forest canopy. In B. A. Hutchison and B. B. Hicks, eds., *The forest-atmosphere interaction*: 421–42. Dordrecht: Reidel.

Denmead, O. T., and I. C. McIlroy. 1970. Measurements of non-potential evaporation from wheat. *Agricultural Meteorology*, 7: 285–302.

Durbin, P. A. 1983. *Stochastic differential equations and turbulent dispersion*. NASA Reference Publication no. 1103. Washington, D.C.

Finnigan, J. J. 1979. Turbulence in waving wheat, II. Structure of momentum transfer. *Boundary-Layer Meteorology*, 16: 213–36.

———. 1985. Turbulent transport in flexible plant canopies. In B. A. Hutchison and B. B. Hicks, eds., *The forest-atmosphere interaction*: 443–80. Dordrecht: Reidel.

Finnigan, J. J., and R. A. Longstaff. 1982. A wind tunnel model study of forced convective heat transfer from a cylindrical grain storage bin. *Journal of Wind Engineering and Industrial Aerodynamics*, 10: 191–211.

Garratt, J. R. 1980. Surface influence upon vertical profiles in the atmospheric near-surface layer. *Quarterly Journal of the Royal Meteorological Society*, 106: 803–19.

Grace, J. 1980. Some effects of wind on plants. In *Plants and their atmospheric environment: The 21st symposium of the British Ecological Society, Edinburgh, 1979*: 31–56. Oxford: Blackwell.

Kumar, A., and N. Barthakur. 1971. Convective heat transfer measurements of plants in a wind tunnel. *Boundary-Layer Meteorology*, 2: 218–27.

Landsberg, J. J., and A. S. Thom. 1971. Aerodynamic properties of a plant of complex structure. *Quarterly Journal of the Royal Meteorological Society*, 97: 565–70.

Legg, B. J., M. R. Raupach, and P. A. Coppin. 1986. Experiments on scalar dispersion within a plant canopy, III. An elevated line source. *Boundary-Layer Meteorology*, 35: 277–302.

Leuning, R. 1983. Transport of gases into leaves. *Plant, Cell and Environment*, 6: 181–94.

McNaughton, K. G., and P. G. Jarvis. 1983. Predicting the effects of vegetation changes on transpiration and evaporation. In T. T. Koslowski, ed., *Water deficits and plant growth* 7: 1–47. New York: Academic.

Monin, A. S., and A. M. Yaglom. 1971. *Statistical fluid mechanics: Mechanics of turbulence* 1, ed., J. L. Lumley. Cambridge, Mass.: MIT Press. [Originally published in Russian (Moscow: Nauka Press, 1965).]

Monteith, J. L. 1965. Evaporation and environment. In G. E. Fogg, ed., *The state amd movement of water in living organisms*: 205–34. Symposia of the Society for Experimental Biology, 19. Cambridge: Cambridge University Press.

———. 1973. *Principles of environmental physics*. London: Arnold.

———, ed. 1975–76. *Vegetation and the atmosphere* 1, 2. London: Academic.

Mulhearn, P. J., and J. J. Finnigan. 1978. Turbulent flow over a very rough, random surface. *Boundary-Layer Meteorology*, 15: 109–32.

Raupach, M. R. 1979. Anomalies in flux-gradient relationships over forest. *Boundary-Layer Meteorology*, 16: 467–86.

———. 1986. A Lagrangian analysis of scalar transfer in vegetation canopies. *Quarterly Journal of the Royal Meteorological Society*, in press.

Raupach, M. R., P. A. Coppin, and B. J. Legg. 1986. Experiments on scalar dispersion within a plant canopy, I. The turbulence structure. *Boundary-Layer Meteorology*, 35: 21–52.

Raupach, M. R., A. S. Thom, and I. Edwards. 1980. A wind tunnel study of turbulent flow close to regularly arrayed rough surfaces. *Boundary-Layer Meteorology*, 18: 373–97.

Schuepp, P. H. 1972. Studies of forced-convection heat and mass transfer of fluttering realistic leaf models. *Boundary-Layer Meteorology*, 2: 263–74.

———. 1973. Model experiments of free convection heat and mass transfer of leaves and plant elements. *Boundary-Layer Meteorology*, 3: 454–67.

———. 1982. Laboratory studies on dry deposition of submicron-size particles on coniferous foliage. *Boundary-Layer Meteorology*, 24: 465–80.

Seginer, I., P. J. Mulhearn, E. F. Bradley, and J. J. Finnigan. 1976. Turbulent flow in a model plant canopy. *Boundary-Layer Meteorology*, 10: 423–53.

Shaw, R. H., R. H. Silversides, and G. W. Thurtell. 1974. Some observations of turbulence and turbulent transport within and above plant canopies. *Boundary-Layer Meteorology*, 5: 429–49.

Shaw, R. H., J. Tavangar, and D. P. Ward. 1983. Structure of the Reynolds stress in a canopy layer. *Journal of Climate and Applied Meteorology*, 22: 1922–31.

Stewart, J. B., and A. S. Thom. 1973. Energy budgets in pine forest. *Quarterly Journal of the Royal Meteorological Society*, 99: 154–70.

Sutera, S. P. 1965. Vorticity amplification in stagnation point flow and its effect on heat transfer. *Journal of Fluid Mechanics*, 21: 515–34.

Taylor, G. I. 1921. Diffusion by continuous movements. *Proceedings of the London Mathematical Society*, ser. A., 20: 196–211.

Tennekes, H., and J. L. Lumley. 1972. *A first course in turbulence*. Cambridge, Mass.: MIT Press.

Thom, A. S. 1972. Momentum, mass and heat exchange of vegetation. *Quarterly Journal of the Royal Meteorological Society*, 98: 124–34.

———. 1975. Momentum, mass and heat exchange of plant communities. In J. L. Monteith, ed., *Vegetation and the atmosphere* 1: 57–109. London: Academic.

Thomson, D. J. 1984. Random walk modelling of diffusion in inhomogeneous turbulence. *Quarterly Journal of the Royal Meteorological Society*, 110: 1107–20.

Willmarth, W. W. 1975. Structure of turbulence in boundary layers. *Advances in Applied Mechanics*, 15: 159–254.

Wilson, J. D., D. P. Ward, G. W. Thurtell, and G. E. Kidd. 1982. Statistics of atmospheric turbulence within and above a corn canopy. *Boundary-Layer Meteorology*, 24: 495–519.

Wilson, N. R., and R. H. Shaw. 1977. A higher-order closure model for canopy flow. *Journal of Applied Meteorology*, 16: 1198–1205.

Wyngaard, J. C. 1982. Boundary layer modeling. In F. T. M. Nieuwstadt and H. van Dop, eds., *Atmospheric turbulence and air pollution modelling*: 69–106. Dordrecht: Reidel.

19

Breeding for Stomatal Characters

Hamlyn G. Jones

Over the past 15 years or so there has been increasing interest among plant breeders in the development of varieties of crop plants with particular stomatal characteristics. This interest in stomata has arisen because of the widespread perception of their crucial role in the control of water loss (and hence of drought tolerance) and of CO_2 uptake (and hence of productivity), and because they provide a simple anatomical character that can be used in selection.

Since the initial enthusiasm, however, it has become clear that, as with other characters, the situation for stomata is not so straightforward as had been hoped. The complications include innate compensation mechanisms that operate when selection is applied to one character in isolation, and the need to consider stomatal responses to environment. In addition, much variation in stomatal characters is phenotypically rather than genetically determined. More important, perhaps, has been the realization that our understanding of the physiological and ecological roles of stomata is incomplete, particularly in relation to drought tolerance. For example, it is necessary to achieve a balance between the need for water conservation and survival (few, small, or closed stomata) and the desire for productivity (many, large, or open stomata). Unfortunately, the optimum depends, in a complex and imperfectly understood way, on other plant characters (such as leaf size and number) and on the environment in which the plant is to be grown.

This review outlines approaches that can be used in breeding for stomatal characters, summarizes what has been achieved, and identifies those characters and methods that hold most potential for the future.

Objectives in Breeding for Stomatal Characters

There are two main reasons for attempting to manipulate stomata by breeding. The first is to maximize productivity, and hence yield, by increasing assimilation rates. An increase in stomatal conductance would be expected to increase CO_2 partial pressure within the leaf, and thus to increase net photosynthesis in those plants (particularly C_3 species) in which the photosynthetic apparatus is not saturated with CO_2. In practice, co-adaptation of the different components of the photosynthetic system tends to keep the intercellular CO_2 partial pressure stable (see Morison, this volume). This may negate any improvement in assimilation, so that the best approach is likely to involve parallel increases in stomatal and intracellular conductances.

The second reason is to improve drought tolerance. Water conservation mechanisms, such as a low stomatal conductance, provide an important way for plants to survive drought, but they also limit potential productivity. For this reason, stomata that close only in response to severe stress may be better for an agricultural crop than stomata that maintain a consistently low conductance. Similarly, stomatal closure in response to increasing humidity deficit (see Jones, 1983) would also tend to enhance water-use efficiency and drought tolerance by restricting stomatal opening to the morning and evening hours of low evaporative demand. Other advantages of a low conductance are that the resulting lowered transpiration rate decreases the depression of leaf water potential and that water-use efficiency can be improved (see Jones, 1976).

There are several other possible purposes for which selection for particular stomatal characters may be of value, such as winter hardiness (very low conductance may prevent winter desiccation) or high temperature tolerance (high conductances would increase evaporative cooling). In addition, there is some evidence that, as might be expected, genotypes with few or sensitive stomata may be more tolerant of atmospheric pollutants (e.g. Butler and Tibbitts, 1979), though other factors may be more important. It has even been suggested that it might be possible to breed trees that effectively absorb and reduce atmospheric pollutants by selecting for large stomatal apertures and a low tendency for stomatal closure (Smith and Dochinger, 1976)! Since many fungal diseases, such as rusts and downy mildews, infect leaves through the stomatal pores, there has been interest in, and some success with, selection for stomatal closure or few stomata as a means for improving resistance to disease (e.g. Gill and Nandpuri, 1978).

In the remainder of this review only stomatal effects on drought tolerance and productivity will be considered.

The Genetics of Stomata

An essential requirement in any breeding program is that appropriate genetic variation exists or can be induced mutagenically for the characters under consideration. It is well known that there are large heritable differences between species in stomatal dimensions, distribution, and morphology (see e.g. Meidner and Mansfield, 1968), but these will become available to breeders only when techniques for interspecific gene transfer are developed. There have also been many studies of the genetics of different stomatal characters (particularly of frequency and size) within a wide range of species, but these studies have been confounded by the large environmental component of variation. For example, dimensions, and especially frequencies, can change

Breeding for Stomatal Characters   433

more than twofold in response to radiation (e.g. Gay and Hurd, 1975), in response to water status (Rawson, Constable, and Howe, 1980), or according to developmental stage (see Meidner and Mansfield, 1968).

Morphological characters. A particular problem with genetic analysis of stomatal characters is that they are often linked to each other and to other characters. For example, stomatal frequency per unit leaf area is commonly related inversely to the size of the guard cells (e.g. Rajendra, Mujeeb, and Bates, 1978) or to the size of the leaf (Jones, 1977a), so that the total pore area per unit leaf area may remain nearly constant (Pallardy and Kozlowski, 1979). Apart from environmental factors that increase cell size and general vigor, genetic factors such as heterosis and increased chromosome numbers at higher ploidy levels also tend to decrease stomatal frequency. The effect of ploidy on stomatal size and frequency is particularly well documented for many genera, including wheat (in which the stomatal frequency of 6n species tends to be about half that of 2n species: Dunstone, Gifford, and Evans, 1973), *Bromus* (Tan and Dunn, 1975; Lea, Dunn, and Koch, 1977), *Vaccinium* (Chandler and Lyrene, 1982), and alfalfa (Setter, Schrader, and Bingham, 1978).

Information on intraspecific genetic variation for several stomatal characters is presented in Table 19.1. It is clear that in several species stomatal frequency can vary by more than twofold, and stomatal size by a factor of more than one and a half. The genetic basis for these variations is not well understood and seems to depend on the species. For example, Dylenok, Khotyleva, and Yatsevich (1981), using monosomic lines of wheat, reported multigenic control of stomatal frequency and size, with the genes involved being on several chromosomes. Yet in maize stomatal frequency appears to be under the control of relatively few genes (Heichel, 1971), and in sorghum it is controlled by one locus or block of loci (Suh et al., 1976). Use of diallel crosses (in which different lines are crossed together in pairs) in a range of species has shown that stomatal frequency and size are highly heritable (e.g. Miskin, Rasmusson, and Moss, 1972; Wilson, 1971; Liang et al., 1975; Tan and Dunn, 1976). Although these workers have found most of the variation to be additive, others have found dominance (Walton, 1974) or overdominance (Suh et al., 1976) to be the main types of variation.

Stomatal conductance. The breeder attempting to improve productivity or drought tolerance should be more concerned with stomatal conductance than with morphological features, which are at best only indicative of physiological functioning. Although frequency, for example, is sometimes correlated with conductance (e.g. Nerkar, Wilson, and Lawes, 1981), in many cases no such relationship can be detected (e.g. Jones, 1977a). This is often because of the inverse relationship between frequency and size, two of the main determinants of conductance per unit area. Thus, conductance may be more closely related to the product of size and frequency. In practice, however, the most important

TABLE 19.1. Some examples of intraspecific variation for stomatal characters

Character and species	Range	Reference
Frequency (stomata mm^{-2})		
Apple		
(abaxial surface)	350–600	Beakbane & Mujamder, 1975
Barley		
(abaxial surface, flag leaf)	39–96	Miskin & Rasmusson, 1970
Bromus inermis		
(abaxial surface, leaf 3)	25.2–36.7	Tan & Dunn, 1975
Panicum antidotale	78–165	Dobrenz et al., 1969
Lolium perenne		
(adaxial surface)	48–129	Wilson, 1971
Soybean		
(abaxial surface)	242–385	Ciha & Brun, 1975
Vicia faba		
(abaxial surface)	50–87	Singh, Singh & Singh, 1982
Stomatal length (μm)		
Barley		
(abaxial surface, flag leaf)	40–56	Miskin & Rasmusson, 1970
Bromus inermis		
(abaxial surface, leaf 3)	56.6–65.9	Tan & Dunn, 1975
Soybean	19.2–21.7	Ciha & Brun, 1975
Lolium perenne		
(adaxial surface)	21.1–36.1	Wilson, 1971
Vicia faba	24.9–26.6	Singh, Singh & Singh, 1982
Conductance (mol m^{-2} s^{-1})[a]		
Bromus inermis		
(mean daytime)	0.044–0.100	Lea, Dunn & Koch, 1977
Vicia faba		
(abaxial surface)	0.050–0.097	Nerkar, Wilson & Lawes, 1981
Indian Mustard		
(total conductance at bud initiation)	0.16–0.38	Singh, Singh & Singh, 1982

[a] Calculated from the original units, assuming that 1 mm s^{-1} = 0.04 mol m^{-2} s^{-1}.

determinant of stomatal conductance is aperture, which varies extremely with environment and development. In spite of this variability, genetic differences in conductance have been demonstrated in many species (e.g. Miskin, Rasmusson, and Moss, 1972; Shimshi and Ephrat, 1975), though in many cases the ranking of cultivars for conductance has been found to change with time of day (e.g. Lea, Dunn, and Koch, 1977; Roark and Quisenberry, 1977), stage of development (e.g. Jones, 1977b; Singh, Singh, and Singh, 1982), or with year (Gaskell and Pierce, 1983).

The inheritance of stomatal conductance is not well understood, except for certain single-gene mutants (see below). One study, based on crosses between two contrasting cotton lines, revealed both additive and dominance components of variance, with a low conductance being dominant (Roark and Quisenberry, 1977). Unfortunately, in this study, effects on conductance were not clearly separable from those on leaf water potential.

Response to stress. Although consistently high conductance (for high photosynthetic rate) or consistently low conductance (for water conservation) may be advantageous in some cases, it has been pointed out that breeding for drought tolerance would more often involve selection of genotypes having an appropriate stomatal response to stress (Jones, 1979). Genotypic differences in stomatal response to water stress have been demonstrated in several species, including sorghum (Henzell et al., 1976), wheat (Jones, 1977b; 1979; Quarrie and Jones, 1979), and millet (Henson et al., 1981). Unfortunately, it has not been possible in these studies to compare many genotypes, because of the large number of measurements required (at the very least, conductances are needed for both control and stress plants). In addition, it is necessary to measure plant water status in order to separate real differences in stomatal response at equal water potentials from differences arising from other factors (such as leaf area) that affect the rate of development of stress and hence affect stomatal closure. Some information on genetic variation in stomatal response has been obtained from measuring the time of decline of transpiration rate of excised leaves and the relative water content at which this occurs (see e.g. Quisenberry, Roark, and McMichael, 1982).

It is well known that the plant growth regulator abscisic acid (ABA) is involved in stomatal closure in response to stress (see Raschke, this volume), and there is strong evidence that differences in stomatal response to stress are at least partially determined by genetic differences in ABA accumulation (for a review, see Quarrie, 1983). For example, the millet cultivar Serere 39, which has a high capacity for ABA accumulation, was found to have a significantly greater stomatal sensitivity to water stress than did two low-ABA accumulators (Henson et al., 1981). Transgressive segregation in a cross between high- and low-ABA-accumulating lines of wheat indicated the involvement of more than one gene. Although evidence for maternal inheritance of the ability to accumulate ABA has been reported for maize (Larqué-Saavedra and Rodriguez, 1979), in wheat and millet reciprocal crosses between high- and low-ABA accumulators give F_1 offspring intermediate between the parents, suggesting predominantly nuclear genes (Quarrie and Henson, 1982).

Of particular interest to breeders are single-gene mutations affecting stomatal behavior. Three X-ray-induced "wilty" mutants, *flacca*, *sitiens*, and *notabilis*, are known in tomato (see Tal and Nevo, 1973). In these a deficiency in ABA synthesis leads to a failure of the stomata to close in response to water deficits. Genetic analysis of these mutants (Taylor and Tarr, 1984) indicates that *notabilis* involves a lesion on the ABA biosynthetic pathway at a point different from that involved in the other two mutants. A similar mutant, "droopy," is known in potato (Simmonds, 1965; Quarrie, 1982), and a recently described wilty mutant of pea may also involve the same lesion (Donkin, Wang, and Martin, 1983). In *Capsicum annuum*, the mutant *scabrous diminutive* also shows abnormal stomatal behavior, but here excessive opening appears to result from a high concentration of ions in the guard cells (Tal, Eshel, and Witztum, 1976).

Contribution of Stomatal Characters to Breeding

At this stage it is useful to consider the possible improvements in yield, or in other plant functions, that could be achieved by altered stomatal characteristics.

Photosynthesis. The main reason for attempting to increase photosynthetic rate, by breeding for stomatal characters, for example, is to improve crop yield. Unfortunately, although production of dry matter often limits crop yields, high photosynthetic rates per unit leaf area do not generally lead to high yield, because on a crop basis many other factors, such as carbohydrate partitioning and leaf-area development, are involved and may compensate for changes in assimilation per unit leaf area (see Jones, 1983). In fact, in wheat (Dunstone, Gifford, and Evans, 1973), as in other genera, assimilation rate per unit leaf area tends to be negatively related to grain yield.

It is not surprising, therefore, that yields are not usually closely related to stomatal characters (e.g. Ledent and Jouret, 1978). Those cases in which yield is positively correlated with stomatal characters such as stomatal length (e.g. Walton, 1974; Ceulemans, Impens, and Steenackers, 1984) may even prove coincidental. There is more reason to expect a close relationship between photosynthesis and stomatal characters, but again any association with morphology tends to be slight, and even conductance can be a poor indicator of photosynthetic capacity (e.g. Miskin, Rasmusson, and Moss, 1972; Dwelle, Hurley, and Pavek, 1983; Gaskell and Pierce, 1983). Although discouraging, these findings do not rule out the possibility that stomatal characters may provide a means of improving productivity; but they do indicate the need to avoid compensating changes by selecting for a range of characters together.

It is appropriate to consider what improvements are theoretically possible. The main considerations are the maximum capacity of the photosynthetic system when CO_2 supply is not limiting (i.e. when stomatal conductance is infinite) and the extent to which stomata in fact do limit photosynthetic rate (see Farquhar and Sharkey, 1982; Jones, 1985). As Farquhar and Sharkey (1982) and Jones (1985) point out, in many species the maximum net photosynthetic rate with nonlimiting CO_2 is only 10 to 20 percent higher than the rate with natural stomatal conductances and ambient CO_2 levels, and in C_4 species this difference tends to be even smaller. Although the increase in net photosynthesis that results from eliminating the stomatal restriction indicates the upper limit to improving assimilation through stomatal changes, the intracellular photosynthetic system is known to adapt very effectively to changed conditions (see Jones, 1983), thus tending to maintain a balance between stomatal and intracellular processes. It is difficult, however, to predict the magnitude of these compensatory changes, which may allow much larger increases in photosynthesis when stomatal conductance is increased.

Drought tolerance. The ability of plants to survive, grow, or yield during periods of drought is determined, at least partially, by how the stomata act to control water loss from the plant canopy and by stomatal effects on water-use efficiency. Stomatal control over evaporation depends on the relative sizes of the boundary-layer conductance (g_a) and the physiological conductance (g_l), largely determined by the stomatal behavior, and is discussed in detail by McNaughton and Jarvis (1983). For a single leaf, g_l tends to be much smaller than g_a, so that the total water-vapor conductance and the evaporation rate are determined largely by changes in stomatal conductance. For crop canopies, the sensitivity of evaporation rate to changing leaf conductance depends on the boundary-layer conductance, as is illustrated for typical conditions in Figure 19.1: in crops with aerodynamically smooth canopies and low g_a (such as grass and dwarf cereals) very large changes in stomatal conductance have only a small effect on transpiration (E), whereas in crops with aerodynamically rough canopies and high g_a (such as orchards or forests), stomata exert more control over E.

A further factor that can be important where large areas of one crop are grown is the feedback effect of any change in E on the humidity deficit of the ambient air. This tends to minimize any change in E, such as one resulting from stomatal closure. In addition, other negative feedbacks between evaporative demand and evaporation tend to result in fairly constant evaporation rates for a range of forest types in different environments (see Roberts, 1983).

It can be concluded, therefore, that breeding for stomatal changes to minimize water loss is likely to be most significant in tall or well-spaced crops, and in crops grown in small areas. Because of the effects described above, major effects on water conservation or water-use efficiency are likely only where breeders can alter the timing of complete stomatal closure, and not where small quantitative changes in conductance are achieved. In this context, it is also worth noting that changes in crop leaf area, including those resulting from changes in length of growing season, are probably more important to total water use over the life of a crop than are changes in stomatal conductance (see Jones, 1983). Further discussion of stomatal effects on water-use efficiency may be found elsewhere (Jones, 1976).

The identification of varieties with a high capacity for accumulating ABA in response to drought has been suggested as an indirect method for detecting those genotypes that have stomata responsive to water stress and that are therefore likely to tolerate drought (Quarrie, 1983). Although varietal differences in capacity for ABA accumulation have been reported for several species, relationships with drought tolerance have been variable (see Quarrie, 1983). In

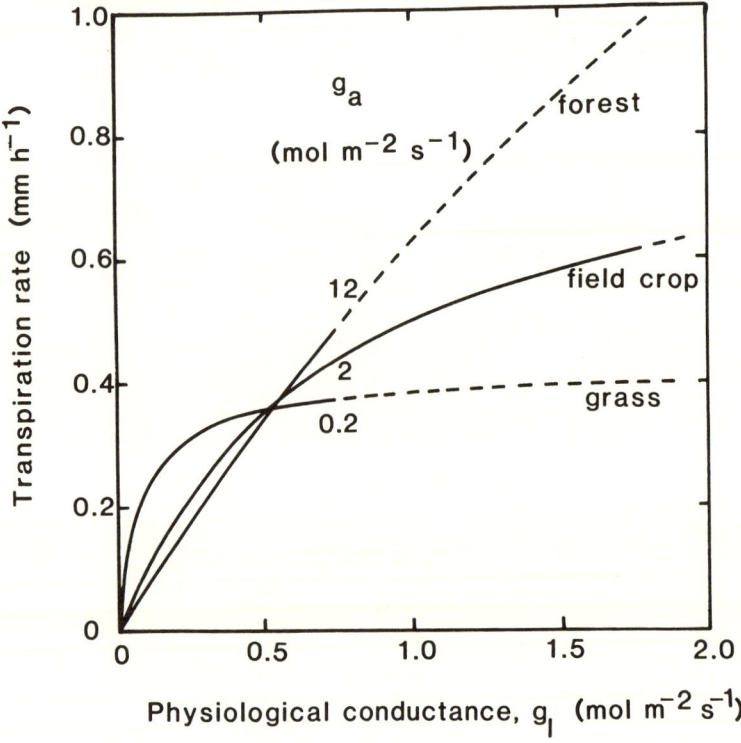

Fig. 19.1. Dependence of transpiration rate on canopy boundary-layer conductance (g_a) and physiological conductance (g_l, calculated using the combination equation for evaporation: see Jones, 1983) for various types of canopies. Conditions: temperature, 15° C; humidity deficit, 1kPa; available energy, 400 W m^{-2}. (Redrawn from Jarvis, 1981.)

order to reduce the background genetic variation that exists in such varietal comparisons, Innes, Blackwell, and Quarrie (1984) compared the field performance of high- and low-ABA-accumulating selections from a single spring wheat cross. In this experiment, the high-ABA selections tended to have a greater grain yield than the low-ABA selections. Furthermore, the high-ABA lines had a better water-use efficiency, which was probably due to a combination of their slightly lower leaf conductances and smaller leaves.

A particularly interesting method for the selection of genotypes having a high water-use efficiency has recently been proposed (Farquhar, O'Leary, and Berry, 1982; Farquhar and Richards, 1984). This method depends on the observations that the naturally occurring carbon isotope ^{13}C is discriminated against in favor of ^{12}C during photosynthesis and that the level of this discrimination decreases as the intracellular partial pressure of CO_2 decreases. Among other factors, a decrease in stomatal conductance would have such an effect. Since water-use efficiency also tends to increase with decreasing intercellular CO_2, measurement of the carbon-isotope ratio should provide an indirect method for selecting for stomatal control of water-use efficiency. Recent data for wheat (Farquhar and Richards, 1984) and groundnut (Hubick, Farquhar, and Shorter, 1986) have confirmed that the technique can detect intraspecific differences in water-use efficiency, at least when plants are water-stressed.

Problems of Screening for Stomatal Characters

Many of the difficulties inherent in screening for stomatal characters have been outlined above, and have been reviewed in detail by Jones (1979). Here I will concentrate on the requirement for large numbers of samples to distinguish genotypes with differing stomatal morphology or conductance. This statistical requirement severely limits the number of lines that can be compared (see Jones, 1979; Carlson et al., 1981). An estimate of the sample numbers required can be obtained from statistical theory if the relevant components of variance are known.

In blue panic grass, for example, the variance components (here reported as coefficients of variation to permit comparisons with other data) were found to be 11.1 percent for variation between culms, 9.2 percent for within-culm variation, and 6.8 percent for variation between microscope fields containing about 73 stomata each (Dobrenz et al., 1969). Similar results for alfalfa were 9.9 percent between stems and 8.9 percent between microscope fields containing approximately 23 stomata (Carlson et al., 1981). From this information the experimenter can optimize resources by balancing the number of stems sampled and the number of replicate counts per stem. In general, however, it is best to increase the number of stems or plants at the expense of replicate microscope fields. In a breeding program it is also necessary to balance the risks of discarding genetically good lines against those of retaining genetically bad lines (see Keuls and Sieben, 1955). Thus, for the alfalfa data, when counting stomata in 4 microscope fields per stem, one would have to sample 97 stems of each genotype in order to detect those genotypes differing by 5 percent (equivalent to counting about 9,000 stomata of each genotype); but when counting stomata in 40 microscope fields per stem, one would still have to sample 81 stems (about 75,000 stomata)! To detect even a 20 percent difference would require sampling 7 stems of each genotype.

These results seem typical of many species and are comparable to the variability found for stomatal conductance. Jones and Cumming (1984), for example, reported a coefficient of variation between different apple trees of 8.9 percent, and between similar leaves on the same tree the figure was 12 percent. Hence, the number of samples required for conductance screening would be similar to that required for morphological characters. This means that neither stomatal morphology nor conductance is likely to provide techniques for rapidly screening the large number of lines found in the early generations of a breeding program. The carbon-isotope method, however, does hold some promise for screening, because it integrates stomatal and photosynthetic behavior over the life of the plant.

Conclusions

Although in theory stomatal characters are of major importance in controlling many plant processes, including photosynthesis and water use, and in spite of the existence of what should be adequate genetic variability for such characters, there has yet been little progress toward their successful incorporation into improved varieties of crop plants. The reasons why include the plant's innate capacity for compensation, which tends to negate the effect of alteration in one character by changes in another, and the fact that screening for stomatal characters is slow and tedious, involving many measurements. Perhaps even more important has been the lack of a complete understanding of the ideal combination of characters actually required in any situation.

For drought tolerance, for example, it is difficult to determine the optimal balance between characters that conserve water (low conductance) and those needed to maximize production (high conductance). This balance depends in a complex manner on the agricultural situation and on the climate (in particular, on the probability of future rainfall; see Jones, 1981). Although it is often argued that stomata that close in response to stress would be ideal in many cases, some researchers have found that an insensitive stomatal apparatus permitting continued assimilation during drought may be preferable (e.g. Henzell et al., 1976).

In spite of such difficulties, there is scope for improving crop-plant performance by appropriate manipulation of stomatal characters; but these improvements are only likely to be achieved when the role of stomata in the adaptation of plants to natural environments is better understood. The use of mutants and isogenic lines with differing stomatal characters provides a particularly powerful technique for elucidating the physiological role of the stomata.

REFERENCES

Beakbane, A. B., and P. K. Mujamder. 1975. A relationship between stomatal density and growth potential in apple rootstocks. *Journal of Horticultural Science*, 50: 285–89.

Butler, L. K., and T. W. Tibbitts. 1979. Stomatal mechanisms determining genetic resistance to ozone in *Phaseolus vulgaris* L. *Journal of the American Society for Horticultural Science*, 104: 213–16.

Breeding for Stomatal Characters 441

Carlson, J. R., R. L. Ditterline, J. M. Martin, and R. E. Lund. 1981. Sampling stomatal density in alfalfa. *Crop Science*, 21: 467–69.
Ceulemans, R., I. Impens, and V. Steenackers. 1984. Stomatal and anatomical leaf characteristics of 10 *Populus* clones. *Canadian Journal of Botany*, 62: 513–18.
Chandler, C. K., and P. M. Lyrene. 1982. Relationship between guard cell length and ploidy in *Vaccinium*. *HortScience*, 17: 53–54.
Ciha, A. J., and W. A. Brun. 1975. Stomatal size and frequency in soybeans. *Crop Science*, 15: 309–13.
Dobrenz, A. K., L. N. Wright, A. B. Humphrey, M. A. Massengale, and W. R. Kneebone. 1969. Stomate density and its relationship to water-use efficiency of blue panic grass (*Panicum antidotale* Retz.). *Crop Science*, 9: 354–57.
Donkin, M.E., T. L. Wang, and E. S. Martin. 1983. An investigation into the stomatal behavior of a wilty mutant of *Pisum sativum*. *Journal of Experimental Botany*, 34: 825–34.
Dunstone, R. L., R. M. Gifford, and L. T. Evans. 1973. Photosynthetic characteristics of modern and primitive wheat species in relation to ontogeny and adaptation to light. *Australian Journal of Biological Sciences*, 26: 295–307.
Dwelle, R. B., P. J. Hurley, and J. J. Pavek. 1983. Photosynthesis and stomatal conductance of potato clones (*Solanum tuberosum* L.). *Plant Physiology*, 72: 172–76.
Dylenok, L. A., L. V. Khotyleva, and A. P. Yatsevich. 1981. (Use of monosomic lines of spring wheat for studying the genetic control of stoma frequency and size in the flag leaf.) *Doklady Akademii nauk BSSR*, 25: 753–55. [In Russian.]
Farquhar, G. D., M. H. O'Leary, and J. A. Berry. 1982. On the relationship between isotope discrimination and the intercellular carbon dioxide concentration in leaves. *Australian Journal of Plant Physiology*, 9: 121–37.
Farquhar, G. D., and R. A. Richards. 1984. Isotopic composition of plant carbon correlates with water-use efficiency of wheat genotypes. *Australian Journal of Plant Physiology*, 11: 539–52.
Farquhar, G. D., and T. D. Sharkey. 1982. Stomatal conductance and photosynthesis. *Annual Review of Plant Physiology*, 33: 317–45.
Gaskell, M. L., and R. B. Pierce. 1983. Stomatal frequency and stomatal resistance of maize hybrids differing in photosynthetic capability. *Crop Science*, 23: 176–77.
Gay, A. P., and R. G. Hurd. 1975. The influence of light on stomatal density in the tomato. *New Phytologist*, 75: 37–46.
Gill, S. S., and K. S. Nandpuri. 1978. The downy mildew incidence as affected by a number of stomata in muskmelon cultivars. *Science and Culture*, 44: 372–73.
Heichel, G. H. 1971. Stomatal movements, frequencies, and resistances in two maize varieties differing in photosynthetic capacity. *Journal of Experimental Botany*, 22: 644–49.
Henson, I. E., V. Mahalaskshmi, F. R. Bidinger, and G. Alargarswamy. 1981. Stomatal responses of pearl millet (*Pennisetum americanum* [L.] Leeke) genotypes, in relation to abscisic acid and water stress. *Journal of Experimental Botany*, 32: 1211–21.
Henzell, R. G., K. J. McCree, C. H. M. van Bavel, and K. F. Schertz. 1976. Sorghum genotype variation in stomatal sensitivity to leaf water deficit. *Crop Science*, 16: 660–62.
Hubick, K. T., G. D. Farquhar, and R. Shorter, 1986. Correlation between water-use efficiency and carbon isotope discrimination in diverse peanut (*Arachis*) germplasm. *Australian Journal of Plant Physiology*, in press.

Innes, P., R. D. Blackwell, and S. A. Quarrie. 1984. Some effects of genetic variation in drought-induced abscisic acid accumulation on the yield and water use of spring wheat. *Journal of Agricultural Science*, 102: 341–51.

Jarvis, P. G. 1981. Stomatal conductance, gaseous exchange and transpiration. In J. Grace, E. D. Ford, and P. G. Jarvis, eds., *Plants and their atmospheric environment*: 175–204. Oxford: Blackwell.

Jones, H. G. 1976. Crop characteristics and the ratio between assimilation and transpiration. *Journal of Applied Ecology*, 13: 605–22.

———. 1977a. Transpiration in barley lines with differing stomatal frequencies. *Journal of Experimental Botany*, 28: 162–68.

———. 1977b. Aspects of the water relations of spring wheat (*Triticum aestivum* L.) in response to induced drought. *Journal of Agricultural Science*, 88: 267–82.

———. 1979. Stomatal behaviour and breeding for drought resistance. In H. Mussell and R. C. Staples, eds., *Stress physiology of crop plants*: 408–28. New York: Wiley.

———. 1981. The use of stochastic modelling to study the influence of stomatal behaviour on yield-climate relationships. In D. A. Charles-Edwards and D. A. Rose, eds., *Quantitative aspects of plant physiology*: 231–44. London: Academic.

———. 1983. *Plants and microclimate: A quantitative approach to environmental plant physiology*. Cambridge: Cambridge University Press.

———. 1985. Partitioning stomatal and non-stomatal limitations to photosynthesis. *Plant, Cell and Environment*, 8: 95–104.

Jones, H. G., and I. G. Cumming. 1984. Variation of leaf conductance and leaf water potential in apple orchards. *Journal of Horticultural Science*, 59: 329–36.

Keuls, M., and J. W. Sieben. 1955. Two statistical problems in plant selection. *Euphytica*, 4: 34–44.

Larqué-Saavedra, A., and G. M. T. Rodriguez. 1979. Maternal inheritance of abscisic acid (ABA) in *Zea mays* L. In *Abstracts of the 10th International Conference on Plant Growth Substances*: 23. Madison, Wis.

Lea, H. Z., G. M. Dunn, and D. W. Koch. 1977. Stomatal diffusion resistance in three ploidy levels of smooth bromegrass. *Crop Science*, 17: 91–93.

Ledent, J. F., and M. F. Jouret. 1978. Relationship between stomatal frequencies, yield components and morphological characters in collections of winter wheat cultivars. *Biologia plantarum*, 20: 287–92.

Liang, G. H., A. D. Dayton, C. C. Chu, and A. J. Casady. 1975. Heritability of stomatal density and distribution on leaves of grain sorghum. *Crop Science*, 15: 567–70.

McNaughton, K. G., and P. G. Jarvis. 1983. Predicting effects of vegetation changes on transpiration and evaporation. In T. T. Kozlowski, ed., *Water deficits and plant growth* 7: 1–47. New York: Academic.

Meidner, H., and T. A. Mansfield. 1968. *Physiology of stomata*. London: McGraw-Hill.

Miskin, K. E., and D. C. Rasmusson. 1970. Frequency and distribution of stomata in barley. *Crop Science*, 10: 575–78.

Miskin, K. E., D. C. Rasmusson, and D. N. Moss. 1972. Inheritance and physiological effects of stomatal frequency in barley. *Crop Science*, 12: 780–83.

Nerkar, Y. S., D. Wilson, and D. A. Lawes. 1981. Genetic variation in stomatal characteristics and behaviour, water use and growth in five *Vicia faba* L. genotypes under contrasting soil moisture regimes. *Euphytica*, 30: 335–45.

Pallardy, S. G., and T. T. Kozlowski. 1979. Frequency and length of stomata of 21 *Populus* clones. *Canadian Journal of Botany*, 57: 2519–23.

Quarrie, S. A. 1982. Droopy: A wilty mutant of potato deficient in abscisic acid. *Plant, Cell and Environment*, 5: 23–26.

_____. 1983. Genetic differences in abscisic acid physiology and their potential uses in agriculture. In F. T. Addicott, ed., *Abscisic acid*: 365–419. New York: Praeger.
Quarrie, S. A., and I. E. Henson. 1982. Biparental inheritance of drought-induced accumulation of abscisic acid in wheat and pearl millet. *Annals of Botany*, 49: 265–68.
Quarrie, S. A., and H. G. Jones. 1979. Genotypic variation in leaf water potential, stomatal conductance and abscisic acid concentration in spring wheat subjected to artificial drought stress. *Annals of Botany*, 44: 323–32.
Quisenberry, J. E., B. Roark, and B. L. McMichael. 1982. Use of transpiration decline curves to identify drought-tolerant cotton germplasm. *Crop Science*, 22: 918–22.
Rajendra, B. R., K. A. Mujeeb, and L. S. Bates. 1978. Relationships between 2x *Hordeum* sp., 2x *Secale* sp. and 2x, 4x, 6x *Triticum* spp. for stomatal frequency, size and distribution. *Environmental and Experimental Botany*, 18: 33–37.
Rawson, H. M., G. A. Constable, and G. N. Howe. 1980. Carbon production of sunflower cultivars in field and controlled environments, II. Leaf growth. *Australian Journal of Plant Physiology*, 7: 555–73.
Roark, B., and J. E. Quisenberry. 1977. Environmental and genetic components of stomatal behavior in two genotypes of upland cotton. *Plant Physiology*, 59: 354–56.
Roberts, J. 1983. Forest transpiration: A conservative hydrological process? *Journal of Hydrology*, 66: 133–41.
Setter, T. L., L. E. Schrader, and E. T. Bingham. 1978. Carbon dioxide exchange rates, transpiration and leaf characters in genetically equivalent ploidy levels of alfalfa. *Crop Science*, 18: 327–32.
Shimshi, D., and J. Ephrat. 1975. Stomatal behaviour of wheat cultivars in relation to their transpiration, photosynthesis and yield. *Agronomy Journal*, 67: 326–31.
Simmonds, N. W. 1965. Mutant expression in diploid potatoes. *Heredity*, 20: 65–72.
Singh, D. P., P. Singh, and M. Singh. 1982. Screening of genotypes of *Brassica juncea* L. for leaf conductance under field conditions. *Journal of Experimental Botany*, 33: 381–87.
Smith, W. H., and L. S. Dochinger. 1976. Capability of metropolitan trees to reduce atmospheric contaminants. In *Better trees for metropolitan landscapes: Proceedings of a symposium at the U.S. National Arboretum*: 49–59. Upper Darby, Pa.: USDA Forest Service.
Suh, H. W., A. D. Dayton, A. J. Casady, and G. H. Liang. 1976. Diallel cross analysis of stomatal density and leaf-blade area in grain sorghum, *Sorghum bicolor*. *Canadian Journal of Genetics and Cytology*, 18: 679–86.
Tal, M., A. Eshel, and A. Witztum. 1976. Abnormal stomatal behaviour and ion imbalance in *Capsicum scabrous diminutive*. *Journal of Experimental Botany*, 27: 953–60.
Tal, M., and Y. Nevo. 1973. Abnormal stomatal behaviour and hormonal imbalance in three wilty mutants of tomato. *Biochemical Genetics*, 8: 291–300.
Tan, G.-Y., and G. M. Dunn. 1975. Stomatal length, frequency, and distribution in *Bromus inermis* Leyss. *Crop Science*, 15: 283–86.
_____. 1976. Genetic variation in stomatal length and frequency and other characteristics in *Bromus inermis* Leyss. *Crop Science*, 16: 550–53.
Taylor, I. B., and A. R. Tarr. 1984. Phenotypic interactions between abscisic acid deficient tomato mutants. *Theoretical and Applied Genetics*, 68: 115–19.
Walton, P. D. 1974. The genetics of stomatal length and frequency in clones of *Bromus inermis* in the relationships between these fruits and yield. *Canadian Journal of Plant Science*, 54: 749–54.
Wilson, D. 1971. Selection response of stomatal length and frequency, epidermal ridging, and other leaf characteristics in *Lolium perenne* L. "Grasslands Ruanui." *New Zealand Journal of Agricultural Research*, 14: 761–71.

20

Calculations Related to Gas Exchange

J. Timothy Ball

This paper considers leaf gas-exchange experiments and the calculation of results to obtain fluxes of carbon dioxide and water vapor, stomatal conductance to these gases, and the concentrations of CO_2 and H_2O in the intercellular air spaces and at the leaf surface. The intercellular CO_2 concentration (c_i) is considered a good estimate of the CO_2 available to the photosynthetic biochemistry, and it is presumed to play a role in feedback regulation of stomatal conductance (see Chap. 10). Diffusion through the stomates mediates exchange between the intercellular spaces and the air at the leaf surface (at the interior of the aerodynamic boundary layer). Since diffusion through, and mixing within, the boundary layer may be quite variable—especially under natural conditions—it is important to distinguish boundary layer processes from physiologically regulated diffusion through the stomata. This is most easily accomplished by considering the responses of stomata to the calculated concentration of CO_2 (c_s) and water vapor (w_s) at the leaf surface (see Ball and Berry, 1982). In a well-stirred gas-exchange cuvette there may be only a small difference between the gas concentrations in the bulk air and the concentrations at the leaf surface, but these differences may be very significant in considering the physiological feedback mechanisms that regulate responses to air humidity or CO_2. Furthermore, it may be argued that an understanding of stomatal responses to the local environment of the leaf surface should be general, not specific to the particular boundary-layer characteristics of the measuring system. Having information about conditions at the leaf surface should provide a good starting point for understanding the interactions between stomata and the boundary layer. Therefore, the procedures and problems of calculating the CO_2 and water vapor concentration at the leaf surface are given major consideration in this paper.

The treatment of diffusion in this paper takes into account the fact, pointed out by Jarman (1974), that diffusion of CO_2 and water through the stomata and

---

The author wishes to thank Graham Farquhar and Chris Field, who discussed and reviewed many of the equations; Joe Berry, Robert Guy, and Ian Woodrow for helpful discussions regarding this paper; and Elizabeth and Hannah C. Ball for their support and patience. Financial assistance provided by the Carnegie Institution of Washington and the McKnight Foundation is appreciated. This paper is CIW-DPB Publication 904.

through the boundary layer is more complex than a simple model based upon Fick's law would indicate. In essence, the commonly used treatment of diffusion, based on Fick's law, neglects the fact that the flux of water vapor out of a leaf tends to sweep CO_2 and air with it. I have drawn upon the expressions of von Caemmerer and Farquhar (1981) for diffusion in a ternary system. This material is discussed in more detail because of its important implications for calculation of the CO_2 concentration in the intercellular air spaces of the leaf and the surface concentrations of water vapor and CO_2.

It is important to realize that there are uncertainties regarding the validity both of some of the assumptions made and of some of the constants used in these calculations. I have attempted to point out assumptions that are made and approaches that may be taken to minimize potential errors. A good example of where assumptions are important is in consideration of the transport of material across the leaf aerodynamic boundary layer: usually one applies constants developed by engineers from analysis of systems that tend to be much simpler than leaves enclosed in gas-exchange cuvettes. Virtually all of the uncertainties about these gas-exchange calculations are present and compound each other in the estimates of the mole fraction CO_2 in leaf intercellular air spaces (c_i). Experiments in which c_i was measured in a manner not dependent on the calculations described here (Sharkey et al., 1982; Mott and O'Leary, 1984) showed reasonable agreement between the estimates of c_i. It is safe to say that the entire set of calculations presented here provides "good engineering results," that is, they are internally consistent and they can be used to formulate predictive relationships that do not violate our understanding of the ways leaves function.

Good gas-exchange measurements require a proper experimental setup, including a cuvette for enclosing leaves, an infrared gas analyzer (IRGA) for CO_2 exchange measurements, a dew point– or relative humidity–measuring device, and ancillary equipment for mixing, humidifying, and measuring the flow of gases. Accurate pressure and temperature measurements are also important. It is beyond the scope of this paper to consider the details of design and construction of such a system. This kind of information is available in a manual of methods (Sestak, Catsky, and Jarvis, 1971), in original articles such as Bloom et al. (1980), Field, Berry, and Mooney (1982), Björkman and Holmgren (1963), Parkinson, Day, and Leach (1980), Field and Mooney (1987), and in papers cited therein.

The equations presented here pertain to "open flow" gas-exchange systems, whether they measure photosynthesis by differential or by compensating means (Fig. 20.1). The calculations do not apply to "transient" gas-exchange systems (often called "closed" systems; see Jarvis et al., 1971). In transient systems a leaf is exposed to a fixed volume of gas for a short period of time, and changes in gas concentrations within that volume per unit time are determined. Open-flow systems are generally able to impose defined environmental conditions on leaves. They are, thus, well suited for the intensive steady-state

Calculations Related to Gas Exchange

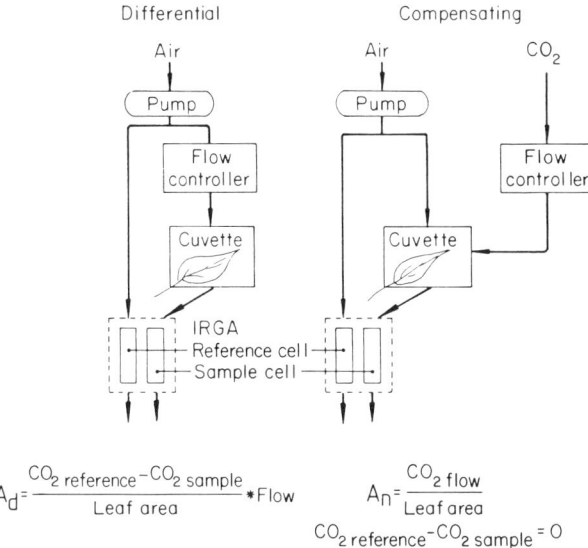

Fig. 20.1. A simplified schematic diagram illustrating the difference between the two basic types of open-flow gas-exchange systems and the difference in the ways that net photosynthetic rate is calculated with each. In differential systems the removal of CO_2 from the air stream by the leaf results in a difference in CO_2 concentrations between the two cells of the infrared gas analyzer (IRGA). In compensating systems the CO_2 uptake by the leaf is compensated by the addition of CO_2 to the cuvette air stream, so that the CO_2 concentration is the same in the two IRGA cells. As is explained in the text, the total flow through a system increases in the cuvette so that the net CO_2 assimilation rate from a compensating system is always the sum of the compensating term ($CO_{2\ flow}$/leaf area) and a differential term [($CO_{2\ reference}$ − $CO_{2\ sample}$) · flow/leaf area]. The differential term can be kept at a value that is quite small relative to the compensating term.

and kinetic measurements typically needed in the study of stomatal responses and their relation to photosynthesis.

The physical differences between differential and compensating open-flow systems are small (Figs. 20.1 and 20.2). Both types pass a known flow rate of air containing a known amount of CO_2 and water vapor across a leaf. After the air passes through the cuvette, its CO_2 and water content are measured. In a differential system, the flow (mass per unit time) of CO_2 into the chamber minus the flow of CO_2 out of the chamber is divided by the projected leaf area to yield the photosynthetic rate per unit area (Fig. 20.1). A similar calculation is made with respect to water vapor for determination of the transpiration rate per unit area. In a compensating or null-balance system, a small, variable amount of CO_2 is metered into the air stream entering the cuvette. When this flow of compensating CO_2 matches the rate of CO_2 uptake by the leaf, the IRGA indicates that there is no difference in CO_2 concentration between the air entering and that leaving the cuvette. With a small adjustment for increased flow through the system (due to the water vapor transpired by the leaf and the carrier gas added with the compensating CO_2), the photosynthetic rate is the measured flow of compensating CO_2 divided by the leaf area (Fig. 20.1). In practice, the compensating flow rarely exactly matches the rate of CO_2 uptake by the leaf, so the photosynthetic rate is usually calculated from a null-balance component plus a differential component. Among the special types of compensating systems are closed recirculating loops to which CO_2 is added and from which water is removed.

Compensating systems have three advantages over differential systems. First, null-balance measurements depend to a much smaller extent on calibration and stability of the IRGA sensitivity. Such minimal dependence is helpful when analyzer sensitivity is prone to drift and when calibration is difficult, as in the field. Second, because the flow rates can be varied widely while a particular ambient CO_2 is maintained, it is possible to use compensating systems without humidification of the ingoing air stream. This is very helpful in the field (Field, Berry, and Mooney, 1982). Third, and of particular interest to stomatal physiologists, a compensating system eliminates interaction between control of ambient CO_2 and control of humidity in the chamber. For example, perhaps the easiest way to fine-tune humidity in a cuvette is with small changes in flow rate. When assimilation rate is compensated by addition of CO_2, the humidity can be changed by changing the flow rate, with almost no effect on the ambient mole fraction of CO_2 in the cuvette.

The equations that are used to calculate transpiration rate (Eq. 12) and photosynthetic rate (Eq. 20) are mass balance equations for water and carbon across the leaf cuvette. To calculate photosynthetic rates from measurements with a differential system, one need know only the flow rate and mole fraction CO_2 of the gas stream at one point before the chamber and one point after the chamber. I show how to calculate the mole fraction of CO_2 and the flow rate at each point in the system because one may occasionally want to know these values at several points in a system. A good way to speed the development of a set of equations for gas-exchange analysis, avoiding errors in the process, is to write out the equations accounting for mass balance of all gases at each point in the gas-exchange system. A table of sample data and calculated results is included in an appendix to assist in developing and checking the output of a gas-exchange program. These data were obtained on a CO_2-compensating openflow system similar to that described by Field, Berry, and Mooney (1982).

The units used here are those recommended by Cowan (1977), and the system of symbols (including the use of lower case symbols for most scalar quantities) is based on those of Cowan (1977) and von Caemmerer and Farquhar (1981). Basic units (mole, meter, etc.) are scaled with exponents rather than with prefixes (milli-, micro-, etc.). This eliminates the need to reconcile the scaling units during arithmetic operations and allows a computer or calculator to handle scaling automatically with minimal confusion during programming. It is simple to convert to commonly reported units scaled with prefixes after all calculations have been completed.

Preliminaries and Useful Subroutines

Pressure. Total pressure (P, in units of pascals) is important in the interconversion of moles and partial pressure. To convert from some commonly reported units of pressure to pascals, one multiplies as follows: bars $\times$ 10^5; atmospheres $\times$ 101325; inches of mercury $\times$ 3386.386; millimeters of mercury $\times$ 133.322.

Thermocouples. The most common device for measuring temperature in gas-exchange systems is the copper-constantan (T-type) thermocouple. The following polynomial fits empirical data (U.S. National Bureau of Standards, NBS Monograph 125) for T-type thermocouples with reference junction at 0° C, to within 0.1° C over the range of most biological interests:

$$°C = \{[(2.84041 \cdot 10^7 \text{ V}) - 7.1535 \cdot 10^5] \text{ V} + 2.59195 \cdot 10^4\} \text{ V}, \quad (1)$$

where V = volts.

Not all commercially available copper and constantan wire meet these standards. Information useful in checking the calibration of thermocouples is available in the booklet *Application note 290: Practical temperature measurements* (Hewlett-Packard Corporation, 1980). This booklet also contains polynomial coefficients for voltage-to-temperature conversion for other thermocouple types, formulas, constants, circuit diagrams, references to other material, and a good discussion of thermometry, much of which is useful information for plant physiologists. It is particularly important that temperature-measuring devices within a gas-exchange system give internally consistent results.

Saturation mole fraction water vapor. The saturation mole fraction water vapor at a particular temperature and total pressure (w_{sat}, mol mol^{-1}) must be found several times in the course of gas-exchange calculations. The World Meteorological Organization adopted the Goff-Gratch equation as the standard fit of saturation vapor pressure as a function of temperature. This rather cumbersome equation is the source of the temperature-vapor pressure table in the Smithsonian Meteorological Tables (List, 1957). The equation as written here yields vapor pressure (v_{sat}) in pascals:

$$\eta = -7.9298 \left(\frac{T_s}{T_K} - 1\right) + 5.02808 \cdot \log_{10}\left(\frac{T_s}{T_K}\right) -$$

$$1.3816 \cdot 10^{-7} \left\{-1 + 10^{\left[11.344\left(1 - \frac{T_K}{T_s}\right)\right]}\right\} +$$

$$8.1328 \cdot 10^{-3} \left\{-1 + 10^{\left[-3.49149\left(\frac{T_s}{T_K} - 1\right)\right]}\right\} + \log_{10}(1013.246) + 2$$

$$v_{sat} = 10^\eta, \quad (2)$$

where

T_K = temperature of interest, °K
T_s = the steam temperature at standard pressure, 373.16° K
η = $\log_{10}$ of v_{sat}
v_{sat} = saturation vapor pressure at T_K.

A much simpler equation, which matches the values from the Goff-Gratch equation to within 1.5 Pa over the physiologically interesting range 0–50° C, has been provided by Richards (1971):

$$v_{sat} = 101325 \cdot e^{\{[(-0.1299t - 0.6445)t - 1.976]t + 13.3185|t\,|\}} \tag{3}$$

where $t = 1 - (T_s/T_K)$. This accuracy is greater than that of humidity or temperature measurements in most gas-exchange systems. The saturation vapor pressure for a particular temperature from either Equation 2 or Equation 3 is converted to mole fraction water vapor (w_{sat}) by

$$w_{sat} = \frac{v_{sat}}{P}, \tag{4}$$

where

w_{sat} = saturation vapor concentration, mole fraction
P = total pressure.

If the pressure at the point in the gas-exchange system where v_{sat} is established exceeds atmospheric pressure, then the additional pressure must be included in the total pressure for this conversion.

Calculations of Flux Rates

Flow rates and gas concentrations before the cuvette. In many gas-exchange systems, air of known mole fraction CO_2 (c_r, mol mol^{-1}) is produced by mixing dry CO_2-free air with nitrogen that contains CO_2 at relatively high mole fraction (typically 10^{-2} to 10^{-1} mol mol^{-1}), using precision mixing pumps, flow meters, or mass flow controllers. By varying the ratio in which these gas streams are mixed, c_r is varied over a physiologically interesting range from 0 up to perhaps $2 \cdot 10^{-3}$ mol mol^{-1}. Alternative sources of air are gas cylinders of compressed or ambient air. Both alternatives have disadvantages. Using gas from cylinders restricts the range of CO_2 concentration at which experiments can be done; furthermore, because CO_2 concentration varies between air tanks, it is difficult to conduct a series of experiments at the same CO_2 concentration. Using atmospheric air that has not been scrubbed of CO_2 and H_2O is difficult, because the concentration of both gases tends to fluctuate over the course of a single experiment. Pearcy and Calkin (1983) used large plastic containers (waterbed bags) filled with ambient air as a stable source of air for gas-exchange measurements in the field.

A small portion of the mixed air (the amount need not be known in most configurations) is sent to the reference cell of an IRGA (Fig. 20.2). The

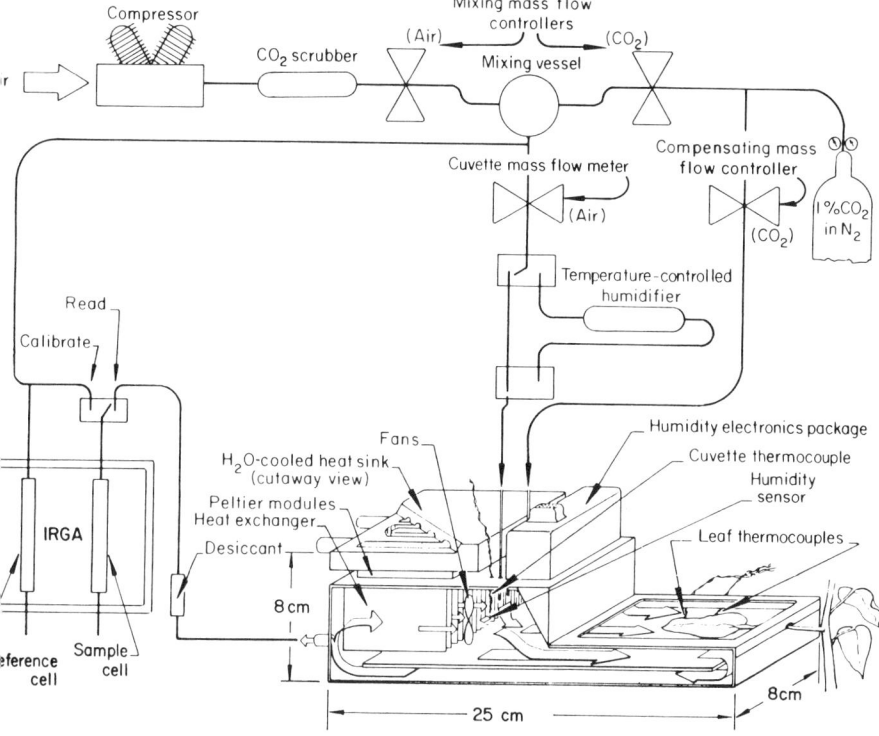

Fig. 20.2. A schematic diagram showing the major components of an open-flow gas-exchange system. The system is drawn as a CO_2-compensating system, but if the flow through the compensating mass flow controller is shut off, the system becomes a differential system. Note that a humidification subsystem can be optionally switched into the air-flow path leading to the cuvette. In this particular cuvette, air is circulated by fans so that each "parcel of air" passes over the leaf a number of times while in the chamber. The fans also insure that the air speed is sufficient to keep the aerodynamic boundary layer around the leaf small.

remaining flow of air, u_r (mol s^{-1}), is measured with a mass flow meter or a capillary/pressure-transducer combination. As mentioned, the gas stream is usually dry at this point, so that the mole fraction water vapor of the mixed air, w_r (mol mol^{-1}), equals zero. It should not be assumed that scrubbed air or air from gas cylinders is dry; this should be checked. If any water is present, it should be included in the mass-balance calculations.

In many systems the source gas is humidified to a known mole fraction water vapor (Fig. 20.2). Oversaturation followed by condensation to a controlled dew point is the usual, and best, way to produce a constant and defined vapor concentration in a stream of gas (Jarvis et al., 1971). Condensation leads to equilibrium; evaporation does not. In practice, the air stream headed toward the cuvette is passed through a small amount of water heated to slightly above the desired dew point, then passed through a condenser of uniform and carefully controlled temperature. The temperature of water at the point of condensation, within the condenser, is used in Equations 2 or 3, followed by Equation 4 to calculate the mole fraction water vapor in the air stream leaving the condenser (w_h). The pressure in many condensers tends to be above ambient atmospheric pressure. As mentioned earlier, the total pressure used in Equation 4 to obtain w_h should include any additional pressure. It is a good idea to calibrate the hygrometers in a system using the temperature measured in the system's condenser as a standard, so that humidity and temperature measurements are all internally consistent. In many systems, w_h is measured by a hygrometer, which may be calibrated independently, providing a check on the humidifier/condenser system.

The flow rate of air through a system is increased by the addition of water vapor in the humidifier. This new flow rate, u_h, is calculated from the general equation for changes in flow rate on addition or removal of component gases:

$$u_{out} = \frac{u_{in}(1 - x_{in})}{(1 - x_{out})}, \tag{5}$$

where x_{in} and x_{out} are the mole fractions of component gases being changed, and u_{in} and u_{out} are the flow rates in mol s^{-1}. In the present case, $u_h = u_{out}$, $u_r = u_{in}$, $x_{out} = w_h$, and $x_{in} = w_r$ (recall that w_r often equals zero). An alternative to calculating the flow correction after humidification is to measure the flow rate at that point (Fig. 20.2). If the flow is measured after humidification, then the flow that entered the humidifier (u_r) can be calculated by rearranging Equation 5. Measuring the flow rate after humidification requires that the flow-measuring device be heated to a temperature above the dew point of the air passing through it. It is important to realize that changing temperature may alter the calibration of many flow-measuring devices. If the flow meter measures volume per unit time, it is also important to take into account any change in pressure across the flow meter, because the calculations require accurate accounting of mass balances.

The mole fraction CO_2 in the air stream is altered by the addition or removal of water vapor at several points in any gas-exchange system. A general equation for handling changes in mole fraction of species x on addition or removal of species y is

$$x_{out} = \frac{x_{in}(1 - y_{out})}{1 - y_{in}}. \tag{6}$$

For the air stream headed toward the cuvette the mole fraction CO_2 after humidification (c_h) is calculated from Equation 6, using the mole fractions CO_2

for x and the mole fraction H_2O for y. Obviously, if there is no humidification step, then $u_h = u_r$, $c_h = c_r$, and $w_h = w_r$ (recall that w_r is usually zero). When an open-flow system is operating in differential mode, these most recently obtained values for flow rate, mole fraction CO_2, and mole fraction H_2O will be the values of these parameters entering the cuvette, so that $u_e = u_h$, $c_e = c_h$, and $w_e = w_h$.

In a system operating in CO_2 compensating mode, an additional small flow (u_n, mol s^{-1}) of dry, concentrated CO_2 (c_n), again usually between 10^{-2} and 10^{-1} mol mol^{-1} CO_2 in N_2, is added to the stream of air passing through the system (Fig. 20.2). The rate of this addition is varied so that the product $u_n c_n$ (= mol CO_2 s^{-1}) closely matches the rate of CO_2 uptake by the leaf. The compensating flow may be added with flow-metering devices such as a capillary/pressure-transducer combination or a mass flow controller. In the mass-balance scheme used here, one considers the compensating CO_2, c_n, as a species separate from the CO_2 in the main gas stream entering the chamber (c_e). This scheme makes the difference between compensating and differential systems clear, and it allows one to calculate the amount of photosynthesis accounted for by null-balance separately from that accounted for by differential measurement. Measuring the photosynthetic rate by two methods is a good way to check for proper functioning of a system. The compensating flow of CO_2 need not be added directly to the cuvette; rather, it may be added at any point after the reference gas (with mole fraction $CO_2 = c_r$) is sent to the IRGA, up to and including the leaf chamber (Fig. 20.2). If the compensating flow is added before the humidification step, it will cause a delay in the system response as the additional CO_2 passes through the volume of the humidifier, which can be disadvantageous in some experiments. The delay is a few tens of seconds when the volume of water in the humidifier is 30 ml and the flow of air is about 10^{-3} mol s^{-1}. One can speed changes in CO_2 concentrations across the condenser by slightly acidifying the water in the bubbler with phosphoric acid, which is nonvolatile. This keeps the $CO_2 \rightleftharpoons HCO_3^-$ equilibrium far in the direction of CO_2. If the compensating gas is added before humidification, the mole fraction CO_2 in the dry air stream (c_r) will be diluted by the added flow of compensating gas, and the total flow u_r will be increased by u_n. (Recall that the compensating CO_2 and the CO_2 in the original mixture are accounted separately here.) These corrections can be made by substituting c_r for c_h and u_r for u_h in Equations 8 and 9, below. Upon humidification the new flow rate and mole fraction CO_2, which will then enter the chamber, will be calculated with Equations 5 and 6, as discussed above.

In the most common system configurations, the compensating flow is added after the humidifier; this includes systems in which the nulling gas is added directly to the cuvette. If the CO_2 concentration in the compensating gas (c_n) is sufficiently high, the flow of this gas for null balance (u_n) will have only a very small impact on the overall flow rate (u_h) and on the mole fraction water vapor

(w_h). One of the advantages of a compensating system is that when c_n is large there is no significant interaction between the control of CO_2 and that of humidity. Nonetheless, correcting u_h, c_h, and w_h for the addition of u_n is simple, and should be done in order to keep the mass-balance calculations correct. To do this, one first corrects the mole fraction water vapor with

$$w_e = \frac{w_h u_h}{u_h + u_n}, \tag{7}$$

then the mole fraction CO_2 with

$$c_e = \frac{c_h u_h}{u_h + u_n}, \tag{8}$$

and, finally, the flow rate is corrected with

$$u_e = u_h + u_n. \tag{9}$$

These new mole fractions of water and CO_2, and the new flow rate, will be the values for these parameters at the entrance to the cuvette.

Transpiration rate. Water loss from a leaf is determined by the mass balance across the cuvette:

$$E \cdot L_a = u_o w_o - u_e w_e, \tag{10}$$

where
  E = transpiration rate per unit projected leaf area, mol m^{-2} s^{-1}
  L_a = projected leaf area, m^2
  u_o = air flow out of the cuvette, mol s^{-1}
  w_o = mole fraction water in the outgoing air.

w_o is often obtained from either relative humidity plus air temperature or the dew point of the outgoing air, using Equations 2 or 3, and Equation 4 to obtain w_{sat}. IRGAs set up to measure the partial pressure of water in a gas stream are available, so w_o can be determined in a manner similar to that for determination of mole fraction CO_2 in an airstream, described below. The flow rate of air changes across the cuvette. The small quantity of CO_2 taken out of the air stream by the leaf is essentially balanced by the efflux of O_2 but the efflux of water from a leaf is not balanced by an influx. The air flow out of the chamber is, therefore, greater than the air flow into the chamber by the amount of water vapor transpired:

$$u_o = u_e + (u_o w_o - u_e w_e) \tag{11a}$$

or, rearranging,

$$u_o = \frac{u_e (1 - w_e)}{(1 - w_o)}. \tag{11b}$$

Equations 10 and 11 are combined to calculate the transpiration rate per unit projected leaf area (mol m^{-2} s^{-1}):

$$E = \frac{u_e (w_o - w_e)}{L_a (1 - w_o)}. \tag{12}$$

The term $(1 - w_o)$ in the denominator of Equation 12, which adjusts the flow rate for the moles of water vapor added by the leaf, typically increases the calculated value of E by 3 to 4 percent over that calculated when the change in the flow rate of air across the cuvette is not taken into account (von Caemmerer and Farquhar, 1981).

Ambient mole fraction water vapor. Gas-exchange cuvettes in which an entire leaf can be placed are usually stirred by some type of fan so that the air passes over the leaf a number of times (Fig. 20.2). When chambers are well stirred and each "parcel of air" passes over the leaf several times, the mole fraction H$_2$O leaving the chamber (w_o) is taken as the ambient mole fraction water vapor (w_a):

$$w_a = w_o. \tag{13a}$$

Some cuvettes are designed to clamp onto leaves, so that the entire chamber is filled with only a portion of the leaf surface. Some chambers of this type allow air to pass over the leaf only once during its residence in the chamber. If the flow in a single-pass cuvette is laminar, then w_a is sometimes taken as

$$w_a = 0.5 (w_e + w_o). \tag{13b}$$

If the flow in such a chamber is turbulent, w_a is sometimes taken as

$$w_a = 0.33 w_e + 0.67 w_o. \tag{13c}$$

Determining whether the flow in any cuvette is laminar or turbulent can be a problem. Particularly if the cuvette volume is small, direct measurement of wind speed and its variation may be difficult when the cuvette is in its normal closed (leaf in place) configuration. Measurements of wind speed made when a chamber is not in its normal closed configuration are probably of little value, because flow patterns are certain to be different when the cuvette is in use. One approach to determining the type of flow through a leaf chamber is to analyze the kinetics of CO$_2$ or water-vapor pulses moving through the cuvette. Such an analysis is beyond the scope of this paper. In unstirred chambers, errors caused by uncertainty about the flow pattern and about which of Equations 13a–c to use can be minimized by keeping the flow rate to the cuvette high, so that the differences in gas concentrations across the cuvette are small (i.e. c_o and w_o are not far from c_e and w_e, respectively). It is generally advisable to ensure that the air in a cuvette is well stirred, so that there are no significant concentration gradients through the cuvette and, thus, the gases leaving the cuvette are certain to be at the average ambient concentration. Stirring the air well will also minimize uncertainties in the measurement of leaf temperature and in estimation of the effects that the aerodynamic boundary layer surrounding the leaf has on gas exchange.

Assimilation rate. Calculation of the mass balance of CO_2 across the cuvette (the net photosynthetic rate) begins by determining the partial pressure of CO_2 in the sample or measuring cell of the IRGA (p_m). IRGAs respond to the molar density of the gas being measured. Thus, it is advisable to keep the pressure in the reference and measuring cells equal and as close to ambient pressure as possible. IRGAs should generally be calibrated in terms of partial pressure rather than of mole fraction, because partial pressure is more closely related to molar density. We generally use and calibrate our IRGAs by the method of Bloom et al. (1980), although not all IRGAs can be calibrated this way, and it may not always be convenient to do so (see also Thorpe, 1978; Parkinson and Legg, 1978; Sinclair, Buzzard, and Knoerr, 1976). This method results in a single calibration applicable over quite a broad range of CO_2 concentrations, provided the partial pressure of CO_2 in the reference cell is known. The calibration curve consists of a plot of inverse sensitivity (the change in partial pressure of CO_2 divided by the change in output voltage) as a function of the mean partial pressure of CO_2 in the two IRGA cells. The partial pressure of CO_2 in the IRGA reference cell, p_r, is obtained from the mole fraction CO_2 in the source air (C_r) and the total pressure (P):

$$p_r = c_r P. \tag{14}$$

The slope (m) and the Y-axis intercept (b) of the calibration line, the voltage signal (s) from the IRGA, and p_r are needed to calculate the partial pressure of CO_2 in the measured (sample) cell:

$$p_m = \frac{p_r[1 + (m\ s/2)] + b\ s}{1 - (m\ s/2)}. \tag{15}$$

The mole fraction CO_2 in the IRGA measuring cell (c_m) is calculated from p_m and the total pressure:

$$c_m = \frac{p_m}{P}. \tag{16}$$

Cross-sensitivity between water vapor and CO_2 is a potential problem with IRGAs. Some instruments are fitted with interference filters, which reduce but do not eliminate cross-sensitivity. Another way to keep the cross-sensitivity problem small is to reduce water vapor to a constant low level by condensing water from the returning gas stream in a cold trap. If the stream of air passing the reference side of the IRGA is humidified to the same low dew point the effect of cross-sensitivity will be eliminated. The mole fraction CO_2 in the humidified reference gas can be calculated using Equation 6. A third option is to dry the returning gas completely by passing it through a small amount of magnesium perchlorate, $Mg(ClO_4)_2$; caution is needed here because some commercially available $Mg(ClO_4)_2$ tends to adsorb and desorb CO_2 (Samish, 1978). Removing water from the measured air stream returning from the chamber makes c_m greater than the mole fraction CO_2 leaving the cuvette (c_o). The mole fraction CO_2 leaving the chamber is related to the mole fraction CO_2 after drying by rearranging and substituting into Equation 6:

$$c_o = \frac{c_m (1 - w_o)}{(1 - w_m)}, \tag{17}$$

where w_m is the mole fraction water vapor in the measured air stream after drying.

The CO_2 assimilation rate per unit leaf area found by differential measurement (A_d) is calculated from

$$A_d = \frac{(u_{in} c_{in} - u_{out} c_{out})}{L_a}. \tag{18}$$

For this calculation, flow rates and mole fraction CO_2 in and out may be taken from any point upstream or downstream, respectively, of the cuvette (Fig. 20.2): for example, $(u_e c_e - u_o c_o)$. One must always be careful to match the mole fraction CO_2 being used to the flow rate at the same point. As an example of how important this is, if A_d is calculated as $(u_e c_e - u_e c_o)/L_a$, failing to take into account the added flow due to transpiration, assimilation will typically be overestimated by 5 to 15 percent, depending on the transpiration rate.

As mentioned earlier, this calculation scheme requires that the upstream mole fraction CO_2 not include the compensating CO_2, because these molecules would then be counted twice. When a system has a flow of null-balancing CO_2 added to the air entering the cuvette (Figs. 20.1, 20.2), the measured air stream returning from the chamber will contain nearly the same CO_2 concentration as the stream that is going to the reference side of the IRGA. Thus, the analyzer will show little or no differential. The portion of the assimilation rate per unit leaf area measured by null balance is then calculated from the rate of CO_2 addition as measured by the compensating flow-metering device:

$$A_n = \frac{u_n c_n}{L_a}, \tag{19}$$

where A_n is the assimilation rate (mol m^{-2} s^{-1}) accounted for by null-balance measurements, L_a is the leaf area (m^2), u_n is the flow rate of the nulling gas, and c_n is the mole fraction CO_2 in that gas. A zero IRGA reading is not quite the balance point for assimilation, because addition of the compensating carrier gas, as well as the addition of water to the air stream by both the humidifier and the leaf, increases the flow rate and dilutes the initial CO_2 in the air stream, c_r. Therefore, photosynthesis measured by a compensating system always includes a small differential term. Because the equation for assimilation in differential mode takes into account changes in flow rate and positive or negative IRGA readings, a compensating system need not be perfectly balanced.

Total CO_2 assimilation rate per unit leaf area (A, mol m^{-2} s^{-1}) is the sum of the rates determined by compensating and differential measurements:

$$A = A_d + A_n. \tag{20}$$

Ambient mole fraction CO_2. Just as in the case of water vapor, the ambient mole fraction CO_2 in the cuvette (c_a) may not be equal to c_o in single-pass cuvettes. Rather c_a is often taken as the mean of c_o and c_e for cuvettes in which the flow is laminar, and as the mole fraction two-thirds of the way between c_o and c_e for cuvettes with turbulent flows (see Equation 13a–c). Again, keeping the flow rate to the chamber high reduces the uncertainty about the real value of c_a.

Conductances and Concentration Gradients

Knowing the ambient concentration of water vapor and CO_2 and having determined their flux rates, we now go on to calculate several derived variables. There are three major goals in this section. The first is to calculate stomatal conductance (inverse resistance) to gas movement between the air just outside the stomatal pore and the intercellular air spaces of the leaf. This is a functional measure of stomatal aperture that we shall carefully define below. There are several reasons for using stomatal conductance rather than stomatal resistance (Burrows and Milthorpe, 1976; Cowan, 1977), the most important being that fluxes of CO_2 and water vapor are approximately linearly related to conductances, whereas they are inversely related to resistances. The second goal is to obtain a value for the mole fraction CO_2 in the intercellular air spaces (c_i). This is an estimate of CO_2 available to the cells within the leaf. Farquhar and von Caemmerer (1982) assume that differences between c_i and the mole fraction CO_2 at the site of carboxylation are often small. However, the intracellular conductance is not infinite, and a significant additional concentration gradient may occur when photosynthetic rates are high (Evans et al., 1986). Nonetheless, the intercellular mole fraction CO_2 reflects the balance between the drawdown of CO_2 by photosynthesis and the supply through varying conductances. Thus, as environmental conditions change, changes in c_i may reflect changes in the partitioning of environmental control of photosyn-

Calculations Related to Gas Exchange

thesis between influences acting primarily at the biochemical level and those related primarily to stomata. The third goal is to obtain estimates of the mole fractions of CO_2 and water vapor at the leaf surface in order the better to define the external conditions to which stomata might respond.

Conductances and gradients based on binary diffusion. Equations related to Fick's first law of diffusion were applied to the movement of CO_2 and water vapor to and from leaves by Penman and Schofield (1951). This concept has been quite useful. The Fick's law formulation is correct to a first approximation, but a more extensive treatment is discussed below. There are three conductances that influence the exchange of water and CO_2 between the intercellular air of a leaf and the bulk surrounding air. The first conductance is stomatal and is primarily a function of stomatal aperture and areal density: diffusion is the major means of mass transport through stomatal pores. The second conductance is due to water loss from the leaf epidermis other than via the stomata and is termed cuticular conductance. There is no routine way to distinguish between cuticular and stomatal conductance. In fully expanded leaves, rates of water loss and carbon uptake from surfaces that are nearly devoid of stomata and/or in which stomatal aperture is thought to be small, are generally very low (e.g. Holmgren, Jarvis, and Jarvis, 1965; Whiteman and Koller, 1967; Jarvis, 1971). This is the basis for the common assumption that cuticular conductance is insignificant relative to stomatal conductance. It is not certain that this is always true, but the assumption is necessary in order to proceed. The third and outermost conductance, the boundary-layer conductance, is due to the still and slowly moving air immediately adjacent to the leaf. Both diffusion and bulk transport play roles in the movement of gases across boundary layers. Using current units and terminology, total conductance (stomatal plus boundary-layer) to water vapor, g_{tw} (mol m^{-2} s^{-1}), is calculated from

$$E = g_{tw} \Delta w_a, \qquad (21)$$

where

E = transpiration rate, mol m^{-2} s^{-1}
$\Delta w_a = w_i - w_a$, the mole fraction vapor gradient from inside the leaf to the bulk air.

Measured parameters here are E and w_a. The mole fraction water vapor in the leaf intercellular air spaces, w_i, is calculated as w_{sat} (from Eq. 2 or 3 and Eq. 4) at the measured leaf temperature. The mole fraction vapor gradient, Δw_a, is the vapor pressure gradient (often called VPD, although this is not the same as the meteorological term "vapor pressure deficit") divided by the total pressure. The total conductance to CO_2 (g_{tc}) is sometimes related to total conductance to water vapor by

460   J. Timothy Ball

$$g_{tc} = \frac{g_{tw}}{1.6}, \tag{22}$$

which is based on the ratio of the binary diffusivities of CO_2 and air and of water and air. As we shall see below, this ratio only applies for diffusion through still air. The mole fraction CO_2 in the leaf intercellular spaces, c_i, is calculated from an equation analogous to Equation 21 but for CO_2 rather than water vapor:

$$A = g_{tc}(c_a - c_i), \tag{23a}$$

rearranged to

$$c_i = c_a - \frac{A}{g_{tc}}. \tag{23b}$$

The calculation of c_i involves two assumptions: first, that the mole fraction water vapor of the air inside the leaf is the saturation mole fraction at leaf temperature; second, that CO_2 diffusing inward meets virtually the same resistances (inverse conductances) that are met by water vapor moving outward. Both of these assumptions have been questioned (e.g. Jarvis and Slatyer, 1970; Meidner, 1975). As leaf water potential falls, the mole fraction water vapor in the air within the leaf falls below saturation. This effect can, however, be shown to be small on physical grounds. The actual vapor pressure (v_i, Pa) at water potential (Ψ, Pa, a negative quantity) is related to the saturation vapor pressure (v_{sat}) as

$$v_i = v_{sat} \cdot e^{(\Psi \bar{V}_w / R T_K)}, \tag{24}$$

where
$\bar{V}_w$ = molal volume of liquid water, $18 \cdot 10^{-6}$ m^3 mol^{-1}
R = the universal gas constant, 8.3143 J K^{-1} mol^{-1}
T_K = temperature, °K.

This amounts to approximately 0.7 percent per 10^6 pascals of water potential. Comparing computed resistances and fluxes with actual diffusion of helium across leaves, Farquhar and Raschke (1978) concluded that the above assumptions were substantially correct. More important, Mott and O'Leary (1984) and Sharkey et al. (1982) confirmed that these assumptions are reasonable by directly measuring the intercellular CO_2 partial pressures. Further, they found that agreement between measured and calculated values was improved when the ternary diffusion considerations discussed below were included.

The need for ternary treatment. Equations 21 and 23 consider only binary diffusion, whereas the problem is actually one of at least ternary diffusion

(Jarman, 1974). That is, these formulas consider only interactions between water vapor and air in the case of transpiration and between CO_2 and air in the case of CO_2 assimilation, neglecting interactions between water vapor diffusing outward and CO_2 diffusing inward. Jarman also considered the influence of outward-diffusing oxygen on the movement of other gases, showing that it generally has an insignificant effect. Leuning (1983) has made a more extensive analysis, including the effect of collisions between molecules diffusing through the stomatal pore and the pore walls. His analysis shows that these collisions have little effect on measured fluxes and calculated values of conductance and c_i until the stomatal aperture approaches 1 μm. Fully open stomata may have diameters approaching 50 μm (Meidner and Mansfield, 1968). When the average stomatal aperture is as low as 1 μm, gas-exchange rates will be quite low, so that calculation of conductances and gradients are likely to be limited by the ability of the gas-exchange system to measure low rates of gas exchange. Von Caemmerer and Farquhar (1981) applied Jarman's analysis to gas-exchange calculations. I have drawn on their treatment of the formulas and system of symbols in developing this section of the paper.

The basis of the ternary treatment is the Stephan-Maxwell equation, which relates the concentration gradient of a particular gas between two points in a system to the molar concentration of each species in the system, the relative diffusion velocity of each species, and the binary diffusivity of each species pair. Simplified expressions for the concentration gradients between leaf intercellular air spaces and the bulk air for the ternary system of CO_2, water vapor, and air (gases other than water vapor and CO_2) are, for water vapor,

$$w_i - w_a = \left(\frac{\bar{a}}{g_{t(wa)}} + \frac{\bar{c}}{g_{t(wc)}} \right) E + \frac{\bar{w}A}{g_{t(wc)}}, \qquad (25)$$

and, for carbon dioxide,

$$c_i - c_a = -\left(\frac{\bar{a}}{g_{t(ca)}} + \frac{\bar{w}}{g_{t(wc)}} \right) A + \frac{\bar{c}E}{g_{t(wc)}}. \qquad (26)$$

In these equations A, E, w_i, w_a, c_i, and c_a are parameters that have been defined above. The spatially averaged mole fractions of water vapor, CO_2, and air within the gradient are approximated as mean mole fractions of each gas at the two ends of each gradient:

$\bar{w} = (w_i + w_a)/2$
$\bar{c} = (c_i + c_a)/2$
$\bar{a} = 1 - (\bar{w} + \bar{c})$.

Since $\bar{c}$ is quite small relative to $\bar{a}$ and $\bar{w}$, the approximation $\bar{a} \approx 1 - \bar{w}$ can be made. The total or leaf (stomatal plus boundary-layer) conductances $g_{t(wa)}$, $g_{t(ca)}$, and $g_{t(wc)}$ are with regard to the binary interaction of water vapor and air, of CO_2 and air, and of CO_2 and water vapor, respectively. A binary conductance involving interactions between substances x and y may be defined as

$$g_{(xy)} = \frac{CD_{(xy)}}{l} \qquad (27)$$

where C is the total molar density of gas, $D_{(xy)}$ is the binary diffusion coefficient for the two substances, and l is the effective length of the gradient over which diffusion is occurring. Note that $D_{(xy)} = D_{(yx)}$. Each of the three terms in Equations 25 and 26 concerns the interaction of water vapor and air, CO_2 and air, and CO_2 and water vapor, respectively. On the right-hand side of Equation 25, the numerators in the first two terms are multiplied by E (the transpiration rate), and the numerator of the third term includes A (the assimilation rate). These multiplications of average concentrations times fluxes represent the interaction of the velocity of water and CO_2 diffusion, respectively, on the movements and gradients of the other gases. There is no term representing the diffusion velocity of air in Equation 25, because there is no net flux of air between the intercellular air spaces and the bulk air. (There is, however, a concentration gradient of air, caused largely by the concentration gradient and flux of water vapor.) Equation 26 lacks the term $\bar{a}$, for the flux rate of air, for the same reason.

Conductances to water vapor flux. In consideration of the flux of water, the concentration ($\bar{c}$) and flux of CO_2 (A) have insignificant effects, so the second and third terms on the right-hand side of Equation 25 can be neglected. Then, substituting $(1 - \bar{w})$ for $\bar{a}$, Equation 25 simplifies to

$$w_i - w_a = \frac{(1 - \bar{w}) E}{g_{t(wa)}}, \qquad (28a)$$

which when rearranged yields

$$g_{t(wa)} = \frac{E(1 - \bar{w})}{(w_i - w_a)} = g_{tw}. \qquad (28b)$$

This equation considers the vapor concentration gradient between the bulk air and the leaf intercellular air spaces and therefore defines the total conductance to water vapor. (Thus the simplifying notation g_{tw}.) This definition yields an estimate of conductance to water vapor that is lower by the factor $(1 - \bar{w})$ than the estimate of Equation 17. The former treatment neglected the existence of a

Calculations Related to Gas Exchange 463

significant concentration gradient of air in the definition of total conductance to water vapor. Total conductance to water vapor must be the first conductance calculated, because values for other conductances as well as concentration gradients are derived from it. Recall that conductances (inverse resistances) in parallel may be summed directly and that resistances in series are summed directly: in other words, the sum of conductances that are in series is the inverse of the sum of the corresponding resistances. For an amphistomatous leaf the stomatal conductances for the two surfaces are in parallel, just as the boundary-layer conductances for the two sides are in parallel with each other. On either side of an amphistomatous leaf and for the lower surface of a hypostomatous leaf, the boundary-layer and stomatal conductances for that side are in series. Thus, for all leaves total conductance to water vapor is the sum of the appropriate boundary-layer conductance and stomatal conductance in series, so that stomatal conductance to water vapor is found as

$$g_{sw} = \frac{1}{\frac{1}{g_{tw}} - \frac{1}{g_{bw}}}. \tag{29}$$

As was mentioned earlier, Equations 25 and 26 contain estimates of the spatially averaged mole fraction of each gas within the gradient. Since the form of all the gradients is unknown, the gas concentrations are assumed to change linearly with distance through their respective gradients. Thus the mean of the mole fraction of each gas at the two ends of each gradient ($\bar{a}$, $\bar{w}$, and $\bar{c}$) is taken as representative of the concentration within each gradient. This assumption is obviously incorrect for the concentration gradients between the bulk air and the intercellular air spaces, because the concentration drop across the boundary layer is rarely the same as the drop across the stomata. In essence, this means that the empirically determined total conductance, the independently determined boundary-layer conductance, and the stomatal conductance (found by difference) rarely (if ever) have a strictly additive relationship. The calculated value of stomatal conductance is, therefore, slightly biased. The discrepancy that this causes is quite small, but it will show up when the concentration of CO_2 at the leaf surface is discussed below. The apparent size of this error can be reduced by iteration to obtain values of the mole fraction water vapor at the leaf surface and g_{sw}. However, other errors in most gas-exchange measurements—for example, in determination of boundary-layer conductance—are likely to be far more serious than this error.

Within the aerodynamic boundary layers that surround leaves several different transport processes occur. The relative importance of these processes is not well understood; moreover, the importance of the various processes shifts as environmental conditions change. For example, as wind speed and freestream turbulence over a leaf increase, turbulent eddies penetrate the leaf boundary layer, and the importance of diffusion transport across the boundary layer is reduced (see Chap. 18). Keeping the wind speed and free-stream turbulence in a gas-exchange cuvette high will keep the boundary-layer conductance large and will minimize the impact of uncertainties, because stomatal conductance will dominate the conductance catena. This is not a totally satisfactory situation, however, because there are interesting questions regarding the impact of wind speed and the boundary layer on photosynthesis and transpiration. For consistency throughout the calculations it is assumed here, as it has been by von Caemmerer and Farquhar (1981) and by Leuning (1983), that ternary interactions are important in transport through the boundary layer.

Boundary-layer conductance to water vapor (g_{bw}) can be determined for any particular leaf size, chamber configuration, and air speed. This is done by measuring the temperature of and evaporation rate from leaf models made of chromatography paper, appropriately positioned in the cuvette and continually wetted (Jarvis et al., 1971). The boundary-layer conductance is calculated in the same way as total conductance, using Equation 28b. Uncertainty as to the temperature of the evaporating surface can be a source of error in determining g_{bw}: there will nearly always be temperature gradients across the leaf model when evaporation is occurring, and good thermal contact between the leaf model and thermocouples may be difficult to achieve. These are, of course, also problems with measurements of real leaves, but in leaf models they may be more critical because of greater evaporative cooling. To minimize these problems, one can make leaf models from two layers of paper, held together with very small pieces of double-sided adhesive tape along the edges. One can then average the readings from three or four thermocouples placed between the sheets of paper. One can also measure the boundary-layer conductance for each side of the leaf independently, by covering the opposite side of the model with a waterproof coating such as adhesive-backed aluminum foil. Such a covering can reduce the evaporative cooling by about 50 percent. Total boundary-layer conductance is then the sum of the parallel conductances from the two sides. This sum may be greater than the value obtained from a leaf model in which evaporation was taking place from both sides. The discrepancy may occur because air that has moved partway across one surface of the leaf model spills over and is swept across the other side (Thom, 1968). Whether boundary-layer conductance should be determined on the basis of the leaf model as a whole or for each side independently depends on the chamber flow characteristics of the chamber used and on the distribution of stomatal conductance between the two leaf surfaces of the species in question.

Conductances to CO_2 flux. The binary diffusivity of CO_2 and water vapor is quite close to that of CO_2 and air (i.e. $D_{(wc)} \approx D_{(ca)}$); thus, we take

$$g_{t(ca)} = g_{t(wc)}. \tag{30}$$

Again, using the fact that $\bar{a} + \bar{w} \approx 1$, the first two terms on the right-hand side of Equation 26 become $1/g_{t(ca)}$, so that Equation 26 can be simplified and rearranged to

$$A = g_{t(ca)} (c_a - c_i) - \frac{(c_a + c_i) E}{2}. \tag{31}$$

In the Fick's-law treatment of the flux and concentration gradient of CO_2 (Eq. 23), the term that includes the effect of the flux of water vapor on the CO_2 gradient ($\bar{c} E$) is not included. Notice in Equation 31 that $g_{t(ca)}$ (hereafter g_{tc}) becomes the familiar flux divided by gradient in the absence of transpiration.

Inasmuch as $D_{(wa)} \approx 1.6 D_{(ca)}$, we can relate stomatal conductance to CO_2 (g_{sc}) to that for water:

$$g_{sc} = \frac{g_{sw}}{1.6}. \tag{32}$$

This applies to diffusion of CO_2 through the "still air" of the stomatal pore but not through the boundary layer, which is a forced-convection regime in virtually all gas-exchange systems.

The ratio of the rates at which two gases are transferred across leaf boundary layers is assumed to be proportional to the two-thirds power of their diffusion coefficients. Thus, the boundary-layer conductance to CO_2 (g_{bc}) is taken as

$$g_{bc} = \frac{g_{bw}}{1.6^{(2/3)}} = \frac{g_{bw}}{1.37}. \tag{33}$$

This transport ratio is obtained from numerical analysis (the Pohlhausen relationship: see Kays and Crawford, 1980) of heat transfer from isothermal flat plates in a flow that is laminar and parallel to the plates. Thom (1968) studied transport of heat, water vapor, and several volatile compounds from leaf models in a wind tunnel and found that the two-thirds power of the ratio of diffusivities explained the data well, especially as the wind speed approached and then exceeded 1 m s^{-1}. The extent to which these engineering experiments adequately approximate the flow and boundary-layer conditions surrounding real leaves is not clear (see Grace, Fasehun, and Dixon, 1980). As the air flow across a leaf increases in speed and turbulence, the ratio of gas transfer approaches unity. The rate at which the ratio approaches unity as a function of wind speed is not well described for leaves under either natural or cuvette conditions (Clark and Wigley, 1975). Leaf boundary layers tend to be laminar in most cuvettes unless high levels of free-stream turbulence are introduced. It is the general practice to assume the value 1.37 for the ratio of water vapor to CO_2 transport through the boundary layer, even under turbulent conditions.

This assumption can be rationalized from the fact that, as turbulence develops, the transport ratio does not go to unity immediately. Further, as wind speed, turbulence, and boundary-layer conductance increase, the conductance catena is increasingly dominated by stomatal conductance. Thus, the error from using 1.37 as the value of the transport ratio becomes quite small. Minimizing this error is one of several arguments made throughout this paper for generally keeping boundary-layer conductance high in gas-exchange cuvettes.

Intercellular CO_2 concentration. Total conductance to CO_2 is the inverse of the sum of the component serial resistances:

$$g_{tc} = \frac{1}{\frac{1}{g_{sc}} + \frac{1}{g_{bc}}}. \tag{34}$$

Having g_{tc}, we can solve Equation 31 for c_i, the intercellular mole fraction CO_2:

$$c_i = \frac{\left(g_{tc} - \frac{E}{2}\right) c_a - A}{\left(g_{tc} + \frac{E}{2}\right)}. \tag{35}$$

By rearranging Equation 31 and comparing it to Equation 23, one can see that the present estimate of the mole fraction CO_2 in the leaf intercellular air spaces is approximately ($\bar{c}\, E/g_{tc}$) less than the estimate of c_i from Equation 23b. ("Approximately," because the values of g_{tc} from Eqs. 22 and 34 are not identical.) This difference in the estimate of c_i can be as much as 20 percent in plants with low ratios of conductance to photosynthesis (C_4 plants) at high Δw_a (low humidity). The discrepancy decreases to the vicinity of 1 percent at low Δw_a in C_3 plants with the highest ratios of conductance to photosynthesis.

The partial pressure of CO_2 in the leaf intercellular air spaces is

$$p_i = c_i P. \tag{36}$$

With regard to the relationship between intercellular CO_2 concentration and photosynthesis, it is most appropriate to consider partial pressure, because the concentration of dissolved gaseous CO_2 in aqueous phase at equilibrium, or the chemical activity of gaseous CO_2, is a function of partial pressure rather than of mole fraction.

Concentrations of water vapor and CO_2 at the leaf surface. Two important factors in photosynthesis, conductance, and their balance, in relation to environmental conditions, are humidity and CO_2 concentration. Among plant

Calculations Related to Gas Exchange   467

physiologists and ecologists the humidity of the air surrounding a leaf is considered in terms of the mole fraction vapor gradient from the leaf to the bulk air (Δw_a), which is the driving gradient for transpiration. The CO_2 concentration that is most often considered is the mole fraction CO_2 in the bulk air (c_a), which is the source end of the concentration gradient established by photosynthesis. Particularly in nature, environmental conditions at the leaf surface vary, from close to those of the bulk air when the wind speed is relatively high to conditions that are quite different from the bulk air as the wind speed over a leaf decreases and as photosynthesis and transpiration increase. In most gas-exchange cuvettes, the boundary-layer conductance is high enough to reduce the difference between bulk air and leaf surface conditions to relatively low levels. The differences that do exist will vary with environmental conditions and leaf responses. The external conditions that are most likely to elicit a physiological response are the conditions at the leaf surface. The differences that do exist between bulk air and leaf surface conditions may be very significant in terms of feedback mechanisms that regulate stomatal responses. The environment of a leaf is, therefore, probably best described by considering conditions at the leaf surface. If environmental and gas-exchange data were routinely expressed in terms of surface conditions, data from different experiments could be more meaningfully compared.

Equation 28 relates the flux of water vapor to concentrations of air and water and to the conductance of the boundary layer and of the stomata in series. At steady state the flux of water through the boundary layer and through the stomata must both equal the total flux. The same is true for the flux of CO_2 through the boundary layer and stomata. These facts make it possible to calculate the mole fraction water vapor (w_s) and CO_2 (c_s) at the leaf surface by substituting stomatal and boundary-layer conductances for the total conductances to water vapor and CO_2 in Equations 28 and 31, respectively.

In the case of water vapor, we must take $\bar{w}$ as ($w_i + w_a$)/2, because we do not know ($w_i + w_s$)/2. This is a very reasonable approximation, because the majority of the concentration drop from w_i to w_a is across the stomata if the boundary-layer conductance is significantly larger than stomatal conductance, as it is in most cuvettes. In other words, in a gas-exchange system w_s is likely to be near w_a, so the point at which the concentration is at the gradient's spatially averaged value is likely to be within the stomatal pore. Making appropriate substitutions in Equation 28, and rearranging,

$$w_s = w_i - \frac{(1 - \bar{w}) E}{g_{sw}}. \tag{37}$$

One can iterate to solve for w_s by revising the estimate of $\bar{w}$. Under most chamber conditions the revisions would be small. Once w_s is calculated, the mole fraction vapor gradient across the leaf epidermis is simply

$$\Delta w_s = w_i - w_s. \qquad (38)$$

Recent work in our laboratory indicates that the relative humidity at the leaf surface may be important to stomatal responses (Ball, Woodrow, and Berry, 1987). It may be of interest to calculate the relative humidity at the leaf surface, which we take as

$$h_s = \frac{w_s}{w_i}. \qquad (39)$$

In calling this "relative humidity" we are assuming that the slow-moving air at the interior of the boundary layer (immediately adjacent to the leaf) is at leaf temperature. Since the usual way of measuring leaf temperature is to appress a thermocouple to the leaf, the temperature that is normally taken to be leaf temperature is actually some average between the leaf temperature and the temperature of air very close to the leaf. These two temperatures are not likely to be very different. Recall also that w_i is w_{sat} at leaf temperature.

Again, because at steady state the flux of CO_2 through the boundary layer must equal the flux through the stomata, one can by analogy to Equations 31 and 35 write

$$c_s \approx \frac{\left(g_{bc} - \dfrac{E}{2}\right) c_a - A}{\left(g_{bc} + \dfrac{E}{2}\right)}. \qquad (40)$$

One can also write

$$c_s \approx \frac{\left(g_{sc} + \dfrac{E}{2}\right) c_i + A}{\left(g_{sc} - \dfrac{E}{2}\right)}. \qquad (41)$$

If one solves both Equations 40 and 41 using actual data, for most experimental situations there will be a difference of something less than 2 μmol mol^{-1} between these two estimates. This small discrepancy comes about because of approximations made regarding the form of the concentration gradient between w_a and w_i and the effect of the approximation when the total conductance to water vapor is partitioned into g_{sw} and g_{bw}. Recall that conductances to CO_2 are calculated as constant proportions of conductances to water vapor. Our understanding of the physiology of leaves is not likely to hinge on differences so small as those between the results of Equations 40 and 41.

Dual-stream or two-sided gas exchange. With a dual-stream gas-exchange system, the activity of the two surfaces of a leaf may be measured independently. In dual- or two-sided systems a portion of a leaf lamina is sealed between two halves of a cuvette and then independent parallel gas streams are provided to each leaf surface. The two gas streams interact with each other only via c_i and w_i, both of which are to some degree under the influence of the physiology of the leaf. It is critical in two-sided gas-exchange systems that there be no significant pressure gradient across the leaf, lest air be swept through the leaf. Although dual-stream systems tend to be substantially more complicated than systems that pass a single stream of air over both sides of a leaf, a significant amount of additional information about the leaf can be obtained with two-sided systems. The flux, gradient, and conductance parameters for each leaf surface are calculated using Equations 28 through 41. The responses of the two separate surfaces can then be compared. To consider the leaf as a whole (that is, for expression on a projected leaf area basis), the fluxes and conductances for the two leaf surfaces are added, because the sides function in parallel. The values of c_i, c_s, and w_s calculated from the summed fluxes and conductances will be the respective means of the values calculated for the two sides independently. One good check of a two-sided gas-exchange system is to measure transpiration and photosynthetic rates of an amphistomatous mesophytic leaf that is not likely to have a CO_2 gradient across the mesophyll. If the system is working well, the values of c_i calculated from each side of the leaf should be quite close.

In two-sided systems, one can assure that conditions to which each surface is exposed are well defined (the same as or different from the opposite side), because adjustments can be made for different rates of addition of water or removal of CO_2 and different conductances on each side of the leaf. This allows the responses of a leaf to environmental conditions to be defined in terms of exact conditions rather than through some average of the conditions over the two leaf surfaces. Using the ability to manipulate environmental conditions at the two surfaces, one can ask whether each side of a leaf responds to the conditions to which it is exposed or whether the two sides of a leaf respond in a coordinated fashion to the average conditions. Two-sided systems may allow important questions to be asked about the relations of stomata on the two surfaces to each other and to the mesophyll. Few such experiments have been done (see Mott and O'Leary, 1984).

An important advantage of two-sided systems is that they require no assumptions about the distribution of conductances between the two surfaces of amphistomatous leaves. In single-stream systems the distribution of conductances cannot be determined, and this can introduce significant artifacts into the calculations. Single-stream systems arrive at the correct value of stomatal conductance only when the boundary-layer and stomatal conductances for each side of the leaf constitute the same proportion of their respective total (summed) conductances. The following example illustrates this potential error. Consider a leaf with a total stomatal conductance to water vapor (g_{sw}) of 0.8 mol m^{-2} s^{-1}, asymmetrically distributed as 0.2 mol m^{-2} s^{-1} and 0.6 mol m^{-2} s^{-1} for the upper and lower surfaces, respectively, in a chamber in which the total boundary-layer conductance to water vapor (g_{bw}) is 2.5 mol m^{-2} s^{-1} distributed as 1.5 mol m^{-2} s^{-1} for the upper side and 1.0 mol m^{-2} s^{-1} for the bottom. The total conductance to water vapor (stomatal plus boundary-layer) for the upper surface (g_{tw}^u) would be

$$g_{tw}^u = \frac{1}{\dfrac{1}{g_{sw}} + \dfrac{1}{g_{bw}}} = \frac{1}{\dfrac{1}{0.2} + \dfrac{1}{1.5}} = .176, \quad (42)$$

whereas for the lower surface the total conductance to water vapor would be

$$g_{tw}^l = \frac{1}{\dfrac{1}{g_{sw}} + \dfrac{1}{g_{bw}}} = \frac{1}{\dfrac{1}{0.6} + \dfrac{1}{1.0}} = .375. \quad (43)$$

The total conductance (stomatal plus boundary-layer) to water vapor for the entire leaf (g_{tw}) is the sum of the two conductances above, 0.551 mol m^{-2} s^{-1}, and would be the value obtained from Equation 28b in an actual experiment. Because the distribution of stomatal conductance can never be determined from this experiment, the combined boundary-layer conductance for the entire leaf cannot be partitioned between the two sides of the leaf. Not knowing the distribution of conductance, we nevertheless go ahead to calculate the stomatal conductance to vapor for the entire leaf from Equation 29, rearranged:

$$g_{sw} = \frac{1}{\dfrac{1}{g_{tw}} - \dfrac{1}{g_{bw}}} = \frac{1}{\dfrac{1}{0.551} - \dfrac{1}{2.5}} = .707. \quad (44)$$

This result is an 11.6 percent error in the value of stomatal conductance. Needless to say, this error in turn causes errors in the value of the derived parameters such as c_i and in the concentrations of CO_2 and water vapor at the leaf surface. The size of the error decreases as the boundary-layer conductance increases, relative to the stomatal conductance; this is another reason to keep the boundary-layer conductance high in gas-exchange cuvettes. During an experiment with a two-sided gas-exchange system, the stomatal conductance for each surface (which would be unknown) would be computed from the measured total conductances for each side (0.176 and 0.375 in this case), and the known boundary-layer conductances (1.5 and 1.0 in this case) by rearranging Equations 42 and 43. The stomatal conductance for the entire leaf would be the sum of the values from each side (0.2 and 0.6 in this case).

Calculations Related to Gas Exchange

Discussion. Stomatal and photosynthetic responses are likely to be more closely related to environmental conditions at the leaf surface than to conditions in the bulk air. At a given flux into or out of a leaf the concentration of the moving gas at the leaf surface depends on the relative magnitude of the boundary-layer and stomatal conductances. The boundary layer separating the leaf surface from bulk conditions varies widely in thickness and therefore in conductance between gas-exchange cuvettes, and even more widely in nature. In cuvettes that are considered well stirred, the differences between bulk air and leaf surface concentrations of both water and CO_2 can be as great as 10 percent. Data expressed in terms of surface conditions are more meaningful and can be more easily compared and applied to other (including natural) situations than data expressed in terms of bulk conditions.

Few studies have attempted to quantify the control ("limitation": see Kacser and Burns, 1973) of photosynthesis and transpiration by the boundary layer. The boundary layer and stomata interact strongly in the control of both processes, not only directly through the flux of CO_2 and water but also through the flux of latent and sensible heat to and from leaves. The boundary layer can constitute a significant proportion of the control ("limitation") of photosynthesis, particularly at the lowered wind speeds and lowered ambient CO_2 concentrations that tend to prevail in crop canopies (Woodrow, Ball, and Berry, 1987; Ball, Woodrow, and Berry, 1987). Separation of the physiologically based control of photosynthesis and transpiration by stomata from the physically based control of these processes by the boundary layer will be important not only in terms of possible increased understanding of the mechanistic connections of stomata with other processes in and around leaves but also in terms of understanding the adaptive significance of leaf form and function. The separation of boundary-layer processes from processes within the leaf is made easier by considering the responses of leaves in terms of conditions at the leaf surface.

LIST OF SYMBOLS

A Net CO_2 assimilation rate, mol $m^{-2} s^{-1}$

$\bar{a}$ mean mole fraction air (excluding CO_2 and H_2O) between two points in a concentration gradient

A_d CO_2 assimilation rate from differential measurements, mol $m^{-2} s^{-1}$

A_n CO_2 assimilation rate from compensating measurements, mol $m^{-2} s^{-1}$

b Y-axis intercept of IRGA calibration regression ($\Delta p_{CO_2}/\Delta V$)

$\bar{c}$ mean mole fraction CO_2 between two points in a concentration gradient

c_a mole fraction CO_2 in bulk air surrounding a leaf

c_e mole fraction CO_2 in air entering a cuvette (excluding c_n)

c_h mole fraction CO_2 in the gas stream following humidification

c_i mole fraction CO_2 in leaf intercellular air spaces
c_m mole fraction CO_2 in gas stream flowing through the IRGA measuring (sample) cell
c_n mole fraction CO_2 in the gas stream added to cuvette to compensate CO_2 uptake
c_o mole fraction CO_2 in gas stream leaving a cuvette
c_r mole fraction CO_2 in the initial air stream and the IRGA reference cell
c_s mole fraction CO_2 in air at the leaf surface (interior to the boundary layer)
$D_{(xy)}$ binary diffusivity of species x and y
E Transpiration rate, mol m^{-2} s^{-1}
g_{bc} conductance of the boundary layer to CO_2, mol m^{-2} s^{-1}
g_{bw} conductance of the boundary layer to water vapor, mol m^{-2} s^{-1}
g_{sc} conductance of stomata to CO_2, mol m^{-2} s^{-1}
g_{sw} conductance of stomata to water vapor, mol m^{-2} s^{-1}
g_{tc} total conductance (stomatal plus boundary-layer) to CO_2, mol m^{-2} s^{-1}
$g_{t(ca)}$ total binary conductance involving CO_2 and air
g_{tw} total conductance (stomatal plus boundary-layer) to water vapor, mol m^{-2} s^{-1}
$g_{t(wa)}$ total binary conductance involving water vapor and air
$g_{t(wc)}$ total binary conductance involving CO_2 and water vapor
h_s relative humidity of the air at the leaf surface
IRGA infrared gas analyzer
L_a projected leaf area, m^2
m slope of IRGA calibration regression (V^{-1})
P total pressure, pascals
p_i partial pressure CO_2 in leaf intercellular air spaces (c_iP)
p_m partial pressure CO_2 in the IRGA measuring (sample) cell
p_r partial pressure CO_2 in the IRGA reference cell
R universal gas constant, 8.3143 J K^{-1} mol^{-1}
s IRGA signal, V
T_K temperature, °K
T_s steam temperature at one atmosphere, 373.16° K
u_e flow rate of air entering a cuvette, mol s^{-1}
u_h flow rate of air leaving a condenser and sent toward a cuvette, mol s^{-1}
u_n flow rate of gas carrying compensating CO_2, mol s^{-1}
u_o flow rate of air leaving a cuvette, mol s^{-1}
u_r flow rate of air sent toward a cuvette before any humidification, mol s^{-1}
v_i vapor pressure of water in leaf intercellular air spaces, Pa
v_{sat} saturation vapor pressure at a given temperature, Pa
$\bar{V}_w$ molal volume of liquid water, $18 \cdot 10^{-6}$ m^3
$\bar{w}$ mean mole fraction water vapor between two points in a concentration gradient
w_a mole fraction water vapor in the bulk air surrounding a leaf
w_e mole fraction water vapor in gas stream entering a cuvette
w_h mole fraction water vapor in gas stream leaving the condenser and sent toward a cuvette
w_i mole fraction water vapor in leaf intercellular air spaces (assumed equal to w_{sat} at leaf temperature)
w_m mole fraction water vapor in air passing through the IRGA measuring (sample) cell
w_o mole fraction water vapor in air leaving the cuvette
w_r mole fraction water vapor in the initially mixed gas (reference gas)
w_s mole fraction water vapor at the leaf surface (interior to the boundary layer)
w_{sat} saturation mole fraction water vapor at a given temperature (v_{sat}/P)
x mole fraction of an air stream component undergoing change due to

Calculations Related to Gas Exchange

addition or removal of another component, y
y mole fraction of an air stream component gas being added or removed
Δw_a mole fraction water vapor gradient from leaf intercellular air spaces to bulk air
Δw_s mole fraction water vapor gradient from leaf intercellular air spaces to air at the leaf surface
η an intermediate result in the calculation of v_{sat}
Ψ leaf water potential, Pa

LIST OF SAMPLE DATA

This list of sample parameter values is provided to aid in developing and checking the output from a computer program for gas-exchange calculations. The table is from an experiment in which the gas exchange of cotton was being studied at a photosynthetically active photon flux area density of $400 \cdot 10^{-6}$ mol m^{-2} s^{-1}. The mean leaf temperature was 24.99° C, and the condenser temperature was 8.07° C. The gas-exchange system was similar in design to that of Field, Berry, and Mooney (1982). Three details of the design are important in interpreting the values: first, the system is a null-balance system, in which the compensating CO_2 is added after humidification; second, cuvette is well stirred, so that w_a is considered to be w_e; third, the gas stream that is sent to the measuring cell of the IRGA is dried completely with magnesium perchlorate.

A	$16.3 \cdot 10^{-6}$ mol m^{-2} s^{-1}	P	$102.269 \cdot 10^3$ Pa
A_d	$0.76 \cdot 10^{-6}$ mol m^{-2} s^{-1}	p_i	22.8 Pa
A_n	$15.58 \cdot 10^{-6}$ mol m^{-2} s^{-1}	p_m	35.3 Pa
b	3.1061 Pa V^{-1}	p_r	35.79 Pa
$\bar{c}$	$280.6 \cdot 10^{-6}$ mol mol^{-1}	R	8.3143 J K^{-1} mol^{-1}
c_a	$338.3 \cdot 10^{-6}$ mol mol^{-1}	S	$-106.0 \cdot 10^{-3}$ V
c_e	$344.3 \cdot 10^{-6}$ mol mol^{-1}	u_e	$527.21 \cdot 10^{-6}$ mol s^{-1}
c_h	$346.3 \cdot 10^{-6}$ mol mol^{-1}	u_h	$524.2 \cdot 10^{-6}$ mol s^{-1}
c_i	$223.1 \cdot 10^{-6}$ mol mol^{-1}	u_n	$2.99 \cdot 10^{-6}$ mol s^{-1}
c_m	$345.2 \cdot 10^{-6}$ mol mol^{-1}	u_o	$532.3 \cdot 10^{-6}$ mol s^{-1}
c_n	$10.11 \cdot 10^{-3}$ mol mol^{-1}	u_r	$518.7 \cdot 10^{-6}$ mol s^{-1}
c_o	$338.3 \cdot 10^{-6}$ mol mol^{-1}	v_i	3164.0 Pa
c_r	$350.0 \cdot 10^{-6}$ mol mol^{-1}	$\bar{w}$	$25.47 \cdot 10^{-3}$ mol mol^{-1}
c_s	$331.5 \cdot 10^{-6}$ mol mol^{-1}	w_a	$20.0 \cdot 10^{-3}$ mol mol^{-1}
E	$2.64 \cdot 10^{-3}$ mol m^{-2} s^{-1}	w_e	$10.47 \cdot 10^{-3}$ mol mol^{-1}
g_{bc}	2.555 mol m^{-2} s^{-1}	w_h	$10.53 \cdot 10^{-3}$ mol mol^{-1}
g_{bw}	3.50 mol m^{-2} s^{-1}	w_i	$30.942 \cdot 10^{-3}$ mol mol^{-1}
g_{sc}	$157.5 \cdot 10^{-3}$ mol m^{-2} s^{-1}	w_m	0.0 mol mol^{-1}
g_{sw}	$252.0 \cdot 10^{-3}$ mol m^{-2} s^{-1}	w_o	$20.0 \cdot 10^{-3}$ mol mol^{-1}
g_{tc}	$148.4 \cdot 10^{-3}$ mol m^{-2} s^{-1}	w_r	0.0 mol mol^{-1}
g_{tw}	$235.1 \cdot 10^{-3}$ mol m^{-2} s^{-1}	w_s	$20.74 \cdot 10^{-3}$ mol mol^{-1}
h_s	0.67 (dimensionless)	Δw_a	$10.95 \cdot 10^{-3}$ mol mol^{-1}
L_a	$19.4 \cdot 10^{-4}$ m^2	Δw_s	$10.21 \cdot 10^{-3}$ mol mol^{-1}
m	$43.969 \cdot 10^{-3}$ V^{-1}		

REFERENCES

Ball, J. T., and J. A. Berry. 1982. The C_i/C_s ratio: A basis for predicting stomatal control of photosynthesis. *Carnegie Institution of Washington Yearbook*, 81: 88–92.
Ball, J. T., I. E. Woodrow, and J. A. Berry. 1987. A model predicting stomatal conductance and its contribution to the control of photosynthesis under different environmental conditions. In J. Biggins, ed., *Proceedings of the 7th International congress on photosynthesis*. The Hague: Junk. [In press.]
Björkman, O., and P. Holmgren. 1963. Adaptability of the photosynthetic apparatus to light intensity in ecotypes from exposed and shaded habitats. *Physiologia plantarum*, 16: 889–914.
Bloom, A. J., H. A. Mooney, O. Björkman, and J. Berry. 1980. Materials and methods for carbon dioxide and water exchange analysis. *Plant, Cell and Environment*, 3: 371–76.
Burrows, F. J., and F. L. Milthorpe. 1976. Stomatal conductance in the control of gas exchange. In T. T. Kozlowski, ed., *Water deficits and plant growth 4, Soil water measurement, plant responses and breeding for drought resistance*: 103–52. New York: Academic.
von Caemmerer, S., and G. D. Farquhar. 1981. Some relationships between the biochemistry of photosynthesis and the gas exchange of leaves. *Planta*, 153: 376–87.
Clark, J. A., and G. Wigley. 1975. Heat and mass transfer from real and model leaves. In D. A. deVries and N. H. Afgan, eds., *Heat and mass transfer in the biosphere*: 413–22. Washington, D.C.: Scripta.
Cowan, I. R. 1977. Stomatal behaviour and environment. *Advances in Botanical Research*, 4: 117–27.
Evans, J. R., T. D. Sharkey, J. A. Berry, and G. D. Farquhar. 1986. Carbon isotope discrimination measured with gas exchange to investigate CO_2 diffusion in leaves of higher plants. *Australian Journal of Plant Physiology*, 13: 281–92.
Farquhar, G. D., and S. von Caemmerer. 1982. Modelling of photosynthetic response to environmental conditions. In O. L. Lange, P. S. Nobel, C. B. Osmond, and H. Ziegler, eds., *Physiological plant ecology* 2, Encyclopedia of plant physiology, n.s., 12B: 549–87. Berlin: Springer-Verlag.
Farquhar, G. D., and K. Raschke. 1978. On the resistance of transpiration of the sites of evaporation within the leaf. *Plant Physiology*, 61: 1000–1005.
Field, C., J. A. Berry, and H. A. Mooney. 1982. A portable system for measuring carbon dioxide and water vapour exchange of leaves. *Plant, Cell and Environment*, 5: 179–86.
Field, C., and H. A. Mooney. 1987. Measurement of photosynthesis under field conditions—Past and present approaches. In P. J. Kramer, B. R. Strain, S. Funada, and Y. Hashimoto, eds., *Scientific instruments in physiological plant ecology*. New York: Academic. [In press.]
Grace, J., F. E. Fasehun, and M. Dixon. 1980. Boundary layer conductance of the leaves of some tropical timber trees. *Plant, Cell and Environment*, 3: 443–50.
Hewlett-Packard Corporation. 1980. *Application note 290: Practical temperature measurements*. Palo Alto, Calif.
Holmgren, P., P. G. Jarvis, and M. S. Jarvis. 1965. Resistance to carbon dioxide and water vapour transfer in leaves of different plant species. *Physiologia Plantarum*, 18: 557–73.
Jarman, P. D. 1974. The diffusion of carbon dioxide and water vapour through stomata. *Journal of Experimental Botany*, 25: 927–36.

Jarvis, P. G. 1971. The estimation of resistance to carbon dioxide transfer. In Z. Sestak, J. Catsky, and P. G. Jarvis, eds., *Plant photosynthetic production: Manual of methods*: 599–631. The Hague: Junk.

Jarvis, P. G., J. Catsky, F. E. Eckardt, W. Koch, and D. Koller. 1971. General principles of gasometeric methods and the main aspects of installation design. In. Z. Sestak, J. Catsky, and P. G. Jarvis, eds., *Plant photosynthetic production: Manual of methods*: 49–110. The Hague: Junk.

Jarvis, P. G., and R. O. Slatyer. 1970. The role of mesophyll cell wall in leaf transpiration. *Planta*, 90: 303–22.

Kacser, H., and J. A. Burns. 1973. The control of flux. In D. D. Davies, ed., *Rate control of biological processes*: 65–104. Symposia of the Society for Experimental Biology, 27. Cambridge: Cambridge University Press.

Kays, W. M., and M. E. Crawford. 1980. *Convective heat and mass transfer*. 2d ed. New York: McGraw-Hill.

Leuning, R. 1983. Transport of gases into leaves. *Plant, Cell and Environment*, 6: 181–94.

List, R. J. 1957. *Smithsonian meteorological tables*. 6th ed. Smithsonian Miscellaneous Collections, vol. 114. Washington, D.C.: Smithsonian Institution Press.

Meidner, H. 1975. Water supply, evaporation, vapour diffusion in leaves. *Journal of Experimental Botany*, 26: 666–73.

Meidner, H., and T. A. Mansfield. 1968. *Physiology of stomata*. London: McGraw-Hill.

Mott, K. A., and J. W. O'Leary. 1984. Stomatal behavior and CO_2 exchange characteristics in amphistomatous leaves. *Plant Physiology*, 74: 47–51.

Parkinson, K. J., W. Day, and J. E. Leach. 1980. A portable system for measuring the photosynthesis and transpiration of graminaceous leaves. *Journal of Experimental Botany*, 31: 1441–53.

Parkinson, K. J., and B. J. Legg. 1978. Calibration of infra-red analysers for carbon dioxide. *Photosynthetica*, 12: 65–67.

Pearcy, R. W., and H. W. Calkin. 1983. Carbon dioxide exchange of C_3 and C_4 tree species in the understory of a Hawaiian forest. *Oecologia*, 58: 26–32.

Penman, H. L., and R. K. Schofield. 1951. Some physical aspects of assimilation and transpiration. In J. F. Danielli and R. Brown, eds., *Fixation of carbon dioxide*: 115–29. Symposia of the Society for Experimental Biology, 5. Cambridge: Cambridge University Press.

Richards, J. M. 1971. Simple expression for the saturation vapour pressure of water in the range -50 degrees to 140 degrees. *British Journal of Applied Physics*, 4: L15–L18.

Samish, Y. B. 1978. Measurement control of CO_2 concentration in air is influenced by the desiccant. *Photosynthetica*, 12: 73–75.

Sestak, Z., J. Catsky, and P. G. Jarvis, eds. 1971. *Plant photosynthetic production: Manual of methods*. The Hague: Junk.

Sharkey, T. D., K. Imai, G. D. Farquhar, and I. R. Cowan. 1982. A direct confirmation of the standard method of estimating intercellular partial pressure of CO_2. *Plant Physiology*, 69: 657–60.

Sinclair, T. R., G. H. Buzzard, and K. R. Knoerr. 1976. A pressure method for frequent differential calibration of CO_2 infra-red analyzers. *Photosynthetica*, 10: 188–92.

Thom, A. S. 1968. The exchange of momentum, mass and heat between an artificial leaf and the airflow in a wind tunnel. *Quarterly Journal of the Royal Meteorological Society*, 94: 44–55.

Thorpe, M. R. 1978. Correction of infra-red gas analyser readings for changes in reference tube CO_2 concentration. *Plant, Cell and Environment*, 1: 59–60.

Whiteman, P. C., and D. Koller. 1967. Species characteristics in whole plant resistances to water vapour and CO_2 diffusion. *Journal of Applied Ecology*, 4: 363–77.

Woodrow, I. E., J. T. Ball, and J. A. Berry. 1987. A general expression for the control of the rate of photosynthetic CO_2 fixation by stomata, the boundary layer and radiation exchange. In J. Biggins, ed., *Proceedings of the 7th International congress on photosynthesis*. The Hague: Junk. [In press.]

Index of Subjects

ABA, see Abscisic acid
ABAGE (abscisyl-β-D-glucopyranoside), 268, 271
ABAGS (1′-O-abscisic acid-β-D-glucopyranoside), 268, 271
Abscisic acid (ABA), 21, 223, 285; and stomatal closing, 130, 135–39 passim, 151–55 passim, 255–63 passim, 267, 272f, 302, 333, 359, 371; and ion fluxes, 151–53, 301; and fluorescence transients, 204; blue-light response, 218–19; CO_2, 232–33, 265–66, 273f, 299ff, 301; and transpiration, 255; and wilting, 255, 267, 270ff, 274; effect on guard cell solutes, 256, 263–65, 301; enantiomers of, 256–58; action of analogues of, 257–58; concentration vs. length of exposure, 258; and pH, 258, 261, 265, 272f, 302; binding sites, 258–59, 261, 263, 265, 272f, 302; action modeled, 259, 272; and K^+, 261ff, 264, 301, 303; inhibition of effects, 262–63, 264f, 273; and respiration, 264; and Ca^{2+}, 265, 302; and temperature, 265f, 269, 302f; and IAA, 266, 294, 297–303 passim; and water stress, 266, 269–74 passim, 299ff, 304ff, 371; long-term effects of, 266–67; biosynthesis of, 267–70, 272; trigger for production, 269–70; metabolism of, 270–71, 272; distribution of, 272ff; and photosynthesis, 273–74, 333; and cytokinins, 281, 284, 286ff, 303; and Na^+, 295, 301; and FC, 301; and stomatal opening in CAM plants, 359; and leaf age, 371; genetic variation in ability to accumulate, 435, 437–38
Acacia tortilis, 324
Acer: campestre, 344; *saccharum*, 344
Achlya, 289
Action spectroscopy, and identification of guard cell photoreceptors, 197–99
Adiantum, 68f, 82f; *capillum veneris*, 49
Adenosine diphosphate (ADP), 134; -glucose pyrophosphorylase, 119–20
Adenosine 5′-triphosphatase (ATPase), 21, 121, 132, 164, 174, 221, 223

Adenosine triphosphate (ATP), 19, 21, 120, 156, 260; and salt accumulation, 131; and light, 134; and proton pump, 134–35, 179–80; and guard cell malate synthesis, 181, 187; and guard cell energy supply, 182–84; in blue light, 185–87; and photophosphorylation, 204–5; and oxidative phosphorylation, 205; and nitrogen distribution, 379
Agathis, 75
Agave: americana, 231, 358f, 361; *deserti*, 231, 361
Aging, see Leaves, aging
Albizzia, 143
Alfalfa, 433, 439
Alisma plantago-aquatica, 50
Alliaceae, 53
Allium, 53, 64, 74, 80f, 83, 170, 221; *cepa* (onion), 133, 215, 230, 260f
Alocasia macrorrhiza, 344
Alsophila, 43
Aluminum, 110
Amaranthus: powelli, 231; *retroflexus*, 231; *hypochondriacus*, 231, 239–40; *palmeri*, 235, 237; *tricolor*, 237, 288
Amaryllis formosissima, 12
Amaryllis-type stomata, 48f
Amphoteric colloid theory, 17
Amyloplasts, 81
Anarthriaceae, 61
Aneimea, 45
Aneimea-type guard cells, 43f
Angiopteris-type guard cells, 44
Angiosperms, 40f, 45, 48–54, 229
Anions, 117, 119, 259ff, 264, 272. See also specific anions by name
Anomodon attenuatus, 39
Anthephora pubescens, 282f, 285f, 289
Anthoceros, 34, 40; *crispulus*, 35; *punctatus*, 49
Anthocerotae (hornworts), 29, 33–34, 35f
Anthocyanin, 288
Anticlinal walls, 59, 69, 76
Antiporter, H^+/K^+, 164f, 261
Aperture, stomatal: size of, and epidermal

Index of Subjects

Aperture (*continued*)
  cells, 91, 96f, 104f, 108; and guard cell walls, 92–93, 94, 101–4; and lumen volume, 97, 103, 105ff; and starch, 116; and K-salt levels, 125, 127; and cell ion contents, 135–37; and osmotic pressure, 143f, 155, 176, 188–89; ion fluxes and proton pumping, 223; measurement of, 323
Apoplast, 272f
Apple, 229f, 241, 434, 439
Aquatic species, 36, 38, 42, 50f, 62
Araceae, 33
Arachis hypogaea (groundnut), 439
Araucaria, 75
Arbutus unedo, 325, 329–34 *passim*, 337
Argyrodendron peralatum, 341ff, 344
Arid-habitat plants: diurnal patterns in leaf conductance, 326–29; midday stomatal closure, 329–37
Asparagales, 53
Aspidiaceae, 43
Aspidium, 43
Aspleniaceae, 43
Asteroxylon, 40; *mackiei*, 41
ATP, *see* Adenosine triphosphate
ATPase, 21, 121, 132, 164, 174, 221, 223
Atriplex halimus, 231
Atropa belladonna, 39
Auxins: early research on, 293–94; and stomatal closure, 293f, 296; synthetic, 293, 296; and transpiration, 293; and photosynthesis, 293; and respiration, 293; and CO_2, 293; and K^+, 294; and starch metabolism, 294; and stomatal opening, 294, 296; and adaxial vs. abaxial stomata, 295–96; and water stress, 299ff, 303. *See also* Indol-3-ylacetic acid
Avena sativa, 230, 282
Avicennia, 67
Azolla, 61, 64, 69f, 79
Azolla-type guard cells, 43

Bacopa caroliniana, 230
Barley, 129, 133, 230, 241, 282, 434
Bean, *see Phaseolus vulgaris*
Beet, 129, 264
Bell pepper, 370, 372
Bending-beam models, 96f
Betacyanin, 288
Beta vulgaris (beet), 129, 264
Betula, 238; *pendula*, 237
Bis-tris-propane chloride, 263
Blechnaceae, 43
Blechnum, 43
Blue light, 17, 19; and hyperpolarization of guard cell plasmalemma, 174, 184f, 221–22, 223; and proton extrusion, 174, 184–85, 215–19, 221–22; and stomatal opening, 184–87, 199–205 *passim*, 224; and intercellular CO_2 levels, 196f, 210–11; and stomatal conductance, 197ff, 210–18 *passim*, 221ff; stomatal responses to, 201ff, 205, 210–11, 214–15; and proton pump, 205, 215, 221–22, 223; and transpiration, 210; and photosynthesis, 210–11; and reciprocity, 212; kinetics model of response, 213–14, 222; light vs. dark reactions, 214–15; acidification response, 216–21 *passim*; and respiration, 217; effect of metabolic inhibitors on, 218–19; and plasmalemma ATPase and proton pump, 221, 223; and PAA, 296
Blue panic grass, 434, 439
Bothriochloa caucasia, 231, 241
Boundary layer, 357, 445f, 471
———conductance, 15, 401f, 405–7, 409, 459, 462; and intercellular CO_2 levels, 298; forced vs. free convection, 402–7 *passim*, 410; laminar vs. turbulent flows, 402–3; to water vapor flux, 462–64, 467; to CO_2 flux, 465–66, 467f; dual-stream measurements of, 469f
Bowenia spectabilis, 40, 46
Brassica: pusna, 230; *campestris*, 282f; *juncea* (Indian mustard), 434
Bromine ions: and stomatal aperture, 138; fluxes, 141, 144–46; intracellular distribution of, 142–44; fluxes in light vs. dark, 147–51, 153; fluxes and ABA, 151–53, 261; fluxes and FC, 155–56
Bromus inermis, 433f
Bryaceae, 36
Bryophyllum daigremontiana, 354ff, 357f
Bryophytes, 29f, 33f, 41. *See also* Anthocerotae; Liverworts; Mosses
Bryum, 37
Buffering capacity, of guard cells, 117
Buxbaumia aphylla, 37

C_3 plants, 204, 329, 353, 466; and CO_2, 229ff, 235, 237ff, 242, 298f, 344; shift to CAM, 357, 361
C_4 plants, 119, 283, 329, 466; and CO_2, 229, 231, 235, 237ff, 242, 298
Cacti, 353f, 357
Calcium, 20, 83; ions and stomatal movement, 129f, 154, 224; iminodiacetate, 132; ions and ABA, 265, 302; ions and cytokinins, 288f
Calla palustris, 50
Callitriche palustris, 51
Callose, 75

Index of Subjects

Calmodulin, 288f, 302
Calvin (-Benson) cycle, 19, 120f, 170, 220, 273
CAM plants, *see* Crassulacean acid metabolism plants
Canna indica, 229f
Canopies, plant: scalar conservation equations for, 386–90; turbulent fluctuation in, 386–87, 390–91; eddy fluxes in, 387–400 *passim*, 403, 405, 419–22; sources and sinks at leaf surfaces, 388–89, 399–400; volume-averaging equations, 388–89; dispersive flux effect, 389f, 419; and molecular flux divergence, 389–90; and wind speed, 391ff; transport in, 393f; intermittency of momentum transport, 393–98 *passim*; heat fluxes in, 393–99 *passim*, 410–11; ejections and sweeps, 394, 397f; momentum transfer, 395, 397f, 418–19; microclimates in, 399–400, 403, 419; exchange processes at leaf surfaces, 400–407; combination equation for concentration fields, 407–12; and radiation fluxes, 407–18 *passim*; single-leaf models, 412–19; second-order closure models, 419–22; Lagrangian models, 422–26; Eulerian models, 422, 424, 426; near-field vs. far-field effects, 424; character of, and stomatal conductance, 437f
Capparis spinosa, 324
Capsicum: annuum, 230, 435; *frutescens* (bell pepper), 370, 372
Caralluma negevensis, 361
Carbon: metabolism of guard cells, 116–20, 259–61; import into guard cells, 170f; balance, and water stress, 328; gain, vs. water-use efficiency and stomatal behavior, 345f; gain, vs. water loss in CAM and C_3 plants, 353
Carbon dioxide, 115, 120; early research on, 9, 14, 17ff, 20f; and cuticle, 29; and guard cell metabolism, 117ff; and stomatal activity, 130, 140, 151, 195–97, 201f, 229–40 *passim*, 243–44, 299, 358, 361ff; and proton pump, 134f; and fluorescence transients, 135, 204; guard cells' ability to fix, 170, 182, 204, 220; and sucrose formation, 182; light, and stomatal opening, 195–97, 201; intercellular, and blue light, 196f, 210–11; concentration, and red- and blue-light thresholds, 199f, 203; and CAM plants, 229, 231, 234, 353–54, 358, 361ff; and conductance, 232–40 *passim*, 345; and stomatal responses in light and dark, 232, 234, 236, 239–40; and water stress, 232, 235,
266, 298ff, 301; and ABA, 232–33, 265–66, 273f, 299ff, 301; and intercellular concentrations and stomatal opening, 233–40 *passim*; assimilation rates, 234–39 *passim*, 242, 318ff, 323f, 326; and humidity, 237, 319f; and temperature, 238, 265f; and leaf age, 238–39, 370, 374–75, 376f; high concentrations of, and stomatal conductance and water-use efficiency, 240–43; mechanisms of stomatal opening, 243–44; and cytokinins, 287f; and auxins, 293; and IAA, 297ff, 301; intercellular concentrations in C_3 and C_4 plants, 298f, 344; intercellular concentrations and wind speed, 298f; uptake, in arid habitats, 328, 332, 334, 335–36; levels, and light intensity in humid climates, 342ff; and photosynthesis, 345, 436, 438, 445, 458, 466; transport in forest canopies, 393f, 399; breeding for improved uptake, 431–32; measurement of, 446–48; experimental control of concentrations, 450, 452–54, 457; calculation of assimilation rates, 456–58; calculation of partial pressures of, 456, 466; cross-sensitivity with water vapor in gas-exchange systems, 456–57; calculation of mole fractions, in gas-exchange systems, 457f; uptake, through nonstomatal surfaces, 459; calculation of conductances based on binary diffusion, 459f; calculation of intercellular, 460, 466; ternary diffusion of, 460–62; calculation of conductances to flux, 464–66, 467; concentration at leaf surface, 467f; dual-stream measurement of, 469
Carbon reductants, 120f
Carbon-reduction pathway, 120
Carbonyl cyanide m-chlorophenyl hydrazone (CCCP), 218
Carboxylation efficiency, 332ff
Carica papaya, 309
Carotenoids, 202
Carrot, 129
Cassia obtusifolia, 230, 241
Cations, and guard cell wall elasticity, 110–11. *See also specific cations by name*
Cell division, 62–65, 83
Cell plates, positioning of, 63f
Ceratopteris thalictroides, 43
Ceratopteris-type guard cells, 43
Cesium chloride, 129
Chara, 133f, 156, 260
Characters, stomatal, breeding for: problems in, 431, 439–40; objectives, 431–32,

480 Index of Subjects

Characters (*continued*)
 436–39; genetics of, 432–36; intraspecific genetic variation in, 433f
Chemiosmotic mechanisms, 165ff, 169–70, 222
Chlorine ions, 53, 68; and K$^+$, 117; fluxes, 129–34 *passim*, 148; and stomatal opening, 164, 169, 176; concentrations in *Vicia faba* guard cells, 169; in chemiosmotic models, 169f; energy costs of uptake, 180; passive uptake of, 222; -H$^+$ cotransport mechanism, 259ff; import into guard cells, 260; and ABA, 261ff, 264
Chlorophyll, 81; in guard cells, 80, 121; and red-light response of stomata, 200–201, 203; and blue light, 201–2, 203f; in guard cells vs. in mesophyll cells, 373; levels during leaf aging, 377
Chlorophytum comosum, 182
Chloroplasts, 45, 80, 82, 118, 121, 173; lack of, in guard cells, 52–53; in guard cells vs. in mesophyll cells, 81, 201, 205, 220; CO$_2$ reduction in guard cells, 135, 204, 220; photophosphorylation in, 135, 172–74, 182ff; and ion transport, 259; and ABA synthesis, 269, 271f
Chrysanthemum, 287
Cibotium-type guard cells, 44f
Citrate, 170
Citric-acid cycle, 260
Citrus: aurantium, 30; *sinensis* (orange), 371
Clivia, 12
Closure, stomatal, 223–24; midday, 15, 20, 328–37 *passim*; and darkness, 130, 134f; and CO$_2$ levels, 130, 298f; and ABA, 130, 135–39 *passim*, 151–55 *passim*, 255–63 *passim*, 267, 272f, 302, 333, 359, 371; initiation of, 146–47, 253, 255; ion fluxes during, 153–54, 259–61; and malate, 172; and PA, 258; and water stress, 266–67, 356–63 *passim*, 371; and auxins, 293f, 296; and temperature, 302, 329ff, 332; and low humidity, 312, 315f, 320, 329; and dry soil, 316ff, 319–20, 329, 333–35; and light intensity, 329, 330–32; and leaf water potential, 329; and photosynthesis, 332–33; diurnal patterns in humid-zone plants, 339; and leaf age, 371
Commelina, 81, 204; *benghalensis*, 283
——*communis*, 21, 50, 178; K$^+$ content of guard cells, 127–28, 136, 140; osmotic pressures in, 129, 137–38, 139, 146; stomatal opening in, 129–30, 131; ion fluxes in, 134, 140–53 *passim*; and ABA, 151–53, 256f, 261f, 266, 297, 299ff;

blue-light responses, 210–16 *passim*; stomatal responses to CO$_2$, 230, 234–40 *passim*, 299ff; and PA, 258; and cytokinins, 283, 285f; FC and stomatal movement, 295; and IAA, 296ff, 303
Commelinaceae, 50, 287
Compositae, 287
Conductance, stomatal, 19, 224; in blue light, 197ff, 210–18 *passim*, 221ff; in red light, 199ff; and intercellular CO$_2$ levels, 232–40 *passim*, 345; and leaf aging, 238–39, 328, 358, 369–79 *passim*; and water-use efficiency in high CO$_2$ concentrations, 240–43; and leaf water potential, 311–19 *passim*, 371; and humidity, 312, 314ff, 319; and soil water potential, 312, 316ff, 319; measurement of, 323–26; diurnal patterns in arid habitats, 326–32 *passim*, 336ff; and photosynthesis, 326–31 *passim*, 334f, 345, 374–80 *passim*; and temperature, 328, 332f, 359; and light intensity, 330–31, 341–45; and water stress, 334, 340–41, 346, 361f, 370f; diurnal patterns in humid climates, 339–40; carbon gain vs. water use, 345f; diurnal patterns in CAM plants, 356, 358, 360; techniques for studying in aging leaves, 367–68; and environmental conditions, 370–71, 379; and leaf-insertion level, 371–72; and ontogenetic changes in guard cells, 372–73; and cuticular depositions, 373; and chlorophyll levels, 377; and concentration fields in plant canopies, 408ff; in single-leaf model of plant canopies, 413–18 *passim*; breeding for improved, 431–39 *passim*; genetics of, 433–34; and character of plant canopies, 437f; vs. resistance, 458; vs. cuticular conductance, 459; vs. boundary-layer conductance, 459; calculations based on binary diffusion, 459–60; calculations based on ternary diffusion, 460–62; to water vapor, 463, 467; to CO$_2$, 465, 467; dual-stream measurements of, 469f
Conifers, 373
Conocephalus ovatus, 53
Corn, *see Zea mays*
Cotton, 229f, 241, 369, 372, 434
Cotyledons, 38f, 45f
Countermicellae elasticity law, 94
Crassula, 360
Crassulaceae, 287
Crassulacean acid metabolism (CAM) plants, 119, 256, 260; stomatal responses to CO$_2$, 229, 231, 234, 353–54, 358, 361ff; ratio of carbon gain to water loss, 353; diurnal

Index of Subjects

patterns of gas exchange, 353–54; stomatal morphology, 353–57, 362; diurnal patterns of stomatal activity, 357; and leaf age, 358; and photoperiod, 359, 361; and thermoperiod, 359; and leaf temperature, 359; and ABA, 359; and cytokinins, 359; and K^+, 359–60; and water, 360ff, 363; mechanism of, 362–63
Crataegus × *macrocarpa*, 371
Crotalaria spectabilis, 230, 241
Cruciferae, 287
Cuticle, 29–30, 33, 79; in Pteridophyta, 40; in gymnosperms, 45f; in angiosperms, 48; formation of, 67f, 70, 82ff; permeability to water, 78–79; depositions during leaf aging, and stomatal conductance, 373; conductance of, 459
Cutin, 29, 31, 33
Cuvettes, use of, 445f, 455–56, 464, 466f, 470f; varieties of, 455; laminar vs. turbulent air flows in, 455, 458, 465–66; dual-stream measurements of, 469
Cyathea capensis, 45
Cyatheaceae, 43
Cyathea-type guard cells, 44
Cyperaceae, 12, 48, 61
Cypripedioideae, 53
Cytokinins, 62–65, 69; and ABA, 281, 284, 286ff, 303; and stomatal movement, 282–88 *passim*, 299, 303, 305–6, 359; and transpiration, 282f, 286; and light intensity, 282–85, 288; and leaf age, 284, 288; and darkness, 285–86; and CO_2, 287ff; mechanism of action, 288–89; synthetic, 299; in CAM plants, 359. *See also* Kinetin; Zeatin
Cytoplasm: ion content of, 142–44, 145f, 150, 153–54, 157; and malate formation, 173; and ABA, 258f, 261; and ion transport, 259f; effect of ABA and CO_2, 266; production of ABA, 269, 271f

Darkness: and stomatal movement, 130, 134ff, 232, 294; and proton efflux, 132, 147; and ion fluxes, 147–51, 153, 156; and CO_2, 232, and ABA, 271f; and cytokinins, 285–86; and IAA, 294
Dark respiration, 378; rate in *Vicia faba* guard cells, 176; energy supply for, 182–83; in aging leaves, 369, 370–71, 373; in young leaves, 370
Daucus carota (carrot), 129
Davalliaceae, 43
DCMU [3-(3,4-dichlorophenyl)-1,1-dimethylurea], 174, 196, 200–205 *passim*, 220, 377–78
Deciduous species, 370

Deformation (of cells), 99f, 110
Desmodium paniculatum, 230
Development, stomatal: in mosses, 37; in ferns, 44–45; in gymnosperms, 46–47; PPBs and cytokinesis, 62–65; wall building, 65–67, 70–76; of cavity and pore, 67–70; of organelles, 79–84. *See also* Evolution, stomatal
Dichodontium pellucidum, 39
Dicksoniaceae, 43
Dicotyledons, 48, 61, 214, 229, 293, 333
Dictyosome vesicles, 65f, 82
Diethylstilbestrol (DES), 218, 223
Diffusion, 14–15; through stomata vs. through boundary layer, 445–46, 459; use of Fick's law, 446, 459, 465; binary, 459–60; ternary, 460–62
Dioon edule, 46
Diplacus aurantiacus, 237
Diplazium celtidifolium, 49
Diplazium-type guard cells, 43, 49
Dipteridaceae, 43
Dipteris-type guard cells, 43
Disease resistance, and breeding for lower stomatal conductance, 432
Dorsal wall, 59f, 63, 66, 71ff, 75, 92ff, 96
DPA (4'-dihydrophaseic acid), 268, 270f
DPAGS (DPA-4'-O-β-D-glucopyranoside), 271
Drymoglossum-type guard cells, 44f
Dryopteris ludoviciana, 45
Dudleya farinosa, 361

Eddy fluxes, 387, 389f, 405; diffusibility, 390, 399, 413–14, 419f, 422; structure of, 390–400; and turbulence, 392–93, 420ff; at leaf boundary layer, 403, 405; in second-order closure models, 419–22
Electrical gradients, across plasmalemma, 166f, 169, 180
Electrochemical potential, 164. *See also* Membrane potential
Electron lucency, 69, 75
Electron transport, 45, 120–21, 373
Encalyptra ciliata, 34
Endoplasmic reticulum (ER), 83, 354f
Environment: and number of stomata, 36; and guard cell walls, 111. *See also* Humidity; Temperature; Water stress; Wind speed
Enzymes, 204f
Epidermal cells/epidermis: in stomata-less plants, 36; in mosses, 36f; in angiosperms, 50; and pore apertures, 91, 96f, 104f, 108; water-extracting powers of, 92; and stomatal mechanics, 94, 96f; interactions with guard cells, 104–9, 111;

Epidermal cells (*continued*)
  mechanical advantage of, 104–5, 108f, 111; lumen of, 109; malate in, 118; ABA content, 272ff; and cytokinins, 284; adaxial vs. abaxial, 294ff; turgor of, and air humidity, 313–14; water potential, 314; waxy, in CAM plants, 354f; water loss through, 459
Equisetatae, 42
Equisetum, 67, 71, 73; *maximum*, 41; *hyemale*, 74
Erithryna caffra, 18
Eucalyptus pauciflora, 196, 230, 236f, 240
Euphorbia, 345
Evaporation, and concentration fields, 409–17 *passim*
Evolution, stomatal: in liverworts, 33; in Anthocerotae, 34; in mosses, 35ff, 38; in Pteridophyta, 40–45; in gymnosperms, 45–47; in angiosperms, 48–54

Fagus sylvatica, 230, 339
Farnesol, 257
Farnesyl pyrophosphate, 267
Ferns, 42–45, 61, 68, 75, 81
Ferocactus acanthodes, 355
Ferrodoxin, 121
Filicatae, 42–45, 61, 68, 75, 81
Finite-element models, 96–97
Flagellariaceae, 61
Flavins, 201f, 205, 222, 296
Floating-leaf plants, 42f, 50–51
Flow rates, 450–54
Fluorescence, 133, 135, 204
Foeniculum officinale, 8
Forests, 390, 393f, 399f, 414, 426
Fragaria virginiana, 379
Front chamber, 59, 68
Funaria, 36ff, 61, 64, 67–84 *passim*; *hygrometrica*, 289
Funariaceae, 36f
Fusicoccin (FC), 131–38 *passim*, 154ff, 183, 218, 262f, 288, 294f, 301

Gametophytes, 33f, 38, 40–41, 43
Gas exchange, 9, 13; measurement of, 323–26; in arid climates, 328–29, 334; in humid climates, 339; of CAM plants, 353–54; shift between C_3 and CAM, 357, 361; binary treatments of, 459–60; ternary treatments of, 460–62; dual stream, 468–70
——systems: controlled-environment, 367–68; open-flow vs. transient, 446; differential vs. compensating open-flow, 447–48, 451, 453, 458; calculation of flow rates and gas concentrations, 450–54; calculation of transpiration, 454–55; ambient mole fraction water vapor, 455–56; calculation of CO_2 assimilation rates, 456–58; ambient mole fraction CO_2 in, 458; dual-stream, 468–69
Geraea canescens, 237
Ginkgo biloba, 45f, 164, 373
Gleicheniaceae, 43
Gluconeogenesis, 119, 170, 172, 259
Glucose, 175
Glyceraldehyde 3-phosphate dehydrogenase, 204
Glycine max, 214, 230, 241, 337, 370ff, 375, 434
Glycolysis, 116, 259
Glycophytes, 129
Gmelina arborea, 237, 339
Gnetinae, 47
Gnetum gnemon, 46f
Gossypium hirsutum (cotton), 229f, 241, 369, 372, 434
Gramineae, 12, 18, 285; stomatal structure, 37, 48, 61, 67–77 *passim*; and blue light, 210, 214; CO_2, 237; and cytokinins, 282–83, 287
Graminean-type stomata, 49
Grana, 80–81, 172
Grasses, *see* Gramineae
Groundnut, 439
Guard-cell parent cells (GPCs), 63ff, 67, 83, 372
Guard cells, 59; early research on, 10–11, 13, 20–21; in hornworts, 34; in mosses, 36–37, 40; in Pteridophyta, 40ff, 43f; lignification of, 41f, 45f, 48, 75f; silicification of, 42; in gymnosperms, 45ff; in angiosperms, 48; dumbbell-shaped, 60–61, 64f, 70ff, 74; kidney-shaped, 61, 63, 65, 71–73, 74; variations in structure, 61–62; shape of, and thickening of walls, 70–73, 82; changes in shape and volume, 72f, 92–93, 131, 253–55; pectins in, 75, 110; historical models of mechanics, 92–93, 96–97; water-extracting power of, 92; expansion of, 92; in non-grass species, 93, 95; wall structure and mechanics, 93f, 96, 101–4; volume of lumen, 94–97, 101–6 *passim*; interactions with epidermal cells, 94, 96f, 104–9, 111; as percent of leaf volume, 115; osmoregulation in, 116, 129, 253–55; behavior, in epidermal strips, 129–31; energy requirements for maintenance, 164, 175, 182f; energy requirements for stomatal opening, 164–72, 178–81, 185; energy sources, 172–74,

182–87; energy demands, 175–82; in CAM plants, 353–57, 362. *See also constituent parts and internal processes by name*
Gymnosperms, 40–50 *passim*, 61f, 72–78 *passim*, 229, 329
Gymnosperm-type stomata, 49

Haemeria discolor, 80
Hakea suaveolens, 52
Halophilic species, 53
Halophytes, 129
Hedera helix, 48
Helianthus: annuus, 230, 282f, 285, 313, 318; *nuttallii*, 314ff
Heliconiaceae, 61
Helleborus, 12
Helleborus-type stomata, 48f
Hexose, 139
Hilaria rigida, 231, 241
Hill oxidant, 121
Hordeum vulgare (barley), 129, 133, 230, 241, 282, 434
Hormones, and stomatal mechanics, 110–11. *See also specific hormones by name*
Hornea, 40
Hornworts, 29, 33–34, 35f
Humid-climate plants: diurnal patterns, 338–39; responses to humidity, 339–40; responses to light fluctuations, 339, 341–45; responses to water potential, 340–41
Humidity, 20; and CO_2, 237, 319ff; and stomatal response, 237, 311–20 *passim*, 345, 370f, 375, 467f; and transpiration, 311f, 315, 320; and turgor, 313–14; and xylem water potential, 313f; and leaf water potential, 314–15; and midday stomatal closure, 329ff, 337; and photosynthesis, 332, 336; and humid-zone plants, 339–40; and aging leaves, 370f, 375; fluxes in canopies, 399, 402, 408; control of, in gas-exchange systems, 448, 451–52, 454; at leaf surfaces, 467f
Hydathodes, 51f, 53
Hydrocharis morsus-ranae, 51
Hydrogen ions, 165ff, 169, 180
Hygrometers, use of, 452
Hymenophyllaceae, 43
Hymenophyllopsidaceae, 43

Indian mustard, 434
Indol-3-ylacetic acid (IAA), 293; and ABA, 266, 294, 297–303 *passim*; and proton extrusion, 266, 301; and temperature, 269, 302–3; and stomatal opening, 294–96,

Index of Subjects 483

303; and Na^+, 295; and CO_2, 297ff, 301; and water stress, 299ff, 303
Infrared gas analyzers (IRGAs), use of, 446ff, 450f, 456–57, 458
Inner walls, 59f, 63, 66, 70, 79, 92f, 109
Ion fluxes: regulation of, 128–29, 140, 156–57; in *Vicia faba*, 140, 147; in *Commelina communis*, 134, 140–53 *passim*; and tonoplasts, 141, 259ff; and external ion concentration, 142ff, 145f, 150, 153–54, 157; transients, 146–47; and light, 147–51; and ABA, 151–53, 301; in stomatal closing, 153–54; in stomatal opening, 154–56; and proton pumping, 223; and CO_2, 301; and IAA, 301. *See also* Proton extrusion
Ion transport, 20–21, 45, 131–35, 145–46, 154, 164–65, 168; and light, 134, 145, 147–51, 156; chemiosmotic mechanism, 166–67, 169–70; regulation of, 167; at steady state, 167–69; energy costs of, 178–81; and ABA, 259–64 *passim*; 287; during stomatal movement, 259–61; and cytokinins, 287ff
Isoetaceae, 42
Isoetes, 42; *malinvernianum*, 41

Juncaceae, 48, 50
Juncus effusus, 50
Juniperus macrocarpa, 49

Kalanchoe: daigremontiana, 256, 283, 286, 359, 363; *blossfeldiana*, 358ff
Kinetin, 282–89 *passim*, 303ff, 359
Kochia indica, 230
Kranz cells, 47

Larix, 47, 79; *leptolepis*, 76
Laurus nobilis, 48
Leaf-surface impressions, 323
Leaves: abaxial surfaces, 294ff, 301, 373, 434; adaxial surfaces, 294ff, 301, 373, 434; and IAA, 294ff, 301; water potential, 311–20 *passim*, 327ff, 330, 334, 340, 371, 460; temperature of, 314, 320, 329f, 333, 336, 338, 341ff, 359, 408, 421; surface and turbulence, 388–89, 399–400; exchange processes, 400–407; water vapor transfer, 400, 405–11 *passim*; molecular diffusion at, 401–3; heat transfer at, 401–11 *passim*; humidity gradients, 402; forced vs. free convection, 402–10 *passim*; effects of clustering, 405f; evaporation, 409–17 *passim*; gas concentrations at surfaces, 445, 466–68; experimental control of temperature, 456,

Leaves (*continued*)
468; dual-stream measurements of, 468–70
———, aging: and stomatal conductance, 238–39, 328, 358, 367–79 *passim*; and CO_2, 238–39, 370, 374–75, 376f; and cytokinins, 284, 288; techniques for studying, 367–68; developmental stages, 368–69; protein content, 369f, 377; RNA content, 369f, 377; and photosynthesis, 369–70, 374–80 *passim*; and environment, 370–71, 379; leaf-insertion level, 371–72; and shading, 372, 375, 378; and reproduction, 372; and ontogenetic changes in guard cells, 372–73; cuticular deposition, 373; and intercellular CO_2 levels, 374–75, 376f; and transpiration, 375–76; and water-use efficiency, 376; and acclimation, 376–80; nitrogen levels, 376, 379–80; chlorophyll levels, 377; RuBPCase levels, 377; regulation of senescence and, 377–78; and plant resource allocations, 378–80
Ledges, 59ff, 62, 68f, 73f, 77, 79
Leguminosae, 64–65, 288
Lemnaceae, 51
Lemna minor, 51f
Lepechinia calycina, 370f, 374f, 379
Light intensity: and stomatal activity, 17ff, 129f, 135–41 *passim*, 145, 195–97, 201f, 204, 288, 331, 345, 370–75 *passim*; and guard cell metabolism, 119ff; and proton pumping, 134–35; and ATP, 134; and ion transport, 134, 145, 147–51, 156; and CO_2, 195–97, 201, 232–40 *passim*; stomatal location and sensitivity to, 196, 294; and ABA production, 271f; and cytokinins, 282–85, 288; and midday stomatal closure, 329–32, 337f; and diurnal patterns in humid climates, 339, 341–45; and photosynthesis, 339, 341–45, 374f; and leaf aging, 370–71
Lignification, 41f, 45f, 48, 75f
Liliaceae, 39, 50
Lilianae, 53
Lindsaya, 43
Lipid globules, 82f
Liquidambar styraciflua, 230, 241
Liverworts, 29f, 33
Lolium perenne (rye grass), 398, 434
Lowiaceae, 61
Loxsomaceae, 43
Lumina, 34, 36, 42, 62f, 69, 71, 94–97, 101–9 *passim*
Lycopersicon: esculentum, 30, 230, 238, 241, 267, 270f, 303, 435; *pennellii*, 63, 81

Lycopodiatae, 42
Lycopodium squarrosum, 41f
Lyellia crispa, 34

Magnesium: ions and ABA, 273; perchlorate, 457
Mahonia aquifolium, 48
Malate, 53, 120; and stomatal activity, 117f, 139–40, 164, 169–72, 176, 181, 205, 259f; and FC, 155; ion concentrations in *Vicia faba*, 169; formation and catabolism, 173, 205; energy costs of, 181–82; and blue light, 187, 203; and red light, 203; and CO_2 fixation, 204, 266; and ABA, 259, 261, 265f; in CAM plants, 362
Malformations, stomatal, 37–38, 39, 45, 47, 51
Malic acid, 131, 260, 362–63
Malus pumila (apple), 229f, 241, 434, 439
Mammalaria dioica, 355
Marantaceae, 61
Marathales, 43
Marattia, 43
Marattiales, 43f
Marchantia, 32f
Marsileales, 43
Matoniaceae, 43
Mechanics, stomatal: historical models of, 92–93, 96–97; and structure of guard cell walls, 93f, 96, 101–4; and volume of lumen, 94–96, 101–6 *passim*; interactions of guard cells and epidermal cells, 94, 96f, 104–9, 111; bending-beam models, 96f; finite-element models, 96–97; volume-expansion models, 97; and polymer elasticity theory, 97–101; and guard cell wall mechanics, 101–4; and water stress, 110; and hormones, 110–11; and minerals, 110
Medicago sativa (alfalfa), 433, 439
Membrane potential, of guard cells, 133, 153, 164, 168f, 221
Meristemoids, 65
Mesophyll, 195f, 273, 319–20, 345, 373; chloroplasts in, vs. in guard cells, 81, 201, 205, 220
Mevalonic acid, 268f
Micelles, 37, 48, 94, 99f
Microbodies, 82, 354
Microfibrils, 99
Microtubules, 62–67 *passim*, 82, 84
Middle lamella, 69f, 75
Millet, 371, 435
Minerals, and stomatal mechanics, 110
Mitochondria, in guard cells, 63, 82, 116, 118; and respiration, 172, 182–83; and malate, 172; oxidative phosphorylation in,

172; and ion transport, 259; in CAM plants, 354f
Mitosis, and cytokinins, 289
Mitotic spindle, 63
Mniaceae, 36
Mnium, 12, 36, 42; *cuspidatum*, 49
Mnium-type stomata, 49
Monocotyledons, 48–50, 61, 214, 229, 293, 333
Mosses, 29, 34–40, 61
Motorphase, 94–95, 101, 103, 111
Movement, stomatal, 15–17, 37, 42, 45, 48, 83–84, 116ff. *See also* Closure, stomatal; Opening, stomatal

N^6-benzyladenine (BA), 282f, 286ff, 289
NADH, 117
NADPH, 121, 205
Naphth-1-ylacetic acid (NAA), 293, 296
Naphth-2-yloxyacetic acid (NOXA), 293, 296
Nectaries, 51–52, 53
Neottia nidus-avis, 39
Nepenthes, 52f
Nerium oleander, 8, 64, 241, 313, 316ff
Nicotiana: tabacum, 60, 80f, 256, 294, 369, 372f, 375; *glauca*, 237
Nitella, 135
Nitrates, 129
Nitrogen, 376, 379–80
Nolina, 354
Notholaena parryi, 235
Nothoscordum inodorum, 50
Nucleus, of guard cells, 60, 62f, 66f, 72, 354f
Nuphar lutea, 51
Nymphaea alba, 51
Nymphaeaceae, 51

Onion, 133, 215, 230, 260f
Opening, stomatal: and K^+, 125, 127ff, 135–40, 164–69 *passim*, 176, 359–60; in intact cells vs. in isolated cells, 127–28, 130f; speed of, 128; in baths of various salts, 129–31; in light vs. in dark, 129f, 134–41 *passim*; and pH, 129f; and turgid subsidiary cells, 130; and proton extrusion, 131–32, 165–66, 167, 176; and ATP, 135; and osmotic pressure, 137–40, 176–78; and sugars, 139; and malate, 139–40, 169–72, 205; energy costs of, 164–72, 178–81, 185; metabolic inhibitors of, 164; chemiosmotic mechanism of, 166–67; ionic steady state during, 167–69; energy supply for, 172–74, 182–87; and influx of water, 175–76; in blue light, 184–87, 199–205 *passim*, 224; and light, 195–97, 201f, 204, 331; and CO_2, 195–97, 201f, 233–40, 243–44, 358; and mesophyll cells, 195; action spectrum of, 198–99; and photoreceptors, 199–200; in red light, 199–204 *passim*; proton pump and ion fluxes, 223; and humidity, 237; and temperature, 238, 331; ion transport during, 259–61; and cytokinins, 282–85, 303, 305–6, 359; and IAA, 294–96, 303; and PAA, 296; and water potential, 313; diurnal patterns in arid habitats, 328; in CAM plants, 353–63 *passim*; and leaf age, 358; and ABA, 359; and water stress, 359ff, 362
Ophioglossales, 43f
Ophioglossum, 42ff; *pendunculosum*, 41
Opuntia: ficus-indica, 231, 354, 358; *basilaris*, 356, 360ff; *inermis*, 360f
Orange, 371
Orchidaceae, 39, 53, 80–81
Orchidoideae, 53
Organelles, in guard cells, 60, 67, 79–84
Orthophosphate, 120
Oryza sativa (rice), 62, 230, 237, 239, 241, 371
Osmoregulation, 53, 116, 120, 129, 253–55
Osmotic potential, 95, 97, 110–11
Osmotic pressures: and K-salts, 137–40, 146; and stomatal opening, 137–40, 176–78; and change in aperture size, 143f, 155, 176, 188–89; and turgor and guard cell volume, 253–55; and ABA, 261, 269f
Osmundales, 43
Osmunda-type guard cells, 43f
Outer wall, 59f, 62, 66, 70, 75, 79, 92f, 109
Oxaloacetate, 119, 362
Oxaloacetic acid, 117, 171
Oxidative pentose phosphate pathway, 116
Oxidative phosphorylation, 135, 172, 182–83, 186, 205, 259
Oxygen, 9, 29, 182–83, 204, 220, 461
Oxyrhynchium swartzii, 39

Palmae, 61
Pandanaceae, 48
Panicum antidotale (blue panic grass), 434, 439
Papaya, 309
Paphiopedilum, 53, 80, 172, 196, 202, 224; *harrissianum*, 214, 220; *leeanum*, 230
Parasitic plants, 38f, 51
Parenchyma, 119f
Parkeriaceae, 43
Paspalum plicatulum, 231, 237, 239, 241

Index of Subjects

Patch clamping, whole-cell, 165f
Pea, see Pisum sativum
Pectins, 75, 110
Pelargonium zonale, 80, 230
Pennisetum: typhoides, 231, 241; *glaucum* (millet), 371, 435
Periclinal walls, 59, 68–69, 71ff
Perilla frutescens, 230
Perityle emoryii, 237
Permeability theory, 17
pH, 17f, 117, 119; and stomatal movement, 117–18, 129f, 132; gradients across plasmalemma, 131; of vacuoles, 132–33, 260; and Cl⁻ fluxes, 133; and ion transport, 134, 154, 156; and proton pumping, 167, 178; and malate, 170; and ABA, 258–59, 261, 263, 265, 272f, 302
Phaeoceros, 34
Phalaris aquatica, 230, 237, 239, 241
Phaseic acid (PA), 258, 268, 270f, 273
Phaseolus vulgaris (bean), 214–15, 230, 266, 293, 302f, 370, 372
Phenylacetic acid (PAA), 296
Phleum, 68, 79
Phosphoenolpyruvate (PEP), 117ff, 170, 259f
Phosphoenolpyruvate carboxylase (PEPCase), 171, 205, 265, 362
Photophosphorylation, 121, 135, 172–74, 182ff, 204f, 220, 301; in red light, 183; in blue light, 186, 205; in guard cells vs. in mesophyll cells, 205, 373
Photoreceptors: use of action spectroscopy to identify, 197–99; and stomatal opening, 199–200; chlorophyll and red light, 200–201; and blue light, 201–2, 205, 222, 224; interaction of red and blue light, 202–4
Photosynthesis, 29; and guard cell metabolism, 120–21; intercellular CO_2 levels, 195f, 298, 376f, 445, 458, 466; and stomatal movement, 195f, 344f; in red light, 200; in blue light, 210–11; and ABA, 273–74, 333; and auxins, 293; in arid climates, 326–37 *passim*; and conductance, 326–31 *passim*, 334f, 345, 374–80 *passim*, 431, 436–37; and midday stomatal closure, 332–33; and water stress, 332, 341, 370; and leaf temperature, 333; in humid climates, 339f; and light intensity, 339, 341–45, 374f; in CAM plants, 363; and leaf aging, 369–70, 374–80 *passim*; and shading of leaves, 370, 372; and transpiration, 370, 375–76, 380; and leaf-insertion level, 371–72; and reproduction, 372; and nitrogen levels, 379–80; and stomatal characters, 436–37; measurement of, 446ff, 453; calculation of, 448, 456, 458; and boundary-layer conditions, 471
Photosynthetically active radiation (PAR), 121, 329f, 332, 336f, 340ff, 343
Photosystems: I, 204; II, 121, 204f
Phragmipedium, 212
Phyllitus scolopendrium, 230
Phylloglossum, 42
Physcomitrium, 38
Picea sitchensis (Sitka spruce), 230, 241, 373
Pinus, 47; *merkusii*, 72; *sylvestris* (Scotch pine), 229f, 232; *ponderosa*, 393f
Pistacia: vera, 313; *lentiscus*, 336, 338
Pisum sativum (pea), 66, 83, 230, 241, 256f, 261, 283ff, 435
Plasmalemma, 65, 67, 77, 79–80, 83, 150, 154; and salt accumulation, 131; measurement of fluxes across, 141; Rb⁺ flux, 141; Br⁻ flux, 141f, 144–46, 149f; APTase in, 164, 174, 221; hyperpolarization of, 165, 169, 174, 178, 184f, 221–22, 233; proton pump at, 165, 221; electrical gradient across, 166f, 169, 180; H⁺ concentration gradient across, 167; in blue light, 221; binding sites of ABA vs. K⁺ channels, 258, 263; ion transport across, 259ff; and ABA, 261, 265
Plasmalemmasomes, 79–80
Plasmodesmata, 76–77, 354
Plasmolysis, 110–11
Plastids, 60, 62f, 66–71 *passim*, 80–82, 117, 354f
Plastoglobuli, 80
Platycerium bifurcatum, 45
Platyhypnidium riparoides, 39
Ploidy, 433
Podocarpus neriifolia, 46
Polar regions, 60, 63, 71f, 95–96, 111
Pollution, and breeding for improved stomatal conductance, 432
Polygonum amphibium, 51
Polymer elasticity theory, 97–101
Polypodiaceae, 43
Polypodium, 67f, 73, 79; *vulgare*, 231
Polysaccharides, 65, 69
Polytrichaceae, 37
Polytrichum, 36f; *strictum*, 36; *alpinum*, 39
Populus: maximowiczii × incrassata, 61; *deltoides*, 231, 241
Pore, 59f, 62f, 79, 81, 115, 461; formation, 67–70, 77, 83; structure, 77–78. *See also* Aperture, stomatal
Porometry, 13–14, 18, 323, 326, 367
Portulacaria afra, 353f, 356f, 359
Potamogeton natans, 51
Potassium, 20–21, 53, 68, 73, 110; and

stomatal movements, 116; in open vs. closed guard cells, 125–31 passim, 135–36; and osmotic changes, 137–40, 146; channels through plasmalemma, 258ff; channels through tonoplast, 261; and stomatal opening in CAM plants, 360
—— bromide, 130f, 136f, 145, 150f
—— chloride, 129ff, 132ff, 136–40, 146, 151, 259, 262–63
—— cyanide, 174, 182f, 186, 204
—— iminodiacetate, 132, 263
—— ions: shuttle, 45; fluxes, 116–17, 118; and stomatal movement, 125, 127ff, 133–40, 164–69 passim, 176, 359–60; uptake, 131f, 178–80, 260; and guard cell membrane potential, 133; and fluorescence transients, 135, 204; uniport, 165; and proton pump, 221–22; and ABA, 261ff, 264, 301, 303; and cytokinins, 285, 288f; and auxins, 294; and IAA, 294, 301, 303
—— malate, 259
—— sulfate, 263
Potato, 267, 435
Poterio-ochromonas, 131
Preprophase band (PPB), 62–65, 67
Pressure: calculation of, 449; in gas-exchange systems, 450, 452
Productivity, plant, 436
Proplastids, 67
Proteins, and leaf aging, 369f, 377
Protocyatheaceae, 43
Proton extrusion, 260; and stomatal opening, 131–32, 165–66, 167, 176; and blue light, 174, 184–85, 215–19, 221–22; energy costs of, 178–80; and ABA, 262ff, 266; and IAA, 266, 301, and cytokinins, 288
Protonmotive force (pmf), 167ff, 171, 179–80
Proton pump, 131–32, 133ff, 155f, 165, 167, 178, 260; and ATP, 134–35, 179–80; and ATPase, 164; and blue light, 205, 215, 221–22, 223; and K^+, 221–22; and ion fluxes, 223; and ABA, 259, 262f, 265, 274, 301; in tonoplast, 260
Protons, 165ff, 169, 180
Protoplasts, 261, 269, 272, 285
Prunus: dulcis, 313; *armeniaca*, 319, 332; *spinosa*, 339, 341, 344, 375
Pseudolarix, 47
Pseudotsuga, 47
Psilophytatae, 40
Psilotatae, 40–41
Psilotum, 40–41, 43; *triquetrum*, 41
Psilotum-type stomata, 49
Pteridiaceae, 43
Pteridophyta, 40–45

Pteris, 43
Pteris-type guard cells, 44
Pulvini, 10, 253
Pyruvate, 119; orthophosphate dikinase, 118

Quercus: ilex, 48, 72, 329, 331ff; *suber*, 325ff, 328, 332, 334ff

Radiation fluxes, in canopies, 407–18 passim
Rapataceae, 61
Rear chamber, 60, 68
Red light, 19, 183, 199–204 passim, 219–20, 288
Reproduction, 372
Resistance, 458
Respiration, 29, 172, 182–83, 205, 217, 264, 293
Restionaceae, 48, 61
Rhynchostegium murale, 39
Rhynia, 40; *major*, 41
Ribes: rubrum, 53; *uva-crispa*, 371
Ribonucleic acid (RNA), 369f, 377
Ribulose-1,5-bisphosphate (RuBP), 120; carboxylase (RuBPCase), 120n, 170, 174, 182, 204, 220, 273, 377
Ribulose 5-phosphate kinase, 204
Rice, 62, 230, 237, 239, 241, 371
Rubidium chloride, 129ff, 136f, 138f, 151
Rubidium ions, 141; fluxes in light vs. dark, 147–51; and ABA, 151–53, 261; and FC, 154, 156
Rumex conglomeratus, 231
Ruscus hypoglossum, 39
Rye grass, 398, 434

Saccharum officinarum (sugarcane), 214
Saliviniales, 43ff
Salvinia natans, 50
Salvinia-type guard cells, 44
Scheuchzeriaceae, 50
Scheuchzeria palustris, 50
Schizaea, 43
Schizaeaceae, 43
Sciadopitys, 47
Sclerophyllous species, 328–29, 334
Scotch pine, 229f, 232
Sedges, 61, 71, 75
Selaginellaceae, 42
Selaginella martensii, 41f
Senescence, *see* Leaves, aging
Sesamum indicum, 231
Setaria viridis, 231
Silicification, of guard cells, 42
Silicon, 110
Sitka spruce, 230, 241, 373
Sodium: chloride, 129f, 151, azide, 182; ions, 285f, 295, 301

488   Index of Subjects

Soil drought: and plant water potential, 311f, 316–18, 320; and stomatal closure, 316ff, 319–20, 329, 333–35
Solanaceae, 39
Solanum tuberosum (potato), 267, 435
Sorghum: 433, 435; *almum*, 231, 241; *bicolor*, 231, 241; *sudanense*, 231, 241
Spannungsphase, 94–95, 101, 110f
Spartina townsendii, 231, 241
Spermatophytes, 45
Sphagnum, 35
Spherosomes, 82
Spinacia oleracea (spinach), 231, 241, 269
Spirodela polyrhiza, 51
Splachnaceae, 36
Sporophytes, 30, 33–40 *passim*
Starch, 260f, 294; in plastids, 62f, 68, 80f; and stomatal movement, 116f, 119; and carbon metabolism, 119f; and malate, 170, 182, 259; synthesis in guard cells, 170–71
Starch-sugar hypothesis, 17
Stomata, 60; structure, 31, 33, 60–62; formation, 37; distribution, 60, 434; cell division in, 62–65; size, 433; frequency, 433f. *See also* constituent parts and internal processes by name
Stomata-less plants, 43
Stylites, 42
Subsidiary-cell parent cells (SPCs), 63ff
Subsidiary cells, 60, 62f, 71ff, 74, 79, 83, 130, 354
Substomatal cavity, 59, 67–70
Sucrose, 118, 139f, 170ff, 182, 263–64
Sugarcane, 214
Sugars, 116, 139f

Taraxacum officinale, 231
Tectonia grandis, 339f
Temperature: and stomatal movement, 15, 19–20, 238, 265f, 328ff, 331ff, 345, 359, 361; and CO_2, 238, 265f; and ABA, 265f, 269, 302f; and IAA, 269, 302–3; and arid-habitat plants, 328ff, 331f; and humid-habitat plants, 345; and CAM plants, 361; and breeding for improved stomatal conductance, 432; measurement of, 449; and saturation mole fraction water vapor, 449–50
Thermocouples, use of, 449, 464, 468
Thermoperiod, 359
3-phosphoglycerate, 120
3-phosphoglyceric acid (3-PGA), 170
Thujopsis, 47
Thurniaceae, 61
Thylakoids, 121, 172
Tonoplasts, 80, 84; and transport processes, 117, 120; and ion fluxes, 141, 259ff; and Br^- fluxes, 142, 144–46, 150; and ABA, 258–59, 261, 264–65
Torreya, 47
Tradescantia, 45, 73, 169; *albiflora*, 285; *nova*, 285; *virginiana*, 313f
Transpiration, 8, 14–15, 16, 467; in blue light, 210; and CO_2, 243, 298; and ABA, 255; and cytokinins, 282f, 286; and auxins, 293; and wind speed, 299; and humidity, 311f, 315, 320; and leaf water potential, 311–18 *passim*; and xylem water potential, 315; measurement of, 323f; in arid habitats, 328, 330f, 338; in CAM plants, 353; and stomatal conductance, 370, 417–18; and photosynthesis, 370, 375–76, 380; and leaf-insertion level, 371–72; and leaf aging, 375–76; calculation of, 447f, 454–55, 459; and boundary-layer conditions, 471
Trapaceae, 51
Trapa natans, 51
Tricarboxylic acid cycle, 116
Tridax procumbens, 283
Triticum aestivum (wheat), 18, 63, 377, 439; ATP levels, 134; and blue light, 202, 212; and CO_2, 231, 240; and cytokinins, 282; and ABA, 286, 371, 435, 438; and water stress, 318, 435, 439; and air flow through canopies, 393, 395, 416–17; genes controlling stomata, 433
Tropaeolum majus, 53
Tulipa gesneriana (tulip), 266
Turbulence (in canopies): fluctuations in, 386–87, 410; variations within, 387, 390–91; and leaf surfaces, 388, 402–3; modeling of, 390, 422–26; and eddies, 392–93, 420ff; random vs. organized flow, 393; and microclimates, 399–400; near-field vs. far-field effects, 424
Turgor, 11, 15–21 *passim*, 34, 176, 313f; in young cells, 68f; and absence of plasmodesmata, 77; and water, 92, 313f; and pressure, 95, 101–11 *passim*; and cell wall structure, 99; and water stress, 110; and cation concentration, 110–11; and PAR, 121; and K^+, 127–28, 139, 301; and ionic fluxes, 128–29; and osmotic pressure, 253–55; and ABA, 269–70, 271, 301f; and IAA, 301; and humidity, 313–14, 320
2,4-dichlorophenoxyacetic acid (2,4-D), 293

Uniporters, 165

Vaccinium, 433
Vacuoles, 60, 63, 66, 72, 83–84, 118; and

Index of Subjects 489

guard cell pH, 132–33, 260; ion content, 142–44, 145, 150f, 157; and ABA, 258f, 261; and stomatal movement, 259f, in CAM plants, 354, 362
Valerianella locusta, 256, 258, 302
Vanadate, 132, 218, 262f
Vapor pressure: deficit, 238f, 360–61; gradients, 459
Ventral wall, 59–72 *passim*, 75, 77, 83, 92
Vesicles, 82–83
Vicia faba, 81, 83, 121, 165f, 172, 239, 434; guard cell mechanics, 94, 97, 102f, 111, 254; and K^+ content, 127–28, 140; osmotic pressures, 129, 139–40, 146; stomatal movement, 129ff, 147; proton extrusion, 132; ion concentrations, 169; chloroplasts in, 172, 174; guard cell bioenergetics, 175–88 *passim*; RuBPCase in, 204, 220; and blue light, 214, 216ff, 221; and red light, 219; and CO_2, 231; and water stress, 255; and ABA, 257f, 261ff, 266, 271; and cytokinins, 283, 286f
Vigna: sinensis, 65; *unguiculata*, 84, 235, 312f, 346; *luteola*, 231, 241
Vitis vinifera, 231, 337
Vittariaceae, 43
Volume-expansion models, 97

Walls, of guard cell, 17; formation of, and microtubules, 65–67; shape, 70–73, 82; thickening, 70–73, 82, 92; structure, 73–75, 93f, 96, 101–4; as polymer structure, 98–101; tensile strength, 104, 106f; and water stress, 110; and minerals and hormones, 110–11; and environmental conditions, 111. *See also specific walls by name*
Water, 16, 115–16, 353, 357; penetration of stomata, 30–31, 46; and stomatal opening, 92, 164, 175–76, 312; pressure, and guard cell volume, 103
Water potential: turgor and osmotic pressures, 253–55; and ABA production, 261; leaf, 311–20 *passim*, 327ff, 330, 334, 340, 371, 460; and stomatal conductance, 311–20 *passim*; soil, 312, 316ff, 319f; xylem, 312ff, 315, 320; epidermal cells, 314, and midday stomatal closure, 329f, 334; and stomatal movement in CAM plants, 360; and leaf aging, 371
Water stress, 345–46, 370; and stomatal mechanics, 110, 232, 235, 266–67; and CO_2, 232, 235, 266, 298ff, 301; and ABA, 266, 269–74 *passim*, 299ff, 304ff, 371; and IAA, 299ff, 303; and cytokinins, 303, 305–6; and stomatal movements in arid habitats, 328; soil, and midday stomatal closure, 329, 333–35, 337; and photosynthesis, 332, 341, 370; and conductance, 334, 340–41, 346, 361f, 370f; in humid habitats, 339ff; in CAM plants, 356–63 *passim*; and leaf age, 371; and breeding for low stomatal conductance, 432f, 437–39; genetics of response to, 435
Water-use efficiency, 240–43, 345f, 375f, 432f, 437–39
Water vapor: loss, 8f, 14, 19f, 29, 115; and leaf water potential, 312f, 316; conductance, 405–11 *passim*; fluxes, in single-leaf models, 412, 415, 417; measurement of, 446f; in gas-exchange systems, 448, 451–57 *passim*; calculation of saturation mole fraction, 449–50, 452; calculation of conductances based on binary diffusion, 459f; intercellular mole fractions, 459f; gradients, 459, 467; ternary diffusion, 460–62; calculation of conductances to flux, 462–64; concentrations at leaf surface, 467–68; dual-stream measurement of, 469f. *See also* Transpiration
Welwitschia mirabilis, 47, 329
Wheat, see *Triticum aestivum*
Wilting: and ABA, 255, 267, 270ff, 274; and CO_2, 265. *See also* Water stress
Wind speed, 298f, 391ff, 402f, 405

Xanthium: strumarium, 19, 198, 231, 233, 265–73 *passim*, 299, 319, 370; *pennsylvanicum*, 256
Xanthoxin, 267f
Xerophytes, 61–62, 78
Xylem, water potential of, 312ff, 315, 320

Yucca schidigera, 357f

Zea mays (corn), 21, 304; stomatal structure, 67f, 70f, 75, 83; and light, 196f, 202; and CO_2, 231f, 235–41 *passim*; and cytokinins, 283, 286, 303ff; and ABA, 302ff, 435; air flow through canopies, 391f, 395; genes controlling stomata, 433
Zeatin, 283, 285ff, 303, 305, 359
Zebrina, 129, 146
Zinc, 110

Index of Authors

Ackerson, R. C., 230, 233, 249, 270, 275
Addicott, F. T., 277, 443
Afgan, N. H., 474
Aizawa, H., 225
Akhmedov, I. S., 291
Akita, S., 231, 245
Alargarswamy, G., 441
Alberts, B., 164, 189
Alden, T., 73, 89
Allaway, W. G., 59, 61, 72, 80–85 *passim*, 92, 111f, 116, 122, 127, 156ff, 172, 189, 196, 206, 209, 225, 231, 245, 260, 275
Allen, J. C., 371, 383
Allen, K. E., 157
Amagasa, T., 119, 122
Amici, G. B., 9, 22
Anderson, J. H., 149, 157
Anderson, L. C., 56, 206
Anderson, L. E., 119, 123, 205, 207
Antonszewski, R., 233, 247
Apostolakos, P., 64–69 *passim*, 82f, 86
Appleby, R. F., 75, 79, 85
Armond, P., 121, 123, 174, 192, 201, 204, 208, 220, 227
Armstrong, J. I., 14, 23
Aslam, M., 378, 380
Assmann, S. M., 165f, 174, 184ff, 189, 196, 202, 205, 208, 214, 216, 220ff, 223f
Astle, M. C., 303, 306
Aubert, B., 353, 363
Austin, R. B., 326, 351
Aylor, D. E., 96f, 112

Baasch, R., 351
Baic, D., 81
Balke, N. E., 218, 224
Ball, J. T., 237, 245, 445, 468, 471, 474f
Ball, M. C., 319f
Ballarin-Denti, A., 132, 158, 217, 226
Bange, G. G. J., 15f, 22
Bannister, P., 105, 111f
Barrs, H. D., 16, 22
Barthakur, N., 406, 427
Bartholomew, B., 349, 361, 363
Bassham, J. A., 170, 190

Batchelor, G. K., 423, 426
Batchelor, S. M., 288, 290
Bates, L. M., 346f
Bates, L. S., 433, 443
Bauer, H., 328, 345, 349
van Bavel, C. H. M., 231, 241, 245, 372, 379, 381, 441
Bazzaz, F. A., 231, 241, 249, 370, 380, 382
Beadle, C. L., 241, 245
Beakbane, A. B., 434, 440
Beardsell, M. F., 14, 22, 272, 275, 326, 347
Begg, J. E., 351
Beilby, M. J., 133, 157
Bell, C. J., 236, 245
Bengtson, C., 282, 290
Berjak, P., 67, 85
Berry, J. A., 237, 245, 326, 344, 347f, 378, 381, 438, 441, 445f, 448, 468, 471, 473ff
Beyschlag, W., 321, 332, 349f
Biddington, N. L., 282, 290
Bidinger, F. R., 441
Bidwell, R. G. S., 370, 372, 381
Bienfait, A., 75, 89
Bierhuizen, J. F., 229, 241, 245
Bierschenk, K., 35ff, 39, 54
Biggins, J., 227, 474f
Bingham, E. T., 433, 443
Bingham, G. E., 249f
Bird, R. B., 403, 427
Bisson, M. A., 129, 158
Björkman, O., 241, 245, 334, 342, 344, 347, 349, 378, 380, 446, 474
Black, C. C., 265, 279, 366
Blackman, P. G., 283, 287, 290, 299, 303ff, 306
Blackwell, R. D., 438, 442
Bloom, A. J., 169, 192, 215, 227, 379f, 446, 456, 474
Boggess, W. R., 231, 241, 249
Bonnet, C., 8–9, 22
Borgeson, C. E., 222, 224
Borle, A. B., 149, 157
Bornman, C. H., 261, 278

Index of Authors

Bosian, G., 324, 347
Bowling, D. J. F., 21, 25, 127, 133, 160, 169, 178, 189, 191, 260, 277
Bowman, B. J., 132, 157, 222, 224
Boyer, J. S., 159, 174, 190f, 207, 225, 277, 334, 349, 353, 364
Boysen-Jensen, P., 293, 306
Bradbury, D., 293, 306
Bradford, J. A., 231, 241, 245
Bradford, K. J., 230, 238, 241, 245, 267, 275, 281, 290, 305f
Bradley, E. F., 393f, 399f, 427f
Brady, C. J., 370, 372, 381
Braun, M., 328f, 331, 348, 350
Bray, D., 189
Bressel, C., 45, 55
Brinckmann, E., 314, 321, 351
Briskin, D. P., 164, 190
Brogårdh, T., 200f, 206, 230, 245
Brown, H., 14f, 22
Brown, J. W., 293, 306
Brown, K. W., 110, 112, 371f, 380, 382
Brown, R., 25, 34, 54, 475
Brown, W. V., 70f, 85
Bruinsma, J., 358, 363
Brulfert, J., 119, 122, 358f, 365
Brun, W. A., 434, 441
Bryant, J. A., 292
Buhr, H., 328, 348
Burden, R. S., 267, 278
Burke, L. L., 132, 158
Burns, D. M., 104, 112
Burns, J. A., 471, 475
Burns, R. H., 250
Burrows, F. J., 368, 371, 373, 380, 458, 474
Burschka, C., 272, 276
Burström, H. G., 53f
Busby, C. H., 61, 64–70 passim, 79, 85
Buschbom, U., 321, 350, 364
Businger, J. A., 386, 427
Butler, L. K., 432, 440
Butler, W. L., 222, 225
Buzzard, G. H., 456, 475

von Caemmerer, S., 370, 378, 380f, 446, 448, 455, 458, 461, 464, 474
Caldwell, M. M., 351, 370, 382
Calkin, H. W., 339, 349, 450, 475
Cameron, L. E., 289, 291
Campbell, N. A., 143, 157
de Candolle, A. P., 7, 9, 22
Canfield, J. E., 350
Carbonneau, A., 326, 347
Carlson, J. R., 439, 441
Carr, D. J., 62, 68f, 73, 79, 85
Carr, S. G., 62, 68f, 73, 79, 85

Casady, A. J., 442f
Catarino, F., 351
Čatský, J., 239, 245, 369f, 380f, 446, 475
Ceulemans, R., 436, 441
Chabot, B. F., 73, 75, 78, 82, 85, 370, 379, 382
Chabot, J. F., 73, 75, 78, 82, 85, 370, 379, 382
Chandler, C. K., 433, 441
Chaney, R. L., 110, 112
Chapin, F. S., III, 380
Charles-Edwards, D. A., 442
Cheesbrough, J. K., 241, 246
Cheung, W. Y., 289f
Cheung, Y. N. S., 99, 112
Christy, A. L., 118, 122, 170, 191
Chu, C., 237, 248
Chu, C. C., 442
Ciferri, O., 160, 277
Ciha, A. J., 434, 441
Clark, J. A., 465, 474
Clark, K. G., 113
Clark, W. C., 240, 245
Clint, G. M., 130, 136f, 141, 154f, 157
Cockburn, W., 231, 234, 245, 283, 286, 357f, 363
Cocucci, M. A., 132, 158
Cohen, D., 272, 275
Collatz, G. J., 334, 349
Colman, B., 220, 225
Comins, H. N., 361, 364
Conde, L. F., 360, 364
Constable, G. A., 369, 372, 374f, 381, 433, 443
Cook, C. S., 250, 351
Cooke, J. R., 96f, 104f, 108, 112
Coombs, J., 119, 122
Cooper, M. J., 282, 286f, 290
Cooper, P. J., 282, 286f, 290
Cooper, S. D., 283, 286, 290
Copeland, E. B., 48
Coppin, P. A., 387, 390, 396ff, 419f, 424, 427f
Cork, R. J., 75, 77, 85
Cornic, G., 273, 275
Cornish, K., 271, 273ff
Corrsin, S., 399, 427
Couot-Gastelier, J., 72, 80, 83ff
Coutinho, L. M., 361, 364
Cowan, I. R., 16, 22, 97, 104, 110, 112, 195f, 208f, 226, 230–40 passim, 245, 249f, 272, 275, 306, 326, 328, 335, 345ff, 351, 365, 368, 374, 376f, 381, 383f, 402, 409, 413, 427, 448, 458, 474f
Coyne, P. I., 231, 241, 245
Cram, W. J., 129, 158, 225f
Craven, C. L., 372f, 383

Index of Authors 493

Crawford, M. E., 465, 475
Creasy, L. L., 191, 226
Creelman, R. A., 267, 269f, 275
Crozier, A., 277
Cumming, I. G., 439, 442
Cummins, W. R., 190, 206, 255, 257, 275
Cure, J. D., 249
Cutler, J. M., 110, 112

Dainty, J., 99, 112, 161
D'Amelio, E., 80f, 172, 190
Danielli, J. F., 25, 475
Darbyshire, B., 303, 307
Darwin, F., 13–18 passim, 22, 195, 206
Das, V. S. R., 283, 287, 290
Davenport, T. L., 303, 307
Davey, J. E., 281, 292, 306, 308
Davidson, B., 14, 22, 326, 347
Davies, D. D., 475
Davies, W. J., 75, 79, 85, 237f, 244, 248, 283, 287, 290, 299, 302–8 passim
Davis, R., 365
Davis, S. D., 239, 245, 372, 374, 379, 381
Day, W., 446, 475
Dayanandan, P., 71, 73f, 82, 85, 360, 364
Dayton, A. D., 442f
Dean, M. L., 358, 366
DeBaerdemaeker, J. G., 112
DeChalain, T., 67, 85
De Journett, R., 79, 82, 88, 354, 366
Delgado, M., 364
Demarty, M., 161
DeMichele, D. W., 92, 96ff, 104f, 112
Denmead, O. T., 393f, 399f, 410, 414, 427
De Silva, D. L. R., 302, 307
deVries, D. A., 474
Di Camelli, C. A., 160, 207, 226
Dickerson, M., 11, 26, 93, 97, 102f, 106f, 111, 113
Dickson, A., 52, 54
Dieter, P., 302, 307
Digby, J., 282, 286f, 290
Dinger, B. E., 361, 364
Ditterline, R. L., 441
Dittrich, P., 204, 207, 260f, 267, 275
Dixon, M., 465, 474
Dobrenz, A. K., 434, 439, 441
Dobrindt, L. A., 75, 87
Dochinger, L. S., 432, 443
Donkin, M. E., 435, 441
Doohan, M., 82, 85
van Dop, H., 427, 429
Dorée, M., 289f
Dörffling, K., 268, 275
Downes, R. W., 231f, 241, 245
Downton, W. J. S., 231, 241, 245, 247, 334, 347

Doyle, W. T., 32f, 35, 54
Dubbe, D. R., 196, 206, 231f, 237, 239f, 245f, 265, 275, 326, 347
Dudley, J. M., 161, 279, 308
Dugger, W. M., Jr., 358f, 366
Dunn, E. L., 328, 347, 349
Dunn, G. M., 433f, 442f
Dunstone, R. L., 433, 436, 441
DuPont, F. M., 132, 158
Durbin, P. A., 426f
Dutrochet, H. J., 10, 22
Dwelle, R. B., 436, 441
Dylenok, L. A., 433, 441

Eamus, D., 266, 275, 302f, 307
Eckardt, F. E., 328, 347, 475
Edwards, A., 169, 178f, 189
Edwards, G. E., 250
Edwards, I., 387, 428
Edwards, M., 15, 24, 79, 85, 92, 104, 110ff
Ehleringer, J., 235, 237, 239, 245, 247, 250, 351, 369, 375, 382
Eidt, D. C., 255, 276
Ekern, P. C., 353, 364
Elbert, C., 119, 123
Eller, B. M., 351
Elliott, D. C., 288ff
Elmore, C. D., 379, 382
English, S. D., 351
Ennis, W. B., 293, 306
Ephrat, J., 434, 443
Epstein, E., 308
Erez, A., 308
Erler, W. L., 353, 364
Erwee, M. G., 77, 86
Esau, K., 60, 62, 72
Escombe, F., 14f, 22
Eshel, A., 436, 443
Evans, J. R., 375, 377, 381, 458, 474
Evans, L. T., 156, 158, 196, 206, 209, 225, 433, 436, 441
Evenari, M., 321, 350, 364

Fahr, A., 75, 86
Falk, S. T., 282, 290
Faraday, C. D., 61, 80–86 passim, 354, 364
Farquhar, G. D., 14, 16, 22, 195f, 206ff, 209, 226, 230ff, 236–41 passim, 245f, 249f, 265, 267, 275, 297, 307, 319f, 326, 328, 335, 345ff, 351, 361, 364, 370, 374–84 passim, 436–41 passim, 446, 448, 455, 458, 460f, 464, 474f
Fasehun, F. E., 339f, 348, 351; 465, 474
Feinleib, M. E., 88, 123, 160, 191, 207, 249, 277
Fellows, M. P., 21, 26, 285, 292

494 Index of Authors

Fernandez, J. M., 167, 192, 222, 226, 258, 260, 278
Ferrar, P. J., 238, 246
Ferri, M. G., 293, 307
Field, C., 197, 202, 208, 224, 227, 244, 251, 326, 348, 371–82 passim, 446ff, 451, 473f
Filion, W. G., 190, 206
Finnigan, J. J., 387ff, 393, 395, 398, 405, 419f, 427f
Firminger, M. S., 75, 87
Firn, R. D., 256, 267f, 278
Fischer, E., 273, 275
Fischer, M., 39, 54
Fischer, R. A., 21f, 116, 122, 125–31 passim, 140, 147, 158, 160, 231, 246
Fisher, D. B., 118, 122, 170, 191
Fitzsimons, P. J., 144, 161, 204, 206, 226, 279, 308
Flint, E. P., 230, 248
Florin, R., 44ff, 47, 54–55
Fogg, G. E., 24, 428
Ford, C. W., 272, 276
Ford, E. D., 442
Forseth, I. N., 250, 351
Forstner, H., 226
Foy, C. D., 110, 112
Franklin, A., 190, 206
Fraser, D. E., 370, 372, 381
Freidman, W. E., 350
French, C. S., 205, 207
Freudenberger, H., 18, 22, 229f, 246
Friedrich, J. W., 377, 381
Fujii, S., 135, 158f
Fujino, M., 21f, 116, 122, 125, 130, 153, 158
Fuller, G. L., 276
Funada, S., 474

Gaastra, P., 230, 246, 376, 381
Galatis, B., 61, 64–73 passim, 79, 82f, 86
Garber, R. C., 143, 158
Gardi, I., 267, 278
Garner, D., 37, 55
Garratt, J. R., 419, 427
Garreau, M., 9, 22
Garrec, J. P., 128, 159
Gaskill, M. L., 434, 436, 441
Gates, D. M., 347, 351
Gay, A. P., 433, 441
Gebel, J., 321, 335f, 348, 350
Gedalovich, E., 75, 86
Gepstein, S., 132, 158, 166, 174, 178, 182f, 190, 193, 204, 208, 216, 218, 220, 225, 227, 261, 275
Gerencser, G. A., 278
Giaquinta, R. T., 170, 181, 190

Gibbs, M., 277, 364f
Giebisch, G., 158
Gifford, R. M., 230f, 237–43 passim, 246, 248, 433, 436, 441
Gill, S. S., 432, 441
Gimmler, H., 269, 271, 276
Givnish, T. J., 381
Glinka, Z., 11, 22
Gnaiger, E., 226
Goatly, M. B., 119, 122
Goebel, K., 35f, 43, 55
Goeschl, J. D., 113
Gollan, T., 311, 313, 317f, 320f
Good, N. E., 284, 290
Goodwin, P. B., 77, 86, 296, 307f
Gotow, K., 174, 183, 192, 207, 215ff, 220, 226f, 261, 276
Govindjee, 159, 191, 207, 225, 277
Graan, T., 174, 190
Grace, J., 298, 307, 340, 348, 405, 427, 442, 465, 474
Graniti, A., 131, 161, 261, 279
Grantz, D. A., 159, 174, 190f, 207, 214, 225, 277
Green, A. E., 326, 349
Gregory, F. G., 14, 16, 18, 23
Grew, N., 8, 23
Griffiths, J. H., 326, 348
Gross, D., 292
Guern, J., 290
Guerrier, P., 289f
Guervin, C., 87
Guettard, J. É., 8, 22
Guinn, G., 233, 249
Gulmon, S. L., 382
Gunar, I. I., 153, 162, 165, 178, 190, 285, 290
Gunning, B. E. S., 61–70 passim, 79, 85f
Gurner, P. J., 326, 351
Gutknecht, J., 129, 158
von Guttenberg, H., 62, 71f, 86, 328, 348
Guyot, M., 20, 23, 84, 86

Haberlandt, G., 13, 23, 35, 37f, 42, 48–55 passim, 93, 112
Hackett, C., 369, 372, 383
Hadley, N. F., 366
Hall, A. E., 209, 226, 230f, 235–40 passim, 246, 249, 316, 321, 323, 328f, 339, 342, 345ff, 350
Hall, A. J., 370, 372, 381
Hall, M. A., 303, 307
Hällgren, J. E., 238, 246
Hambleton, S., 65, 68f, 82, 84, 88
Hamill, O. P., 165, 190
Hampp, R., 116, 122
Hanebuth, W. F., 196, 207, 231, 249

Index of Authors

Hanscom, Z., 356f, 360f, 364
Hansen, U. P., 133, 160
Hardwick, K., 377, 381
Harms, H., 19, 23
Harper, J. L., 370, 380
Harris, G. G., 231, 241, 246
Harrison, A. T., 328, 349
Hartsock, T. L., 358ff, 365
Hartung, W., 256, 258, 267–72 passim, 275f, 302, 307
Hashimoto, Y., 282, 285, 291, 474
Hastings, D. F., 129, 158
Hatch, M. D., 245
Haupt, W., 88, 123, 160, 191, 207, 249, 277
Hayama, T., 159
Hays, R. L., 349
Heagle, A. S., 249
Heald, J. K., 270, 277
Hearn, A. B., 351
Heath, O. V. S., 11, 16ff, 19f, 23ff, 195, 206, 229ff, 232, 246, 326, 348
Heber, U., 170, 190
Heck, W. W., 249
Hedrich, R., 167, 192, 222, 226, 258, 260, 273, 278, 319, 320
Hedwig, J., 9, 23
Heichel, G. H., 370, 381, 433, 441
Heilmann, B., 269, 271, 276
Heim, G., 347
Heldt, H. W., 134, 161, 170, 190
Heller, F. O., 20, 23
Hellkvist, J., 99, 112
Hellmers, H., 230, 241, 248
Hellmuth, E., 329, 348
Hendricks, S. B., 133, 158
Henson, I. E., 270, 272, 276, 371, 382, 435, 441, 443
Henzell, R. G., 435, 440f
Hepler, P. K., 19, 27, 59, 64f, 73, 77, 80, 86f, 161, 169, 192, 196, 208, 215, 227, 289, 292
Hesketh, J. D., 383
Hetherington, A. M., 302, 307
Hewlett-Packard Corporation, 449, 474
Hiatt, A. J., 133, 158
Hicks, B. B., 427
Higgins, T. J. V., 296, 307f
Hildebrand, F., 42, 44, 55
Hillman, J. R., 152, 161, 256, 263, 277, 279, 301, 308
Hiron, R. W. P., 21, 27, 255, 276, 279
Hodge, L. D., 83, 87
Hodges, T. K., 218, 224
Hodgkinson, K. C., 370, 381
Hoff, C. C., 364
Hoglund, H.-O., 202, 206

Holmgren, P., 446, 459, 474
Hopmans, P. A. M., 16, 23
Horgan, R., 267, 270, 277
Hornberg, C., 257f, 263, 269, 274, 276, 279
Horton, R. F., 283, 287, 291
Howarth, G. L., 256, 276, 287, 291, 359f, 364
Howe, G. N., 433, 443
Hrazdina, G., 191, 226
Hsiao, T. C., 115f, 122, 125, 127ff, 147, 156ff, 165, 190, 196, 206, 209, 225, 305f
Huber, B., 16, 23, 324, 348
Hubick, K. T., 439, 441
Huffaker, R. C., 377, 381
Humbert, C., 20, 23, 84, 86
Humble, G. D., 21, 23, 94, 112, 127, 129, 132, 139, 158, 160, 165f, 169, 176, 178f, 183, 190, 192, 216, 226, 260, 262, 278
Humphrey, A. B., 441
Hunt, L. A., 378, 380
Hurd, R. G., 433, 441
Hurley, P. J., 436, 441
Hutchinson, B. A., 427

Idso, S. B., 243, 247
Iino, M., 165, 182, 184f, 190, 192, 210–27 passim
Iljin, W. S., 17, 23
Imai, K., 243, 246, 249, 383, 475
Imamura, S.-I., 20, 23, 116, 122, 125, 129, 139, 146, 159
Imber, D., 267, 278, 308
Impens, I., 436, 441
Incoll, L. D., 256, 261, 276, 282–87 passim, 291, 359f, 364
Innes, P., 438, 442
Ishikawa, F., 282, 291
Ishikawa, H., 159, 165, 178, 190f, 207, 221, 225
Israel, D. W., 249
Issaias, S., 206
Itai, C., 200, 207, 261, 276, 303, 307
Izawa, S., 284, 290

Jackson, S. G., 56, 206
Jacobs, M., 132, 158f, 166, 178, 183, 190, 216, 218, 225, 261, 275
Jahner, D., 328, 334, 351
Jahnke, R., 73, 85
Jamieson, A., 19, 23
Janáček, K., 225f
Janzen, P., 34, 55
Jarman, P. D., 445, 461, 474
Jarvis, M. S., 459, 474
Jarvis, P. G., 14, 22, 26, 87, 99, 112f, 122, 196, 207, 230f, 234–48 passim, 299,

Jarvis, P. G. (*continued*)
307f, 326, 328, 339, 347ff, 373, 382, 410, 427, 437f, 442, 446, 451, 459f, 464, 474f
Jarvis, R. G., 130, 151, 159, 243f, 247, 285, 291, 295, 301, 307
Jeffree, C. E., 373, 382
Jennings, D. H., 162, 347, 381
Jewer, P. C., 244, 256, 261, 276, 283–87 *passim*, 291, 359f, 363f
Jeyaseelan, K., 153, 160
John, A., 307
Johnson, C. B., 277, 308
Johnson, H. B., 353, 357, 360, 365f
Johnson, R. P. C., 373, 382
Johnson, S. C., 70f, 85
Johnsson, A., 200, 206, 208, 210, 214, 225, 230, 245
Johnsson, M., 200, 206
Jones, H. G., 432–43 *passim*
Jones, J. W., 383
Jones, M. B., 358, 364
Jones, P. G., 250
Jones, R. J., 256, 261, 276f
Jordan, W. R., 110, 112, 303, 307, 371, 382
Joshi, M. C., 353, 364
Jouret, M. F., 436, 442
Jurik, T. W., 351, 370, 379, 382

Kacser, H., 471, 475
Kaiser, W. M., 170, 190, 272, 276
Kamiya, A., 205f
Kappen, L., 320f, 350, 364
Kapuya, J. A., 307
Karlsson, P. E., 202, 206, 209, 212, 214, 222, 225
Katsaros, C., 64–69 *passim*, 82, 86
Kaufman, P. B., 65, 69, 71, 73, 77, 85f, 360, 364
Kaufmann, K., 48, 54
Kaufmann, M. R., 231, 238, 246, 339, 348
Kauss, H., 131, 159
Kays, W. M., 465, 475
Kazaryan, A. A., 291
Kenda, G., 373, 382
Kende, H., 255, 275, 282, 291
Kennedy, J., 139, 160, 171, 191
Keuls, M., 439, 442
Khare, P. K., 44, 56
Khotyleva, L. V., 433, 441
Khush, G. S., 48, 50, 56
Kidd, G. E., 429
Kiermayer, O., 86
Kikuyama, M., 134, 159
Kimball, B. A., 243, 246f
Kishimoto, T., 289f

Kishira, H., 190, 225
Klemm, P., 48, 55
Klepper, E., 16, 22
Klockare, R., 202, 206
Kluge, M., 119, 122, 353f, 356f, 364
Kneebone, W. R., 441
Knight, R. C., 16, 18, 24
Knoerr, K. R., 456, 475
Koch, D. W., 433f, 442
Koch, W., 230, 247, 324, 329, 348, 475
Kohl, F. G., 19, 24
Koller, D., 14, 25, 230f, 250, 459, 475
Kolpikov, D. I., 9, 24
Komor, E., 264, 276
Kondo, N., 174, 183, 192, 207, 215ff, 220, 226, 261, 276
Kondo, T., 44, 55
Korbes, R. E., 87
Korelege, S., 153, 160
Körner, C., 345, 348
Kowallik, W. A., 186, 190, 205f, 217, 225
Koziol, M. J., 55
Kozlowski, T. T., 85, 247, 291, 380, 433, 442, 474
Kralovic, J., 307
Krämer, G., 119, 123
Kramer, P. J., 353, 360, 364f, 474
Krarup-Hjort, C. F., 68f, 86
Kraus, F. 34, 42f, 48, 55
Krause, G. H., 135, 159
Kriedemann, P. E., 231, 242, 247, 255, 276
Krikorian, A. D., 96f, 112
Krizek, D. T., 370, 382
Ktitorova, I. N., 290
Ku, M. S. B., 250
Kubik, M., 233, 247
Kuhlbrodt, H., 35f, 38, 55
Kuiper, P. J. C., 19, 24, 196, 206
Kumar, A., 406, 427
Kummer, V., 267f, 271, 276
Kummerow, J., 328, 349
Küppers, M., 237, 246, 339ff, 344–50 *passim*, 370f, 375, 377, 382
Kuraishi, S., 282, 285, 291

Laffray, D., 72, 80, 83ff, 128, 159
Laidlaw, C. G. P., 18, 24
Landré, P., 61, 65, 69, 86–87
Landsberg, J. J., 229f, 241–250, 405, 427
Lange, G., 68ff, 79, 89
Lange, O. L., 113, 249, 307, 311, 313, 320f, 324, 328–33 *passim*, 337, 347–52 *passim*, 361, 364, 380, 474
LaPre, L., 354, 357f, 364
Larcher, W., 328, 349

Index of Authors

Laroche, J., 67, 71, 73, 87
Larqué-Saavedra, A., 214f, 226, 435, 442
Larsson, S., 282, 290
Lawes, D. A., 433f, 442
Lawlor, D. W., 371, 382
Lea, H. Z., 433f, 442
Leach, J. E., 446, 475
Lecoq, C., 87
Ledent, J. F., 436, 442
Lee, D. R., 370, 382
Legg, B. J., 387, 390, 396, 419f, 424, 427f, 456, 475
Le John, H. B., 289, 291
Lendzian, K. J., 30, 55
Leopold, A. C., 276
Letham, D. S., 296, 307f
Lettau, J., 127, 130, 136ff, 159
Leuning, R., 400, 427, 461, 464, 475
Lewis, J., 189
Lex, A., 293, 307
Liang, G. H., 433, 442f
Liebig, M., 19, 24, 200, 206
Liebisch, H.-W., 292
Lightfoot, E. N., 403, 427
Lilley, R. M., 134, 161
Lin, Z. F., 237, 239, 247, 369, 375, 382
Link, H. F., 9, 24, 46, 55
Linsbauer, K., 17f, 24
Lippmann, G., 45, 55
List, R. J., 449, 475
Little, C. H. A., 255, 276
Livingston, B. E., 353, 364
Livnè, A., 282, 291
Lloyd, C. W., 86f
Lloyd, F. E., 13, 15, 17, 24, 52, 55
Loftfield, J. V. G., 15, 19, 24, 302, 307
Long, S. P., 231f, 241, 247
Longstaff, R. A., 405, 427
Loomis, R. S., 110, 112
Lösch, R., 45, 55, 230, 238, 247, 319, 320, 329, 331, 349ff
Louguet, P., 72, 80, 83ff, 128, 159
Loukari, H., 86
Louwerse, W., 230, 237, 247
Loveys, B. R., 268, 270, 276
Lowe, S. B., 378, 380
Lowry, O. H., 127, 139, 160, 170, 175, 191
Lucas, W. J., 75f, 80, 89, 161
Ludewig, M., 275
Ludlow, M. M., 231, 241, 247, 272, 276, 344, 347, 365, 370, 373f, 382
Lumley, J. L., 387, 419, 428
Lund, R. E., 441
Lurie, S., 205f
Lüttge, U., 99, 113, 122, 158, 260, 276

Lyalin, O. O., 285, 291
Lyrene, P. M., 433, 441

Macallum, A. B., 20, 24, 125, 159
McCree, K. J., 110, 112, 239, 245, 372, 374, 379, 381, 441
MacDonald, S. G. G., 104, 112
MacIachlan, G. A., 204, 207
McIlroy, I. C., 410, 427
McMichael, B. L., 435, 443
MacMillan, J., 292
McNaughton, K. G., 243, 247, 410, 427, 437, 442
McPherson, H. G., 231, 241, 247, 326, 349
MacRobbie, E. A. C., 116, 122, 127f, 130, 136ff, 141, 146–55 passim, 158f, 176, 188, 190, 223, 225, 301, 307
Madhavan, S., 204, 206, 220, 225
Mahalaskshimi, V., 441
Maheshwan, P., 47, 55
Mahlert, A., 46, 55
Maier-Maercker, U., 79, 87, 131, 159
Mainzer, S. E., 157
Malpighi, M., 7–8, 24
Manchester, J., 122, 139, 160, 170, 191, 207, 226
Mang, H. A., 112
Mansey, G. M., 226
Mansfield, T. A., 16, 19f, 23f, 87, 93, 113, 122, 129ff, 131, 151, 159, 161, 184, 192, 209, 225, 229ff, 233, 243–51 passim, 256, 261, 263, 266, 276ff, 279, 283ff, 287, 291, 293–302 passim, 307f, 348f, 358, 365, 432f, 442, 461, 475
Marcelle, R., 365, 382
Margaris, N. S., 351, 381
Marigo, G., 260, 276
Marmé, D., 289, 291
Maroti, I., 44, 56
Marrè, E., 131, 139, 159f, 261, 277, 288, 291
Marrè, M. T., 132, 158
Martin, E. S., 16, 24, 45, 56, 59, 67ff, 70, 73, 80ff, 88, 226, 435, 441
Martin, G. E., 45, 56, 204, 206
Martin, J. M., 441
Marty, A., 165, 190
Marx, J. L., 291
Maskell, E. J., 15f, 24
Massengale, M. A., 441
Matthews, M. A., 334, 349
Mawson, B. T., 174, 190, 201, 206, 227
Maxwell, K. M., 132, 161
Mayne, B. C., 122, 191, 207
Mayo, J. M., 53, 56, 172, 191, 230, 248
Medina, E., 350, 361, 364

Index of Authors

Meidner, H., 15f, 18ff, 23ff, 79, 85, 92f, 104f, 110ff, 113, 196, 202, 205, 208f, 225, 229f, 233f, 246ff, 261, 276, 282, 291, 323, 349, 358, 365, 432f, 442, 460f, 475
Meinzer, F. C., 353, 365
Meister, H. P., 351
Melis, A., 121, 123, 135, 159, 174, 182, 190, 192, 201, 204, 207f, 220, 225, 227
Mérida, T., 79, 87
Mertens, R., 257, 271, 277
Methy, M., 347
Meyer, A., 337, 348ff
Miginiac, E., 273, 275
Milborrow, B. V., 257, 267ff, 270f, 276f, 279
Milford, G. F. J., 371, 382
Miller, A. G., 134, 161, 220, 225
Milthorpe, F. L., 59, 61, 72, 80f, 85, 92, 111f, 174, 191, 236, 250, 368, 370f, 373, 380, 382, 458, 474
Miroslavov, E. G., 84, 87
Mishkind, M., 65, 74, 87
Miskin, K. E., 433f, 436, 442
Mitrakos, K., 61, 65–70 *passim*, 73, 79, 82, 86
Mittelheuser, C. J., 21, 25, 255, 277, 282, 286, 291
Miyachi, S., 205ff
Miyaji, K.-I., 370, 382
Mizrahi, Y., 281, 292
von Mohl, H., 10ff, 13, 16f, 25, 35, 56, 91f, 113
Molisch, H., 13, 25, 373, 377, 382
Mollenhauer, H. H., 76, 80, 82, 87
Monin, A. S., 424, 428
Monnier, A., 161
Monteith, J. L., 390, 407, 412, 427f
Moody, W., 133, 153, 159, 161, 165, 178, 191f, 221, 225, 227
Mooney, H. A., 202, 208, 224, 227, 237, 248, 324–29 *passim*, 334, 347f, 349ff, 375f, 379ff, 382, 446f, 451, 473ff
Moreshet, S., 14, 25
Morgan, P. W., 303, 307
Morison, J. I. L., 196, 207, 230ff, 235–48 *passim*, 299, 308, 328, 348, 432
Morrow, P. A., 328, 344, 347, 349
Morse, S. R., 350
Moss, D. N., 231, 234, 245, 248, 433f, 436, 442
Mott, K. A., 446, 460, 469, 475
Mouravieff, I., 11, 17, 19, 25, 209, 225, 230, 233, 248
Mujamder, P. K., 434, 440
Mujeeb, K. A., 433, 443

Mulhearn, P. J., 387, 428
Müller, N. J. C., 10ff, 13f, 25
Munns, R., 318, 320
Muñoz, V., 222, 225
Murata, Y., 243, 246
Mussell, H., 442

Nadel, M., 11, 17, 25
Naegeli, C., 35, 56
Nagarajah, S., 235, 248, 311, 321
Nandpuri, K. S., 432, 441
Napp-Zinn, K., 45, 56, 62, 87
Neales, T. F., 231, 248, 353f, 358f, 365
Neher, E., 165f, 190, 192
Neill, S. J., 267, 270, 277
Neilson, R. E., 241, 245
Nelmes, B. J., 75, 77, 85
Nelson, S. D., 53, 56, 172, 191, 230, 248
Nerkar, Y. S., 433f, 442
Nevo, Y., 435, 443
Ng, P. A. P., 229, 248
Ng, T. T., 272, 276
Nicholls, D. G., 168, 179, 191
Nieuwstadt, F. T. M., 427, 429
Nishida, K., 353, 358, 365
Nobel, P. S., 113, 168, 177, 183, 191, 231, 235, 241, 248f, 307, 321, 347, 349f, 353, 357ff, 360f, 365, 380, 474
Northcote, D. H., 63, 88
Nott, D. L., 360, 365
Nowak, R. S., 370, 382
Nyman, B., 73, 89

Odhnoff, C., 53f
Ogasawara, N., 205, 207
Ogawa, T., 135, 156, 159, 165, 174, 187, 190f, 196f, 200f, 203f, 207, 210ff, 213f, 220–27 *passim*, 273, 277
Ogunkanmi, A. B., 263, 279, 283f, 287, 292, 294, 308
Ohl, B., 267, 271, 276
Okali, D. U. U., 339f, 348, 351
O'Kane, D. J., 87
O'Leary, J. W., 446, 460, 469, 475
O'Leary, M. H., 170, 191, 438, 441
O'Neill, S. D., 262, 277
Oppenheimer, H., 328, 349
Ormrod, D. P., 239, 249
Osmond, C. B., 113, 119, 123, 245, 249, 307, 321, 347, 349f, 357, 360f, 365, 380, 474
Osonubi, O., 237f, 248
Osteryoung, K., 345, 349
Outlaw, W. H., Jr., 56, 115–22 *passim*, 126f, 135, 139, 144, 160f, 170–75 *passim*, 191, 204ff, 207, 216, 220, 226

Paetz, K. W., 19, 25
Paez, A., 230, 241, 248
Palevitz, B. A., 12, 25, 59, 61, 64–87 *passim*
Pallaghy, C. K., 127, 140, 160, 233, 248
Pallardy, S. G., 433, 442
Pallas, J. E., Jr., 45, 57, 76, 80, 82, 87, 265, 279
Panichkin, L. A., 153, 162, 165, 178, 190, 285, 290
Pant, D. D., 44, 56
Paolillo, D. J., Jr., 37f, 55f, 61, 64, 67ff, 70, 73–84 *passim*, 88
Parker, L. L., 233, 249
Parkinson, K. J., 446, 456, 475
Parlange, J.-Y., 96f, 112
Passioura, J. B., 318, 320f
Paterson, N. W., 161, 279, 308
Patten, D. T., 361, 364
Patterson, D. T., 230, 248
Pavek, J. J., 436, 441
Pearce, R. B., 370, 382
Pearcy, R. W., 339, 342ff, 345, 349, 450, 475
Pearman, G. I., 245
Pearse, H. L., 16, 18, 23
Pearson, C. J., 172, 187, 191
Peaud-Lenoel, C., 290
Pekarek, J., 18, 25, 132, 160
Pemadasa, M. A., 153, 160, 209, 226, 293ff, 296, 308
Penman, H. L., 14f, 25, 459, 475
Penning de Vries, F. W. T., 175, 191
Penny, M. G., 21, 25, 127, 133, 160, 178, 191, 260, 277
Penth, B., 134, 160
Percival, J., 12, 25
Pereira, J. S., 329, 349ff
Pertz, D. F. M., 13, 22
Petering, L. B., 86
Peterson, R. L., 65, 68f, 75, 82, 84, 87f
Pickett, J. M., 205, 207
Pickett-Heaps, J. D., 63, 88
Pierce, M. L., 97f, 102f, 106, 111, 113, 256, 265–71 *passim*, 277f
Pierce, R. B., 434, 436, 441
Pike, C. S., 241, 245
Pilet, P. E., 291
Pitman, M. G., 122, 129, 158, 160
Platt-Aloia, K., 61, 80–86 *passim*, 354, 364
Poljakoff, A., 328, 349
Poole, R. J., 164, 180, 190f, 221, 226
Popiela, C. C., 265, 278
Porsch, O., 31, 34ff, 39, 46, 52, 56
Pospíšilová, J., 239, 250, 367, 383
Powell, R. D., 113

Powles, S. B., 365
Prange, R. K., 239, 249
Preiss, J., 120, 123
Probst, W., 41ff, 44f, 56
Proctor, J. T. A., 239, 249
Pyrkosch, G., 19, 25

Quarrie, S. A., 267, 277, 371, 382, 435, 437f, 442f
Queiroz, O., 119, 122, 358f, 365
Quisenberry, J. E., 434f, 443

Rachid, M., 293, 307
Radin, J. W., 230, 233, 249, 270, 275, 379, 382
Radtke, F., 53, 56
Raff, M., 189
Raghavendra, A. S., 165, 182, 191, 283, 287, 290
Raikhel, N. V., 65, 74, 87
Rains, D. W., 110, 112
Raison, J. K., 344, 347
Rajendra, B. R., 433, 443
Ramos, C., 209, 226, 230f, 237, 239f, 249
Rand, R. H., 112
Randall, D., 345, 349
Randall, D. D., 160, 207, 226
Rao, I. M., 119, 123, 205, 207, 283, 287, 290
Rapp, B., 160, 226
Rapp, G., 207
Raschke, K., 11, 14, 19–26 *passim*, 53, 56, 59, 71f, 75, 78, 84, 88, 92ff, 97f, 102f, 106f, 111ff, 115, 118, 123, 127, 129, 132f, 139, 147, 158, 160, 166, 169f, 172–79 *passim*, 183, 190f, 196–207 *passim*, 209, 216, 222, 226, 229–40 *passim*, 243ff, 246, 249f, 254–79 *passim*, 285, 292, 299, 301, 308, 320f, 326, 333, 347, 349, 460, 474
Rasmusson, D. C., 433f, 436, 442
Raupach, M. R., 387, 390, 396, 419f, 424f, 427f
Raven, J. A., 117, 119, 123, 184, 192, 275
Rawlins, S. L., 234, 248
Rawson, H. M., 351, 369–75 *passim*, 381, 383f, 433, 443
Rayder, L., 360f, 365
Regehr, D. L., 231, 241, 249
Rehfous, L., 9, 22, 24f, 46, 56
Renner, O., 15, 26
Resch, A., 20, 23
Revsbech, N. P., 220, 226
Rhine, J. B., 373, 383
Richards, F. J., 371, 383
Richards, G. P., 99, 112

Index of Authors

Richards, J. M., 450, 475
Richards, R. A., 438f, 441
Richmond, A., 303, 307
Rieber, F., 41f, 56
Roark, B., 434f, 443
Robert, D., 87
Roberts, J., 437, 443
Roberts, J. K. M., 134, 160
Roberts, K., 189
Robichaux, R. H., 339, 350
Robinson, N. L., 120, 123
Rodriguez, G. M. T., 435, 442
Rodríguez, J. L., 302f, 308
Rodriguez, M. T., 214, 226
Rogers, C. A., 113
Rogers, H. H., 230, 241, 249, 250
Rollinson, P. L., 326, 349
Rose, D. A., 442
Rosenberg, N. J., 372, 380
Roth-Bejerano, N., 200, 207
Rouschal, E., 328, 350
Roy, J., 329, 350
Rubery, P. H., 303, 306
Rudesch, M. L., 364
Ruhland, W., 246, 321, 350
Rundel, P. W., 350, 353, 365
Running, S. W., 339, 350
Russell, J., 18f, 23, 195, 206, 231f, 246
Rutter, J. C., 53, 56, 80ff, 88
Rybová, R., 225f

Sack, F. D., 38, 56, 61, 64, 67ff, 70, 73-84 *passim*, 88
Saftner, R. A., 133, 160, 169, 192, 264, 278, 285, 292
Sakaki, T., 192
Sakata, M., 190, 225
Sakmann, B., 166, 190, 192
Salisbury, E. J., 13, 26
Samish, Y. B., 457, 475
Sanchez, S., 80, 88
Sanders, D., 133, 160, 217, 226, 260f, 278
San Pietro, A., 290
Sasaki, T., 207
Satler, S. O., 239, 250, 377, 383
Saunders, M. J., 289, 292
de Saussure, H. B., 8f, 26
Sauvezon, R., 347
Saxe, H., 78f, 88
Sayre, J. D., 16f, 19, 26
Scarth, G. W., 17f, 26, 116, 123, 132, 161, 195, 207, 229, 249
Schäfer, E., 210, 225
Schäfer, R. F. C., 104, 113
Scheel, J. A., 345, 348
Schertz, K. F., 441
Schilling, K., 311, 321

Schimper, W. Ph., 35, 56
Schleiden, M. J., 11, 26
Schmerl, R., 347
Schmidt, H., 79, 87
Schmidt, W., 222, 226
Schnabl, H., 53, 56, 85, 88, 119, 123, 170, 192, 260f, 269, 278f
Schneider, E. A., 296, 308
Schneider, G., 292
Schoch, P. G., 372, 383
Schofield, R. K., 15, 25, 459, 475
Schönherr, J., 30, 32, 56, 79, 87
Schrader, L. E., 433, 443
Schroeder, J. I., 165ff, 174, 184ff, 189, 192, 214, 216, 222ff, 226, 258, 260, 278
Schuepp, P. H., 387, 404f, 428
Schulze, E.-D., 235, 237f, 246, 248f, 311-21 *passim*, 323-29 *passim*, 332f, 335, 339f, 342, 345-50 *passim*, 364
Schürmann, B., 36, 56
Schutte, K. H., 361, 365
Schwartz, A., 135, 161, 164f, 174, 182ff, 186, 192f, 199f, 202f, 207, 224, 226, 373, 384
Schwendener, S., 10ff, 13, 26, 36, 42, 48, 50, 56, 92f, 113
Seginer, I., 387, 428
Sembdner, G., 281, 292
Sen, D. N., 88
Senger, H., 192, 207, 227
Šesták, Z., 446, 475
Setter, T. L., 433, 443
Setterfield, G., 80ff, 84f, 116, 122, 172, 189
Sexton, R., 76, 89
Seybold, A., 16, 26
Shackel, K., 314, 321
Sharkey, T. D., 19, 26, 174, 192, 196-202 *passim*, 207, 222, 226, 230ff, 234, 237ff, 240f, 244f, 249f, 258, 267, 275, 278, 297, 307, 319, 321, 368, 376, 381, 383, 436, 441, 446, 460, 474f
Sharp, R. E., 305, 307
Sharpe, P. J. H., 91-113 *passim*, 169, 176, 192
Shaw, J., 256, 261, 276, 283ff, 291
Shaw, M., 204, 207
Shaw, R. H., 389, 392, 395, 421, 428f
Sheriff, D. W., 92, 104, 111f
Shibata, K., 159, 191, 207, 225
Shimada, K., 159, 191, 207, 225
Shimazaki, K., 165, 172, 174, 182ff, 185f, 192, 205, 207f, 215-26 *passim*
Shimmen, T., 135, 158
Shimshi, D., 434f, 443
Shiraishi, M., 282, 285, 291
Shmueli, E., 68ff, 79, 89

Index of Authors

Shoemaker, E. M., 96, 113
Short, K. C., 239, 250, 263, 279, 283, 287, 292f, 308
Shorter, R., 439, 441
Shreve, E. B., 353, 365
Sieben, J. W., 439, 442
Sierp, H., 16, 19, 26
Sigler, K., 225f
Sigworth, F., J., 190
Silversides, R. H., 392, 428
Simmonds, N. W., 435, 443
Simoncini, L., 165f, 174, 184ff, 189, 214, 216, 222ff
Sinclair, T. R., 456, 475
Singh, A. P., 61, 63, 67–84 passim, 88
Singh, D. P., 434, 443
Singh, M., 434, 443
Singh, P., 434, 443
Síonít, N., 249f
Sivakumaran, S., 307
Skaar, H., 200, 208, 210, 214, 226
Skoog, F., 275
Slatyer, R. O., 14, 26, 229, 231, 241, 245, 247, 460, 475
Slavik, B., 326, 350
Slayman, C. L., 217, 226
Slayman, C. W., 132, 157
Small, J., 132, 161
Smith, B. N., 204, 206, 220, 225
Smith, F. A., 117, 119
Smith, H., 119, 122, 192
Smith, J. A. C., 260, 276
Smith, J. M., 249
Smith, M. L., Jr., 371, 383
Smith, W. H., 432, 443
Snaith, P. J., 233, 250, 266, 278, 296ff, 308
Solárová, J., 238f, 245, 250, 367, 369f, 381, 383
Solereder, H., 37, 56
Sondheimer, E., 257, 275
Sorokin, H., 82, 88
Sorokin, S., 82, 88
Spalding, M. H., 234, 250
Spanner, D. C., 15, 25
Spanswick, R. M., 132, 134, 158, 161, 221, 226, 262, 277
Spence, R. D., 101f, 105–13 passim
Sprecher, A., 45, 56
Springer, S. A., 118, 121f
Squire, G. R., 131, 161, 301, 308
Srivastava, L. M., 61, 69–77 passim, 80, 82ff, 88, 96, 113
Stahl, F., 14, 26
Ståhlfelt, M. G., 11–16 passim, 20, 26f, 94, 113, 255, 278, 311, 321
Stanhill, C. S., 14, 25
Staples, R. C., 442

Stebbins, G. L., 48, 50, 56
Steenackers, V., 436, 441
Stemmerman, L., 350
Sternberg, L. O., 231, 234, 245, 357f, 363
Steudle, E., 99, 113
van Steveninck, R. F. M., 21, 25, 255, 277, 282, 286, 291
Stevens, R. A., 45, 56, 59, 67ff, 70, 73, 80ff, 88
Stewart, J. B., 408, 428
Stewart, W. E., 403, 427
Steyn, R., 361, 365
Stitt, M., 134, 161, 273, 275
Stocker, O., 323f, 329, 350
Stoddart, J. L., 369f, 383
Strain, B. R., 230, 241, 243, 248, 250f, 474
Strand, M., 238, 246
Strasburger, E., 35, 42, 48, 57
Streich, J., 275
Stumpf, D. K., 250
Stüning, M., 257, 271, 277
Suh, H. W., 433, 443
Sundbom, E., 238, 246
Surand, K. A., 249
Sutera, S. P., 405, 428
Sward, R. J., 231, 247
Syonō, K., 261, 276
Syvertsen, J. P., 371, 383
Szarek, S. R., 353, 356f, 360ff, 365f

Taiz, L., 100, 113, 131f, 158f, 166, 178, 183, 190, 216, 218, 225, 261, 275
Tal, M., 267, 278, 303, 308, 435f, 443
Tan, G.-Y., 433f, 443
Tanner, C. B., 318, 321
Tanner, W., 264, 276
Tarczynski, M. C., 116–22 passim
Tarr, A. R., 435, 443
Tavangar, J., 395, 428
Taylor, G. I., 423, 428
Taylor, H. F., 267, 278f
Taylor, I. B., 435, 443
Taylor, S., 227
Tazawa, M., 135, 158f
Tenhunen, J. D., 319, 321, 325–34 passim, 348ff, 351, 357, 366
Tennekes, H., 387, 419, 428
Thaler, I., 373, 382
Thellier, H., 161
Thimann, K. V., 239, 250, 377, 383
Thom, A. S., 387, 405, 408, 418, 427f, 464, 468, 475
Thomas, H., 369f, 383
Thomas, J. C., 110, 112, 371, 382
Thomas, J. P., 249
Thomas, T. H., 282, 290
Thompson, M. L., 359, 366

Index of Authors

Thomson, D. J., 426, 428
Thomson, W. W., 61, 79–88 *passim*, 354, 364, 366
Thorpe, M. R., 229f, 241, 250, 456, 475
Thorpe, N., 80ff, 88, 236, 250
Thurtell, G. W., 392, 428f
Tibbitts, T. W., 432, 440
Tichá, I., 239, 245, 367–72 *passim*, 380f, 383
Tietz, D., 275
Ting, I. P., 119, 123, 231, 234, 245, 277, 353–66 *passim*
Tobolsky, A. V., 101, 113
Tosteson, D. C., 158
Travis, A. J., 184, 192, 243f, 247
Treloar, L. R. G., 91, 98f, 113
Treviranus, L. C., 34, 57
Trewavas, A., 288, 292
Troughton, J. H., 364
Tucker, D. J., 256, 279, 283ff, 287, 292, 294, 308
Turner, N. C., 131, 161, 261, 279, 294, 308, 311, 313, 317f, 320f, 337, 351, 369ff, 381, 383
Tyree, M. T., 99, 112f, 234, 250

Uchikawa, T., 149, 157
U. S. National Bureau of Standards, 449
Ussing, H. H., 158

Vaadia, Y., 282, 291f, 303, 307
Vaclavic, J., 369, 372, 383
Van Bel, A. J. E., 77, 86
Van Kirk, C. A., 118, 123, 169f, 172, 176, 192, 261, 279
Van Staden, J., 281, 292, 306, 308
Varela, F., 161, 192, 227
Vasil, V., 47, 55
Vasquez-Yanes, C., 350
Vassilyev, A., 73, 75, 88
Vassilyeva, G., 73, 75, 88
Vaughan, G. T., 257, 271, 279
Vaughn, K. C., 135, 161
Veith, G. M., 160, 207, 226
Vesque, M. J., 48, 57
Vienken, J., 85, 88
Vieweg, G. H., 324, 350
de Vries, H., 11, 27

Wagner, Th., 51f, 57
Walker, D. A., 165, 192, 241, 246
Walker, N. A., 133, 157
Walles, B. B., 73, 89
Wallihan, E. F., 14, 27
Walton, D. C., 267, 277
Walton, P. D., 433, 436, 443
Walz, H., 350

Wang, T. L., 435, 441
Ward, D. P., 395, 428f
Wardle, K., 239, 250, 263, 279, 283, 287, 292f, 308
Wareing, P. F., 278, 290f, 349
Warrit, B., 229f, 241, 250
Wassermann, J., 51, 57
Waterkeyn, L., 75, 89
Watson, J. D., 189
Weber, F., 13, 27, 373, 382
Weber, H., 264, 276
Weber, J. A., 321, 339, 344, 350f
Weigl, J., 134, 160
Weiler, E. W., 257f, 263, 268ff, 274–79 *passim*
Werk, K. S., 244, 250, 345, 351
van der Westhuizen, M., 361, 365
von Wettstein, R., 34, 57
Weyers, J. D. B., 144, 152, 161, 204, 206, 216, 226, 256, 263f, 279, 301f, 308
Whatley, F. R., 55
Whatley, J. M., 80, 89
White, M. C., 110, 112
Whitehead, D., 339, 351
Whitelam, G. C., 282f
Whiteman, P. C., 230f, 250, 459, 475
Wiggans, R. G., 11, 27
Wightman, F., 296, 308
Wigley, G., 465, 474
Wille, A. C., 75f, 80, 89
von Willert, D. J., 329, 351
Williams, B. A., 326, 346, 351
Williams, T., 9, 27
Williams, W. E., 351
Willmarth, W. W., 393, 428
Willmer, C. M., 19, 23, 45, 53, 56f, 59, 76, 80ff, 88f, 129f, 161, 230, 235, 247, 250, 265, 279
Wilson, C. C., 19f, 27
Wilson, D., 433f, 442f
Wilson, G. L., 231, 241, 247, 373f, 382
Wilson, J. A., 262, 265f, 277, 279, 299ff, 309
Wilson, J. D., 391, 429
Wilson, J. M., 275, 299, 302f, 307, 309
Wilson, N. R., 389, 421, 429
Winter, K., 119, 123, 357, 365f
Wittenbach, V. A., 377, 383
Witztum, A., 436, 443
Wong, S. C., 195f, 208f, 226, 230, 232, 236–43 *passim*, 247, 250, 328, 345, 351, 374, 377, 383f, 432
Wood, M., 376, 381
Woodrow, I. E., 468, 471, 474f
Woodward, R. G., 370–75 *passim*, 383f
Woolhouse, H. W., 231f, 241, 247, 369, 376, 381, 384
Wright, L. N., 441

Wright, S. T. C., 21, 27, 255, 276, 279
Wu, H., 91–113 *passim*, 169, 176, 192
Wulff, R. D., 230, 243, 251
Wyngaard, J. C., 420ff, 429
Wyse, R. E., 264, 278

Yaglom, A. M., 424, 428
Yamashita, T., 21, 27, 125, 161
Yatsevich, A. P., 433, 441
Ýcas, J. W., 377, 384
Yianoulis, P., 234, 250
Yocum, C. S., 86

Zeevaart, J. A. D., 267–75 *passim*, 279
Zeiger, E., 19, 27, 80f, 115, 120f, 123, 126, 133ff, 147, 153, 156, 159, 161, 164f, 169, 172, 174, 178, 182ff, 185f, 190ff, 193, 196–208 *passim*, 209–27 *passim*, 244, 251, 285, 288, 292, 301, 309, 373, 384
Zelitch, I., 165, 192, 293, 309
Zemel, E., 174, 182, 193, 204, 208, 220, 227
Zenger, V. E., 122, 191, 207
Ziegenspeck, H., 13, 27, 37, 43, 48, 57, 65, 73, 89–94 *passim*, 100, 113f
Ziegler, H., 30, 32, 53, 56, 68ff, 79, 89, 113, 170, 192, 249, 260f, 278, 307, 321, 347, 349f, 380, 474
Zimmermann, U., 85, 88, 99, 113, 129, 161
Zimmermann, W., 40f, 48, 57
Zinth, P. M., 360, 365
Zlotnikova, I. F., 153, 162, 165, 178, 190, 285, 290

Library of Congress Cataloging-in-Publication Data

Stomatal function.

Includes indexes.
1. Stomata.   I. Zeiger, Eduardo.   II. Farquhar, G. D.,
1947–   .   III. Cowan, I. R.
QK873.S85   1987   581.1'2   86-23162
ISBN 0-8047-1347-2 (alk. paper)